METHODS IN MOLECULAR BIOLOGY

For further volumes:
http://www.springer.com/series/7651

For over 35 years, biological scientists have come to rely on the research protocols and methodologies in the critically acclaimed *Methods in Molecular Biology* series. The series was the first to introduce the step-by-step protocols approach that has become the standard in all biomedical protocol publishing. Each protocol is provided in readily-reproducible step-by step fashion, opening with an introductory overview, a list of the materials and reagents needed to complete the experiment, and followed by a detailed procedure that is supported with a helpful notes section offering tips and tricks of the trade as well as troubleshooting advice. These hallmark features were introduced by series editor Dr. John Walker and constitute the key ingredient in each and every volume of the *Methods in Molecular Biology* series. Tested and trusted, comprehensive and reliable, all protocols from the series are indexed in PubMed.

Plant Genotyping

Methods and Protocols

Edited by

Yuri Shavrukov

College of Science and Engineering, Biological Sciences, Flinders University, Adelaide, SA, Australia

Humana Press

Editor
Yuri Shavrukov
College of Science and Engineering
Biological Sciences
Flinders University
Adelaide, SA, Australia

ISSN 1064-3745 ISSN 1940-6029 (electronic)
Methods in Molecular Biology
ISBN 978-1-0716-3023-5 ISBN 978-1-0716-3024-2 (eBook)
https://doi.org/10.1007/978-1-0716-3024-2

This Humana imprint is published by the registered company Springer Science+Business Media, LLC, part of Springer Nature.
The registered company address is: 1 New York Plaza, New York, NY 10004, U.S.A.

Preface

> "The dear old magnolias!" says the young man, pinching one of my leaves. "I love them all". – Magnolia! Well, wouldn't that--say! Those innocents thought I was a magnolia! What the--well, wasn't that tough on a genuine little old New York rubber plant?"...

This "sentimental dialog" from *The Rubber Plant's Story* (1917) by the renowned American short story writer O'Henry reflects how commonly humans are confused by plant identification.

Our life is always surrounded by plants, and we observe their external traits as plant phenotypes. In contrast, we cannot directly see plant genotypes, the set of genetic material that encodes these phenotypic traits. To make genotypes accessible for research and understanding, various genotyping methods are used. Genotyping can be employed to study the origin of plant characteristics and traits, where each plant has a unique genotype. Many methods of plant genotyping were initially developed for medical research, but all genotyping methods, if they are to be successful, should be suitable for application across the full range of studies within plant biology. This relates particularly to the hundreds of thousands of offspring generated in plant progenies through diverse types of plant propagation (i.e., sexual and clonal, self- and cross-pollination) and across the relatively short life cycles of annual plant species which of course include the majority of crop species. Plant genotyping methods may be based on a variety of assessments, including DNA microarray, with its hundreds of thousands of simultaneous reactions, or separate individual studies of DNA sequencing and fragment analysis, PCR and qPCR, allele-specific molecular probes and primers, digestion with restriction endonucleases, microscopy, and many others. Results of plant genotyping can be easily converted into molecular markers, which can then be used not only for academic study but more importantly for practical application in crop breeding, biodiversity and biosecurity, and analysis of food products from plants. Therefore, methods of plant genotyping and their results are very important for our future and the future of our global biosphere.

The current book represents the wide range of existing methods, from the very popular to the more exotic and rarely used. Each researcher is therefore awarded the opportunity to update their knowledge and choose the most suitable method of plant genotyping for their chosen application.

Adelaide, Australia *Yuri Shavrukov*

Contents

Contributors

IRINA ADONINA • *Institute of Cytology and Genetics, Russian Academy of Sciences, Siberian Branch, Novosibirsk, Russia*

NATALIA CRISTINA AGUIRRE • *Instituto de Agrobiotecnología y Biología Molecular (IABiMo), Unidad Ejecutora de Doble Dependencia Instituto Nacional de Tecnología Agropecuaria (INTA) – Consejo Nacional de Ciencia y Técnica (CONICET), Hurlingham, Argentina*

AIGUL AMANGELDIYEVA • *Kazakh Research Institute of Agriculture and Plant Growing, Almalybak, Almaty, Kazakhstan*

JUNREY C. AMAS • *School of Biological Sciences, University of Western Australia, Perth, WA, Australia*

AKMARAL BAIDYUSSEN • *S.Seifullin Kazakh AgroTechnical University, Astana, Kazakhstan*

KEN BAKER • *General Bioinformatics, Jealott's Hill International Research Centre, Berkshire, UK*

JACQUELINE BATLEY • *School of Biological Sciences, University of Western Australia, Perth, WA, Australia*

LUCA BRAGLIA • *Institute of Agricultural Biology and Biotechnology (IBBA), Milan, Italy*

DIEGO BREVIARIO • *Institute of Agricultural Biology and Biotechnology (IBBA), Milan, Italy*

ANTHONY BRUSA • *Department of Agronomy and Plant Genetics, University of Minnesota, Minneapolis, MN, USA*

KULPASH BULATOVA • *Kazakh Research Institute of Agriculture and Plant Production, Almalybak, Almaty, Kazakhstan*

CATARINA CAMPOS • *MED-Mediterranean Institute for Agriculture, Environment and Development & CHANGE-Global Change and Sustainability Institute, Institute for Advanced Studies and Research, Universidade de Évora, Evora, Portugal*

MARIA DOROTEIA CAMPOS • *MED-Mediterranean Institute for Agriculture, Environment and Development & CHANGE-Global Change and Sustainability Institute, Institute for Advanced Studies and Research, Universidade de Évora, Evora, Portugal*

ALDRIN Y. CANTILA • *School of Biological Sciences, University of Western Australia, Perth, WA, Australia*

HÉLIA CARDOSO • *MED-Mediterranean Institute for Agriculture, Environment and Development & CHANGE-Global Change and Sustainability Institute, Institute for Advanced Studies and Research, Universidade de Évora, Evora, Portugal*

TIRTHARTHA CHATTOPADHYAY • *Department of Plant Breeding and Genetics, Bihar Agricultural College, Bihar Agricultural University, Sabour, Bhagalpur, Bihar, India*

ALINE DA COSTA LIMA MORAES • *Department of Plant Biology, Biology Institute, University of Campinas (UNICAMP), Campinas, Brazil*

JIANGBO DANG • *Key Laboratory of Horticulture Science for Southern Mountains Regions of Ministry of Education, College of Horticulture and Landscape Architecture, Southwest University, Beibei, Chongqing, China*

MARÍLIA DE CASTRO RODRIGUES PAPPAS • *Embrapa Genetic Resources and Biotechnology, Brasilia, Distrito Federal, Brazil*

ANETE PEREIRA DE SOUZA • *Department of Plant Biology, Biology Institute, University of Campinas (UNICAMP), Campinas, Brazil; Center for Molecular Biology and Genetic Engineering (CBMEG), University of Campinas (UNICAMP), Campinas, Brazil*

JOE DOWNES • *Herbicide Bioscience, Syngenta, Jealott's Hill International Research Centre, Berkshire, UK*

CARLA VALERIA FILIPPI • *Instituto de Agrobiotecnología y Biología Molecular (IABiMo), Unidad Ejecutora de Doble Dependencia Instituto Nacional de Tecnología Agropecuaria (INTA) – Consejo Nacional de Ciencia y Técnica (CONICET), Hurlingham, Argentina; Laboratorio de Bioquímica, Departamento de Biología Vegetal, Facultad de Agronomía, Universidad de la República, Montevideo, Uruguay*

MARGARET FLEMING • *Department of Plant, Soil, and Microbial Sciences, Michigan State University, East Lansing, MI, USA*

JUNYI GAI • *Soybean Research Institute, Nanjing Agricultural University, Nanjing, China*

FLORIANA GAVAZZI • *Institute of Agricultural Biology and Biotechnology (IBBA), Milan, Italy*

SILVIA GIANÌ • *Institute of Agricultural Biology and Biotechnology (IBBA), Milan, Italy*

RAFAQAT ALI GILL • *Key Laboratory of Biology and Genetic Improvement of Oil Crops, Ministry of Agriculture and Rural Affairs, Oil Crops Research Institute, Chinese Academy of Agricultural Sciences, Wuhan, China*

QIGAO GUO • *Key Laboratory of Horticulture Science for Southern Mountains Regions of Ministry of Education, College of Horticulture and Landscape Architecture, Southwest University, Beibei, Chongqing, China*

PATRICK HARDINGE • *School of Biosciences, Cardiff University, Cardiff, Wales, UK*

SAMANTHA L. HARVIE • *School of Biological Sciences, University of Western Australia, Perth, WA, Australia*

JIANBO HE • *Soybean Research Institute, Nanjing Agricultural University, Nanjing, China*

MD MOSTOFA UDDIN HELAL • *Key Laboratory of Biology and Genetic Improvement of Oil Crops, Ministry of Agriculture and Rural Affairs, Oil Crops Research Institute, Chinese Academy of Agricultural Sciences, Wuhan, China*

PATRICIO HINRICHSEN • *Instituto de Investigaciones Agropecuarias, INIA La Platina, Santiago, Chile*

MARCUS A. HOOKER • *Department of Plant Pathology, Washington State University, Pullman, WA, USA*

MING HU • *Key Laboratory of Biology and Genetic Improvement of Oil Crops, Ministry of Agriculture and Rural Affairs, Oil Crops Research Institute, Chinese Academy of Agricultural Sciences, Wuhan, China*

SARAH-JANE HUTCHINGS • *Herbicide Bioscience, Syngenta, Jealott's Hill International Research Centre, Berkshire, UK*

HIDETOSHI IKEGAMI • *Fukuoka Agriculture and Forestry Research Center, Buzen Branch, Yukuhashi, Japan*

MIR ASIF IQUEBAL • *Division of Agricultural Bioinformatics, ICAR-Indian Agricultural Statistics Research Institute, New Delhi, India*

SARIKA JAISWAL • *Division of Agricultural Bioinformatics, ICAR-Indian Agricultural Statistics Research Institute, New Delhi, India*

SATYVALDY JATAYEV • *S.Seifullin Kazakh AgroTechnical University, Astana, Kazakhstan*

PALCHAMY KADIRVEL • *ICAR-Indian Institute of Oilseeds Research, Hyderabad, India*

Won-Hee Kang • *Department of Horticulture, Division of Applied Life Science (BK21 Four), Gyeongsang National University, Jinju, South Korea; Institute of Agriculture & Life Science, Gyeongsang National University, Jinju, South Korea*

Shiv Shankhar Kaundun • *Herbicide Bioscience, Syngenta, Jealott's Hill International Research Centre, Berkshire, UK*

Sholpan Khalbayeva • *Kazakh Research Institute of Agriculture and Plant Production, Almalybak, Almaty, Kazakhstan*

Gulmira Khassanova • *A.I. Barayev Research and Production Centre of Grain Farming, Shortandy, Kazakhstan; Faculty of Agronomy, S. Seifullin Kazakh AgroTechnical University, Astana, Kazakhstan*

Nayoung Kim • *Department of Horticulture, Division of Applied Life Science (BK21 Four), Gyeongsang National University, Jinju, South Korea*

Nobuhiro Kotoda • *Faculty of Agriculture, Saga University, Saga, Japan*

Ashutosh Kumar • *Department of Bioengineering and Biotechnology, Birla Institute of Technology, Mesra, Ranchi, Jharkhand, India; Department of Biotechnology, School of Sciences, PP Savani University, Kosamba, Surat, Gujarat, India*

Chippa Anil Kumar • *ICAR-Indian Institute of Oilseeds Research, Hyderabad, India*

Dinesh Kumar • *Division of Agricultural Bioinformatics, ICAR-Indian Agricultural Statistics Research Institute, New Delhi, India; Department of Biotechnology, School of Interdisciplinary and Applied Sciences, Central University of Haryana, Mahendergarh, Haryana, India*

Marzhan Kuzbakova • *S.Seifullin Kazakh AgroTechnical University, Astana, Kazakhstan*

Ji-Su Kwon • *Department of Horticulture, Division of Applied Life Science (BK21 Four), Gyeongsang National University, Jinju, South Korea*

Verónica Viviana Lia • *Instituto de Agrobiotecnología y Biología Molecular (IABiMo), Unidad Ejecutora de Doble Dependencia Instituto Nacional de Tecnología Agropecuaria (INTA) – Consejo Nacional de Ciencia y Técnica (CONICET), Hurlingham, Argentina*

Guolu Liang • *Key Laboratory of Horticulture Science for Southern Mountains Regions of Ministry of Education, College of Horticulture and Landscape Architecture, Southwest University, Beibei, Chongqing, China*

Shengyi Liu • *Key Laboratory of Biology and Genetic Improvement of Oil Crops, Ministry of Agriculture and Rural Affairs, Oil Crops Research Institute, Chinese Academy of Agricultural Sciences, Wuhan, China*

Yan Liu • *Department of Plant Pathology, Washington State University, Pullman, WA, USA*

Melina Cristina Mancini • *Center for Molecular Biology and Genetic Engineering (CBMEG), University of Campinas (UNICAMP), Campinas, Brazil*

Susana Noemí Marcucci Poltri • *Instituto de Agrobiotecnología y Biología Molecular (IABiMo), Unidad Ejecutora de Doble Dependencia Instituto Nacional de Tecnología Agropecuaria (INTA) – Consejo Nacional de Ciencia y Técnica (CONICET), Hurlingham, Argentina*

Karol Marlowe • *USDA-ARS Wheat Health, Genetics and Quality Research Unit, Pullman, WA, USA*

Shynar Mazkirat • *Kazakh Research Institute of Agriculture and Plant Production, Almalybak, Almaty, Kazakhstan*

Marco Meneses • *Instituto de Investigaciones Agropecuarias, INIA La Platina, Santiago, Chile*

MANMODE DARPAN MOHANRAO • *ICAR-Indian Institute of Oilseeds Research, Hyderabad, India*
LAURA MORELLO • *Institute of Agricultural Biology and Biotechnology (IBBA), Milan, Italy*
ARNAB MUKHERJEE • *Department of Plant Breeding and Genetics, Bihar Agricultural College, Bihar Agricultural University, Sabour, Bhagalpur, Bihar, India*
CLAUDIA MUÑOZ-ESPINOZA • *Universidad Andrés Bello, Center for Plant Biotechnology, Santiago, Chile*
MUHAMMAD NAEEM • *Department of Agricultural Genetic Engineering, Nigde Omer Halisdemir University, Nigde, Turkey; Department of Agriculture Extension and Adaptive Research,, Punjab Agriculture Department, Govt. of Pakistan, Attock, Punjab, Pakistan*
SATOSHI NANAMI • *Graduate School of Science, Osaka Metropolitan University, Osaka, Japan*
HUITANG PAN • *Beijing Key Laboratory of Ornamental Plants Germplasm Innovation and Molecular Breeding, National Engineering Research Center for Floriculture, College of Landscape Architecture, Beijing Forestry University, Beijing, China*
DEV MANI PANDEY • *Department of Bioengineering and Biotechnology, Birla Institute of Technology, Mesra, Ranchi, Jharkhand, India*
NORMA BEATRIZ PANIEGO • *Instituto de Agrobiotecnología y Biología Molecular (IABiMo), Unidad Ejecutora de Doble Dependencia Instituto Nacional de Tecnología Agropecuaria (INTA) – Consejo Nacional de Ciencia y Técnica (CONICET), Hurlingham, Argentina*
GEORGIOS JOANNIS PAPPAS JR. • *Department of Cell Biology, University of Brasilia, Brasilia, Distrito Federal, Brazil*
ERIC PATTERSON • *Department of Plant, Soil, and Microbial Sciences, Michigan State University, East Lansing, MI, USA*
WENDELL JACINTO PEREIRA • *School of Forest, Fisheries, and Geomatics Sciences, University of Florida, Gainesville, FL, USA; Department of Cell Biology, University of Brasilia, Brasilia, Distrito Federal, Brazil*
WIRULDA POOTAKHAM • *National Omics Center, National Science and Technology Development Agency (NSTDA), Pathum Thani, Thailand*
GOPAL KUMAR PRAJAPATI • *Department of Bioengineering and Biotechnology, Birla Institute of Technology, Mesra, Ranchi, Jharkhand, India; R & D Biologics Division, Promea Therapeutics Pvt Ltd, Sultanpur, Hyderabad, India*
ANDREA FABIANA PUEBLA • *Instituto de Agrobiotecnología y Biología Molecular (IABiMo), Unidad Ejecutora de Doble Dependencia Instituto Nacional de Tecnología Agropecuaria (INTA) – Consejo Nacional de Ciencia y Técnica (CONICET), Hurlingham, Argentina*
ANIL RAI • *Division of Agricultural Bioinformatics, ICAR-Indian Agricultural Statistics Research Institute, New Delhi, India*
AWAIS RASHEED • *Department of Plant Sciences, Quaid-i-Azam University, Islamabad, Pakistan; Institute of Crop Science, Chinese Academy of Agricultural Sciences (CAAS), & CIMMYT-China Office, Beijing, China*
YARABAPANI RUSHWANTH REDDY • *ICAR-Indian Institute of Oilseeds Research, Hyderabad, India*
TRAVIS M. RUFF • *USDA-ARS Wheat Health, Genetics and Quality Research Unit, Pullman, WA, USA*
YUTARO SAITO • *Yamagata Integrated Agricultural Research Center, Horticultural Research Institute, Yamagata, Japan*

DEVEN R. SEE • *USDA-ARS Wheat Health, Genetics and Quality Research Unit, Pullman, WA, USA; Department of Plant Pathology, Washington State University, Pullman, WA, USA*

SENAPATHY SENTHILVEL • *ICAR-Indian Institute of Oilseeds Research, Hyderabad, India*

DAUREN SERIKBAY • *College of Agronomy, Northwest A&F University, Yangling, Shaanxi, China*

DANILO AUGUSTO SFORÇA • *Center for Molecular Biology and Genetic Engineering (CBMEG), University of Campinas (UNICAMP), Campinas, Brazil*

YURI SHAVRUKOV • *College of Science and Engineering, Biological Sciences, Flinders University, Adelaide, SA, Australia*

FUMITO TADA • *Yamagata Integrated Agricultural Research Center, Horticultural Research Institute, Yamagata, Japan*

TADASHI TAKASHINA • *Yamagata Integrated Agricultural Research Center, Horticultural Research Institute, Yamagata, Japan*

CHAO TAN • *Gentides Biotech Co., Ltd., Wuhan, China*

GITANJALI TANDON • *Division of Agricultural Bioinformatics, ICAR-Indian Agricultural Statistics Research Institute, New Delhi, India*

MINQIANG TANG • *Key Laboratory of Genetics and Germplasm Innovation of Tropical Special Forest Trees and Ornamental Plants, Ministry of Education, College of Forestry, Hainan University, Haikou, China*

WILLIAM J. W. THOMAS • *School of Biological Sciences, University of Western Australia, Perth, WA, Australia*

CHAOBO TONG • *Key Laboratory of Biology and Genetic Improvement of Oil Crops, Ministry of Agriculture and Rural Affairs, Oil Crops Research Institute, Chinese Academy of Agricultural Sciences, Wuhan, China*

MARAL UTEBAYEV • *A.I. Barayev Research and Production Centre of Grain Farming, Shortandy, Kazakhstan*

PABLO ALFREDO VERA • *Instituto de Agrobiotecnología y Biología Molecular (IABiMo), Unidad Ejecutora de Doble Dependencia Instituto Nacional de Tecnología Agropecuaria (INTA) – Consejo Nacional de Ciencia y Técnica (CONICET), Hurlingham, Argentina*

BIANCA BACCILI ZANOTTO VIGNA • *Embrapa Pecuária Sudeste, Brazilian Agricultural Research Corporation, São Carlos, Brazil*

DANIEL VON MAYDELL • *Julius Kuehn-Institute (JKI), Institute for Breeding Research on Horticultural Crops, Quedlinburg, Germany*

HAIYAN WANG • *Key Laboratory of Horticulture Science for Southern Mountains Regions of Ministry of Education, College of Horticulture and Landscape Architecture, Southwest University, Beibei, Chongqing, China*

AAKANKSHA WANY • *Department of Bioengineering and Biotechnology, Birla Institute of Technology, Mesra, Ranchi, Jharkhand, India; Department of Biotechnology, School of Sciences, PP Savani University, Kosamba, Surat, Gujarat, India*

SATOSHI WATANABE • *Faculty of Agriculture, Saga University, Saga, Japan*

CHANGLONG WEN • *Beijing Institute of Vegetable Science, Beijing Academy of Agricultural and Forestry Sciences, Beijing, China*

YOSHIYUKI YAMAGATA • *Faculty of Agriculture, Kyushu University, Fukuoka, Japan*

JINGJING YANG • *Beijing Institute of Vegetable Science, Beijing Academy of Agricultural and Forestry Sciences, Beijing, China*

YANYU YANG • *Gentides Biotech Co., Ltd., Wuhan, China*

SEON-IN YEOM • *Department of Horticulture, Division of Applied Life Science (BK21 Four), Gyeongsang National University, Jinju, South Korea; Institute of Agriculture & Life Science, Gyeongsang National University, Jinju, South Korea*

RAUSHAN YERZHEBAYEVA • *Kazakh Research Institute of Agriculture and Plant Growing, Almalybak, Almaty, Kazakhstan*

MUHAMMAD FARHAN YOUSAF • *Department of Agricultural Genetic Engineering, Nigde Omer Halisdemir University, Nigde, Turkey; Faculty of Technical Sciences, Center for Quantitative Genetics and Genomics, Aarhus University, Aarhus, Denmark*

GIUSI ZAINA • *Department of Agricultural, Food, Environmental and Animal Sciences, University of Udine, Udine, Italy*

JACO D. ZANDBERG • *School of Biological Sciences, University of Western Australia, Perth, WA, Australia*

JIAN ZHANG • *Beijing Institute of Vegetable Science, Beijing Academy of Agricultural and Forestry Sciences, Beijing, China*

YUEQI ZHANG • *School of Biological Sciences, University of Western Australia, Perth, WA, Australia*

YANG ZHOU • *Beijing Key Laboratory of Ornamental Plants Germplasm Innovation and Molecular Breeding, National Engineering Research Center for Floriculture, College of Landscape Architecture, Beijing Forestry University, Beijing, China*

MUHAMMAD ABU BAKAR ZIA • *Department of Plant Breeding and Genetics, College of Agriculture, Bahauddin Zakariya University, Bahadur Sub Campus Layyah, Multan, Pakistan*

Chapter 1

Genotyping by Sequencing (GBS) for Genome-Wide SNP Identification in Plants

Wirulda Pootakham

Abstract

Marker-assisted selection has played a pivotal role in developing several elite varieties in the past two decades. Molecular markers employed in plant breeding programs have recently shifted from microsatellites or simple sequence repeats (SSRs) to single nucleotide polymorphisms (SNPs) due to the ubiquity of SNP markers in the genome and the availability of various high-throughput SNP genotyping platforms. Rapid advances in sequencing technologies and the reduction in sequencing cost have facilitated SNP discovery in several plant species including non-model organisms with little or no genomic resources. Despite the lower cost of sequencing, genome complexity reduction approaches are still useful for SNP identification because many applications do not require every base of the genome to be sequenced. Genotyping-by-sequencing (GBS) is a quick and affordable reduced representation method that can simultaneously identify and genotype a large number of SNPs that has been successfully applied to a wide range of plant species. This chapter describes a robust two-enzyme GBS method for SNP discovery and genotyping that has been verified in non-model plant species.

Key words Genotyping by sequencing (GBS), Reduced representation, Restriction site associated DNA sequencing (RADseq), Single nucleotide polymorphism (SNP)

1 Introduction

The advent of DNA-based genetic markers allowed plant breeders to move from phenotype-based towards genotype-based selection of agronomical traits, accelerating the breeding programs. Marker-assisted selections employ DNA markers that are tightly linked to the target loci (associated with traits of interest) as a substitute for or to assist in phenotypic screenings [1]. In recent years, marker-assisted selections have played a pivotal role in agricultural breeding programs as they have several advantages over traditional breeding approaches. Genotypic screening is often simpler and much less laborious than the phenotypic screening. In addition, the marker-based selections can be performed early at a seedling or juvenile stage. The ability to identify individuals that are likely to exhibit

Yuri Shavrukov (ed.), *Plant Genotyping: Methods and Protocols*, Methods in Molecular Biology, vol. 2638,
https://doi.org/10.1007/978-1-0716-3024-2_1,

desirable traits prior to field trials would reduce the duration and size of the breeding programs considerably. Most importantly, many agriculturally valuable traits such as yield are frequently controlled by more than one gene. The identification of quantitative trait loci (QTLs) based solely on conventional phenotypic evaluation is not possible, and the use of molecular markers creates opportunities for breeders to select for progeny with important quantitative traits.

The dawn of molecular marker development in plants began with the utilization of RFLP (*R*estriction *F*ragment *L*ength *P*olymorphism) and RAPD (*R*andom *A*mplified *P*olymorphic *D*NA) markers [2, 3]. RFLPs and RAPDs are usually among the first types of markers developed because they are relatively easy to generate, and they do not require prior sequence information. However, they are generally not highly polymorphic, and they are often inherited in a dominant manner. Screening a large number of hybridization-based RFLP markers can also be laborious and time-consuming [4]. The next class of markers generated is simple sequence repeats (SSRs) or microsatellites. SSRs typically behave as co-dominant markers, rendering them useful for population genetic studies and mapping. They are also relatively abundant and randomly distributed in the genome.

Recently, attention has been geared toward the use of single nucleotide polymorphisms (SNPs) as molecular markers. Although SNPs are less polymorphic than SSRs due to their bi-allelic nature, the ubiquity of SNPs in plant genomes and their usefulness as genetic markers has been well established over the last decade. SNP markers typically occur at frequencies of one per ~100–500 bp in plant genomes, depending on the species, e.g., 1 SNP/490 bp in soybean [5], 1 SNP/124 bp in maize [6] and 1 SNP/540 bp in pea [7]. SNPs have emerged as the markers of choice in many areas of molecular genetics as they are amenable to high-throughput automated genotyping platforms such as Axiom, Infinium, and MassARRAY iPLEX. Moreover, the cost of SNP genotyping per data point has become cheaper than that of SSRs. With rapid advancement in sequencing throughput together with an overall decrease in sequencing cost, next generation sequencing technologies have been applied to SNP identification in various plant species [8, 9]. However, it remains costly to employ whole-genome sequencing to evaluate several hundred individuals in a mapping population or a germplasm, especially for plant species with large genomes. Reduced representation methods are extremely useful, not only because of their cost-reducing aspects, but also because many research applications do not require every base of the genome to be sequenced. Several techniques have been developed to reduce genome complexity and capture only a fraction of the genome to be sequenced. Restriction-site-associated DNA sequencing (RAD-seq) was introduced by Miller et al. [10] and adapted to incorporate barcoding for multiplexing by Baird et al.

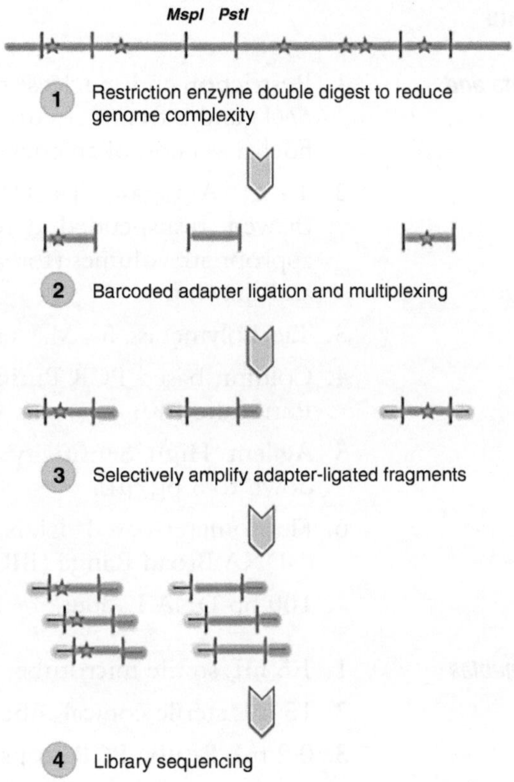

Fig. 1 The flow chart shows the overview of the GBS approach. (**1**) GBS library construction begins with a restriction enzyme double digest followed by (**2**) barcoded adapter ligation (for multiplexing purposes) and (**3**) amplification of adapter-ligated digested fragments prior to sequencing (**4**)

[11]. Elshire et al. [12] proposed a less complicated method for constructing highly multiplexed reduced representation genotyping-by-sequencing (GBS) libraries. GBS is an efficient strategy that can simultaneously discover and genotype tens of thousands of SNP markers. This technique has been demonstrated to be affordable, rapid, and robust across a wide range of species, and it has successfully been applied to marker-trait association analyses, phylogenetic studies, and cultivar identification in several plant species [13–16]. Another key advantage of GBS is that the approach can be applied to species with no reference genomes [17].

This chapter describes a GBS method that employs two restriction enzymes, which are chosen based on preferred locations of SNPs to be identified and downstream applications (*see* details in **Notes** below). For genetic map construction and QTL analysis or other applications that favor genic SNPs or SNPs locating in the vicinity of coding regions, we recommend using a combination of methylation-sensitive enzymes (*Pst*I/*Msp*I or *Aat*II/*Msp*I), and for applications where intergenic SNPs are preferred, we recommend using a pair of methylation-insensitive enzymes (such as *Sph*I/*Mse*I) [18]. The overview of the GBS approach is illustrated in Fig. 1.

2 Materials

2.1 Reagents and Kits

1. Restriction endonuclease of choice: a rare cutter (*Pst*I, *Aat*II or *Sph*I) and a frequent cutter (*Msp*I or *Mse*I). Choose the high-fidelity version of an enzyme, if available (such as *Pst*I-HF).
2. T4 DNA Ligase. T4 DNA Ligase Buffer (10×) should be thawed, resuspended at room temperature, and aliquoted in appropriate volumes (for single use) to avoid repeated freezing and thawing.
3. Taq Polymerase 5× Master Mix.
4. Column-based PCR Purification Kit (such as QIAQuick PCR Purification Kit).
5. Agilent High Sensitivity DNA Kit for DNA quantification down to 5 pg/μL.
6. Fluorometer-based dsDNA quantification Kit, such as Qubit dsDNA Broad Range (BR) Assay Kit.
7. 100 bp DNA Ladder.

2.2 Consumables

1. 1.5 mL sterile microtube
2. 15 mL sterile conical tube
3. 0.2 mL 8-tube PCR strips with attached caps
4. Sterile molecular biology grade water.
5. 1× Elution Buffer (EB): 10 mM Tris-Cl pH 8.0
6. 10× Adapter Buffer (AB): 500 mM NaCl, 100 mM Tris-Cl pH 8.0
7. E-Gel EX Agarose Gels, 2%.

2.3 Equipment

1. PCR Thermal Cyclers.
2. Benchtop centrifuge for microcentrifuge and PCR strips.
3. E-Gel Electrophoresis Apparatus.
4. Bioanalyzer Instrument.

3 Methods

3.1 Adapter Preparation (See Note 1)

1. Resuspend single-stranded adapter oligos to 100 μM in 1× EB.
2. To prepare 10 μM barcoded adapters (Adapter 1), mix 10 μL of each single-stranded oligo (100 μM), 10 μL of 10× AB and 70 μL sterile water (total volume 100 μL).
3. Heat to 95 °C and cool at 1 °C per min to 30 °C. Hold at 4 °C.
4. Dilute Adapter 1 to 3 μM, quantify using Bioanalyzer and normalize to 2.2 ng/μL (~0.1 μM).

5. To prepare the common adapter (Adapter 2), follow **steps 2** and **3** above. Leave Adapter 2 at 10 μM.
6. To prepare 20 μL of "working adapter stock," add the following to 0.2 mL tubes and mix well: 4 μL barcoded adapter (Adapter 1) at ~0.1 μM, 6 μL common adapter (Adapter 2) at 10 μM and 10 μL 1× AB (*see* **Note 2**).

3.2 DNA Normalization and Digestion

1. Quantify DNA samples (*see* **Note 3**) using the Qubit dsDNA Broad Range (BR) Assay Kits and dilute them to 20 ng/μL.
2. This protocol uses a double digestion with one 6-cutter and one 4-cutter (*see* **Note 4**). Prepare a restriction enzyme digest master-mix in a 1.5 mL microtube and aliquot into 0.2 mL PCR strips with attached caps. Each reaction contains: 2 μL of 10× Buffer, 0.4 μL of the 6-cutter (5 units), 0.4 μL of the 4-cutter (5 units), 200 ng of DNA template and sterile water to 20 μL.
3. Incubate the reactions at 37 °C for 2 h.
4. Incubate at 65 °C for 20 min and proceed directly to ligation.

3.3 Adapter Ligation

1. Prepare the ligation master-mix as follows (per reaction): 4 μL NEB 10× Ligation Buffer, 0.5 μL T4 DNA Ligase (200 U), 5 μL working adapter stock and 10.5 μL sterile water.
2. Add 20 μL of the ligation master-mix to the digestion (the ligation is completed in the same tubes as the digestion). The total volume for the ligation is 40 μL.
3. Incubate at 22 °C for 2 h and 65 °C for 20 min.
4. Complete ligation reactions can be stored at −20 °C prior to the multiplex and amplification step.

3.4 Multiplexing (See Note 5)

1. For 12-plex, pool 30 μL from each sample to a single 1.5 mL microtube. For 24-plex, 48-plex and 96-plex, pool 15 μL, 10 μL and 5 μL from each sample, respectively.
2. Due to the large volume of the pooled samples, each library requires two columns of QIAquick or another similar product for a clean-up. Split pooled ligation DNA (360 μL) into two 1.5 mL microtubes and add 900 μL PB (PCR purification) to each tube.
3. Follow the instruction in the PCR purification manual and resuspend the pooled sample in 60 μL.
4. Combine the elution from two columns. The total volume should be 120 μL.

3.5 PCR Amplification (See Note 6)

1. Prepare 100 μL of PCR cocktail mix as follows: 20 μL of pooled library, 20 μL 5× Master-mix, 8 μL 10 μM forward and reverse primers and 52 μL sterile water.
2. Aliquot the PCR mix into four PCR tubes and run the following cycle: 95 °C (30 s); [95 °C (30 s), 62 (20 s), 68 °C (1 min)] × 16 cycles; 72 °C (5 min); and hold at 4 °C.
3. Prior to PCR clean-up, pool amplified products from four PCR tubes and reserve 1 μL of the PCR product for Bioanalyzer tests.
4. Perform PCR clean-up following a PCR Purification Kit and elute in 50 μL Elution Buffer.

3.6 Size Selection and Library Quantification

1. For each library, use four lanes of 2% E-gel Agarose Gel for the size selection step. Mix 10 μL of 100 bp DNA ladder with 30 μL sterile water and load 20 μL of the mixture in wells #1 and #4.
2. Load 25 μL of column-purified library in wells #2 and #3 (25 μL each; 50 μL total).
3. Collect the samples when the appropriate size has entered the well (*see* **Note 7**).
4. Rinse each well with 10 μL sterile water and collect the wash.
5. Quantify the cleaned, size-selected library concentration using the Bioanalyzer High Sensitivity DNA Kit. It is a good practice to analyze 1 μL of the cleaned PCR products before the size selection to ensure that the amplification was successful.
6. The cleaned, size-selected library is ready to be loaded in the sequencer. The amount that should be loaded depends on the sequencing platform used and can be calculated based on the library concentration obtained from the Bioanalyzer (*see* **Note 8**).

4 Notes

1. Adapters are ordered as standard desalting oligos. For each adapter, two single-stranded oligos must be ordered in complementary pairs and must be annealed to form double-stranded adapters prior to use. After annealing, adapters should be stored at −20 °C.
2. Final concentrations of Adapter 1 and Adapter 2 in "working adapter stock" are 0.02 μM and 3 μM, respectively. Extra care should be taken to make sure that the working adapter stocks are kept at 4 °C or lower.
3. It is important to quantify and normalize DNA samples to the same concentration to ensure that each barcoded sample will

produce an even number of sequence tags. It is recommended that DNA be quantified using a fluorescence-based approach such as the Qubit Fluorometric System.

4. A combination of a rare cutter (6-base cutter such as *Pst*I, *Aat*II, *Sph*I) and a frequent cutter (4-base cutter such as *Msp*I and *Mse*I) is used to perform a double digestion. The 6-cutter overhang corresponds to Adapter 1 (barcoded adapter), and the 4-cutter overhang corresponds to the common Adapter 2. The choice of restriction endonucleases used depends on the desired number and location of SNP markers. If genic SNPs or SNPs locating in the vicinity of coding regions are preferred, methylation sensitive enzymes are recommended (*Pst*I/*Msp*I or *Aat*II/*Msp*I). On contrary, if SNPs in the intergenic regions are preferred, methylation-insensitive enzymes such as *Sph*I/*Mse*I are recommended. Please see Pootakham et al. [18] for a thorough evaluation on effects of methylation-sensitive enzymes on the enrichment of genic SNPs.
5. The level of multiplexing depends on the number of sequence reads desired per sample and the sequencing output of the platform used.
6. PCR amplification is performed in four separate reactions (25 μL × 4) to avoid accumulating mutations that arise during this step.
7. The appropriate library size depends on the sequencing platform used. For Ion Torrent S5 sequencers, we select amplicons of approximately 270 bp. For Illumina sequencers, the preferred average amplicon size is ~350 bp.
8. The protocol detailed here has been used to generate genetic linkage maps in oil palm and rubber tree using two methylation-sensitive enzymes, *Pst*I and *Msp*I [19, 20].

References

1. Lande R, Thompson R (1990) Efficiency of marker-assisted selection in the improvement of quantitative traits. Genetics 124:743–756. https://doi.org/10.1093/genetics/124.3.743
2. Rafalski JA, Tingey SV (1993) Genetic diagnostics in plant breeding: RAPDs, microsatellites and machines. Trends Genet 9:275–280. https://doi.org/10.1016/0168-9525(93)90013-8
3. Ragot M, Hoisington DA (1993) Molecular markers for plant breeding: comparisons of RFLP and RAPD genotyping costs. Theor Appl Genet 86:975–984. https://doi.org/10.1007/BF00211050
4. Mohan M, Nair S, Bhagwat A, Krishna TG, Yano M, Bhatia CR et al (1997) Genome mapping, molecular markers and marker-assisted selection in crop plants. Mol Breed 3:87–103. https://doi.org/10.1023/A:1009651919792
5. Choi IY, Hyten D, Matukumalli L, Song Q, Chaky J, Quigley C et al (2007) A soybean transcript map: gene distribution, haplotype and single-nucleotide polymorphism analysis. Genetics 176:685–696. https://doi.org/10.1534/genetics.107.070821
6. Ching A, Caldwell KS, Jung M, Dolan M, Smith O, Tingey S et al (2002) SNP frequency, haplotype structure and linkage disequilibrium

in elite maize inbred lines. BMC Genet 3:19. https://doi.org/10.1186/1471-2156-3-19

7. Leonforte A, Sudheesh S, Cogan N, Salisbury P, Nicolas M, Materne M et al (2013) SNP marker discovery, linkage map construction and identification of QTLs for enhanced salinity tolerance in field pea (*Pisum sativum* L.). BMC Plant Biol 13:161. https://doi.org/10.1186/1471-2229-13-161
8. Mammadov J, Aggarwal R, Buyyarapu R, Kumpatla S (2012) SNP markers and their impact on plant breeding. Int J Plant Genomics 2012:728398. https://doi.org/10.1155/2012/728398
9. Ganal M, Altmann T, Röder M (2009) SNP identification in crop plants. Curr Opin Plant Biol 12:211–217. https://doi.org/10.1016/j.pbi.2008.12.009
10. Miller MR, Dunham JP, Amores A, Cresko WA, Johnson EA (2007) Rapid and cost-effective polymorphism identification and genotyping using restriction site associated DNA (RAD) markers. Genome Res 17:240–248. https://doi.org/10.1101/gr.5681207
11. Baird NA, Etter PD, Atwood TS, Currey MC, Shiver AL, Lewis ZA et al (2008) Rapid SNP discovery and genetic mapping using sequenced RAD markers. PLoS One 3:e3376. https://doi.org/10.1371/journal.pone.0003376
12. Elshire RJ, Glaubitz JC, Sun Q, Poland JA, Kawamoto K, Buckler ES et al (2011) A robust, simple genotyping-by-sequencing (GBS) approach for high diversity species. PLoS One 6:e19379. https://doi.org/10.1371/journal.pone.0019379
13. D'Agostino N, Taranto F, Camposeo S, Mangini G, Fanelli V, Gadaleta S et al (2018) GBS-derived SNP catalogue unveiled wide genetic variability and geographical relationships of Italian olive cultivars. Sci Rep 8:15877. https://doi.org/10.1038/s41598-018-34207-y
14. Poland JA, Brown PJ, Sorrells ME, Jannink JL (2012) Development of high-density genetic maps for barley and wheat using a novel two-enzyme genotyping-by-sequencing approach. PLoS One 7:e32253. https://doi.org/10.1371/journal.pone.0032253
15. He J, Zhao X, Laroche A, Lu Z, Liu H, Li Z (2014) Genotyping by sequencing (GBS), an ultimate marker-assisted selection (MAS) tool to accelerate plant breeding. Front Plant Sci 5:484. https://doi.org/10.3389/fpls.2014.00484
16. Favre F, Jourda C, Besse P, Charron C (2021) Genotyping-by-sequencing technology in plant taxonomy and phylogeny. In: Besse P (ed) Molecular plant taxonomy: methods and protocols. Methods in molecular biology, vol 2222. Humana, New York, pp 167–178. https://doi.org/10.1007/978-1-0716-0997-2_10
17. Berthouly-Salazar C, Mariac C, Couderc M, Pouzadoux J, Floc'h JB, Vigouroux Y (2016) Genotyping-by-sequencing SNP identification for crops without a reference genome: using transcriptome based mapping as an alternative strategy. Front Plant Sci 7:777. https://doi.org/10.3389/fpls.2016.00777
18. Pootakham W, Sonthirod C, Naktang C, Jomchai N, Sangsrakru D, Tangphatsornruang S (2016) Effects of methylation-sensitive enzymes on the enrichment of genic SNPs and the degree of genome complexity reduction in a two-enzyme genotyping-by-sequencing (GBS) approach: a case study in oil palm (*Elaeis guineensis*). Mol Breed 36:154. https://doi.org/10.1007/s11032-016-0572-x
19. Pootakham W, Jomchai N, Ruang-Areerate P, Shearman JR, Sonthirod C, Sangsrakru D et al (2015) Genome-wide SNP discovery and identification of QTL associated with agronomic traits in oil palm using genotyping-by-sequencing (GBS). Genomics 105:288–295. https://doi.org/10.1016/j.ygeno.2015.02.002
20. Pootakham W, Ruang-Areerate P, Jomchai N, Sonthirod C, Sangsrakru D, Yoocha T et al (2015) Construction of a high-density integrated genetic linkage map of rubber tree (*Hevea brasiliensis*) using genotyping-by-sequencing (GBS). Front Plant Sci 6:367. https://doi.org/10.3389/fpls.2015.00367

Chapter 2

Genotyping by Multiplexed Sequencing (GMS) Using SNP Markers

Travis M. Ruff, Karol Marlowe, Marcus A. Hooker, Yan Liu, and Deven R. See

Abstract

SNP-based genotyping has become the most effective approach to generate target-specific data for use in genetic studies. In this chapter, we will describe a high-throughput genotyping method that multiplexes hundreds to thousands of SNP markers in a two-step PCR protocol that can be customized to fit the specific needs of a study.

Key words Genotyping by multiplexed sequencing, Genotyping, Single nucleotide polymorphism (SNP), Next-generation sequencing (NGS), Amplicon sequencing

1 Introduction

Over the last decade, genotyping has been moving away from low-throughput, uniplex assays to high-throughput, multiplexed technologies [1]. This shift was possible with advances in next-generation sequencing capacities and their incorporation of single nucleotide polymorphism (SNP) based assays [1, 2]. SNPs make an ideal molecular marker as they are typically co-dominant and are the largest source of genetic variability in eukaryotic genomes [3]. SNP markers can be used to track important agronomic traits in crops, follow diagnostic SNPs linked to genes of interest or be used as anchor points for a variety of genetic studies including genomic selection (GS), association studies and marker-assisted selection [4]. SNP-based assays have allowed researchers to obtain more data points per assay at a lesser cost and in a shorter amount of time [5]. Genotyping by multiplexed sequencing (GMS) is a high-throughput method that pools hundreds to thousands of SNP markers into a single assay that track specific loci throughout a genome [6]. GMS was developed to be used with GS but can be

Yuri Shavrukov (ed.), *Plant Genotyping: Methods and Protocols*, Methods in Molecular Biology, vol. 2638,
https://doi.org/10.1007/978-1-0716-3024-2_2,

adapted to fit a researcher's specific needs. The flexibility of this technology is in the primer pool(s) as a researcher can customize primer pairs in a pool to fit the specific needs of a given experiment. The GMS protocol begins with two PCRs. The first PCR generates the targeted amplicons and adds the Illumina read 1 and 2 adaptors. The second PCR incorporates the sample index and Illumina sequencing adapters. These PCRs are followed by four library purification and two quantification steps, before sequencing on an Illumina platform.

2 Materials

2.1 Consumables

1. Standard 96-well PCR plates.
2. Silicone sealing mats for 96-well PCR plate.
3. 1.5 mL microcentrifuge tubes.
4. 2.0 mL microcentrifuge tubes.
5. 5.0 mL microcentrifuge tubes.
6. 0.2 mL 12-well strip tubes with caps.
7. Qubit assay tubes.

2.2 Kits

1. PCR Purification Kit.
2. Gel Extraction Kit.
3. Qubit dsDNA Quantitation, high sensitivity (ThermoFisher or similar product).
4. High Sensitivity DNA Kit (Agilent Technologies or similar product).

2.3 Reagents

1. Primer pool (250 nM) (*see* **Notes 1–4**).
2. HoTaq DNA polymerase (5 U/μL) (MCLAB or similar Hot Start DNA polymerase).
3. $MgCl_2$ (25 mM).
4. dNTPs (5 and 100 mM).
5. Molecular-grade water.
6. Illumina forward sequencing primer (10 μM).
7. Illumina i7 indexes (2 μM).
8. Ethanol (80%) (freshly prepared).
9. Isopropanol (100%).
10. AMPure XP beads (Beckman Coulter or similar product).
11. E-Gel SizeSelect II Agarose Gels, 2% (ThermoFisher or similar).
12. E-Gel EX Agarose Gels, 4% (ThermoFisher or similar).
13. 50 bp DNA ladder.

2.4 Equipment

1. Thermal cycler.
2. Centrifuge with plate rotor.
3. Benchtop centrifuge for microcentrifuge tubes.
4. Water bath.
5. Magnetic 1.5 mL tube rack.
6. E-Gel™ Power Snap Electrophoresis Device (ThermoFisher or similar).
7. Invitrogen™ Qubit™ 4 Fluorometer (ThermoFisher or similar).
8. 2100 Bioanalyzer Instrument (Agilent Technology or similar).

3 Methods

3.1 SNP Primer Design for Primer Pools

1. Determine the SNPs that will be included in the new primer pool (*see* **Note 1**).
2. To design primers, choose a design program that was developed for multiplexing primer pairs, as these programs analyze primer interactions within a given design (*see* **Note 2**).
3. Order the primers in a 96-well plate with each forward and reverse primer pair combined into a single well. Request each well be normalized to a concentration of 100 μM with standard purification (*see* **Note 3**).

3.2 Creating Primer Pools

1. Thaw the primer plate(s) to be used in creating the primer pool.
2. Determine the concentration and volume of the primer pool and note how many primers will be in it.
3. Calculate the volume to add from each primer well to create a 250 nM primer pool with 300 primer pairs, in a total volume of 1000 μL (*see* **Note 4**).
4. Using a 12-channel pipette, aspirate 2.5 μL from each well of the primer plate and dispense into a new 12-well strip tube, changing tips after each dispense.
5. With a single-channel pipette, remove the volume from each well of the strip tube and dispense into a new 1.5 mL tube.
6. Cap the tube, vortex briefly and spin down (*see* **Note 5**).

3.3 GMS PCR-1

1. Normalize the sample plate DNA so the concentration of each well is the same (*see* **Note 6**).
2. Prepare PCR-1 master mix for the number of sample libraries to be created in the run: In a new 1.5 mL tube combine 10× Taq PCR Buffer (no dNTPs), $MgCl_2$, dNTPs, primer pool, Hot start Taq polymerase and molecular-grade water according to Table 1 (*see* **Note 7**).

Table 1
GMS PCR-1 master mix guide

PCR-1 Master Mix		
Reagent	**Per sample (μL)**	**×110 (μL)**
MCLAB 10× Taq PCR Buffer (no dNTPs)	1	110
$MgCl_2$ (25 mM)	0.45	49.5
dNTPs (5 mM)	1	110
Primer pool (250 nM)	0.5	55
MCLAB HoTaq (5 U/μL)	0.2	22
Molecular-grade water	2.85	313.5
Total	6	660

Table 2
Thermal cycler PCR-1 conditions

Step	Temperature (°C)	Time	Note
1	94	10 min	
2	94	20 s	
3	56	2 min	
4	68	30 s	
5			Go to step 2, 35×
6	72	3 min	
7	4	∞	

3. Cap the tube, vortex briefly and spin down. Dispense 55 μL of PCR-1 master mix to each well of a new 12-well strip tube.
4. Using a 12-channel pipette, dispense 6 μL of PCR-1 master mix to each well of a new 96-well PCR plate (*see* **Note 8**).
5. Add 4 μL of normalized sample DNA to its corresponding well in the PCR-1 plate with a 12-channel pipette and mix. Change tips between dispenses of sample DNA.
6. Seal the 96-well PCR plate with a silicone mat, vortex briefly and spin down.
7. Place the PCR-1 plate on a thermal cycler and follow the profile in Table 2.
8. Remove the PCR-1 plate and spin down.

3.4 PCR-1 Dilution

1. In a new 96-well PCR plate make a 1:1 dilution of PCR-1 product with molecular-grade water.
2. Add 9 μL of molecular-grade water to each well of the PCR-1 dilution plate using a 12-channel pipette.
3. Aspirate 9 μL of PCR-1 product from each well and dispense into the corresponding well of the new PCR-1 dilution plate. Change tips between dispenses of PCR-1 product (*see* **Note 9**).

3.5 GMS PCR-2

1. Create the PCR-2 master mix according to the number of libraries generated in PCR-1. In a new 1.5 mL tube combine 10× Taq PCR Buffer (no dNTPs), $MgCl_2$, dNTPs, Illumina-F, Hot start Taq polymerase and molecular-grade water according to Table 3 (*see* **Note 10**).
2. Cap the tube, vortex briefly and spin down. Aliquot 18 μL of PCR-2 master mix to each well of a new 12-well strip tube.
3. With a 12-channel pipette, aliquot 2 μL of PCR-2 master mix to each well of a new 96-well PCR plate. Add 2 μL of diluted PCR-1 product to its corresponding well in the PCR-2 plate, changing tips between dispenses.
4. Add 2 μL of each i7 index to its corresponding well in the PCR-2 plate using a 12-channel pipette, changing tips between dispenses of indexes (*see* **Note 11**).
5. Seal the plate with a silicone mat, vortex briefly and spin down.
6. Place the PCR-2 plate on a thermal cycler and use the conditions as shown in Table 4.
7. Remove the PCR-2 plate and spin down.

Table 3
GMS PCR-2 master mix guide

PCR-2 Master Mix		
Reagent	**Per sample (μL)**	**×110 (μL)**
MCLAB 10× Taq PCR Buffer (no dNTP)	0.6	66
$MgCl_2$ (25 mM)	0.2	22
dNTPs (100 mM)	0.03	3.3
Illumina-F (10 μM)	0.2	20
MCLAB HoTaq (5 U/μL)	0.2	20
Molecular-grade water	0.77	84.7
Total	2	216

Table 4
PCR-2 thermal cycler conditions

Step	Temperature (°C)	Time	Note
1	94	10 min	
2	94	20 s	
3	60	30 s	
4	72	1 min	
5			Go to **step 2**, 15×
6	72	3 min	
7	4	∞	

8. With a 12-channel pipette, remove 5 μL from each well of the plate and dispense into a new 12-well strip tube. Remove all volume from each well of the strip tube and dispense into a new 5.0 mL tube to pool the indexed libraries. Vortex briefly and spin down (*see* **Note 12**).

3.6 GMS Library Pool Cleanup

1. To clean the indexed GMS libraries, follow the QIAquick PCR purification kit protocol with the following modifications (*see* **Note 13**).
2. Use a single column per plate to keep the library concentration as high as possible.
3. Elute the column with 30 μl of EB buffer at 60 °C and incubate for 1 min before centrifuging.

3.7 GMS Library Pool Bead Cleanup

1. Place a new 12-well strip tube on a thermal cycler at 37 °C, add 35 μL of molecular-grade water to a well and cap the strip tube. Prepare 200 μL of 80% ethanol (prepare fresh each time).
2. Ensure the AMPure XP beads are at room temperature (~22 °C) and mix thoroughly. Measure the volume in the 1.5 mL tube from the GMS library pool cleanup and add beads at a 1:1 ratio. Mix the libraries and beads until homogeneous.
3. Incubate the tube at room temperature for 5 min.
4. Place the tube onto a magnetic 1.5 mL tube rack and incubate at room temperature for 5 min.
5. Slowly aspirate all supernatant from the tube, without bead carryover (*see* **Note 14**).
6. Dispense 200 μL of 80% ethanol above the bead cluster in the tube and incubate at room temperature for 30 s.
7. Aspirate and discard all ethanol from the tube.

8. Remove tube from the magnetic rack and dispense 30 μL of molecular-grade water (37 °C) to the tube above the bead cluster. Mix the water and beads until homogenous. Incubate at room temperature for 1 min.
9. Place the tube onto the magnetic tube rack for 5 min.
10. Keep the tube on the magnetic rack and transfer the eluate into a new 1.5 mL tube. Avoid bead carryover.

3.8 4% Gel Size Selection

1. Place a 4% E-Gel cassette into the E-Gel Power Snap Electrophoresis Device (EPSED).
2. Load 25 μL of GMS library to the sample lane.
3. Load 15 μL of E-Gel 50 bp DNA ladder to both sides of the sample lane.
4. On the EPSED, select the program E-Gel EX 4% and click "Start run."
5. Use the backlight to track library band migration until the 100 bp ladder fragment is at the top of the lower sticker (*see* Fig. 1).
6. Once the 100 bp ladder fragment has reached the top of the bottom sticker, stop the program and remove the E-Gel cassette (*see* **Note 15**).

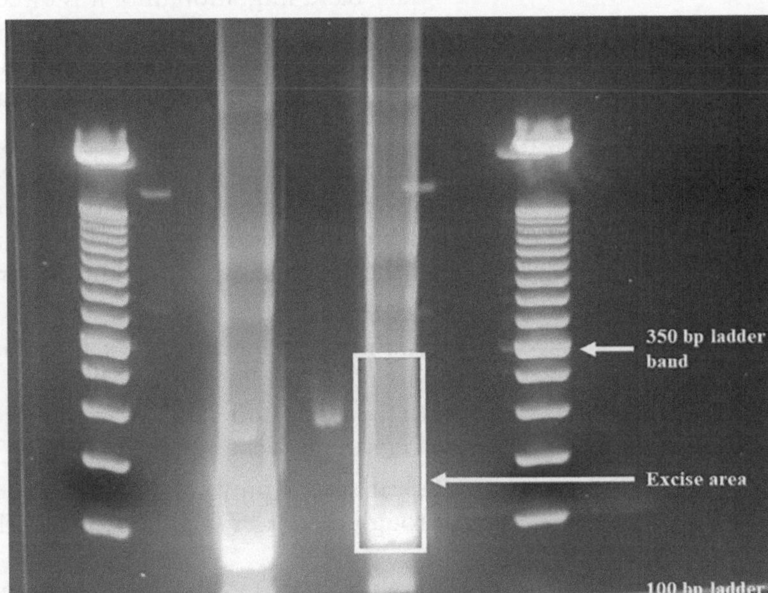

Fig. 1 GMS library migration on a 4% E-Gel

7. Using a small, flathead screwdriver with a thin blade, carefully pry open the plastic cassette to expose the gel for excision (*see* **Note 16**).
8. Excise the GMS library band with a scalpel and place it into a new 2.0 mL tube. The excise area is typically 150–350 bp (*see* Fig. 1 and **Note 17**).

3.9 Gel Purification

1. To purify the excised gel, follow the Qiagen QIAquick Gel Extraction Kit protocol with the following exceptions (*see* **Note 18**):
2. Add 6 volumes of QG buffer to the tube with the excised gel slice.
3. Elute DNA with 30 μL EB buffer at 60 °C.

3.10 2% Gel Size Selection

1. Place a 2% E-Gel into the EPSED. Select program "SizeSelect 2%."
2. Fill lower wells with 25 μL of molecular-grade water.
3. Load 15 μL of 50 bp ladder to the top, middle well.
4. Lightly vortex the 1.5 mL tube from the gel purification step and spin down. Add 25 μL of GMS library to the top well adjacent to the ladder, close the lid and press "Start run."
5. While the program is running, use the backlight to track the library band migration until it is directly above the lower sample well. Pause the program and remove the molecular-grade water from the lower sample well. Add 25 μL of fresh molecular-grade water to the lower sample well.
6. Close the lid and resume the run. Turn on the backlight to visualize the library band migrate into the lower sample well. Pause the program, aspirate the liquid from the lower sample well and dispense into a new 1.5 mL tube. Add 25 μL of fresh molecular-grade water to the lower sample well and close the lid.
7. Resume the program for 15 s, then pause the run. Open the lid and aspirate the liquid from the lower sample well and dispense into the tube. Add 25 μL of fresh molecular-grade water to the lower sample well.
8. Repeat **step 7** until the library band has completely migrated through the lower sample well. Cap the tube and spin down.

3.11 GMS Library Pool Cleanup

1. To clean the GMS libraries, follow the QIAquick PCR purification kit protocol with the following modifications (*see* **Note 13**):
2. Use a single column per plate to keep the library concentration as high as possible.
3. Elute the column with 30 μl of EB buffer at 60 °C and incubate for 1 min before centrifuging.

3.12 GMS Library Pool Quantification and Length Assays

1. To quantify the GMS library pool, follow the Qubit dsDNA HS Assay Kit user guide with the following exception (*see* **Note 19**).
2. Use 2 μL of GMS library pool for quantification.
3. Determine GMS library length by following the Agilent High Sensitivity DNA Kit Guide.
4. The GMS libraries are ready for sequencing on an Illumina platform. Check with your sequencing center of choice to determine the concentration and volume required for submission (*see* **Note 20**).

4 Notes

1. Limit the number of primer pairs in a single primer pool to 300 or fewer. It was found primer pools that contain more than 300 primer pairs yield diminishing returns.
2. The GMS protocol uses the MassArray Assay Design 4.0 software (Agena Bioscience, San Diego, CA) for primer design, but other design software can be used. The important parameters for GMS primer design are the melting temperature (T_m) and amplicon length. The T_m range must be 56–60 °C with the optimum temperature of 56 °C. The amplicon length should be 100–150 bp with the optimum length of 135 bp. After primer design, Illumina Read adapters must be added to the 5′ end of each forward and reverse primers as shown below. The Nextera Read 1 and 2 adapter sequences are in italics. The forward and reverse locus-specific primer sequences are underlined.

   GMS forward primer with Read 1 adapter:

   5′-*TCGTCGGCAGCGTCAGATGTGTATAAGAGACAG*CCTGTTAGTAGTGATGGTCC

   GMS reverse primer with Read 2 adapter:

   5′-*GTCTCGTGGGCTCGGAGATGTGTATAAGAGACAG*CCAGAGCACACCAAAGTA

   For additional adapter and index sequences refer to "Illumina Adapter Sequences" Document # 1000000002694 v16. If sequencing GMS libraries on alternative sequencing platforms, be sure to use platform specific adaptor sequences.
3. Have the primers synthesized at the lowest scale available. The GMS protocol has the primers resuspended in molecular-grade water. Ordering GMS primers in a 96-well plate makes the process of constructing primer pools simpler, as a 12-channel pipette can be used and will limit pipetting errors. Store primers at −20 °C.

4. Use the following equation when making a primer pool: (concentration of new pool) (volume of new pool) = (concentration of primer well) (X volume of each primer well to add to the new primer pool).

   To calculate the volume to add from each primer well see the following example:

   $$(250 \text{ nM})\ (1000\ \mu\text{L}) = (100{,}000 \text{ nM})\ (X\ \mu\text{L})$$

   X = 2.5 μL per primer well, with 300 total primer pairs the primer volume would be 750 μL. Add 250 μL of molecular-grade water to the primer pool to bring the concentration to 250 nM.

5. To reduce the number of freeze-thaw cycles a primer pool experiences, make smaller aliquots and store them at −20 °C. Freeze-thaw cycles degrade primers over time and can lead to non-specific amplification.
6. Use an extraction method that yields high-quality DNA suitable for next-generation sequencing. DNA concentrations were determined by fluorescence on a plate reader. Spectrophotometry can be used for quantification, but the results are not as accurate as fluorescence. DNA normalization was performed by a liquid handling system. Normalizing sample DNA in this manner is recommended to reduce cross-contamination by human error. When normalizing your sample plate DNA use the standard concentration for PCR in your lab. For reference, when performing PCR with wheat our lab uses a DNA concentration of 20 ng/μL and 6.6 ng/μL for barley.
7. To create the PCR-1 master mix for 96 samples, the GMS protocol multiplies the Table 1 "Per sample" volume of each reagent by 110 to account for pipette error. Keep all reagents and sample DNA on ice until **step 6**.
8. Using a 12-channel pipette for dispensing master mixes, sample DNA and i7 indexes saves time and will reduce cross-contamination by human error.
9. Evaporation from wells may occur during PCR, especially in the outer wells. If evaporation occurred, determine which well has the lowest volume and use this volume to make your dilution from each well. Do not use less than 6 μL from PCR-1 to make the dilution. Try to keep well volumes similar during this step as it will help limit sample bias when sequencing.
10. The Illumina-F reagent in Table 3 is the forward primer for PCR-2. In the "Illumina Adapter Sequences" document, its sequence is the same as the Index 2 read without an i5 index (*see* **Note 2**). The Index 2 Read sequence is in italics and the PCR-1 overhang sequence is in bold:

Illumina-F: 5′*AATGATACGGCGACCACCGAGATC-TACAC***TCGTCGGCAGCGTC**

If sequencing on a paired-end flow cell, incorporate an i5 index sequence into the Index 2 read forward primer between the italicized and underlined sequences. Keep all reagents and diluted PCR-1 products on ice until **step 5**.

11. To create the i7 indexes follow the "Index 1 (i7) Adapters" example sequence in the "Illumina Adapter Sequences" document (*see* **Note 2**). The following is an example of an i7 index primer that would be used in PCR-2. The Index 1 Read adaptor sequence is in italics, the underlined sequence is a unique i7 index and the PCR-1 overhang sequence is in bold. 5′*CAAG-CAGAAGACGGCATACGAGAT*CGCTCAGTTC**GTCT CGTGGGCTCGG**

    If sequencing on a paired-end flow cell, add an i5 index sequence to the forward primer in PCR-2 (*see* **Note 10**). The "Illumina Adapter Sequences" document has 384 unique i7 and i5 index sequences which allows for a multitude of plate multiplexing options depending on the type of flow cell chosen for sequencing.

12. Each well has a unique index so there is no need to change tips during pooling. It is important to remove all liquid from each well to ensure each sample is represented as equally as possible during sequencing.

13. The GMS protocol uses the QIAquick PCR purification kit, but other PCR purification kits can be used.

14. If beads are aspirated with the supernatant, dispense back into the tube and repeat **step 4**. We do not want any bead carryover as the GMS libraries are bound to the beads.

15. If the 100 bp ladder band is not at the top of the lower sticker at the end of the program, run the same program again tracking the band with the backlight. This allows for maximum separation between the GMS library and primer dimer band for gel excision.

16. Each side of the cassette has a groove to insert the blade of the screwdriver. Place the cassette onto a flat surface, gently push the blade into the groove and slide it toward the cassette's top corner. You will be able to hear and see the top and bottom of the cassette separating. If you cannot slide the tip any further turn the handle of the screwdriver a ¼ turn or until you see the plastic separate. Continue to slide and turn the blade of the screwdriver until all four sides of the cassette are unsealed. While removing the cassette top the gel may stick and tear. If a tear occurs through the excise area, carefully complete the tear, remove the gel from the cassette top and replace to the bottom gel into its correct orientation for the excision. See "E-Gel™ Technical Guide" for other methods to open the gel cassette.

17. The goal of this size selection is to exclude the primer dimer band at ~120 bp (*see* Fig. 1). Excluding primer dimer increases the amount of useful data received from sequencing, as the GMS libraries occupy more of a flow cell's oligonucleotide lawn instead of primer dimer. If using the backlight with the EPSED lid up to visualize the excision area, wear the supplied Safe Imager Viewing Glasses that are included with the EPSED kit. Weigh the new 2.0 mL tube before placing the excised gel into it, as this weight will be needed in the gel purification step.
18. The GMS protocol uses the Qiagen QIAquick Gel Extraction Kit, but other gel purification kits can be substituted.
19. A minimum GMS library concentration of 1.0 ng/μL is needed for high-quality sequencing results. Sequencing with less concentrated GMS libraries has been attempted but the sequencing results were inconsistent.
20. Amplicon sequencing methods such as GMS typically do not create nucleotide diverse libraries as they are targeted genotyping methods. Nucleotide diversity of libraries needs to be considered when sequencing on newer Illumina platforms as they have transitioned away from using a different fluorescent dye for each nucleotide to using two fluorescent dyes. If only sequencing a GMS project on a flow cell, additional PhiX control will need to be loaded on the run to increase nucleotide diversity. If a GMS project is to be spiked into a sequencing run with other libraries of high diversity (such as exome capture or randomly fragmented sample libraries) increasing the amount of PhiX is not needed if the GMS libraries do not exceed 50% of the libraries sequenced on a flow cell or single lane. Let your sequencing center know you will be submitting amplicon libraries with low nucleotide diversity and discuss how to proceed with sequencing. For further information about library nucleotide diversity see the Illumina bulletin "What is nucleotide diversity and why is it important?"

Acknowledgments

The GMS manuscript was funded by the USDA-ARS.

References

1. Thomson M (2014) High-throughput genotyping to accelerate crop improvement. Plant Breed Biotech 2:195–212. https://doi.org/10.9787/PBB.2014.2.3.195
2. Bhat J, Ali S, Salgotra R, Mir ZA, Dutta S, Jadon V et al (2016) Genomic selection in the era of next generation sequencing for complex traits in plant breeding. Front Genet 7:221. https://doi.org/10.3389/fgene.2016.00221
3. Agarwal M, Shrivastava N, Padh H (2008) Advances in molecular marker techniques and their applications in plant sciences. Plant Cell Rep 27:617–631. https://doi.org/10.1007/s00299-008-0507-z

4. Bernardo A, Wang S, St. Amand P, Bai G (2015) Using next generation sequencing for multiplexed trait-linked markers in wheat. PLoS One 10:e0143890. https://doi.org/10.1371/journal.pone.0143890
5. Rasheed A, Wen W, Gao F, Zhai S, Jin H, Liu J et al (2016) Development and validation of KASP assays for genes underpinning key economic traits in bread wheat. Theor Appl Genet 129:1843–1860. https://doi.org/10.1007/s00122-016-2743-x
6. Ruff T, Marston E, Eagle J, Sthapit SR, Hooker MA, Skinner DZ et al (2020) Genotyping by multiplexed sequencing (GMS): a customizable platform for genomic selection. PLoS One 15: e0229207. https://doi.org/10.1371/journal.pone.0229207

Chapter 3

Computational Protocol for DNA Methylation Profiling in Plants Using Restriction Enzyme-Based Genome Reduction

Wendell Jacinto Pereira, Marília de Castro Rodrigues Pappas, and Georgios Joannis Pappas Jr.

Abstract

Epigenetics can be described as heritable phenotype changes that do not involve alterations in the underlying DNA sequence. Having widespread implications in fundamental biological phenomena, there is an increased interest in characterizing epigenetic modifications and studying their functional implications. DNA methylation, particularly 5-methylcytosine (5mC), stands out as the most studied epigenetic mark and several methodologies have been created to investigate it. With the development of next-generation sequencing technologies, several approaches to DNA methylation profiling were conceived, with differences in resolution and genomic scope. Besides the gold standard whole-genome bisulfite sequencing, which is costly for population-scale studies, genomic reduced representation methods emerged as viable alternatives to investigate methylation loci. Whole-genome bisulfite sequencing provides single-base methylation resolution but is costly for population-scale studies. Genomic reduction methods emerged as viable alternatives to investigate a fraction of methylated loci. One of such approaches uses double digestion with the restriction enzymes *Pst*I and one of the isoschizomers, *Msp*I and *Hpa*II, with differential sensitivity to 5mC at the restriction site. Statistical comparison of sequencing reads counts obtained from the two libraries for each sample (*Pst*I-*Msp*I and *Pst*I-*Hpa*II) is used to infer the methylation status of thousands of cytosines. Here, we describe a general overview of the technique and a computational protocol to process the generated data to provide a medium-scale inventory of methylated sites in plant genomes. The software is available at https://github.com/wendelljpereira/DArTseqMet.

Key words DNA methylation, Restriction enzymes, Methylation sensitivity, Next-generation sequencing, Differential expression, Computational protocol

1 Introduction

Genotyping aims to catalog heritable differences that discriminate between individuals at the DNA sequence level (genetics), which is the blueprint enacting the phenotype. However, in multicellular organisms, this information is decoded differently across the cells, in a phenomenon governed by the accessibility of the DNA

Yuri Shavrukov (ed.), *Plant Genotyping: Methods and Protocols*, Methods in Molecular Biology, vol. 2638, https://doi.org/10.1007/978-1-0716-3024-2_3,

molecule in its intimate association with proteins (chromatin). The dynamic and reversible modulation of the chromatin structure imposes a new layer of cellular regulation that can be reshaped in response to internal and external stimuli. This is the domain of epigenetics, where distinct cellular phenotypes arise from the same underlying DNA sequence. Over the years, it was established that epigenetic phenomena play pivotal roles in modulating gene expression and genome organization, intertwined with numerous physiological processes such as growth, development, response to biotic and abiotic stresses, and eventually adaptive evolution [1–5]. It is established that epigenetic states can be transmitted over generations and, for sessile organisms like plants, this can expand the repertoire of adaptive phenotypic responses to environmental changes [1]. In this scenario, DNA sequence-only surveys may not completely explain the emergence and mechanisms of phenotypic plasticity and it has been proposed that epigenetic modifications may explain part of the missing heritability found in genome-wide association studies [6].

The main epigenetic effectors are histone modifications, small RNAs, and reversible chemical modifications to the DNA molecule, without affecting the base identity. In the latter case, the most prevalent of such modifications is the methylation of cytosine at position 5′, resulting in 5-methylcytosine (5mC). A sizable portion of eukaryotic genomes bears this modification, particularly in the local sequence context of CG dinucleotide, but also CHG and CHH (where H is any base other than G) [7]. 5mC is normally recognized as a repressive mark, directing chromatin insulation and resulting in the silencing of gene expression and transposable element mobilization. Highly methylated spans in DNA may offer circumstantial evidence for regional chromatin organization and transcriptional activity, without resorting to mRNA abundance measurements.

The looming prospects of climate change impose realistic concerns on food production and global sustainability [8]. For many important food crops, it is observed a narrow genetic variation because of the use of inbred lines and clonal propagation. In this setting, it is of utmost importance to identify genotypes capable of adapting to environmental challenges, as well as the loci contributing to favorable phenotypic characters [9–11]. Epigenetic profiling offers an attractive strategy to pinpoint this facet in a limited genetic variation context. Nonetheless, it is still difficult to establish a clear causality between epigenetics and phenotypic plasticity [1].

Given the far-reaching implications of epigenetics in several aspects of the cellular regulatory landscape and genotype-environment interactions, various methodologies have been developed for the discovery and typing of different epigenetic marks [12, 13]. There are alternatives in terms of resolution and costs, and methods vary from site-specific targeted methylation analysis, sampling a few sites, to whole-genome single-base resolution, sampling millions of methylation loci. Consequently, it is possible to accommodate different types of studies and species. Our focus will be on genome-wide DNA methylation profiling in plants.

Unlike DNA sequence-based genotyping, 5mC snapshots vary within the same organism. Therefore, methylation profiling should be conducted in an experimental design akin to transcriptional profiling, contrasting treatments, tissues, or developmental stages. This feature per se makes the methylation analysis more complex than DNA genotyping, since samples comprise an aggregate of different cell types, each with its methylation pattern, resulting in an averaged signal.

Several approaches were created and/or adapted since the early global methylation quantification with HPLC-UV (High-performance liquid chromatography-ultraviolet) adopted in 1980 [14]. New strategies flourished, especially after the establishment of next-generation sequencing technologies, allowing methylation detection at different levels [15–17]. The list can sum up dozens of techniques, each with advantages and drawbacks, so that the most appropriate assay will vary according to the research needs. For example, important factors are the required sensitivity and specificity, quantity and quality of DNA samples, and if the goal is to discover or assess methylation in target regions, to name a few.

Considered the gold-standard method for DNA methylation detection, the whole-genome bisulfite sequencing technique (WGBS, BS-seq), might survey the entire landscape of 5mC in the genome at single-base resolution, often referred to as the methylome [18, 19]. Recently, third-generation sequencing technologies proved able to distinguish the cytosine methylation status directly, without chemical treatments or amplification, as in single-molecule sequencing techniques such as Oxford Nanopore or Pacific Biosciences [17, 20].

Notwithstanding, the aforementioned technologies are still costly and prohibitive to population-scale experiments, especially for species with poorly characterized genomes. To remedy that, strategies relying on reducing the genome complexity were devised, either by restriction enzyme (RE) digestion or affinity enrichment by immunoprecipitation [20]. Those represent cost-effective strategies at the expense of sampling a smaller fraction of methylated loci in a genome.

In terms of DNA digestion techniques, the strategies are based on the differential sensitivity to methylation by some REs that cleave the same restriction site (isoschizomers). The isoschizomers *Hpa*II and *Msp*I (5′-C/CGG-3′) are a common combination of restriction enzymes used for DNA methylation detection: methylation blocks *Hpa*II nuclease activity at the internal cytosine, whereas *Msp*I is unaffected [20].

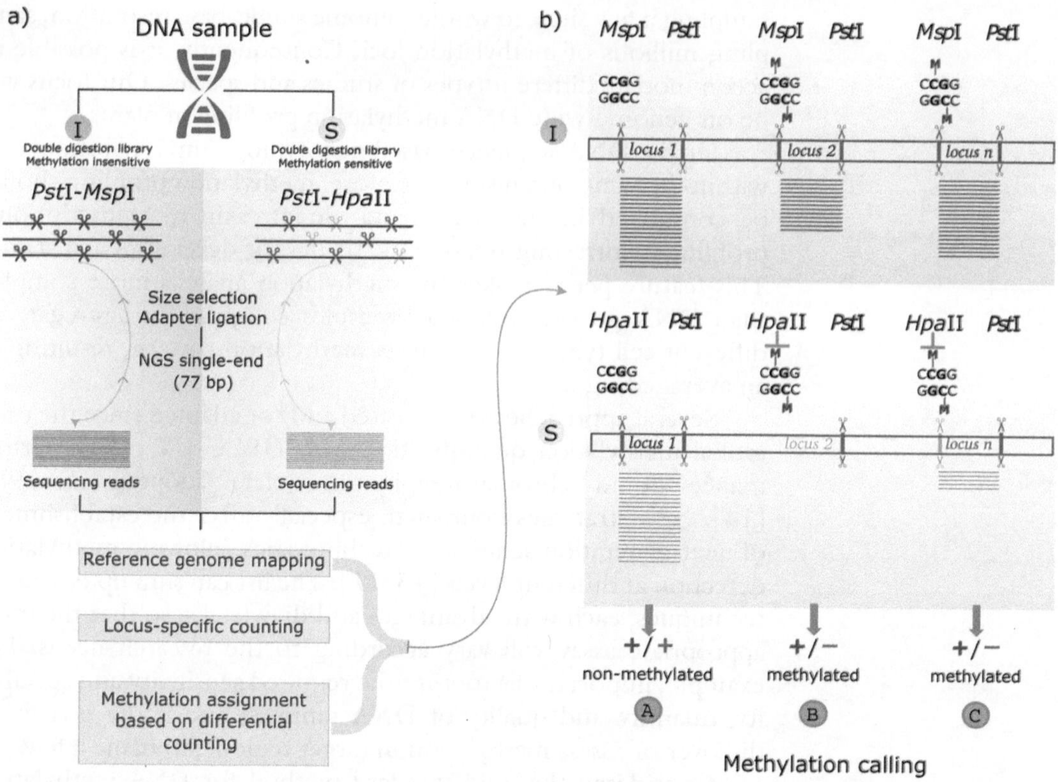

Fig. 1 DNA methylation detection via double digestion using restriction enzymes with contrasting methylation sensitivities. (**a**) Two double-digestion libraries are created per DNA sample, using the enzyme *Pst*I and one of the isoschizomers *Msp*I or *Hpa*II. The DNA fragments are then submitted to size selection and adapter ligation for PCR amplification. Next, the libraries are sequenced producing single-end reads and mapped to a reference genome. The mapped loci are filtered by checking restriction site boundaries and read counts defined for each sample in the context of the libraries indicated by (I), methylation-insensitive [*Msp*I], and (S), methylation-sensitive [*Hpa*II]. (**b**) Methylation assignment by contrasting fragment counts. Three profiles emerge (A) *non-methylated*: counts are statistically similar in both libraries, indicating the absence of DNA methylation in the restriction site; (B) *methylated*: fragments are detected only in the *Pst*I-*Msp*I library, implying the presence of DNA methylation that inhibits *Hpa*II cleavage; and (C) *methylated*: read counts are detected in both libraries but are significantly higher in *Pst*I-*Msp*I library. This may happen due to differing methylation states of the cell types from tissues/organs. Therefore, even if DNA methylation is prevalent in a sample, the same locus can be unmethylated in some cells and generate reads, to a lesser extent, in the *Pst*I-*Hpa*II library

The general principle comprises constructing two parallel digestion libraries, using each isoschizomer, from aliquots of the same DNA sample, followed by next-generation sequencing (Fig. 1a). To determine DNA methylation calls, the sequencing reads from both libraries are compared. For a genomic locus, the presence of reads in both libraries is an indication that the locus is not methylated. If reads are produced in the methylation-insensitive library (*Pst*I-*Msp*I) and absent in the methylation-sensitive counterpart (*Pst*I-*Hpa*II), then the internal cytosine of the restriction site (5′-C$\check{C}$GG-3′) is considered being methylated [21, 22]. However, for methylated loci, the expected behavior of no reads in the methylation-sensitive library (*PstI-Hpa*II) does not always hold because DNA samples represent a mix of cell types of a tissue or organ. Thus, it is expected that methylation signals are averaged out for some loci, resulting in sequencing reads in both libraries, but in differing quantities. If counts are significantly more abundant in the *PstI-Msp*I library compared to the *Pst*I-*Hpa*II library, then this is interpreted as being the consequence of DNA methylation at the internal cytosine for the majority of cells, as illustrated in Fig. 1b. To account for these fluctuations in locus-specific read counts, statistical models applied to differential gene expression in transcriptomic analyses are used for methylation calling [37], as implemented by the widely adopted programs for RNA-seq analysis, edgeR [42] and DESeq2 [43].

Several methodologies are based on this approach as methyl-sensitive cut counting (MSCC) [21, 23], Methyl-seq [24], MRE-seq [25–27], HELP-seq [28], MSAP-seq [22], and MCSeEd [29]. A common characteristic of these techniques is a bias towards hypomethylated regions, unlike methods based on affinity enrichment that usually present bias for hypermethylated regions, commonly associated with transposable elements [26]. Another alternative in this category is an adaptation of the DArTseq™ (Diversity Arrays Technology, Australia) genotyping method [30, 31], in which double digestion with the enzymes *Pst*I and *Hpa*II is used to construct a genomic library with reduced complexity. To allow methylation profiling, the DArTseqMet variant produces another double-digestion library for the same sample, with the methylation-insensitive isoschizomer *Msp*I (Fig. 1a).

Incorporating *Pst*I in the double-digestion approach of DArTseqMet introduces another selectivity level for low complexity regions, given that *Pst*I is also sensitive to methylation within its restriction site (5′-CTGCA/G-3′), comprising a CHG context [7]. The rationale is to provide another level of selectivity for low copy regions since CHG methylation is a typical silencing signal of repetitive elements in plant genomes. In traditional genotyping experiments, it was shown that fragments generated from *Pst*I-*Msp*I libraries were significantly enriched in the genic space of *E. guineensis*, which potentially increases the chances of finding informative markers for adaptative phenotypic responses [30, 32].

Arguably, the computational processing of DArTseqMet data is the most convoluted part of the methodology. The software package msgbsR [35] is an option for this type of analysis, providing a comprehensive solution for methylation-sensitive restriction enzyme sequencing data. However, it is not fully automated and was conceived for single digestion datasets. This prompted us to develop an open-source workflow to analyze DArTseqMet data, automating all required steps and capable of working with double digestion libraries. The analytical protocol was applied for DArT-seqMet data generated for *Eucalyptus grandis*, uncovering thousands of differentially methylated loci across tissues from an individual tree [36] and for different genotypes grown in contrasting environments [37]. The methylation calls generated by this computational protocol from DArTseqMet data were validated by BS-seq, confirming the concordance between the techniques and reinforcing the usefulness of this approach [36].

This chapter will cover a simplified experimental procedure to generate double-digestion libraries for DNA methylation profiling that can be adopted for population studies and illustrate a computational protocol to process the generated data.

2 Materials

2.1 Reagents and Kits

1. Restriction endonucleases: *Pst*I, *Msp*I, and *Hpa*II.
2. T4 DNA Ligase.
3. PCR amplification kit.
4. PCR primers.
   (a) *Pst*I-compatible adapter: AATGATACGGCGACCACC GAGATCTACACTCTTTCCCTACAC GACGCTCTTCCGATCT
   (b) *Hpa*II-compatible adapter (reverse): CAAGCAGAA GACGGCATACGAGATCGGTCTCGG CATTCCTGCTGAACCGCTCTTCCGATCTCGG

2.2 Equipment

1. Agarose gel electrophoresis apparatus.
2. PCR Thermal cycler.
3. Desktop computer with at least 8 Gb RAM.

2.3 Software

1. Preferably Linux operating system (OS), but any OS is supported.
2. Conda (Miniconda) – https://docs.conda.io/en/latest/miniconda.html
3. Bioconda – https://bioconda.github.io
4. Snakemake – https://snakemake.readthedocs.io
5. DArTseqMet pipeline – https://github.com/wendelljpereira/DArTseqMet

2.4 Data

1. Reference genome for the studied species, in "FASTA" format.

3 Methods

3.1 Profiling Methylation by Double-Digestion with Restriction Enzymes

Library preparation and digestion/ligation reaction. Digestion and adapter ligation are performed simultaneously in parallel for up to 96 samples using microtiter plates or individually in PCR microtubes (*see* **Note 1**) in a 10 μL aqueous solution containing:

1. 75 ng of high-quality genomic DNA.
2. Using 2 Units of each restriction enzyme perform concurrent digestion using a mix of *Pst*I-*Msp*I and *PstI*-*Hpa*II (*see* **Note 2**).
3. 80 Units of T4 DNA Ligase.
4. 0.05 μM of each adapter.
5. Incubate at 37 °C for 2 h, followed by 2 h at 60 °C.

3.2 PCR Amplification

1. Take 1 μL of digestion/ligation product as a template for PCR amplification (*see* **Note 3**) in a 50 μL reaction with the following cycling parameters:
   - 94 °C for 1 min.
   - 30 cycles including:
   - 94 °C for 20 s.
   - 58 °C for 40 s.
   - 72 °C for 1 min.
   - Final extension at 72 °C for 7 min.
2. PCR products are visually inspected by loading 5 μL of the amplification product in a 1.2% agarose gel stained with ethidium bromide.
3. Equimolar amounts of amplification products from each sample of the 96-well microtiter plate were bulked together, purified, quantified, and sequenced (see below).

3.3 Next-Generation Sequencing (NGS)

1. The resulting libraries should be sequenced using a short-read NGS platform of choice, using single-end sequencing and at least 75 bp per read (*see* **Notes 4** and **5**).
2. Raw sequencing reads in "FASTQ" format is the expected output.

3.4 Computational Workflow

The processing of sequencing data obtained from methylation profiling using the restriction-based family of methods, including the particularities of DArTseqMet, involves the stepwise execution of an assortment of computer programs, as shown in Fig. 2.

To process DArTseqMet data, we developed an analytical pipeline enacted by Snakemake, a workflow composition framework that provides a reproducible and simplified execution environment [38]. The pipeline is freely available on GitHub (*see* Materials) and

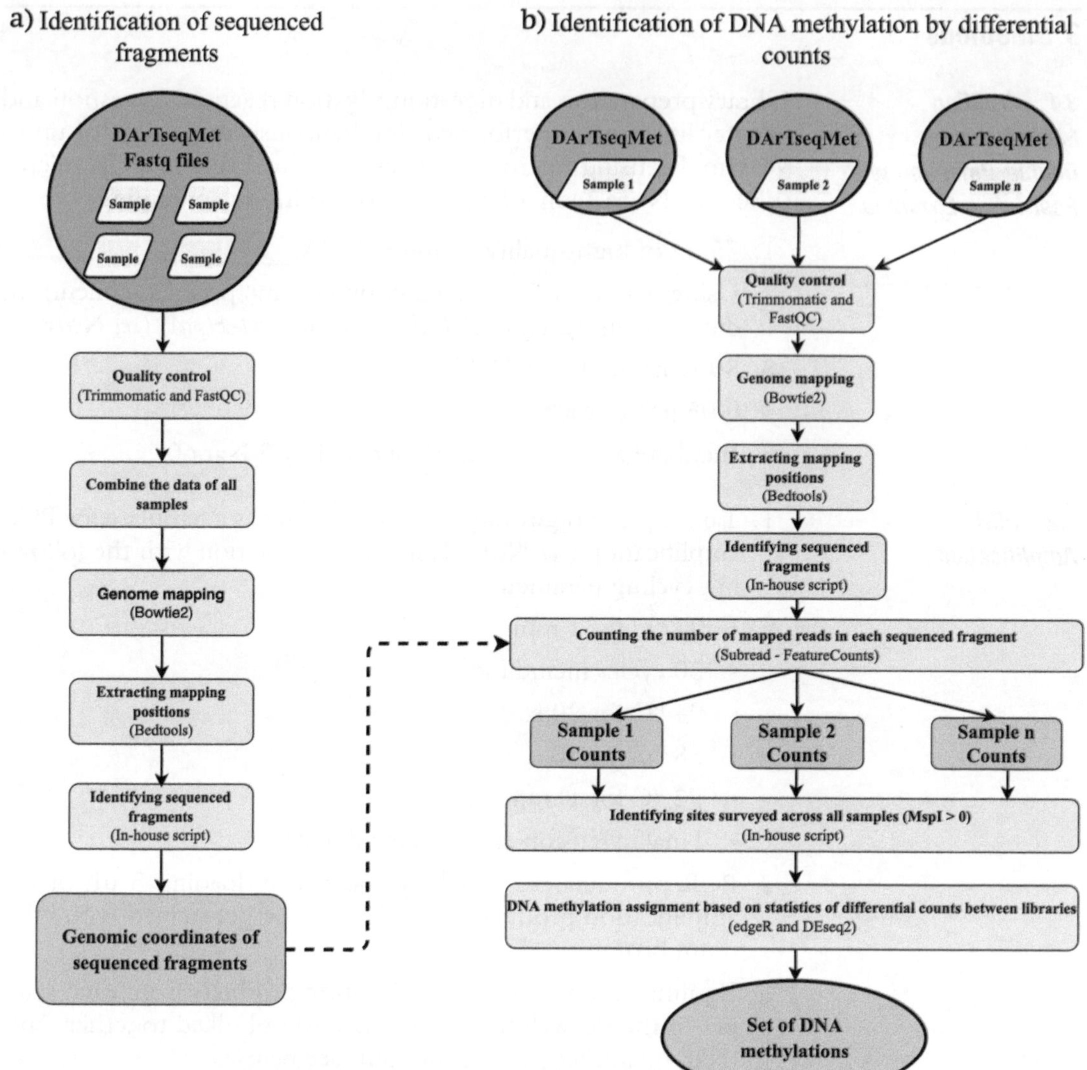

Fig. 2 Computational protocol designed for DNA methylation identification from DArTseqMet data. (**a**) Identification of potentially methylated loci from all samples based on the mapping of the sequencing reads to a reference genome. The mapped positions are used to identify sites subjected to DNA methylation by reconstructing the sequenced fragments in both libraries, *Pst*I-*Msp*I and *Pst*I-*Hpa*II. (**b**) For each sample, the read counts in each library (methylation-sensitive and insensitive) are computed and compared using statistical tests to infer significantly larger counts in the *Pst*I-*Msp*I library (insensitive), which serves as an internal control for normalization. The lack of a significant number of read counts in the *Pst*I-*Hpa*II library indicates DNA methylation at the restriction site for the locus. The result is a collection of methylated loci per sample

can be installed in any operating system. The default pipeline execution parameters should be a good starting point for most plant species (*see* **Note 6**). Nonetheless, they can be readily modified to suit other species (*see* **Note 3**).

3.5 Software Installation

1. Conda/Bioconda:

   wget https://repo.anaconda.com/miniconda/Miniconda3-latest-Linux-x86_64.sh

   bash Miniconda3-latest-Linux-x86_64.sh

   conda config --add channels bioconda

2. DArTseqMet pipeline:

   git clone https://github.com/wendelljpereira/DArTseqMet.git

   cd DArTseqMet

   conda env create --file dartseqmet.yaml

   conda activate dartseqmet

3.6 Pipeline Execution

To execute all analytical steps stated in Fig. 2, users should invoke the Snakemake command inside the DArTseqMet directory, where the pipeline descriptor file (Snakefile) is located. Finally, all pipeline steps are executed with the following command line (*see* **Note** 7 for a description of the parameters):

snakemake -p -c 2 --use-conda all

3.7 Pipeline Steps

As described in the previous point, from the user standpoint, a single command line enacts all the pipeline steps and suffices to generate all methylation calls from the raw sequence data. Below we provide a general description of the main pipeline stages (Fig. 2), with their respective programs.

3.8 Quality Control

The first analytical step is the quality control of the raw sequencing reads. For that, the software Trimmomatic [39] is used in the single-end mode to remove any barcoded adaptors and low-quality segments. Next, FastQC [40] is used to access the data quality after sample barcode removal. Only the data passing the quality control is used for the next steps.

3.9 Identifying Potential DNA Methylation Sites

The workflow detects DNA methylation at sites in the genome, by contrasting read counts in the *Pst*I-*Msp*I library against the counts obtained for the *Pst*I-*Hpa*II library for the same site. To achieve this goal, the method relies on mapping the sequenced reads against the reference genome and verifying if the mapped region is compatible with a fragment generated by the double digestion with the enzymes *Pst*I and *Hpa*II (*Msp*I).

Read mapping against the reference genome is produced by bowtie2 [41] with stringent parameters for reducing the number of allowed mismatches and excluding indels within each read (*see* **Note 8**). The mapped reads are combined and their genomic coordinates are obtained using BEDTools [42]. Next, an in-house R script using the mapped positions, in conjunction with the position of each restriction site of the enzymes *Pst*I and *Msp*I, reconstructs the sequenced fragments. A sequence fragment is then defined as the region between the *Pst*I site, where the reads are mapped, to the following *Msp*I site in the same strand.

3.10 Counting Sequencing Reads

To generate the counts per sample for each sequenced fragment, the software featureCounts [43] is used. Only restriction sites accessible to the insensitive RE in all samples are considered, that is, sites with at least one count in the *Pst*I-*Msp*I libraries. Using the *Pst*I-*Msp*I library for this filtering is justified because these libraries are not affected by DNA methylation in the internal cytosine, and therefore make up a baseline for the sequenced loci in each sample.

3.11 Identification of DNA Methylation by Differential Counts Analysis

For each sample, the read counts for the methylation-sensitive and insensitive libraries are compared using edgeR [33] and DESeq2 [34]. Only loci in which counts differ with fold change larger than 2, and q-value <0.05 (p-value corrected by False Discovery Rate) in both edgeR and DESeq2 are considered as true methylations.

After this step, users are provided with a file containing the genomic coordinates of all DNA methylation loci identified in the dataset. In addition, a file containing the presence/absence of the DNA methylations per sample is also made available.

3.12 Biological Inference Using Detected DNA Methylation

The set of identified DNA methylation loci can be used in downstream analyses to evaluate the potential functional implications or associations with desired traits. The protocol described here focuses on DNA methylation detection and does not provide the means to perform downstream analyses, which should be tailored for each experimental design and depending on the available genomic resources for the species being studied (*see* **Note 9**).

4 Notes

1. The *Pst*I adapter can be tagged with up to 96 different barcodes enabling multiplexing in a 96-well microtiter plate.
2. Our computational protocol will work with different enzymes if a similar experimental design is employed. A replacement for the enzyme *Pst*I is possible if adapters are used to guarantee that the sequencing starts at the restriction site of the new enzyme. When this constraint is observed, the REs *Hpa*II and *Msp*I can be replaced by any other pair of isoschizomers that differ in their sensibility to DNA methylation, to generate the two contrasting libraries. To make the computer code compatible with the new enzymes, users need only to provide the restriction site of the REs as a FASTA file.

3. The *Pst*I and *Hpa*II primers internal to the adapters are proprietary (Diversity Arrays Technology Pty Ltd, Australia) and this step is outsourced but can be adapted by the user.
4. Read lengths exceeding 75 bp would not improve methylation resolution but will benefit the read mapping against a reference genome.
5. Because of the particularities of the DArTseqMet library preparation, only fragments generated by the cleavage with both enzymes *Pst*I and *Hpa*II (or *Msp*I) are sequenced. Each read starts at the restriction site of the enzyme *Pst*I and is extended by up to 77 bp. Shorter reads can also emerge when fragments are generated from nearby restriction sites.
6. For the usage of the workflow described here, a reference genome for the species (or close relative) being analyzed is required.
7. The rule name "all" triggers the execution of pipeline steps. It is also possible to execute the workflow step-by-step, which can be useful for better understanding the intermediate files generated during the analyses. These steps are named in the file "Snakefile." In the command line, we used Snakemake parameters, such as the option "-p," which informs the software to print the executed commands in the terminal. The option "-c" is required by Snakemake and informs the number of processor cores to be used. Increasing the number of cores, up to the maximal number of cores available in the computer, will typically accelerate the execution. The execution of the whole pipeline using 24 cores (option "-c 24") and 60 Gb of RAM will take less than a day for a few samples. However, the time will be increased when using dozens of samples.
8. We selected parameters that are like the defined for the "-very-sensitive option in bowtie2," with some modifications to increase the sensitivity and precisely control for gaps and mismatches. The first change is in the parameter "-L," which controls the size of the seed used to start the alignment, the value was replaced from 20 to 15. This change makes the alignment slower, but more sensitive. In addition, the option "-end-to-end" is used to guarantee that only instances where the whole read was aligned would be accepted. Finally, the option "-score-min" is used to control the number of gaps and mismatches allowed in each step of the mapping.
9. Example scripts for exploring the methylated loci in the context of available *E. grandis* genome annotations are available at the GitHub repositories: https://github.com/wendelljpereira/techs_ms_dart_seq and https://github.com/wendelljpereira/ms-dart-seq.

References

1. Lloyd JPB, Lister R (2022) Epigenome plasticity in plants. Nat Rev Genet 23:55–68. https://doi.org/10.1038/s41576-021-00407-y
2. Brodie ED, Gregory B, Lisch D, Riddle NC (2022) The epigenome and beyond: how does non-genetic inheritance change our view of evolution? Integr Comp Biol 61:2199–2207. https://doi.org/10.1093/icb/icab084
3. Higo A, Saihara N, Miura F, Higashi Y, Yamada M, Tamaki S et al (2020) DNA methylation is reconfigured at the onset of reproduction in rice shoot apical meristem. Nat Commun 11:4079. https://doi.org/10.1038/s41467-020-17963-2
4. Xu G, Lyu J, Li Q, Liu H, Wang D, Zhang M et al (2020) Evolutionary and functional genomics of DNA methylation in maize domestication and improvement. Nat Commun 11:5539. https://doi.org/10.1038/s41467-020-19333-4
5. Dowen RH, Pelizzola M, Schmitz RJ, Lister R, Dowen JM, Nery JR et al (2012) Widespread dynamic DNA methylation in response to biotic stress. Proc Natl Acad Sci U S A 109: E2183–E2191. https://doi.org/10.1073/pnas.1209329109
6. Noshay JM, Springer NM (2021) Stories that can't be told by SNPs; DNA methylation variation in plant populations. Curr Opin Plant Biol 61:101989. https://doi.org/10.1016/j.pbi.2020.101989
7. Sarkies P (2022) Encyclopaedia of eukaryotic DNA methylation: from patterns to mechanisms and functions. Biochem Soc Trans 50: 1179–1190. https://doi.org/10.1042/BST20210725
8. McGuigan K, Hoffmann AA, Sgrò CM (2021) How is epigenetics predicted to contribute to climate change adaptation? What evidence do we need? Philos Trans R Soc B Biol Sci 376: 20200119. https://doi.org/10.1098/rstb.2020.0119
9. Springer NM, Schmitz RJ (2017) Exploiting induced and natural epigenetic variation for crop improvement. Nat Rev Genet 18:563–575. https://doi.org/10.1038/nrg.2017.45
10. Gahlaut V, Zinta G, Jaiswal V, Kumar S (2020) Quantitative epigenetics: a new avenue for crop improvement. Epigenomes 4:25. https://doi.org/10.3390/epigenomes4040025
11. Varotto S, Tani E, Abraham E, Krugman T, Kapazoglou A, Melzer R et al (2020) Epigenetics: possible applications in climate-smart crop breeding. J Exp Bot 71:5223–5236. https://doi.org/10.1093/jxb/eraa188
12. Zentner GE, Henikoff S (2014) High-resolution digital profiling of the epigenome. Nat Rev Genet 15:814–827. https://doi.org/10.1038/nrg3798
13. Mehrmohamadi M, Sepehri MH, Nazer N, Norouzi MR (2021) A comparative overview of epigenomic profiling methods. Front Cell Dev Biol 9:714687. https://doi.org/10.3389/fcell.2021.714687
14. Kuo KC, McCune RA, Gehrke CW, Midgett R, Ehrlich M (1980) Quantitative reversed-phase high performance liquid chromatographic determination of major and modified deoxyribonucleosides in DNA. Nucleic Acids Res 8: 4763–4776. https://doi.org/10.1093/nar/8.20.4763
15. Kurdyukov S, Bullock M (2016) DNA methylation analysis: choosing the right method. Biology 5:E3. https://doi.org/10.3390/biology5010003
16. Umer M, Herceg Z (2013) Deciphering the epigenetic code: an overview of DNA methylation analysis methods. Antioxid Redox Signal 18:1972–1986. https://doi.org/10.1089/ars.2012.4923
17. Zeng H, He B, Yi C (2019) Compilation of modern technologies to map genome-wide cytosine modifications in DNA. Chembiochem 20:1898–1905. https://doi.org/10.1002/cbic.201900035
18. Lister R, O'Malley RC, Tonti-Filippini J, Gregory BD, Berry CC, Millar AH et al (2008) Highly integrated single-base resolution maps of the epigenome in *Arabidopsis*. Cell 133: 523–536. https://doi.org/10.1016/j.cell.2008.03.029
19. Kawakatsu T, Huang SSC, Jupe F, Sasaki E, Schmitz RJ, Urich MA et al (2016) Epigenomic diversity in a global collection of *Arabidopsis thaliana* accessions. Cell 166:492–505. https://doi.org/10.1016/j.cell.2016.06.044
20. Yong WS, Hsu FM, Chen PY (2016) Profiling genome-wide DNA methylation. Epigenetics Chromatin 9:26. https://doi.org/10.1186/s13072-016-0075-3
21. Colaneri A, Staffa N, Fargo DC, Gao Y, Wang T, Peddada SD et al (2011) Expanded methyl-sensitive cut counting reveals hypomethylation as an epigenetic state that highlights functional sequences of the genome. Proc Natl Acad Sci U S A 108:9715–9720. https://doi.org/10.1073/pnas.1105713108
22. Chwialkowska K, Korotko U, Kosinska J, Szarejko I, Kwasniewski M (2017) Methylation sensitive amplification polymorphism

sequencing (MSAP-Seq) – a method for high-throughput analysis of differentially methylated CCGG sites in plants with large genomes. Front Plant Sci 8:2056. https://doi.org/10.3389/fpls.2017.02056

23. Ball MP, Li JB, Gao Y, Lee JH, Le Proust EM, Park IH et al (2009) Targeted and genome-scale strategies reveal gene-body methylation signatures in human cells. Nat Biotechnol 27: 361–368. https://doi.org/10.1038/nbt.1533
24. Brunner AL, Johnson DS, Kim SW, Valouev A, Reddy TE, Neff NF et al (2009) Distinct DNA methylation patterns characterize differentiated human embryonic stem cells and developing human fetal liver. Genome Res 19:1044–1056. https://doi.org/10.1101/gr.088773.108
25. Maunakea AK, Nagarajan RP, Bilenky M, Ballinger TJ, D'Souza C, Fouse SD et al (2010) Conserved role of intragenic DNA methylation in regulating alternative promoters. Nature 466(7303):253–257. https://doi.org/10.1038/nature09165
26. Wischnitzki E, Sehr EM, Hansel-Hohl K, Berenyi M, Burg K, Fluch S (2015) How to isolate a plant's hypomethylome in one shot. Biomed Res Int 2015:570568. https://doi.org/10.1155/2015/570568
27. Wischnitzki E, Burg K, Berenyi M, Sehr EM (2016) Selecting hypomethylated genomic regions using MRE-Seq. In: Hehl R (ed) Plant synthetic promoters: methods in molecular biology, vol 1482. Springer, New York, pp 83–102. https://doi.org/10.1007/978-1-4939-6396-6_6
28. Oda M, Glass JL, Thompson RF, Mo Y, Olivier EN, Figueroa ME et al (2009) High-resolution genome-wide cytosine methylation profiling with simultaneous copy number analysis and optimization for limited cell numbers. Nucleic Acids Res 37:3829–3839. https://doi.org/10.1093/nar/gkp260
29. Marconi G, Capomaccio S, Comino C, Acquadro A, Portis E, Porceddu A et al (2019) Methylation content sensitive enzyme ddRAD (MCSeEd): a reference-free, whole genome profiling system to address cytosine/adenine methylation changes. Sci Rep 9: 14864. https://doi.org/10.1038/s41598-019-51423-2
30. Kilian A, Wenzl P, Huttner E, Carling J, Xia L, Blois H et al (2012) Diversity arrays technology: a generic genome profiling technology on open platforms. In: Pompanon F, Bonin A (eds) Data production and analysis in population genomics: methods and protocols, vol 888. Springer, New York, pp 67–89. https://doi.org/10.1007/978-1-61779-870-2_5
31. Sansaloni C, Franco J, Santos B, Percival-Alwyn L, Singh S, Petroli C et al (2020) Diversity analysis of 80,000 wheat accessions reveals consequences and opportunities of selection footprints. Nat Commun 11:4572. https://doi.org/10.1038/s41467-020-18404-w
32. Pootakham W, Sonthirod C, Naktang C, Jomchai N, Sangsrakru D, Tangphatsornruang S (2016) Effects of methylation-sensitive enzymes on the enrichment of genic SNPs and the degree of genome complexity reduction in a two-enzyme genotyping-by-sequencing (GBS) approach: a case study in oil palm (*Elaeis guineensis*). Mol Breed 36:154. https://doi.org/10.1007/s11032-016-0572-x
33. Robinson MD, McCarthy DJ, Smyth GK (2009) edgeR: a Bioconductor package for differential expression analysis of digital gene expression data. Bioinformatics 26:139–140. https://doi.org/10.1093/bioinformatics/btp616
34. Love MI, Huber W, Anders S (2014) Moderated estimation of fold change and dispersion for RNA-seq data with DESeq2. Genome Biol 15:550. https://doi.org/10.1186/s13059-014-0550-8
35. Mayne BT, Leemaqz SY, Buckberry S, Rodriguez Lopez CM, Roberts CT, Bianco-Miotto T et al (2018) msgbsR: an R package for analysing methylation-sensitive restriction enzyme sequencing data. Sci Rep 8:2190. https://doi.org/10.1038/s41598-018-19655-w
36. Pereira WJ, de Castro Rodrigues Pappas M, Grattapaglia D, Pappas GJ (2020) A cost-effective approach to DNA methylation detection by methyl sensitive DArT sequencing. PLoS One 15:e0233800. https://doi.org/10.1371/journal.pone.0233800
37. Pereira WJ, Pappas MCR, Camargo Campoe O, Stape JL, Grattapaglia D, Joannis Pappas Jr G (2020) Patterns of DNA methylation changes in elite *Eucalyptus* clones across contrasting environments. For Ecol Manag 474:118319. https://doi.org/10.1016/j.foreco.2020.118319
38. Köster J, Rahmann S (2012) Snakemake – a scalable bioinformatics workflow engine. Bioinformatics 28:2520–2522. https://doi.org/10.1093/bioinformatics/bts480
39. Bolger AM, Lohse M, Usadel B (2014) Trimmomatic: a flexible trimmer for Illumina sequence data. Bioinformatics 30:2114–2120. https://doi.org/10.1093/bioinformatics/btu170

40. Andrews S (2010) FastQC a quality control tool for high throughput sequence data. https://www.bibsonomy.org/bibtex/2b6052877491828ab53d3449be9b293b3/ozborn. Accessed: 15 July 2022
41. Langmead B, Salzberg SL (2012) Fast gapped-read alignment with Bowtie 2. Nat Methods 9:357–359. https://doi.org/10.1038/nmeth.1923
42. Quinlan AR, Hall IM (2010) BEDTools: a flexible suite of utilities for comparing genomic features. Bioinformatics 26:841–842. https://doi.org/10.1093/bioinformatics/btq033
43. Liao Y, Smyth GK, Shi W (2014) featureCounts: an efficient general purpose program for assigning sequence reads to genomic features. Bioinformatics 30:923–930. https://doi.org/10.1093/bioinformatics/btt656

Chapter 4

Double Digest Restriction-Site Associated DNA Sequencing (ddRADseq) Technology

Natalia Cristina Aguirre, Carla Valeria Filippi, Pablo Alfredo Vera, Andrea Fabiana Puebla, Giusi Zaina, Verónica Viviana Lia, Susana Noemí Marcucci Poltri, and Norma Beatriz Paniego

Abstract

Double digest restriction-site associated DNA sequencing (ddRADseq) technology combines genome reduced representation by digestion with two restriction enzymes and next generation sequencing (NGS) to obtain thousands of markers (SNP, SSR, and InDels) and genotype tens to hundreds of samples simultaneously. In this chapter, we describe a 96-plex derived ddRADseq protocol that can be set up to obtain different depth of coverage per locus and can be exploited to model and non-model plant species.

Key words Reduced representation libraries, Next generation sequencing, RADseq, Genotyping by sequencing, SNP, SSR, InDels, Molecular markers

1 Introduction

As first envisaged by Luikart et al. [1] and then highlighted by Davey et al. [2]: "The ideal molecular approach for population genomics should uncover hundreds of polymorphic markers that cover the entire genome in a single, simple and reliable experiment. Unfortunately, at present there is no such approach."

Nowadays, with the great advancement of Next Generation Sequencing (NGS) technologies, we might think that the ideal genotyping technique should be whole genome sequencing. This strategy offers the power to uncover all sorts of polymorphisms, from short ones such as Single Nucleotide Polymorphisms (SNPs), Insertion/Deletions (InDels), and Simple Sequence Repeats (SSR) to the large Copy Number Variants (CNVs), Presence/Absence Variants (PAVs), and others Structural Variants (SVs). Nevertheless, it is still too expensive and time-consuming to apply to large numbers of individuals, especially for non-model species [3].

Yuri Shavrukov (ed.), *Plant Genotyping: Methods and Protocols*, Methods in Molecular Biology, vol. 2638,
https://doi.org/10.1007/978-1-0716-3024-2_4,

Reduced representation sequencing (RRS) strategies have been developed to reduce the amount of genomic sequence to be analyzed and simplify its handling. Some of them require prior knowledge of the genome sequences, such as exome capture [4, 5], while others are capable to overcome this limitation, such as RNA sequencing [6]. A possible drawback of these approaches is that they both focus only on coding regions, ignoring the large proportion of non-coding sequences present in most genomes [7]. Because of their potential neutrality, non-coding sequences are of uttermost interest for population genetics studies. In addition, they may provide a link to phenotype since they also comprise regulatory regions and non-coding RNAs [8].

RRS techniques based on restriction enzymes are approaches that combine genome reduction, sampling both coding and non-coding regions, without requiring prior genomic information [2, 9]. This allows their application to model and non-model organisms, particularly in plants [10].

In this approach, only a small fraction of the genome is sequenced (usually less than 5%), reducing the cost per sample while developing thousands of markers distributed throughout the genome. These features make them suitable in a wide range of applications, from genetic diversity, linkage disequilibrium analysis, conservation genetics and genome assembly improvement to QTL mapping, association studies and genomic selection in plant breeding programs [7].

The angular stone and first restriction-enzyme RRS strategy was restriction-site associated DNA sequencing technology (RAD-seq) [11]. When the strategy was published, Illumina high-throughput sequencing was still in its early beginning, and sequencing costs were still high, so the strategy did not generate a genotyping revolution. Moreover, from a methodological point of view, the strategy had some complexities, such as a high requirement of DNA per sample (up to 1 μg) and many hands-on steps, including a random sharing step [9, 12]. Nevertheless, after the publication of RADseq, several protocols have been developed aiming to keep the brand-new advantages and improving or eliminating the disadvantages [13]. The common steps of all these methods are DNA digestion (with one or two restriction enzymes, methylation-sensitive or not), adapter ligation (with barcodes or indexes), sample pooling, selection or reduction of DNA fragment population, amplification of DNA fragments (by short polymerase chain reaction (PCR) in most protocols, but some avoid this step), and short-read high throughput sequencing (generally Illumina sequencing by synthesis). Among the strategies, Genotyping by Sequencing (GBS) [14] and Double digest restriction-site associated DNA sequencing (ddRADseq) [12] have emerged as the most popular. There are reports of optimization of these techniques for different species (from fungi to animals and plants and either wild or agricultural species), breeding systems, genome size, population diversity or heterozygosity level, etc.

Compared to other protocols, ddRADseq has emerged as a suitable alternative for species with low inbreeding, highly heterozygous genotypes and natural populations because it permits tuning the number of regions to be analyzed and the depth of coverage per locus [9, 12]. For this reason, it can provide high coverage per locus with the same number of reads as other protocols, so that it can be exploited to well distinguish homozygous from heterozygous genotypes, or to approach polyploids, which needs a large depth of coverage.

Here we describe an improved 96-plex version of the optimized and derived ddRADseq protocol for *Eucalyptus dunnii* published by Aguirre et al. [15], which potentially allows multiplexing more than 70,000 samples in the same run. This method has been already applied to a few plant species, diploids and polyploids, with or without reference genome or other genomic information, providing the simultaneous discover and genotyping of many molecular markers, not only SNPs, but also InDels and SSR [15–18].

2 Materials

2.1 Plasticware and Consumables

1. Sterile tips p10, p100, p200, p1000 with filter.
2. Sterile microcentrifuge tubes (0.5, 1.5, and 0.2 mL).
3. Silicone mat or adhesive film for sealing PCR microplates (96 Round Well).
4. PCR 96 well microplates.
5. Microplates 96 well V-shape RP.
6. Sterile deionized water.
7. TE buffer (1×: 10 mM Tris-HCl pH 7.5, 1 mM EDTA), sterile.
8. Elution buffer (EB; 10 mM Tris-HCl, pH 8.5).
9. Annealing buffer stock (10×: 100 mM Tris-HCl, pH 8, 500 mM NaCl, 10 mM EDTA).
10. 80% ethanol (freshly made).
11. 1% agarose gel in TAE buffer (1×: 40 mM Tris, pH 7.6, 20 mM acetic acid, 1 mM EDTA).
12. 0.5 μg/mL ethidium bromide staining solutions in 1 × TAE.
13. 0.2 N NaOH (sodium hydroxide).

2.2 Enzymes and Kits

1. Qubit high sensitivity dsDNA kit.
2. DNF-474-0500 HS NGS Fragment Kit (1-6000 bp).
3. Absolute ethanol.
4. *Sph*I-HF (20,000 units/mL).

5. *Mbo*I (5000 units/mL).
6. rCutSmart buffer or other suitable.
7. AMPure XP reagent (Beckman Coulter) or similar other brands.
8. T4 DNA ligase (5 Unit/μL).
9. Quant-iT picogreen dsDNA or similar other brands.
10. 2 × agarose gel cassette for SAGE ELF or similar.
11. Phusion HF DNA polymerase 2 U/μL or similar.
12. Adapters and primers (*see* Tables 1 and 2).
13. PhiX control v3 kit.
14. NovaSeq 6000 SP Reagent Kit or other kit for Illumina sequencing.

2.3 Equipment

1. Micropipettes: p10, p100, p1000, multichannel p200.
2. Microvolume spectrophotometer (e.g., Nanodrop ND-1000 or similar).
3. Microplate Fluorometer (e.g., FluorStar Óptima or similar).
4. Fluorometer (e.g., Qubit or another brand).
5. Horizontal gel electrophoresis unit with power supply.
6. Fragment Analyzer System or similar.
7. Magnetic bed, MB (e.g., DynaMag or similar) for microplates and 1.5 mL microcentrifuge tubes.
8. Vortex platform.
9. Benchtop centrifuge with plate rotor.
10. Benchtop centrifuge for microcentrifuge tubes.
11. Thermocycler for PCR plates (96 wells).
12. SpeedVac or similar.
13. SAGE ELF (Sage Science) or similar.
14. NovaSeq 6000 or another Illumina sequencer.

3 Methods

3.1 General Description of ddRADseq Method

The ddRADseq procedure consists in six main steps: (1) double digestion, (2) adapters ligation, (3) sample pooling, (4) size selection, (5) PCR amplification, and (6) NGS. In this chapter, a 96-plex version of the optimized and derived ddRADseq protocol for *Eucalyptus dunnii* is presented based on published paper by Aguirre et al. [15], where the optimal enzymes, *Sph*I and *Mbo*I, were used. To optimize the protocol for other species, it is recommended to follow the specifications detailed on this publication [15] (*see* **Note 1**).

Table 1
Oligonucleotides for adapter preparation [15], based on [20, 21]

Oligo name	Barcode sequence	Oligo sequence
SphIA1	AAGTGA	ACACTCTTTCCCTACACGACGCTCTTCCGATCTAAGTGACA TG
SphIA2	ACAGA	ACACTCTTTCCCTACACGACGCTCTTCCGATCTACAGACATG
SphIA3	ACCA	ACACTCTTTCCCTACACGACGCTCTTCCGATCTACCACATG
SphIA4	AGAATGA	ACACTCTTTCCCTACACGACGCTCTTCCGATCTAGAATGACA TG
SphIA5	AGTGTTAA	ACACTCTTTCCCTACACGACGCTCTTCCGATCTAGTG TTAACATG
SphIA6	ATCATACCT	ACACTCTTTCCCTACACGACGCTCTTCCGATCTATCATACC TCATG
SphIA7	CACGACCA	ACACTCTTTCCCTACACGACGCTCTTCCGATC TCACGACCACATG
SphIA8	CCACTGG	ACACTCTTTCCCTACACGACGCTCTTCCGATCTCCACTGGCA TG
SphIA9	CCATCCACT	ACACTCTTTCCCTACACGACGCTCTTCCGATCTCCATCCAC TCATG
SphIA10	CCTG	ACACTCTTTCCCTACACGACGCTCTTCCGATCTCCTGCATG
SphIA11	CGACG	ACACTCTTTCCCTACACGACGCTCTTCCGATCTCGACGCATG
SphIA12	CTCACT	ACACTCTTTCCCTACACGACGCTCTTCCGATCTCTCACTCA TG
SphIA13	GACTCGG	ACACTCTTTCCCTACACGACGCTCTTCCGATCTGACTCGGCA TG
SphIA14	GATGA	ACACTCTTTCCCTACACGACGCTCTTCCGATCTGATGACATG
SphIA15	GGAG	ACACTCTTTCCCTACACGACGCTCTTCCGATCTGGAGCATG
SphIA16	GGCCGA	ACACTCTTTCCCTACACGACGCTCTTCCGATCTGGCCGACA TG
SphIA17	GTATA	ACACTCTTTCCCTACACGACGCTCTTCCGATCTGTATACATG
SphIA18	GTGCACCA	ACACTCTTTCCCTACACGACGCTCTTCCGATCTG TGCACCACATG
SphIA19	TATTCCACT	ACACTCTTTCCCTACACGACGCTCTTCCGATCTTATTCCAC TCATG
SphIA20	TCTCA	ACACTCTTTCCCTACACGACGCTCTTCCGATCTTCTCACATG
SphIA21	TGCGAGA	ACACTCTTTCCCTACACGACGCTCTTCCGATCTTGCGAGACA TG
SphIA22	TGGC	ACACTCTTTCCCTACACGACGCTCTTCCGATCTTGGCCATG

(continued)

Table 1
(continued)

Oligo name	Barcode sequence	Oligo sequence
SphIA23	TTGACCAG	ACACTCTTTCCCTACACGACGCTCTTCCGATC TTTGACCAGCATG
SphIA24	TTGTAG	ACACTCTTTCCCTACACGACGCTCTTCCGATCTTTGTAGCA TG
MboIA1	GTGAC	GTGACTGGAGTTCAGACGTGTGCTCTTCCGATCTGTGAC
MboIA2	CACGTA	GTGACTGGAGTTCAGACGTGTGCTCTTCCGATCTCACGTA
SphIB1	TCACTT	TCACTTAGATCGGAAGAGCGTCGTGTAGGGAAAGAGTGT
SphIB2	TCTGT	TCTGTAGATCGGAAGAGCGTCGTGTAGGGAAAGAGTGT
SphIB3	TGGT	TGGTAGATCGGAAGAGCGTCGTGTAGGGAAAGAGTGT
SphIB4	TCATTCT	TCATTCTAGATCGGAAGAGCGTCGTGTAGGGAAAGAGTGT
SphIB5	TTAACACT	TTAACACTAGATCGGAAGAGCGTCGTGTAGGGAAAGAGTGT
SphIB6	AGGTATGAT	AGGTATGATAGATCGGAAGAGCGTCGTGTAGGGAAAGAGTGT
SphIB7	TGGTCGTG	TGGTCGTGAGATCGGAAGAGCGTCGTGTAGGGAAAGAGTGT
SphIB8	CCAGTGG	CCAGTGGAGATCGGAAGAGCGTCGTGTAGGGAAAGAGTGT
SphIB9	AGTGGATGG	AGTGGATGGAGATCGGAAGAGCGTCGTGTAGGGAAAGAGTGT
SphIB10	CAGG	CAGGAGATCGGAAGAGCGTCGTGTAGGGAAAGAGTGT
SphIB11	CGTCG	CGTCGAGATCGGAAGAGCGTCGTGTAGGGAAAGAGTGT
SphIB12	AGTGAG	AGTGAGAGATCGGAAGAGCGTCGTGTAGGGAAAGAGTGT
SphIB13	CCGAGTC	CCGAGTCAGATCGGAAGAGCGTCGTGTAGGGAAAGAGTGT
SphIB14	TCATC	TCATCAGATCGGAAGAGCGTCGTGTAGGGAAAGAGTGT
SphIB15	CTCC	CTCCAGATCGGAAGAGCGTCGTGTAGGGAAAGAGTGT
SphIB16	TCGGCC	TCGGCCAGATCGGAAGAGCGTCGTGTAGGGAAAGAGTGT
SphIB17	TATAC	TATACAGATCGGAAGAGCGTCGTGTAGGGAAAGAGTGT
SphIB18	TGGTGCAC	TGGTGCACAGATCGGAAGAGCGTCGTGTAGGGAAAGAGTGT
SphIB19	AGTGGAATA	AGTGGAATAAGATCGGAAGAGCGTCGTGTAGGGAAAGAGTGT
SphIB20	TGAGA	TGAGAAGATCGGAAGAGCGTCGTGTAGGGAAAGAGTGT
SphIB21	TCTCGCA	TCTCGCAAGATCGGAAGAGCGTCGTGTAGGGAAAGAGTGT
SphIB22	GCCA	GCCAAGATCGGAAGAGCGTCGTGTAGGGAAAGAGTGT
SphIB23	CTGGTCAA	CTGGTCAAAGATCGGAAGAGCGTCGTGTAGGGAAAGAGTGT
SphIB24	CTACAA	CTACAAAGATCGGAAGAGCGTCGTGTAGGGAAAGAGTGT
MboIB1	GTCAC	GATCGTCACAGATCGGAAGAGCGAGAACAA
MboIB2	TATCGTG	GATCTACGTGAGATCGGAAGAGCGAGAACAA

Oligonucleotides A must be annealed with the corresponding oligonucleotide B to have a double-strand adapter for the ligation step. All oligonucleotides B must have a 5Phos modification at the 5′ end (must have an available phosphate group to be used by the ligase enzyme)

Table 2
Dual-index primers [22]

Index1-Reverse and Index2-Forward	Index sequence	Primer sequence
R_8	TACTCTAC	CAAGCAGAAGACGGCATACGAGATTACTCTACG TGACTGGAGTTCAGACGTG
R_9	ATAAGCTA	CAAGCAGAAGACGGCATACGAGATATAAGCTAG TGACTGGAGTTCAGACGTG
R_20	TACCGGTG	CAAGCAGAAGACGGCATACGAGATTACCGGTGG TGACTGGAGTTCAGACGTG
R_21	ATCGAAAC	CAAGCAGAAGACGGCATACGAGATATCGAAACG TGACTGGAGTTCAGACGTG
R_31	ATAGGAAT	CAAGCAGAAGACGGCATACGAGATATAGGAATG TGACTGGAGTTCAGACGTG
R_32	ATTAGTTG	CAAGCAGAAGACGGCATACGAGATATTAGTTGG TGACTGGAGTTCAGACGTG
R_43	TAAAGCTA	CAAGCAGAAGACGGCATACGAGATTAAAGCTAG TGACTGGAGTTCAGACGTG
R_44	ATCGATTA	CAAGCAGAAGACGGCATACGAGATATCGATTAG TGACTGGAGTTCAGACGTG
R_55	ATGGAACT	CAAGCAGAAGACGGCATACGAGATATGGAACTG TGACTGGAGTTCAGACGTG
R_56	ATTGACAT	CAAGCAGAAGACGGCATACGAGATATTGACATG TGACTGGAGTTCAGACGTG
R_67	ATATTATA	CAAGCAGAAGACGGCATACGAGATATATTATAGTGAC TGGAGTTCAGACGTG
R_68	ATTCGGGA	CAAGCAGAAGACGGCATACGAGATATTCGGGAG TGACTGGAGTTCAGACGTG
R_79	TATGACAT	CAAGCAGAAGACGGCATACGAGATTATGACATG TGACTGGAGTTCAGACGTG
R_80	TAGGACGG	CAAGCAGAAGACGGCATACGAGATTAGGACGGG TGACTGGAGTTCAGACGTG
R_91	TAGTCGTC	CAAGCAGAAGACGGCATACGAGATTAGTCGTCG TGACTGGAGTTCAGACGTG
R_92	TAGCTGTA	CAAGCAGAAGACGGCATACGAGATTAGCTGTAG TGACTGGAGTTCAGACGTG
F_01	TAGATCGC	AATGATACGGCGACCACCGAGATCTACACTAGA TCGCACACTCTTTCCCTACACGA
F_02	CTCTCTAT	AATGATACGGCGACCACCGAGATCTACACCTCTCTA TACACTCTTTCCCTACACGA

(continued)

Table 2
(continued)

Index1-Reverse and Index2-Forward	Index sequence	Primer sequence
F_03	TATCCTCT	AATGATACGGCGACCACCGAGATCTACACTATCCTC TACACTCTTTCCCTACACGA
F_04	AGAGTAGA	AATGATACGGCGACCACCGAGATCTACACAGAG TAGAACACTCTTTCCCTACACGA
F_05	GTAAGGAG	AATGATACGGCGACCACCGAGATCTACACG TAAGGAGACACTCTTTCCCTACACGA
F_06	ACTGCATA	AATGATACGGCGACCACCGAGATCTACACACTGCA TAACACTCTTTCCCTACACGA
F_07	AAGGAGTA	AATGATACGGCGACCACCGAGATCTACACAAGGAG TAACACTCTTTCCCTACACGA
F_08	CTAAGCCT	AATGATACGGCGACCACCGAGATCTACACCTAAGCC TACACTCTTTCCCTACACGA
F_09	TGAACCTT	AATGATACGGCGACCACCGAGATCTACACTGAACC TTACACTCTTTCCCTACACGA
F_10	TGCTAAGT	AATGATACGGCGACCACCGAGATCTACACTGCTAAG TACACTCTTTCCCTACACGA
F_11	TAAGTTCC	AATGATACGGCGACCACCGAGATCTACACTAAG TTCCACACTCTTTCCCTACACGA
F_12	ATAGAGGC	AATGATACGGCGACCACCGAGATCTACACA TAGAGGCACACTCTTTCCCTACACGA
F_13	GGCTCTGA	AATGATACGGCGACCACCGAGATCTACACGGCTC TGAACACTCTTTCCCTACACGA
F_14	AGGCGAAG	AATGATACGGCGACCACCGAGATC TACACAGGCGAAGACACTCTTTCCCTACACGA
F_15	TAATCTTA	AATGATACGGCGACCACCGAGATCTACACTAATC TTAACACTCTTTCCCTACACGA
F_16	CAGGACGT	AATGATACGGCGACCACCGAGATCTACACCAGGACG TACACTCTTTCCCTACACGA

Sixteen Index 1-Reverse primers (R) of the 96 published by Lange et al. [22] and 16 Index2-Forward (F)

After each enzyme reaction (digestion, ligation and PCR steps), it is recommended to purify the DNA to avoid salt, enzymes, or remaining oligonucleotides interfering with the next reaction (*see* **Note 2**). For this reason, there are several intermediate steps of DNA purification using magnetic beads. Besides, it is better to check as many steps as possible, either by quantifying the DNA or by examining the DNA profiles.

3.1.1 Double Digestion

The double digestion step involves using two restriction enzymes to obtain a complete digestion of the genomic DNA of each sample. The enzymes are usually a frequent cutter enzyme (four bases in the recognition site) and a rare cutter enzyme (six or more bases in the recognition site). However, for species with a large genome size it is better to use a combination of two rare cutter enzymes. This complete digestion must start with a high-quality genomic DNA and result in a fragment population less than 3 Kbp in size (*see* **Note 3**), rich in fragments between 200 and 800 bp (fragment size able to be sequenced on a second generation NGS instrument).

3.1.2 Adapters Ligation

The adapters ligation step consists in covalently binding double-stranded adapters to the DNA fragments, a reaction mediated by a T4 ligase. The sequences of the adapters have three main regions: a sticky end complementary to the sticky end that has been left by the restriction enzyme on the target DNA fragment, a specific short sequence (from 4 to 9 bp) that identifies each sample (barcode), and a complementary sequence to the PCR and sequencing primers. There are two different types of adapters: one with a sticky end for the rare cutter enzyme (adapter 1 or A1), and another with a sticky end for the frequent cutter enzyme (adapter 2 or A2). Usually, each type of adapter must be prepared by an annealing step from two different oligonucleotides. One of the oligonucleotides must be phosphorylated at its 5′ end (this oligo modification must be required, since regular oligonucleotide synthesis does not include the 5′ phosphate group), to allow covalent union with 3′ end of the target DNA fragment from each sample. The adapter 2 shows a "Y" form, where the two oligonucleotides, which are used to prepare it, have only a partial complementary region in-between the sticky end and the mid part of the adapter (*see PCR step*, **Notes 4–6**, and Table 1).

3.1.3 Sample Pooling

The sample pooling step consists of combining equimolar aliquots of all the final inactivated ligation reactions in a single microcentrifuge tube and then concentrating them. Here, it is proposed to divide the samples into groups of 48 items (*see* **Notes 3**, 7, and **8**).

3.1.4 Size Selection

Here it is proposed to use automatic size selection equipment as in the original protocol [12] and give an example of such an instrument. This step could also be performed by manual size selection using a conventional electrophoresis in stained agarose gel, a molecular weight ladder, and a scalpel [15, 19]. In this protocol, to reduce the genomic fraction to be sequenced, it is selected a 70 bp wide window (range specified by the equipment) of target DNA fragments that are between 330 and 400 bp long. Because selection is performed after the addition of adapters, their length (between 75 and 85 bp, depending on the length of the barcodes) must be considered when estimating the actual size selection windows to be applied to each pool of libraries. For this reason, the size selection step is set to an average of 450 bp (between 415 and 485 bp, corresponding to the length of the target DNA plus both adapters) (*see* **Note 9**).

3.1.5 PCR Amplification

The PCR step includes only ten amplification cycles, which allows not only to obtain a higher concentration of the library, but also to increase its length. The primers have a portion complementary to each adapter and add an index sequence (short sequence of 8 bp) and an additional sequence complementary to the oligonucleotides on the flow cell of the Illumina sequencers (*see* **Note 10** and Table 2). For each pool of 48 samples, two indexes (at each end of the library) are added, allowing each pool of libraries to be distinguished from one another. Additionally, because of the A2 "Y" design, PCR allows amplification of only those target DNA fragments that have both the rare and frequent cutting enzyme ends, resulting in a second reduction in the subpopulation of DNA fragments (corresponding to genomic regions) to be sequenced.

3.1.6 Next Generation Sequencing

Finally, as described in this procedure, it is needed to perform an equimolar mixture of both 48-plex pools of libraries, validate their concentration and fragment size distribution and sequence them on a second-generation sequencer (*see* **Notes 11** and **12**). The present adapter and primer sequences and libraries are specifically designed for Illumina sequencers. Then, there is a range of read length that can be chosen (from 75 to 300 bp) and in single-end read (SE, sequencing A1 extreme only) or paired-end (PE, sequencing both extremes) format. The expected sequencing depth depends on the species. For example, for reaching a 10× mean coverage in a diploid and highly heterozygous species, with ~30K genomic regions (estimated in silico) the recommend number of sequenced reads per sample would be between 1 and 2 million (M; more than three times the estimated depth of coverage), to minimized missing data points. Therefore, between 96 and 192 M reads are required for the following 96-plex protocol.

3.2 Preparing of Genomic DNA

1. Verify DNA integrity for all samples by electrophoretic run in 1% agarose gel. Optimal genomic DNA must be a band of around 20 Kbp, without smear (*see* Fig. 1).
2. Verify DNA quality by estimating absorbance with a spectrophotometer (e.g., NanoDrop). Optimum values are 260/280 and 260/230 > 1.8.
3. Quantify DNA concentration by measuring it with a microplate fluorometer (e.g., FluorStar Óptima and Quant-iT Picogreen dsDNA), following manufacturer instructions.

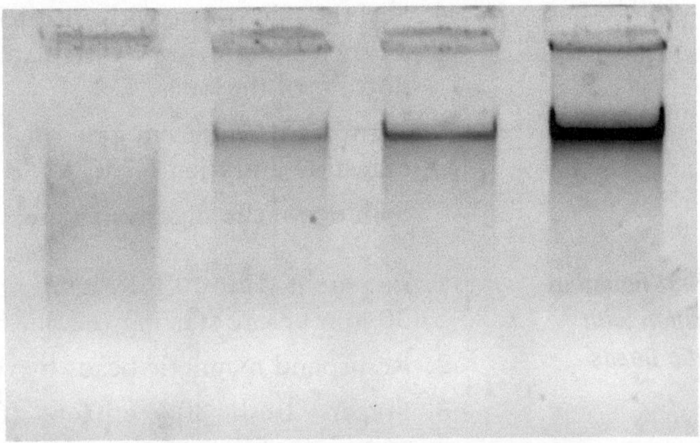

Fig. 1 Evaluation of genomic DNA integrity. Examples of different DNA integrity and concentration. 1% Agarose gel, from left to right: degraded genomic DNA and optimal integrity of genomic DNA at three different increasing concentrations

Table 3
Digestion mix per one reaction

Mix per one digestion reaction per sample	1× (μL)
H_2O	11.40
*Sph*I-HF (20 U/μL)	0.12
*Mbo*I (5 U/μL)	0.48
rCutsmart buffer (10×)	3.00
Final mix volume per sample	*15.00*

4. Normalize all DNA samples with ultrapure sterile water or buffer EB. Per reaction, it requires 150 ng of DNA. The concentration suggested to use is 10 ng/μL. For processing 96 samples, aliquot 15 μL of each one in a sterile PCR microplate (digestion plate, 0.2 mL/well).

3.3 DNA Digestion

1. Prepare reaction-enzyme master mixes for each group of 48 samples (consider four extra reactions in setting up each master mix; *see* Table 3 for quantities per one reaction) in a 1.5 mL microtube as follow: 592.8 μL of deionized sterile water, 156 μL of rCutsmart Buffer (10×), 6.24 μL of *Sph*I-HF (20 U/μL), 24.96 μL of *Mbo*I (5 U/μL). Final volume 780 μL (*see* **Note 1**). A total of two mixes are required to process 96 samples.
2. Spin down the mix on a benchtop centrifuge.
3. Add 15 μL of reaction enzymes master mix to each well of the sample/digestion plate.

4. Seal the microplate with an autoclaved silicone mat and spin on a benchtop centrifuge with plate rotor to settle the liquid to the bottom of the wells.
5. Incubated samples in a thermal cycler at 37 °C for 90 min for digestion and then 65 °C for 20 min for enzyme inactivation.
6. Spin down the digestion plate.

3.4 First Round of Purification with Magnetic Beads

1. Put the AMPure XP Reagent at room temperature for at least 30 min before starting the purification.
2. Resuspend magnetic beads by vortexing.
3. Prepare fresh 80% ethanol (for 96 samples, 15 mL are required).
4. Open digestion plate carefully, add 1.5 volumes of AMPure XP Reagent (45 μL) to each sample, seal the plate, mix by vortexing and spin at low speed (250 × *g*).
5. Incubate at room temperature (RT) for 5 min.
6. Precipitate beads on magnetic bed (MB) for microplates during 5 min.
7. Remove supernatant.
8. Keeping the plate on the MB, wash each well twice with 80% ethanol (Add 150 μL per 30 s and remove it). Then, air dry (*do not over-dry*) (*see* **Note 2**).
9. Add 20 μL of EB and seal the digestion plate with a silicone mat.
10. Mix, incubate 5 min at room temperature and spin.
11. Place-back the digestion plate on the MB for 5 min.
12. Transfer 15 μL of purified sample into a new PCR microplate (ligation plate). Keep the remaining 5 μL for digestion quantification and checking (*see* **Note 3**).
13. *Safe Stop: Protocol can be stopped here and holds samples overnight at 4 °C.*

3.5 Ligation

1. Slowly defreeze adapters in ice (For adapter design, annealing procedure and oligonucleotide sequences *see* **Notes 4–5** and Table 1).
2. Prepare on ice two ligation mixes for 48 samples each (having a total of 96 samples) in a 1.5 mL microcentrifuge tube. As indicated in Subheading 3.3, consider four extra reaction volumes in the set-up of reaction mix (*see* Table 4 for quantities per one reaction), as follows: 109.2 μL of ultra-pure water, 312 μL of T4 DNA ligase Buffer (5×) and 124.8 μL of T4 DNA ligase (5 Weiss Unit/μL) (*see* **Note 6**). Final volume 546 μL.

Table 4
Ligation mix per one reaction

Mix per one ligation reaction per sample	1× (μL)
T4 DNA ligase (5 Weiss unit/μL)	2.4
T4 DNA ligase buffer (5×)	6.0
H_2O	2.1
Final mix volume per sample	*10.5*

3. Spin each tube and dispense 10.5 μL of ligation mix to each well of ligation plate, at RT.
4. Add to each sample: 2 μL of *Sph*I 1 μM adapter (with barcodes—2 pM) and 2.5 μL of *Mbo*I 2 μM adapter (with barcodes—5 pM) (*see* **Note 7**). Final volume of each reaction: 30 μL.
5. Seal the ligation plate with a silicone mat and spin.
6. Incubate the reaction for 1 h at 23 °C and 1 h at 20 °C; inactivate it for 20 min at 65 °C.
7. *Safe Stop: Protocol can be stopped here and hold samples overnight at 4 °C.*

3.6 Pooling Libraries

1. Quantify the digestions using a microplate fluorometer (e.g., FluorStar Optima and Quant-iT Picogreen dsDNA), following manufacturer instructions (*see* **Note 3**).
2. Pool together 48 libraries (two pools for 96 samples). Pay attention, group only libraries with different barcoded-adapter (Adapter 1 and Adapter 2) combinations (*see* **Note 7**).
3. Aliquot equimolar quantities of each ligation in one single 1.5 mL microcentrifuge tube (e.g., use the total ligation volume of the less concentrated sample as reference and add equimolar quantities of the remaining samples).
4. Concentrate each pool to 150–200 μL by using SpeedVac (e.g., 90 min at 45 °C).

3.7 Second Round of Purification with Magnetic Beads

1. Follow the same steps as in the first purification with magnetic beads (*see* Subheading 3.4), with the following modifications.
2. Use 1 × AmpureXP beads proportion: add the appropriate volume to each pool of libraries.
3. Use a MB for microcentrifuge tubes.
4. Add 45 μL of EB.
5. Transfer 43 μL of purified sample into a new 1.5 mL tube.

3.8 Automated Size Selection

1. Set SAGE ELF instrument (Sage Science) to collect fraction containing fragments 450 bp long on average and use a 2% agarose cassette.
2. Mix 30 μL of one pool of libraries and 10 μL of marker mix on a microcentrifuge tube, then sow it on a cassette and run an automatic size selection in the SAGE ELF following manufacturer instructions. Run one SAGE ELF size selection per pool of libraries.
3. Collect 30 μL of the sample of the desired fragment size range from the elution well and transfer it into a new 1.5 mL microcentrifuge tube. Repeat this step for each pool of libraries.

3.9 Third Round of Purification with Magnetic Beads

1. Follow the same steps as in the second purification with magnetic beads (*see* Subheading 3.7) for each pool of libraries, with the following modifications.
2. Use 0.8 × AmpureXP beads proportion: add 24 μL to each pool.
3. Add 32 μL of EB buffer.
4. Transfer 30 μL of purified sample into a new 1.5 mL tube.
5. Quantify each pool with Qubit 2.0 fluorometer (High Sensitivity dsDNA kit) (*see* **Note 9**).

3.10 PCR Enrichment of Libraries

1. For 96 libraries, in two 48-plex pools, prepare PCR mix for five reactions (two amplifications each pool and one extra volume). On ice and in a 0.5 mL microcentrifuge tube add: 117.5 μL of ultra-pure water, 50 μL of HF buffer (5×), 5 μL of dNTPs (10 μM), 2.5 μL of Phusion High-Fidelity DNA polymerase Phusion HF DNA polymerase (2 U/μL) (*see* Table 5 for quantities per one reaction).
2. Mix by pipetting slowly up and down and then spin down mixes with benchtop microcentrifuge.
3. Dispense 35 μL of mix in each 0.2 mL microcentrifuge tube.

Table 5
PCR mix per one reaction

Mix per one PCR reaction (half of pool of libraries)	1× reaction (μL)
HF buffer (5×)	10.0
dNTPs (10 μM)	1.0
H_2O	23.5
Phusion HF DNA polymerase (2 U/μL)	0.5
Final mix volume per PCR	*35.0*

4. Add 13 μL of each pool of libraries in each microcentrifuge tube (in duplicate for each pool).
5. Add 1 μL of Primer Forward (12.5 μM) and 1 μL Primer Reverse (12.5 μM) to each tube. Use a pair of primers with dual indexes identifying each pool of libraries (*see* **Note 10** and Table 2). Final volume per reaction is 50 μL.
6. Spin down.
7. Place the microcentrifuge tubes inside the thermal cycler and set it with the following parameters: 3 min at 95 °C for initial denaturation; 10 cycles of 30 s at 95 °C, 30 s at 60 °C and 45 s at 72 °C for amplification; and 2 min at 72 °C for final extension, hold at 8 °C.
8. Run the PCR cycles, use a slow ramp if possible.

3.11 Fourth Round of Purification with Magnetic Beads

1. Mix both PCRs from the same pool of 48 samples/libraries.
2. Follow the same steps as in the third purification with magnetic beads (*see* Subheading 3.9) with the following modifications:
3. Use 1 × AmpureXP beads proportion: Add 100 μL to the combined PCR products.
4. Add 52 μL of EB.
5. Transfer 50 μL of purified sample into a new 1.5 mL tube.
6. *Safe Stop: Protocol can be stopped here and hold samples overnight at 4 °C.*

3.12 Library Validation

1. Check the resulting library size distribution in the Fragment Analyzer system (DNF-474-0500 HS NGS Fragment Kit 1-6000 bp) (*see* **Note 11**).
2. Quantify each pool of libraries wit Qubit 2.0 fluorometer (High sensitivity dsDNA kit) (*see* **Note 11**).

3.13 Preparing Libraries for Sequencing

1. Estimate molarity for each library with the following formula: *[concentration (ng/μL)/660 (g/mol)* × *library_mean_fragment_size (bp)]* × *10*^{6}.
2. Mix both pools together in equimolar ratio and dilute to 10 nM by using EB.
3. Dilute the final pool of 96 libraries to 4 nM by using EB.
4. Prepare 20 pM denature libraries (1 mL in total):
   (a) Mix 5 μL of 4 nM pool of libraries with 5 μL of 0.2 N NaOH (freshly prepared) and incubate 5 min at RT.
   (b) Add 5 μL 200 mM Tris-HCl (pH 7), mix by pipetting.
   (c) Finally, add 985 μL HT1 (provided by the sequencing kit), and mix by pipetting.

5. Dilute library to 1.5 pM (97 μL library 20 pM + 1203 μL HT1).
6. Prepare the final sample for sequencing as follow: 1235 μL of the 1.5 pM library and 65 μL of 1.5 pM PhiX (PhiX final concentration: 5%).
7. Place into an Illumina Next generation sequencer (e.g., NovaSeq 6000) and start a run by using a sequencing kit (e.g., NovaSeq 6000 SP Reagent Kit) (*see* **Note 12**).

4 Notes

1. Before starting to apply this protocol to a new species, it is highly recommended to do a screening of some enzyme combinations and a range of size selection windows, both in silico and in vitro, to achieve a complete genome digestion (*see* Fig. 2) and a population of DNA fragment enriched in the size selection region [15].
2. For every round of purification with magnetic beads, be careful and try to not touch beads with the tip when remove supernatant, ethanol or EB. Before resuspending with EB, make sure that the ethanol has evaporated. However, avoid leaving the microplate uncapped for too long, as it is very difficult to resuspend beads completely if samples were over-dry.
3. At this point, complete digestion of DNA (at least for some samples per master mix) should be assessed by electrophoresis in a Fragment Analyzer system or using a 1.5% agarose gel. A homogeneously distributed fragment population below 3K bp

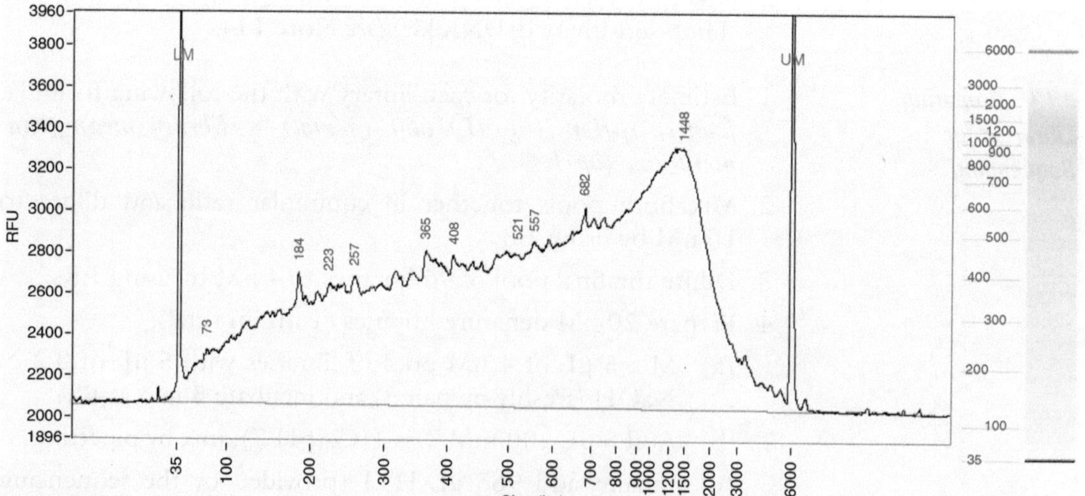

Fig. 2 Evaluation of digestion efficiency of *E. dunnii* DNA using *Sph*I-*Mbo*I. Fragment Analyzer electrophoresis (DNF-486-High Sensitivity NGS Fragment Analysis kit, 35–6000 bp)

is expected (*see* Fig. 2). In addition, 2 μL of DNA digestion should be quantified (between 5 and 10 ng/μL are expected per sample) to perform equimolar pooling of ligations.

4. As detailed in Aguirre et al. [15] adapter design was based on double-stranded oligonucleotides published by Peterson et al. [20]. Adapter 2 (A2) had a "Y" form for the specific amplification of fragments with different cut site endings. Adapter 1 (A1) and A2 were modified by replacing their sticky ends for *Sph*I and *Mbo*I, respectively. Besides, 24 variable-length (4–9 bp) barcodes designed by Poland et al. [21] were added to avoid low-complexity in the sequence of the first fragment bases due to the restriction site and to be able to pool the libraries before PCR as the original ddRADseq protocol by Peterson et al. [12]. The concentration of adapters (2 and 5 pMol for A1 and A2 respectively) was chosen based on protocol applied by Scaglione et al. [19]. The concentration of the stock of single-stranded oligonucleotides must be 100 μM (*see* Table 1), and it is recommended to store them in a 96-well PCR microplate. To obtain a double-stranded adapter, proceed as **Note 5**.

5. Adapter Annealing steps are as follows.

   (a) Prepare 24 double-stranded adapter 1 (*Sph*I) and two adapter 2 (*Mbo*I) stocks in a 96-well PCR microplate. To prepare the stocks, add to each well 50 μL of water, 20 μL of barcoded oligonucleotide A (e.g., oligo *Sph*IA1 at well A1) (*see* Table 1), 20 μL of barcoded oligonucleotide B (e.g., oligo *Sph*IB1 at same well A1) (*see* Table 1) and 10 μL Annealing buffer (10×). The final volume per each reaction is 100 μL and the final concentration of each double-stranded adapter is 20 μM.

   (b) Incubate in a thermocycler at 94 °C for 2.5 min, and then cool at a rate of no more than 1 °C per min until the solution reaches a temperature of 21 °C. Hold at 4 °C.

   (c) Prepare working solutions of adapters in a new 96-well PCR microplate. Final concentrations are 1 μM for barcoded double-stranded adapters for the rare-cutter enzyme (*Sph*I-1 to *Sph*I-24) and 2 μM for barcoded double-stranded adapters for the frequent-cutter enzyme (*Mbo*I-1 and *Mbo*I-2).

6. This protocol is based on T4 ligase enzyme concentration in Weiss units. Pay attention on the T4 ligase enzyme units to be used. Depending on the brand, concentration is in Cohesive end ligation or Weiss-units. Look at which unit the supplier manual uses and calculate the needed quantity of enzyme.

7. For a straightforward and unambiguous bioinformatic identification of each sample, add a unique combination of adapters with barcodes per sample. For example, for sample 1, use A1 *Sph*I-1 and A2 *Mbo*I-1, for sample 2, use A1 *Sph*I-2 and A2 *Mbo*I-1, and so on. With the adapters designed here, a maximum of 48 libraries can be pooled by combining 24 adapters 1 (*Sph*I) with two adapters 2 (*Mbo*I). Be careful and pay attention to pool only samples/libraries with different barcodes combinations. Use two 48-plex pools for 96 libraries. Each 48-plex pool is identified with specific dual-index primers added in PCR step.
8. In this step, it is suggested considering the concentration obtained from the quantification of each purified DNA digestion to merge ligations reaction (*see* **Note 3**). That is why if pooling step is based on ligations concentration, the remaining adapter dimers could affect the actual values. On the other hand, if pooling is based on the initial quantity of DNA, this value after the digestion and purification step could be less concentrated and could vary among samples.
9. Quantifying the libraries allows to know the quantity of DNA added in the PCR enrichment step (*see* **Note 11**). If this protocol is being performed for the first time, it is also recommended that you analyze the libraries by electrophoresis in agarose gel or with the Fragment Analyzer to determine if the size selection was performed correctly (*see* Fig. 3). In this step, for each pool of libraries is expected a band or pick between 415 and 485 bp (with a mean at 450 bp) at mean concentration of 0.5 ng/μL.

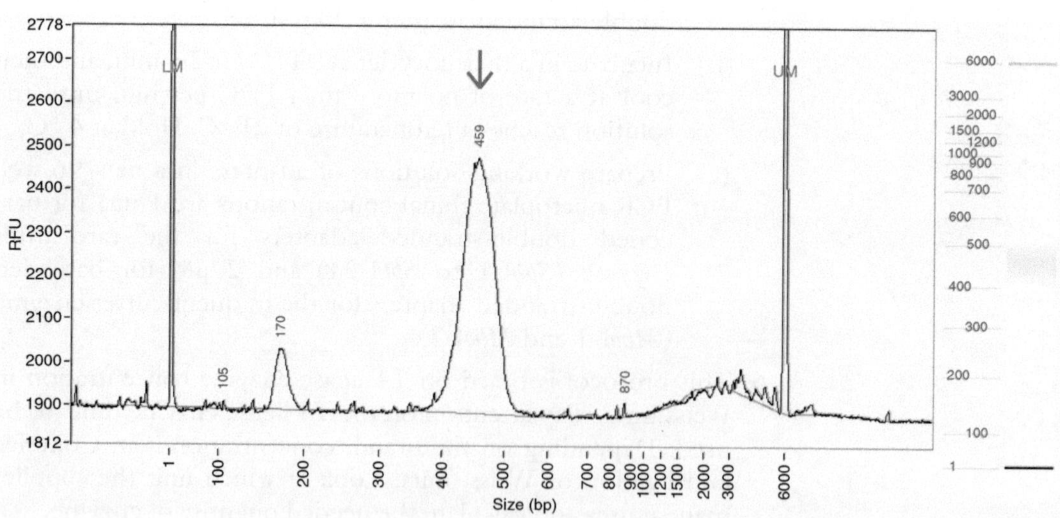

Fig. 3 Evaluation of the library generated after automatic size selection with ELF (Sage). Library with a mean size fragment population of 459 bp (pointed by red arrow). Fragment Analyzer electrophoresis (DNF-486-High Sensitivity NGS Fragment Analysis kit, 35–6000 bp)

10. Dual-indexed primers designed by Lange et al. [22] are used for the PCR reactions (*see* Table 2). The oligonucleotides have one part complementary to P5/P7 oligonucleotides of the Illumina sequencing platforms, another complementary to the adapters, and an index (8 bp) that allows identification of each pool of the library. With these primers, a maximum of 1536 pools could be sequence at the same time in a sequence platform (a combinatorial of 16 primers forward and 96 primers reverse). As a result, a total of 73,728 samples (1534 pools of 48 samples each) might potentially be sequence together.
11. A mean fragment population size of ~519 bp (~70 pb width) and a PCR product concentration that varies between around 2 and 17 ng/μL are expected (*see* Fig. 4), which represent an increase in both parameters compared to the values obtained in the previous step (Subheading 3.8, **step 2**: 450 bp and around 0.5 ng/μL, respectively). If an additional peak appears in the electropherogram in the low molecular weight region, it could be free adapters. It is recommended to remove them by another round of AMPureXP purification with 0.7 × beads and elution in 52 μL EB.
12. Here it can be chosen between Illumina sequencers such as MiSeq, NextSeq, HiSeq, NovaSeq and between SE or PE reads of different length (from 75 to 300 bp). In this 96-plex protocol, it is suggested to use a NovaSeq 6000 SP Reagent Kit, which provides between 1.3 and 1.6 billion (B) reads. With this kit, it is possible to run around 800 samples (eight PCR plates) with an average of 2 M reads each one, either SE (1 × 300 bp) or PE (2 × 150 bp).

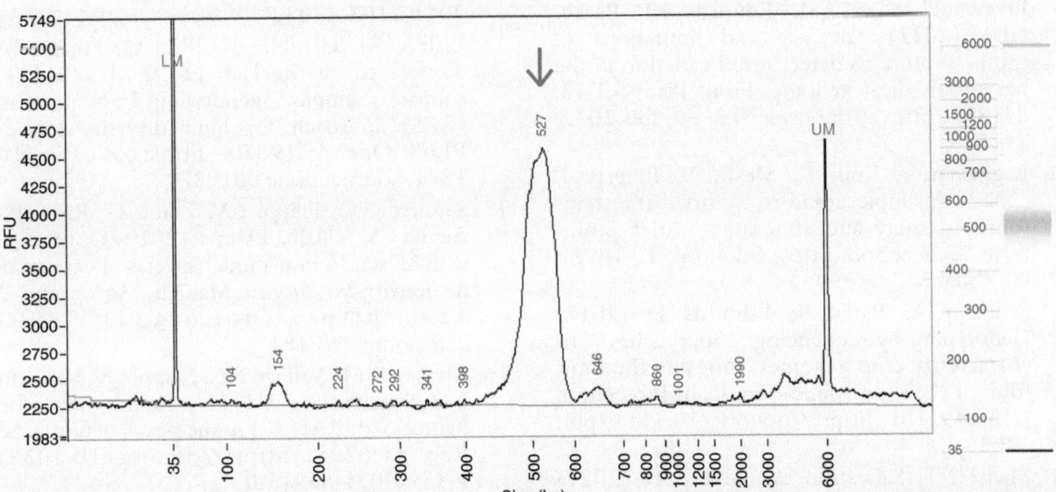

Fig. 4 Evaluation of the library generated after PCR step. Library with a mean size fragment population of 527 bp (pointed by red arrow). Fragment Analyzer electrophoresis (DNF-486-High Sensitivity NGS Fragment Analysis kit, 35–6000 bp)

Acknowledgements

We would like to thank to IGA technology services srl (Udine, Italy) for the support in setting up the ddRADseq approach in *Eucalyptus*, the Genomic Unit of INTA (Hurlingham, Argentina) for technical support, and to the National *Eucalyptus* Improvement Programme of INTA for provide the genetic material.

References

1. Luikart G, England PR, Tallmon D, Jordan S, Taberlet P (2003) The power and promise of population genomics: from genotyping to genome typing. Nat Rev Genet 4:981–994. https://doi.org/10.1038/nrg1226
2. Davey JW, Hohenlohe PA, Etter PD, Boone JQ, Catchen JM, Blaxter ML (2011) Genome-wide genetic marker discovery and genotyping using next-generation sequencing. Nat Rev Genet 12:499–510. https://doi.org/10.1038/nrg3012
3. Fuentes-Pardo AP, Ruzzante DE (2017) Whole-genome sequencing approaches for conservation biology: advantages, limitations and practical recommendations. Mol Ecol 26: 5369–5406. https://doi.org/10.1111/mec.14264
4. Bayer M, Morris JA, Booth C, Booth A, Uzrek N, Russell JR et al (2019) Exome capture for variant discovery and analysis in barley. In: Harwood W (ed) Barley, Methods in molecular biology, vol 1900. Humana Press, New York, pp 283–310. https://doi.org/10.1007/978-1-4939-8944-7_18
5. Burridge AJ, Winfield MO, Wilkinson PA, Przewieslik-Allen AM, Edwards KJ, Barker GLA (2022) The use and limitations of exome capture to detect novel variation in the hexaploid wheat genome. Front Plant Sci 13: 841855. https://doi.org/10.3389/fpls.2022.841855
6. Raghavan V, Kraft L, Mesny F, Rigerte L (2022) A simple guide to de novo transcriptome assembly and annotation. Brief Bioinform 23:bbab563. https://doi.org/10.1093/bib/bbab563
7. Scheben A, Batley J, Edwards D (2017) Genotyping-by-sequencing approaches to characterize crop genomes: choosing the right tool for the right application. Plant Biotechnol J 15:149–161. https://doi.org/10.1111/pbi.12645
8. Micheel J, Safrastyan A, Wollny D (2021) Advances in non-coding RNA sequencing. Noncoding RNA 7:70. https://doi.org/10.3390/ncrna7040070
9. Andrews KR, Good JM, Miller MR, Luikart G, Hohenlohe PA (2016) Harnessing the power of RADseq for ecological and evolutionary genomics. Nat Rev Genet 17:81–92. https://doi.org/10.1038/nrg.2015.28
10. Deschamps S, Llaca V, May GD (2012) Genotyping-by-sequencing in plants. Biology 1:460–483. https://doi.org/10.3390/biology1030460
11. Baird NA, Etter PD, Atwood TS, Currey MC, Shiver AL, Lewis ZA et al (2008) Rapid SNP discovery and genetic mapping using sequenced RAD markers. PLoS One 3:e3376. https://doi.org/10.1371/journal.pone.0003376
12. Peterson BK, Weber JN, Kay EH, Fisher HS, Hoekstra HE (2012) Double digest RADseq: an inexpensive method for *de novo* SNP discovery and genotyping in model and non-model species. PLoS One 7:e37135. https://doi.org/10.1371/journal.pone.0037135
13. Campbell EO, Brunet BMT, Dupuis JR, Sperling FAH (2018) Would an RRS by any other name sound as RAD? Methods Ecol Evol 9: 1920–1927. https://doi.org/10.1111/2041-210X.13038
14. Elshire RJ, Glaubitz JC, Sun Q, Poland JA, Kawamoto K, Buckler ES et al (2011) A robust, simple genotyping-by-sequencing (GBS) approach for high diversity species. PLoS One 6:e19379. https://doi.org/10.1371/journal.pone.0019379
15. Aguirre NC, Filippi CV, Zaina G, Rivas JG, Acuña CV, Villalba PV et al (2019) Optimizing ddRADseq in non-model species: a case study in *Eucalyptus dunnii* Maiden. Agronomy 9: 484. https://doi.org/10.3390/agronomy9090484
16. Aballay MM, Aguirre NC, Filippi CV, Valentini GH, Sánchez G (2021) Fine-tuning the performance of ddRAD-seq in the peach genome. Sci Rep 11:6298. https://doi.org/10.1038/s41598-021-85815-0
17. Gutiérrez AV, Filippi CV, Aguirre NC, Puebla AF, Acuña CV, Taboada GM et al (2021) Development of novel SSR molecular markers

using a next-generation sequencing approach (ddRADseq) in *Stetsonia coryne* (Cactaceae). An Acad Bras Cienc 30:e20201778. https://doi.org/10.1590/0001-3765202120201778

18. Molina C, Aguirre NC, Vera PA, Filippi CV, Puebla AF, Marcucci Poltri SN et al (2022) ddRADseq-mediated detection of genetic variants in sugarcane. Plant Mol Biol. https://doi.org/10.1007/s11103-022-01322-4
19. Scaglione D, Fornasiero A, Pinto C, Cattonaro F, Spadotto A, Infante R et al (2015) A RAD-based linkage map of kiwifruit (*Actinidia chinensis* Pl.) as a tool to improve the genome assembly and to scan the genomic region of the gender determinant for the marker-assisted breeding. Tree Genet Genomes 11:115. https://doi.org/10.1007/s11295-015-0941-3
20. Peterson GW, Dong Y, Horbach C, Fu YB (2014) Genotyping-by-sequencing for plant genetic diversity analysis: a lab guide for SNP genotyping. Diversity 6:665–680. https://doi.org/10.3390/d6040665
21. Poland JA, Brown PJ, Sorrells ME, Jannink JL (2012) Development of high-density genetic maps for barley and wheat using a novel two-enzyme genotyping-by-sequencing approach. PLoS One 7:e32253. https://doi.org/10.1371/journal.pone.0032253
22. Lange V, Böhme I, Hofmann J, Lang K, Sauter J, Schöne B et al (2014) Cost-efficient high-throughput HLA typing by MiSeq amplicon sequencing. BMC Genomics 15:63. https://doi.org/10.1186/1471-2164-15-63

Chapter 5

Whole Genome Wide SSR Markers Identification Based on ddRADseq Data

Gitanjali Tandon, Sarika Jaiswal, Mir Asif Iquebal, Anil Rai, and Dinesh Kumar

Abstract

The advent of advanced NGS technologies have led to the generation of enormous amount of sequence data which further aid in the discovery of the various type of markers such as SSRs, SNPs, InDels, etc. Among all these markers, microsatellite SSR markers can be mined from the ddRADseq data as certain properties of SSR markers make them ideal markers for study. These assist researchers and breeders in diversity analysis and producing new varieties with desired traits. To extract the markers, first, the ddRAD-seq data is assembled into consensus sequences using STACKS program which are further assembled for mining microsatellites using QDD along with MISA tool.

Key words SSR markers, ddRADseq, STACKS, MISA, QDD, consensus sequences

1 Introduction

Studies related to genetic variations have always facilitated the plant breeders in developing new improved varieties with desirable traits. For this, markers play an important role. With the advent of new sequencing technologies, preferably next generation sequencing (NGS) that produces millions of sequences at reasonable costs also pave the way of discovery of large number of genetic markers [1, 2] (*see* **Note 1**). One of the widely used sequencing approaches used is RADseq that targets the sequences around the restriction sites in the genome. RAD sequencing is usually followed when the reference genome of the organism is not available. Each sample of study is attached to an adapter with a unique barcode for executing multiplexed sequencing, thus leading to reduction of financial costs [3]. A small variation in RADseq during which double digestion of the DNA site is carried out via restriction enzymes leads to double digest restriction site associated sequencing (ddRADseq). This double digestion aids in improving the size selection step more

Yuri Shavrukov (ed.), *Plant Genotyping: Methods and Protocols*, Methods in Molecular Biology, vol. 2638,
https://doi.org/10.1007/978-1-0716-3024-2_5,

accurately [4] (*see* **Notes 2** and **3**). The DNA libraries obtained are bonded to specific adapters which further helps in reduced libraries representation. By plummeting, the segment of the sequenced genome, leads in generating a wide set of SNP markers that can be used to deduce the genetic diversity and population structure of germplasm collections very precisely [5] (*see* **Note 4**). Moreover, the double digestion by the restriction enzymes facilitates paired-end sequencing of the identical loci across all the samples thus providing high accuracy during mapping of the reads. Apart from all above ddRADseq also assists in expansion of number of markers from thousands to millions. This technique also provides the wide range of selection of the restriction enzymes which is selected based on the genome [6].

Significance of microsatellites and their use as genetic markers, amongst the wide diversity of available markers, microsatellites have become the preferred markers for study over time due to their specific properties of rapid amplification via PCR and larger extent of allelic variations on each locus [7, 8]. Moreover, these markers are highly polymorphic, reproducible, codominant, and are highly abundant throughout the genome of the eukaryotes [9, 10, 11].

SSRs are spread throughout the whole genome and are specifically located on the euchromatin of eukaryotes (both coding as well as non-coding regions) [12]. SSR distribution has been observed to be highly organized and can be varying in several regions as studied under the comparative analysis of rice and *Arabidopsis thaliana* [13]. These markers have been studied extensively as they are intensely informative about the mutation rates that occur per locus for each generation (i.e., 10^{-7}–10^{-3}) [14]. Furthermore, being co-dominant in nature, SSRs assist in direct measurement of heterozygosity [15–18].

2 Materials

2.1 Reagents Used

1. Restriction endonucleases (RE) and a buffer that is compatible with both the enzymes (e.g., MspI and EcoRI).
2. Adapter P1 and P2 are ligated using T4 endonuclease in Annealing buffer. The composition of annealing buffer is:
   (a) 100 mM of Tris HCl.
   (b) pH 8500 mM of NaCl.
   (c) 10 mM of EDTA.
3. For clean-ups after PCR amplification: *AMPore XP kit* [4].

2.2 Software Used

1. *FASTQC:* Quality check of the reads.
2. *STACKS:* Pre-processing of the reads and designing of consensus sequences.
3. *QDD:* Filtering of polymorphic sequences.
4. *MISA:* SSR mining.
5. *Primer3:* Primer designing.

3 Methods

3.1 Library Preparation and Sequencing

1. Before proceeding, DNA quality must be checked on an agarose gel and make sure that DNA should be of high molecular weight. Thereafter, digest DNA with two restriction enzymes simultaneously via PCR.
2. DNA repair and adenylation of 3′-ends.
3. Ligating the adapters with the sticky ends of the digested DNA using ligase.
4. Further enriching the DNA fragments.
5. Normalizing and pooling of the libraries.

Figure 1 represents the ddRAD sequencing methodology. After checking the quality and quantity of the libraries, these are loaded on the flow cell of the sequencer [19]. The output is generated in the form of FASTQ files which are quality files and assessed as discussed below.

3.2 Data Pre-processing and Construction of Consensus Sequences

The ddRADseq data is assessed for the quality with the help of FASTQC (http://www.bioinformatics.babraham.ac.uk/projects/fastqc/). After that, STACKS v2.41 [20] is employed for pre-processing of the reads (*see* **Note 5**). The reads whose score of being correct is less than 90% probability should be discarded and the read length limit is set (generally 135) using *process_radtags* program available in STACKS. This program handles barcoded and unbarcoded data, help in adapter removal and filtering the reads on the basis of quality. Further, *ustacks* script available in STACKS v2.41 is used to create consensus sequences for each sample [21]. These consensus sequences are created by aligning the short reads having exact match into stacks (*see* **Notes 6** and 7).

3.3 Microsatellite Identification

For identifying the polymorphic microsatellite loci in the consensus sequences for all the samples, QDD_v3 is used [22] (*see* **Note 8**). Pipe1.pl script in QDD_v3 is used to detect the microsatellites having di to hexa-nucleotide motifs in the sequences for all the samples. The sequences with microsatellites are aligned via pipe2.pl QDD script to identify the polymorphic microsatellite loci in all the samples. The polymorphic consensus sequences for all these

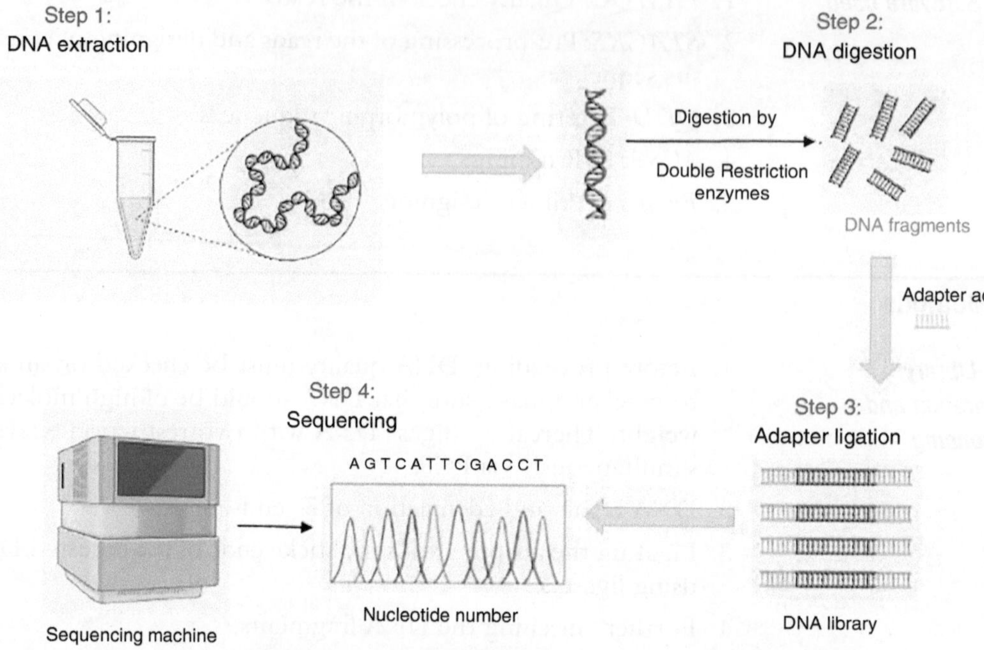

Fig. 1 Main steps of the ddRAD sequencing technology

individuals are further used for microsatellite screening in MIcro-SAtellite (MISA: http://pgrc.ipkgatersleben.de/misa/) tool Perl script (*see* **Note 9**). This script identifies di- to hexa-nucleotide motifs.

3.4 Primer Design

Once the SSR markers are mined, Primer v3 program [23] is implemented for generating the primers with parameters, namely, annealing temperature – min: 57 °C, optimal: 60 °C, maximum: 63 °C, GC content: 50%; primer size – min: 15, optimal: 18, maximum: 28 oligonucleotides (*see* **Note 10**). These parameters may be altered as per the requirement and choice. Figure 2 demonstrates the SSR mining and primer generation protocol from ddRAD sequences.

3.5 Whole Genome SSR Markers Identification with Example of ddRADseq Data

The example datasets have reads from forty samples, which were 9,151,583 in total. Out of these, 249,726 being of low quality as checked by FASTQC, were discarded while 9,126,610 were retained. After alignment and comparing these reads using *ustacks*, the data further reduced to 2,591,028 reads.

Further, after using QDD_v3, 4235 reads containing microsatellites were obtained with pipe1.pl script. These reads contained microsatellites from di to hexa-nucleotides motifs. Taking these

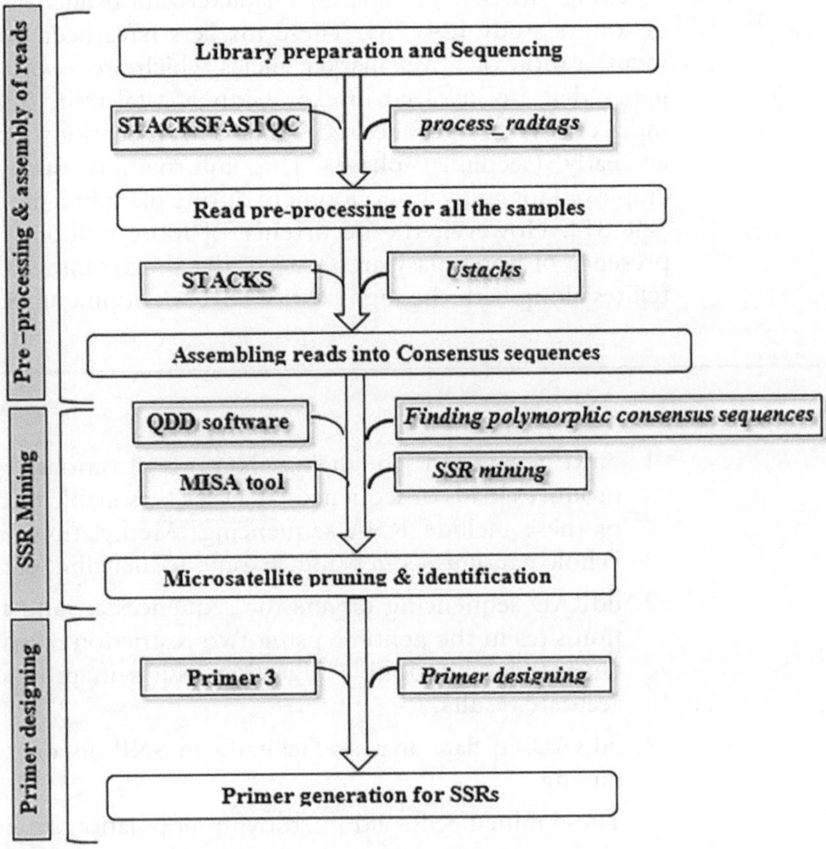

Fig. 2 SSR mining and primer generation protocol from ddRAD sequences

reads, further pipe2.pl script of QDD_v3 consensus sequences were constructed and a total of 2770 sequences were obtained.

These polymorphic consensus sequences were later on screened with the help of MISA tool for microsatellites mining and thus 2043 sequences contained microsatellites which composed of c & c*, p1, p2, p3, p4, p5 and p6 (here p1 refers to mono-nucleotide repeats; p2 means di-nucleotide repeats; p3 represents tri-nucleotide repeats; p4 is tetra-nucleotide repeats; p5 refers to penta-nucleotide repeats and p6 to hexa- nucleotide repeats while c means compound microsatellites). Using these, 2043 primers were designed with the help of Primer v3 and total 1576 primers were obtained.

3.6 Utility of ddRAD Derived SSR Markers

Microsatellites have been widely implemented in the study of genetics of a crop/ animal which includes genetic diversity, construction of linkage map between genes and the markers, marker assisted selection (MAS) for desired traits, genome-wide association study (GWAS), haplotype determination, marker assisted breeding (MAB), germplasm characterization and various other genomics study [24–28]. These markers have been used for the identification of novel marker alleles which are associated to the genes that are involved in expression of vital trait, thus allowing indirect selection of desired traits at segregations occurring at an early (seedling) phase. This information can be further employed for cultivar development during plant breeding programs [29–31]. However, the occurrence of more null alleles, and the presence of homoplasy are some of the shortcomings of microsatellites along with the high cost of SSR development [32].

4 Notes

1. Next generation sequencing (NGS) is of various types, and it produces loads of sequences at a very reasonable pricing. Some of these include RNA sequencing, Methylation sequencing, Whole genome sequencing, Exome sequencing, etc.
2. ddRAD sequencing targets the sequences around the restriction sites in the genome using two restriction enzymes simultaneously. This double digestion assists in producing more accurate results.
3. ddRADseq data analysis facilitates in SNP discovery and SSR mining,
4. These mined SSRs aid in studying population differentiation, phylogeography and crop improvement.
5. One of the most common tools used for consensus sequence generation is STACKS program. It allows the usage of both the approaches for obtaining the desired results.
6. For ddRADseq data, SSRs are discovered from consensus sequences.
7. These consensus sequences can be obtained by using two approaches either reference based or de novo based approach. Reference approach is implemented if the genome sequence of the species under studied is available whereas in the absence of whole genome sequence, de novo approach is used.
8. Furthermore, for pruning of polymorphic SSRs sequences, QDD – version 3 is very suitable software.
9. For the filtered polymorphic sequences, SSRs could be mined with the MISA tool.
10. The primers, for the mined SSRs, are designed using PRIMER version 3 tool.

References

1. Kalia RK, Rai MK, Kalia S, Singh R, Dhawan AK (2011) Microsatellite markers: an overview of the recent progress in plants. Euphytica 177: 309–334. https://doi.org/10.1007/s10681-010-0286-9
2. Taheri S, Lee AT, Yusop MR, Hanafi MM, Sahebi M, Azizi P et al (2018) Mining and development of novel SSR markers using next generation sequencing (NGS) data in plants. Molecules 23:399. https://doi.org/10.3390/molecules23020399
3. Varala K, Swaminathan K, Li Y, Hudson ME (2011) Rapid genotyping of soybean cultivars using high throughput sequencing. PLoS One 6:e24811. https://doi.org/10.1371/journal.pone.0024811
4. Peterson BK, Weber JN, Kay EH, Fisher HS, Hoekstra HE (2012) Double digest RADseq: an inexpensive method for *de novo* SNP discovery and genotyping in model and non-model species. PLoS One 7:e37135. https://doi.org/10.1371/journal.pone.0037135
5. Esposito S, Cardi T, Campanelli G, Sestili S, Díez MJ, Soler S et al (2020) ddRAD sequencing-based genotyping for population structure analysis in cultivated tomato provides new insights into the genomic diversity of Mediterranean 'da serbo' type long shelf-life germplasm. Hortic Res 7:134. https://doi.org/10.1038/s41438-020-00353-6
6. Shirasawa K, Hirakawa H, Isobe S (2016) Analytical workflow of double-digest restriction site-associated DNA sequencing based on empirical and in silico optimization in tomato. DNA Res 23:145–153. https://doi.org/10.1093/dnares/dsw004
7. Ab Razak S, Ghazalli MN, Azman NHEN, Abd Majid AM, Ismail SN (2021) RAD sequencing for the development of microsatellite markers for identification of Malaysian taro cultivars. Biotechnol Equip 35:1284–1290. https://doi.org/10.1080/13102818.2021.1969278
8. McCormack JE, Hird SM, Zellmer AJ, Carstens BC, Brumfield RT (2013) Applications of next-generation sequencing to phylogeography and phylogenetics. Mol Phylogenet Evol 66:526–538. https://doi.org/10.1016/j.ympev.2011.12.007
9. Miller MR, Dunham JP, Amores A, Cresko WA, Johnson EA (2007) Rapid and cost-effective polymorphism identification and genotyping using restriction site associated DNA (RAD) markers. Genome Res 17:240–248. https://doi.org/10.1101/gr.5681207
10. Wang JY, Yan SY, Hui WK, Gong W (2020) SNP discovery for genetic diversity and population structure analysis coupled with restriction-associated DNA (RAD) sequencing in walnut cultivars of Sichuan Province, China. Biotechnol Equip 34:652–664. https://doi.org/10.1080/13102818.2020.1797531
11. Miah G, Rafii MY, Ismail MR, Puteh AB, Rahim HA, Islam KN et al (2013) A review of microsatellite markers and their applications in rice breeding programs to improve blast disease resistance. Int J Mol Sci 14:22499–22528. https://doi.org/10.3390/ijms141122499
12. Phumichai C, Phumichai T, Wongkaew A (2015) Novel chloroplast microsatellite (cpSSR) markers for genetic diversity assessment of cultivated and wild Hevea rubber. Plant Mol Biol Rep 33:1486–1498. https://doi.org/10.1007/s11105-014-0850-x
13. Lawson MJ, Zhang L (2006) Distinct patterns of SSR distribution in the *Arabidopsis thaliana* and rice genomes. Genome Biol 7:R14. https://doi.org/10.1186/gb-2006-7-2-r14
14. Buschiazzo E, Gemmell NJ (2006) The rise, fall and renaissance of microsatellites in eukaryotic genomes. BioEssays 28:1040–1050. https://doi.org/10.1002/bies.20470
15. Oliveira EJ, Pádua JG, Zucchi MI, Vencovsky R, Vieira MLC (2006) Origin, evolution and genome distribution of microsatellites. Genet Mol Biol 29:294–307. https://doi.org/10.1590/S1415-47572006000200018
16. Selkoe KA, Toonen RJ (2006) Microsatellites for ecologists: a practical guide to using and evaluating microsatellite markers. Ecol Lett 9: 615–629. https://doi.org/10.1111/j1461-0248.2006.00889.x
17. Fan L, Zhang MY, Liu QZ, Li LT, Song Y, Wang LF et al (2013) Transferability of newly developed pear SSR markers to other Rosaceae species. Plant Mol Biol Rep 31:1271–1282. https://doi.org/10.1007/s11105-013-0586-z
18. Mason AS (2015) SSR genotyping. In: Batley J (ed) Plant genotyping. Methods and protocols. Humana, New York, pp 77–89. https://doi.org/10.1007/978-1-4939-1966-6_6
19. Davey JW, Hohenlohe PA, Etter PD, Boone JQ, Catchen JM, Blaxter ML (2011) Genome-wide genetic marker discovery and genotyping using next-generation sequencing. Nat Rev Genet 12:499–510. https://doi.org/10.1038/nrg3012

20. Catchen J, Hohenlohe PA, Bassham S, Amores A, Cresko WA (2013) Stacks: an analysis tool set for population genomics. Mol Ecol 22:3124–3140. https://doi.org/10.1111/mec.12354
21. Catchen JM, Amores A, Hohenlohe P, Cresko W, Postlethwait JH (2011) Stacks: building and genotyping loci de novo from short-read sequences. G3 Genes Genom Genet 1:171–182. https://doi.org/10.1534/g3.111.000240
22. Meglécz E, Costedoat C, Dubut V, Gilles A, Malausa T, Pech N et al (2010) QDD: a user-friendly program to select microsatellite markers and design primers from large sequencing projects. Bioinformatics 26:403–404. https://doi.org/10.1093/bioinformatics/btp670
23. Untergasser A, Nijveen H, Rao X, Bisseling T, Geurts R, Leunissen JA (2007) Primer3Plus, an enhanced web interface to Primer3. Nucleic Acids Res 35:W71–W74. https://doi.org/10.1093/nar/gkm306
24. Zargar SM, Raatz B, Sonah H, Bhat JA, Dar ZA, Agrawal GK et al (2015) Recent advances in molecular marker techniques: insight into QTL mapping, GWAS and genomic selection in plants. J Crop Sci Biotechnol 18:293–308. https://doi.org/10.1007/s12892-015-0037-5
25. Gao H, Jiang K, Geng Y, Chen XY (2012) Development of microsatellite primers of the largest seagrass, *Enhalus acoroides* (Hydrocharitaceae). Am J Bot 99:e99–e101. https://doi.org/10.3732/ajb.1100412
26. Jain SM, Brar DS, Ahloowalia BS (eds) (2009) Molecular techniques in crop improvement. Springer, Dordrecht
27. Antiqueira LMOR (2013) Application of microsatellite molecular markers in studies of genetic diversity and conservation of plant species of Cerrado. J Plant Sci 1:1–5. https://doi.org/10.11648/j.jps.20130101.11
28. Vieira MLC, Santini L, Diniz AL, Munhoz CDF (2016) Microsatellite markers: what they mean and why they are so useful. Genet Mol Biol 39:312–328. https://doi.org/10.1590/1678-4685-GMB-2016-0027
29. Cyriac A, Paul R, Anupama K, Sheeja TE, Nirmal Babu K, Parthasarathy VA (2016) Isolation and characterization of genomic microsatellite markers for small cardamom (*Elettaria cardamomum* Maton) for utility in genetic diversity analysis. Physiol Mol Biol Plants 22:219–229. https://doi.org/10.1007/s12298-016-0355-1
30. Gupta PK, Varshney RK (2000) The development and use of microsatellite markers for genetic analysis and plant breeding with emphasis on bread wheat. Euphytica 113: 163–185. https://doi.org/10.1023/A:1003910819967
31. Garris AJ, Tai TH, Coburn J, Kresovich S, McCouch S (2005) Genetic structure and diversity in *Oryza sativa* L. Genetics 169: 1631–1638. https://doi.org/10.1534/genetics.104.035642
32. Nadeem MA, Nawaz MA, Shahid MQ, Doğan Y, Comertpay G, Yıldız M et al (2018) DNA molecular markers in plant breeding: current status and recent advancements in genomic selection and genome editing. Biotechnol Equip 32:261–285. https://doi.org/10.1080/13102818.2017.1400401

Chapter 6

High-Throughput Association Mapping in *Brassica napus* L.: Methods and Applications

Rafaqat Ali Gill, Md Mostofa Uddin Helal, Minqiang Tang, Ming Hu, Chaobo Tong, and Shengyi Liu

Abstract

Oil seed rape (*Braasica napus* L.) is ranked second among oil seed crops cultivated globally for edible oil for human, and seed cake for animal consumption. Recent genetic and genomics advancements highlighted the diversity that exists within *B. napus*, which is largely discovered using the most promising genetic markers called single nucleotide polymorphism (SNP). Their calling rate is also enhanced to ~100 folds after the continuous advancements in the next generation sequencing (NGS) technologies. As the high throughput of NGS resulted in multi-Giga bases data, the detailed quality control (QC) prior to downstream analyses is a pre-requisite. It mainly involved the removal of false positives, missing proportions, filtering of low-quality SNPs, and adjustments of minor-allele frequency and heterozygosity. After marker-trait association, for conformation of target SNPs, validations of SNPs can be performed using various methods, especially allele-specific PCR assay-based methods have been utilized for SNP genotyping of genes targeting agronomic traits and somaclonal variations occurred during transgenic studies. In the present study, the authors mainly argue on the genotypic progress, and pipelines/methods that are being used for detection, calling, filtering, and validation of SNPs. Also, insight is provided into the application of SNPs in linkage and association mapping, including QTL mapping and genome-wide association studies targeting mainly developmental traits related to the root system and plant architecture, flowering time, silique, and oil quality. Briefly, the present study provides the recent information and recommendations on the SNP genotyping methods and its applications, which can be useful for marker-assisted breeding in *B. napus* and other crops.

Key words Agronomic traits, *Brassica napus* L., GWAS, QTL mapping, SNP genotyping

1 Introduction

Single nucleotide polymorphisms (SNPs) are one of the most popular markers for the fine mapping of heritable traits, especially for diploid species. SNPs are the excellent markers for high-density genetic map construction, physical ordering of chromosome contigs, association and linkage disequilibrium (LD) studies, and comparative and evolutionary genomics analyses. The low mutation

Yuri Shavrukov (ed.), *Plant Genotyping: Methods and Protocols*, Methods in Molecular Biology, vol. 2638,
https://doi.org/10.1007/978-1-0716-3024-2_6,

rate of SNPs makes them valuable for understanding the complex genetic traits and genome evolution. With the increasing development in high-throughput sequencing technologies started from first generation Sanger DNA sequencing, to second generation Illumina/Solexa (including MiSeq, NextSeq 500, and the HiSeq series such as HiSeq 2000, 2500, 3000, and Xten), and third generations including Pacific Bioscience (PacBio) and Oxford Nanopore (ONP) enabled researchers around the globe to sequence 100 s of organisms. Improvements in the efficiency of sequencing platforms not only reduce the cost but also enhance the throughput to ~100 folds. The above advancements enabled the resequencing of diverse genetic populations and later on the development of an efficient and convenient method for calling millions of SNPs to dissect the genetic mechanism of agronomic traits in both plants and animals including polyploids. Among polyploid plant species, *Brassica napus* L. is a classical example, where researchers observed the whole genome duplication (WGD) event occurred more recently (~7500 years) [1]. *B. napus* (AACC, $2n = 38$) is an allopolyploid plant species that was produced during the post-Neolithic speciation from a small number of naturally occurring hybridization events between the diploid progenitor species *B. rapa* (AA, $2n = 20$) and *B. oleracea* (CC, $2n = 18$). The above crop provides an ideal system for researchers to gain deep insight into the mechanisms of de novo and recurrent polyploidization and their following consequences [1, 2]. Domestication of *B. napus* has been started since ~700 years ago, and as a result of evolution, it is divided into three ecotypes such as winter oilseed rape (WOR), spring oilseed rape (SOR), and semi-winter oilseed rape (SWOR) [3–5].

Among the mapping approaches, quantitative trait loci (QTL) and Genome-wide association study (GWAS) are the two most popular and widely used methods in population genetics and genomics research for identifying SNPs associated with different qualitative and quantitative agronomic traits in almost all crops. For example, the QTL and GWAS combined approaches identified 86 significant QTLs and 103 significant SNPs as associated with root system architecture (RSA)-related traits [6, 7]. In cases of oil quality, four ecotype-specific haplotype blocks (HBs) identified were significantly associated with glucosinolate (GSL) contents through GWAS of 307 *B. napus* accessions [8]. A novel QTL related to Oleic acid content from a panel of 23,168 SNPs was detected using a panel of 375 rapeseed germplasms [9]. Flowering is a key component of plant life cycles, marking its transition from the vegetative to reproductive phase. Traits related to flowering time (FT) are the most promising agronomic traits that directly impact the seed yield and oil quality of *B. napus.* Developing early flowering and maturity rapeseed varieties is one of the most important breeding objectives in *B. napus.* A combined QTL mapping

(2795 SNPs) and GWAS (3,85,691 SNPs) approaches were used to dissect the genetic mechanism of FT, growth period (GP) and maturity time (MT)-related traits in *B. napus* [10]. They reported that 83 QTLs and 146 SNPs were found to be associated with days to initial flowering (DIF), days to final flowering (DFF), flowering period (FP), GP and MT.

In *B. napus*, plant height (PH) is not only one of the most important traits related to plant architecture but also an important agronomic trait associated with crop yield, lodging resistance, and mechanized harvesting [11]. Silique is a critical part of the yield performance of *B. napus*, which serves as both a source organ for seed development and a sink organ for absorbing and storing photosynthetic substances produced by leaves [12]. The yield of the *B. napus* is determined by the number of the siliques per plant (NSPP), the number of seeds per silique (NSPS), and 1000-seed weight (TSW) [13], in which NSPP is mainly influenced by the number of the branches per plant (NBPP) and number of siliques on the main inflorescence (NSMI) [14, 15]. There were 38 significant SNPs were identified associated with plant height (PH), branch initiation height (BIH), and NBPP through GWAS using 32,297 SNPs called from a diverse panel of 333 OSR germplasms [16]. Moreover, 101 and 77 SNPs identified were significantly associated with NSPP and TSW, respectively [17]. Similarly, GWAS analysis was performed in *B. napus* and seven SNPs distributed on five different chromosomes were associated with seed oil and protein content [18]. Among candidate genes targeting the above two traits, five genes were orthologous to the model plant *Arabidopsis thaliana* and were considered as candidates for future breeding programs.

In this chapter, the authors' main objectives are to describe the detailed methods used for SNP genotyping in polyploid *B. napus*, and its global application in identifying the genes targeting agronomic traits in *B. napus*.

2 Materials and Methods

2.1 QTL Analysis

2.1.1 Plant Material

For QTL mapping either through segregation or association or both, mapping population either can be alone or in a combination, including F_2 population (it is achieved by selfing of F_1 hybrid, here segregation for co-dominant marker is 1:2:1 and for dominant marker is 3:1, it is recommended for preliminary mapping), $F_{2:3}$ population (achieved through selfing of F_2 plants, it is recommended for mapping of quantitative traits, and mapping of recessive genes), double haploid (DH) population (it is achieved through chromosomal doubling of pollen/egg derived haploid plants from F_1 generation, here marker segregation ration is just like F_1 plants, 1:1 and widely used for mapping of both quantitative and qualitative traits), backcross population (BC, it is achieved through the crossing of F_1 with one of its parent, usually recessive parent, and mostly used for getting the elite genotypes), recombination inbred lines (RILs) population (it is achieved through the continuous selfing/sibmating progeny of individual F_2 plant until optimum homozygosity, here segregation ration is 1:1, can be replicated in multiple locations, widely used in QTL mapping, especially for identification of tightly linked markers) and near isogenic lines (NILs) population (it is achieved through the repeated selfing/BC of F_1 with recurrent parent, here marker segregation ratio is 1:1, irrespective of marker is dominant or co-dominant).

2.1.2 Phenotyping

To associate the phenotypic traits with genotypic markers, biometrical measurement of target trait/s is a very crucial step in genetic studies. Phenotypic traits including environmental (abiotic or biotic stress) and developmental traits (i.e., root system architecture, plant architecture, flowering time, oil quality and quantity) are measured accordingly as each trait has its recording method. Generally, data should be recorded from at least ten plants in each plot (replicated twice/thrice using the randomized block design method) and are grown in the normal growing season in single or multiple locations for several years (usually 2–10, but the more, the better to get more stable markers). Phenotypic trait analyses including analysis of variance (ANOVA) and Pearson's correlation analysis can be performed using Statistical Analysis System (SAS, Institute Inc., Cary, NC, USA), and broad sense heritability in the mapping population can be calculated in the biological replications grown at single location or multiple locations using "lmerTest," an R package [19].

2.1.3 Genotyping

Genomic DNA can be extracted from fresh leaf samples of the mapping population (DH, F_2, RILs and NILs) using the CTAB method (for example, Sigma-Aldrich) following the manufacturer's protocol. DNA quality should be assessed using electrophoresis on 1% agarose gels and quality can be evaluated using NanoDrop one spectrophotometer (for example, Thermo Fisher Scientific). For QTL mapping, there are several types of molecular markers have been widely used in *B. napus* such as simple sequence repeat (SSR), amplified fragment length polymorphism (AFLP), and single nucleotide polymorphism (SNP). For the first case, the analyses can be performed by following the procedure described earlier [20], and also can get from relevant (1) public sources [i.e., http://www.brassica.info/resource/markers/ssr-exchange.php and http://brassicadb.org/brad/geneticMarker.php]; (2) directly from published articles, i.e., Refs. [21–27]; and (3) can be designed based on the sequences of known genes. For the AFLP marker, the analysis should be performed as described by Vos et al. [28]. Other details like giving the name and labeling of primers for AFLP and set band size should be followed by Liu et al. [29]. The amplified product of both markers (SSR and AFLP) can be detected using a capillary reaction-based sequencer (for example, ABI Applied Biosystems). Lastly, for SNP genotyping, a microarray approach so-called "Brassica 60K Infinium Array" developed by Illumina (San Diego, USA) should be used as it is widely accepted and has been used in multiple crop species. The produced markers can be filtered by setting a criterion such as missing data <30%, minor allele frequency >0.05, and genotype × missing data <20%. After that, further filtering such as inter-homeologous polymorphism can be performed according to Trick et al. [30]. Finally, physical position can be assigned to clean SNPs by aligning with reference genome of *B. napus* [1]. To evaluate the expected segregation ratio for each marker (e.g., AFLP) in terms of "goodness to fit," the v2 test can be performed as suggested by Liu et al. [29].

2.1.4 QTL Mapping

Linkage analysis for SSR, AFLP and SNP makers can be performed using JoinMap 4.0 (https://www.kyazma.nl/index.php/JoinMap). The threshold of recombination frequency of markers can be set ≤0.25 and a minimum logarithm of odds (LOD) ration for grouping loci which can be set as ≥6 [31]. The genetic distance between two loci produced by either above mentioned markers can be recorded using Kosambi's function [32]. To confirm the makers orders and quality, the linkage map can be drawn using "Mapchart 2.1" as described by Voorrips [33], and also in R using the "R/qtl", an R statistical package as described earlier [34, 35]. Further, a name can be assigned to each linkage group based on nomenclature as recommended by "the Multinational *Brassica* Genome Project Steering Committee" (http://www.brassica.info/resource/maps/lg-assignments.php). Alternatively, QTL mapping can also be performed using the composite interval mapping (CIM) method using "WinQTLcart 2.5" as described by Wang [36] (http://statgen.ncsu.edu/qtlcart/WQTLCart.htm). In this method, the threshold for the QTL scoring should be determined using a 1000-time permutation test (can be performed using "Zmapqtl program") and LOD score values should be ≥3. Moreover, CIM additionally associate the phenotypic trait data with single marker intervals and also considered the additional markers as co-variates to minimize the residual variance and control over the genetic diversity of both linked and unlinked markers even beyond interval of interest as described by Broman et al. [34]. Thus, CIM has an advantage over interval mapping (IM, scanone function in "R/qtl" as described by Lander and Botstein [37]) approach in terms of better mapping resolution. However, for a smaller QTL region IM method is better as in CIM, model complexity sometimes showed relatively less sensitivity. To determine the region (a confidence interval) for each QTL, a change in a one LOD from the peak position should be recorded. In case of multiple identified peak positions that are close to each other (≤ one-half LOD distance), the peak with the highest LOD score should be considered as a QTL region. On the other hand, if peak positions are close to more than one-half LOD score distance then each peak should be considered as a separate QTL. However, in case of multiple environments, peaks that are present within overlapped confidence intervals should be treated as a same QTL as mentioned before [38, 39].

The epistatic values between markers should be conducted using a 2D genome scan in "QTLNetwork 2.0" with a threshold level set at 0.05 [40]. To estimate the real QTLs (stable in multiple environments) among several overlapping QTLs, meta-analyses can be performed using software "BioMercator 2.1" as described by Goffinet and Gerber [41]. Moreover, in this approach, multiple overlapping QTLs can be integrated and then a real QTL can be estimated, as well as confidence interval (CI). Under a single location (environment), there could be a several candidate positions localized in a linkage group (LG). In such QTL, five different models can be extracted and best one could be determined based on the minimum Akaike Information Criteria (AIC, lower the value more likely the model) value, which can be considered as a real QTL. Consensus QTL can be determined by considering real QTL (determined based on AIC values) along with other overlapping QTLs using tool meta-analysis in software "BioMercator 2.1." Further, a consensus QTL is the one that contained at least two original QTLs on a LG, and CI for consensus QTL should be shrunk by a wide margin. Mean R^2 values of the original QTLs in region would explain this consensus QTL as suggested by Sun et al. [42].

2.1.5 Identification of Candidate Genes in the Target Region

Identification of candidate genes with the CI region in a LG, a gene cloning strategy can be applied as described earlier [24]. For example, coding sequence of a known gene homolog to Arabidopsis can be cloned in the mapping population. After that, SNP markers can be developed based on nucleotide diversity between parents of the mapping population. Lastly, SNP primers can be made for the characterization of all genotypes of the mapping population as well as for the linkage analysis. To find the physical position of markers, BLASTn approach should be followed using the sequence information of primer associated with target QTL/s. This source is maintained in the Brassica genome sequence assembly information related to QTLs and putative genes targeting phenotypic traits (e.g., flowering time). Another strategy to find the candidate genes is to use the physical position of a marker aligned with a reference genome. For example, in case of SNP markers called from "Brassica 60 K SNP Array" that have determined position after mapping to the reference genome (*B. napus*; http://www.genoscope.cns.fr/brassicanapus). So, based on the SNP markers position in the CI of QTLs can be aligned with the reference genome (using gff file) and genes in that region will be considered as the potential candidates.

2.2 Genome Wide Association Analysis

2.2.1 Plant Material

In the GWAS studies, population can be either less diverse as DH population and RIL population, or more diverse (so called the GWAS population), which is consists of material collected from all growing regions and crop types (ecotypes) around the globe. If the purpose is to dissect the genetic architecture of a single trait, then a combination of two populations (DH/RIL and GWAS) will be better to find the consensus QTLs.

2.2.2 Phenotypic Data Analysis

The phenotyping data recording method is similar as in case of QTL mapping. However, for GWAS analysis there are three types of methods have been reported for each studied trait. First method is to directly use the mean values of 10 plants for each accession/genotype. Second one is the calculation of best linear unbiased prediction (BLUP) values that potentially discard the outliers in 10 plant data. This method is generally used for data collected from genotypes grown at a single location with normally >2 years. The BLUP values can be calculated using an Rscript (www.eXtension.org/pages/61006) [43]. The third one is the best linear unbiased estimates (BLUE, generally it is recommended when data is collected from multiple environments/multiple locations) values and can be performed using "lme4", an R package (r-project.org) and lsmeans [44]. One of the above data such as mean, BLUP, and BLUE values can be directly used to perform GWAS analysis.

2.2.3 *Brassica* 60K SNP Array Analysis

The 60K SNP array is most widely used genotyping method for both QTL and GWAS analyses. This method is described above in Sect. 2.1.3.

2.2.4 DNA Resequencing

Genomic DNA can be extracted by following the similar approach as described above in the Subheading 2.1.3. After the quality check, for DNA sequencing there are several platforms have been used in different plant species such as Hi-Seq 2000, 2500, X10, and Sequel II, etc. to get high quality reads. Generally, for genotype by sequencing or population genomic studies Hi-Seq 2000 PE product is enough. For filtering, adapter sequences and low-quality reads (>5% of non-sequenced (N) bases or more than 20% $Q \leq 20$) can be removed using SOAPnuke [45]. After that, clean reads should be aligned to the reference genome of *B. napus* (Darmou-bzh_V5, http://brassicadb.org) using HISAT2 and BWA tool [46].

2.2.5 mRNA Sequencing

Total RNAs from leaf tissues (preferably more sensitive or growing tissue will be better, if later purpose is to link mRNA expression analysis with GWAS results) should be isolated using an RNA Prep Pure Plant Kit, which later can be used to build the RNA-seq library. Pre-mRNAs of these tissues should be subjected for sequencing using the Illumina HiSeq 2000 PE150 (BGI). Here, filtering procedure will be the same as described in the above Subheading 2.2.4.

2.2.6 Calling of SNPs

For SNP calling, the product of "Stringtie-1.3.3" as a substrate (or directly from HISAT2 product) can be subjected to Genome Analysis Toolkit (GATK, version = 4.0.3.0) as described earlier [47].

2.2.7 Filtering of SNPs

There are several filters were described in many studies. However, few standards during filtering should be keep in mind such as (1) first filtering of SNPs should be done using vcftools with the following parameters min-alleles 2 –max-alleles 2 –maf 0.05 –max-missing 0.9. Second, impute the missing values using Beagle (http://faculty.washington.edu/browning/beagle/beagle.html).

2.2.8 Association of SNPs with Phenotypic Traits

To date, there are several single and multi-locus methods have been used in many plant species. However, mixed linear model (MLM) in the "EMMAX" (beta version) platform [48] is most widely accepted and used to associate the SNPs with phenotypic data of the agronomic traits. The matrix of pairwise genetic distances should be calculated using EMMAX as the variance-covariance matrix of random effects to test the trait-SNP associations [49, 50]. Significant P-value thresholds can be determined by dividing the 0.05 with total number of clean SNPs (0.05/ n, n = total SNP) to control the type 1 error rate.

2.2.9 Identification of Candidate Genes

For the identification of potential genes targeting agronomic traits, the first step is to determine the QTL region (up-down stream region from a significant marker). In the literature, there are several suggested and measured indicators, which could help to set the QTL region. For example, in some studies, QTL region ± from a significant marker is determined based on the linkage disequilibrium (LD, r^2) value calculated using all SNPs on all chromosomes. However, for high quality SNP data called from the NGS data), r^2 values is very less compared to "Brassica 60 K SNP Array" data. In the literature, generally the range is from 100 KB to 1 MB. After determining the ± KB region from the marker, next step is to find candidate genes from "gff" files of *B. napus* reference genome (http://www.genoscope.cns.fr/brassicanapus) using custom script.

3 Application of SNPs in Association Mapping Targeting Agronomic Traits in *Brassica napus*

3.1 Root Architecture-Related Traits

Plants depend on their RSA due to the three-dimensional spatial configurations of complex distribution of root system in the changing soil environment (root growth, lateral and deep in search of water, and gravity) to uptake water and mineral nutrients to enhance the productivity and yield stability [51–53]. An increased in lateral root (LR) number and other genetic modifications to root architecture enable plants to increase overall crop yield and improve stress tolerance capability [51, 54]. For example, in *B. napus* [55] using "shovelomics" approach, 216 diverse genotypes for five RSA-related traits were phenotyped in normal field conditions for 2 years (2015 and 2016). These phenotypic traits were soil-level taproot diameter (R_1Dia), below ground taproot diameter (R_2Dia), primary root branches (PRB), root angle (RA), and root score (RS). A genotypic panel consisting of 30,262 SNPs was used for GWAS. Their mapping data showed that six marker loci for each R_1Dia and R_2Dia, seven for PRB, and eight for each RA, and RS. Further, they reported that same seven markers which were targeting root diameter-related traits (R_1Dia and R_2Dia), were also targeting PRB and RS. Thus, they suggested that in canola, taproot diameter has been appeared as a major detrimental trait among RSA traits. Therefore, it can be used as a proxy for probing the impact on other traits. Lastly, they identified 15 candidate genes targeting the RSA-traits. The distance of these genes from the associated markers was ±100 Kb. So, these associated markers can be utilized for MAS for RSA-traits in *B. napus* breeding program. Similarly, a panel of 37,500 SNPs was exploited against the seven phenotypic RSA-traits including root length (RL), RA, PRB, root dry weight (RDW), root vigor score (RVS), and R1Dia and R2Dia [56]. Their GWAS data showed that ten significant makers were targeting RL, 11 RA, nine each PRB and RDW, seven RVS, and six makers for R1Dia and R2Dia. Interestingly, the majority of their identified markers were localized on only five chromosomes in two subgenomes, i.e., A01, A02, A04 (A subgenome), C03, and C04 (C subgenome). Within ±50 Kb region, they identified in a total of 22 candidate genes targeting RSA-related traits. Importantly, it was also reported that several makers co-localized on chromosomes A01, A02, A04, and C03 as the physical distance between them was very short.

Exploring the genetic variation in RSA is essential for enhancing plant capability to adapt well in a dynamic environment. For this, a sophisticated system called "non-destructive gel-based mini-rhizotron" was used for phenotyping of RSA-related traits in *B. napus* [57]. The important feature of the above system is to facilitated the visualization of the phenotypic data of large genotypic data sets including 94 DH (produced from the cross of Express 617 × V8) and 439 inbred lines. First, based on the CIM, authors constructed a high-density genetic map for the identification of QTLs in DH population. Second, 6K SNPs data were used for association mapping in the inbred lines. In the result, 11 QTLs were detected in DH population and 38 significant markers were identified in inbred lines targeting the RSA-related traits. More recently, the connection between RSA and shoot system architecture (SSA) was explored to gain further insight on the genetic architecture of root growth [58]. Seven RSA and SSA-related traits were phenotyped at five consecutive vegetative growth stages in a diverse collection of 280 *B. napus* accessions. Further, a persistent and growth-specific genetic mechanism was established by identifying 16 persistent and 32 growth-specific QTLs through GWAS. Moreover, the root dynamics (slow and fast growth) was noticed during all five vegetative growth stages and subjected to the variations for transcriptome analysis. The mRNA data showed that a total of 367 differentially expressed genes (DEGs) with persistent expression were identified during the root developmental stages. These DEGs were significantly enriched in GO terms, such as response to biotic and abiotic stresses and energy metabolism. Similarly, a total of 485 stage-specific DEGs were found, and enriched in GO terms including nitrogen metabolism. Moreover, the integrated results of GWAS, weighted gene co-expression network (WGCNA) and DEGs analyses discovered common four and eight persistent and stage-specific candidate genes, respectively, potentially involved in the RSA development. Finally, the distance of these genes was less than 100 Kbp from SNP peaks. These above findings can be utilized in the genetic improvement and MAS for breeding program with the objective of improved RSA in *B. napus*.

3.2 Plant Architecture-Related Traits

In *B. napus*, the plant architecture (PA)-related traits mainly include PH, BIH, branch angle (BA), NBPP, number of the branches on the main inflorescence (NBMI), leaf and inflorescence morphology, and stem length. There are several insights from studies on the genetic mechanism of these above-mentioned traits using different association mapping approaches. For example, the genetic mechanism of PH in *B. napus* was dissected using a panel of 60 K Brassica Infinium SNP array from 476 worldwide collected inbred lines grown under six different environments [59]. The GWAS results were produced using Anderson–Darling, A–D test methods, which are (i) robust; (ii) novel; and (iii) non-parametric. In the result, 24 loci were detected targeting PH, overlapping with selective sweep signals data. The above mentioned finding clearly indicated the signatures of semi-dwarf breeding in the evolutionary history of *B. napus*. Moreover, the LD decay (up- and down-stream of 65 loci) data detected the candidate genes, the orthologs in Arabidopsis, which are involved in PH regulation. Additionally, a locus that co-localized with the established PH locus "*BnRGA*" in *B. napus* was described. More recently, a genetic linkage map of F_2 population (including 200 individuals in Yangluo-2018 derived from ZS11-HP × *sdw-e*) was constructed using whole-genome resequencing data to detect the QTLs targeting PH [60]. This linkage map consists of 4323 bins with markers, and it covers a total distance of 2026.52 centimorgan (cM, ~equal to 1 Mb) with an average marker interval of 0.47 cM. In detail, in linkage group 10, a major QTL qPHA10 targeting PH was identified. This QTL was consistent in the presented QTL-seq data. Further, authors integrated the results of variation sites with changes in the expression of genes in the region of the QTL qPHA10 to identify the candidate genes. As suggested, a major QTL along with other candidate genes can be used in MAS for rapeseed breeding program. To expand the scope, PH was investigated along with other two major PA traits such as BIH and NBPP, and associated with genetic markers through GWAS using a panel 60 K Brassica Infinium SNP array in 333 *B. napus* accessions [16]. Seven loci in total were described targeting PH, four BIH, and five for BN. Subsequently, 31, 15, and 17 candidate genes, respectively, were identified using LD decay of 38 significant SNPs. Further, a strong correlation was revealed between PH and BIH traits in this study. Among the candidate genes, *BnRGA* (GA-signaling gene) and *BnFT* (flowering time regulatory gene), located in chromosome A02, were associated with PH regulation, whereas *BnLOF2* (meristem initiation gene) and *BnCUC3* (NAC domain transcriptional factor) were locilized in A07, most likely the candidate genes associated with BN. The above findings on the genetic mechanism of three important traits and their correlation dissected the novel genetic mechanism of PA, and it may facilitate the researchers in designing molecular marker-assisted breeding targeting the studied traits.

Besides, BA is an important agronomic trait that determine the ideotype of a plant. Unfortunately, there is less information available that underlying its genetic mechanism. A DH population was developed from a cross including one *B. napus* introgression line derived from *Capsella bursa-pastoris* contains compressed branches and wooden stem [61]. A high-density genetic map was constructed covering a total genetic distance of 2242.14 cM and the average distance between two markers was 0.73 cM. In case of phenotypic data, they phenotyped all DH population across six environments and constructed the "inclusive composite interval mapping algorithm" to analyze the QTLs associated with BA. In the results based on a single environment, 17 QTLs were identified in total, mainly localized on four chromosomes, A01, A03, A09 and C03. Among them, three major QTLs including qBA.A03-2, qBA.C03-3 and qBA.C03-4, were highly significant, and each QTL explained >10% of the total phenotypic variation under at least two environments. Moreover, 10 QTLs were detected using the mapping data for QTL by environment interactions (QEI), and seven of them were also detected in a single environment analysis. Lastly, 27 candidate genes were identified using SNP position with their functional annotation based on genome of *Arabidopsis thaliana*. Based on this study, prominent genes were indicated as responsive to early auxin-response, small auxin-up RNA, auxin/indoleacetic acid, and Gretchen Hagen 3.

Similarly, lodging-related traits such as stem strength (SS), stem breaking resistance (SBR), stem diameter (SD), and lodging coefficient (LC) are very crucial among PA traits, which contributed to yield productivity and seed quality. To dissect the genetic mechanism underlying these above mentioned traits, a comprehensive

GWAS study was performed using a panel of Brassica 60K SNP array of 472 *B. napus* accessions [62]. Results showed that, a total of 67 significantly associated QTLs and 71 candidate genes were detected. Besides, a WGCNA was also performed, and a significant module was found that associated with cellulose biosynthesis. Their integrated GWAS and WGCNA results detected three candidate genes such as BnaC08g26920D (*Eskimo1, ESK1*), BnaA09g06990D (*Cellulose synthase 6, CESA6*), and BnaC04g39510D (*Fragile fiber 8, FRA8*). These above findings look into the genetic mechanism underlying stem lodging resistance, and both candidate QTLs and genes can be used for revealing the genetic basis underlying stem lodging and provided the promising QTLs harboring genes for the genetic improvement of stem lodging resistance traits in *B. napus.*

3.3 Flowering Time-Related Traits

Flowering is a key component of a plant's life cycle, marking its transition from the vegetative to the reproductive phase. Opportune FT is a crucial for the survival of a plant in a specific environment. It regulates life-cycle duration, and contribute to high yield, seed quality, disease resistance, and crop rotation. Thus, it can be said that traits related to FT are the most promising agronomic traits that directly impact on seed yield and oil quality in *B. napus.* In the literature, several FT-related traits have been reported so far, such as budding, bolting, DIF, DFF, and FP. The genetic mechanism of FT was studied in related traits through GWAS using a panel of "*Brasica* 60K Illumina Infinium SNP array" [63]. Their results showed that, a total of 41 SNPs distributed on 14 chromosomes were significantly associated with DIF. Among these, 12 SNPs were located in the confidence intervals (CI) of a QTL. Interestingly, twenty-five candidate genes were orthologous to *A. thaliana* flowering-related genes. Moreover, to dissect the genetic mechanism of FT, they performed GWAS on two derived phenotypic traits including the environment- and temperature-sensitivity. Interestingly, they detected a promising SNP marker localized near "Bn-scaff_16362_1-p380982," that just 13 kb away from an important photoperiod pathway-related gene BnaC09g41990D (orthologs to an Arabidopsis *Constans, CO*). The integration of two or more approaches can give better insight than a single approach to dissect the genetic mechanism of agroinomic traits, for instance FT. For this purpose, a joint QTL mapping and RNA-Seq were used to reveal the candidate genes targeting DIF [64]. In case of phenotypic material, recombinant inbred lines were used including lines with significant variation in their FT while grown under six environments. Results showed that, a total of 27 QTLs distributed on eight chromosomes were detected from six environments including a major QTL on C02, which was stable across all environments and alone explained the 11–25% of phenotypic variation. Besides, mRNA gene expression data revealed 105 DEGs targeting FT were involved in four

FT-related pathways such as circadian-clock or photoperiod pathway, autonomous pathway, hormone and vernalization pathway. Their integrated results (genes localized on the candidate QTL regions and DEGs from mRNA) revealed that eight genes/DEGs were common including important FT targeting genes such as *Pseudo Response Regulator 7* (*PRR7*), a temperature-sensitive circadian system, and *FY*, an mRNA processing factor is involved in regulation of FT through affecting FCA mRNA processing. Interestingly, these both genes were located in a major QTL region on C02 [64].

Among FT-related traits, GP-related traits such as DIF, DFF, and FP are very crucial as they not only impact on the yield but also adaptation to the changing environment. To elucidate the genetic mechanisms of above traits in *B. napus*, a combinatorial approach was used including linkage mapping, GWAS and mRNA-Seq analyses [10]. The detailed results showed that, a total of 146 SNPs and 83 QTLs were identified through GWAS and linkage mapping using RIL population. Of these, 19 SNPs were pleotropic and six including q18DFF.A03-2, q18MT.A03-2, q17DFF.A05-1, q18FP.C04, q17DIF.C05, and q17GP.C09 were detected in both methods. Moreover, integration of GWAS and linkage mapping with RNA-Seq results of early and late RILs identified 12 candidate genes were associated with the above traits. Finally, their resequencing data showed that among the above 12 genes, seven have polymorphic sites located 2 Kb upstream region from their coding sequence. Lastly, pleotropic haplotypes such as "BnaSOC1.A05-Haplb" and "BnaLNK2.C06-Hapla" that target multiple phenotypic traits and the candidate genes can be used in MAS for rapeseed breeding program focusing on the productivity, yield stability, and environmental adaptation. More recently, the genetic mechanism of FT was further investigated through the integration of SNP-GWAS, and with haplotype GWAS analysis results produced from a panel of 60K SNP array of 373 diverse world widely collected *B. napus* accessions were phenotyped under four different environments [65]. In the results, a total of 15 and 37 QTLs were identified from both above-mentioned methods. Of these, three (SNP-GWAS) and eight (haplotype-GWAS) were environmentally stable QTLs. Moreover, their integrated results showed that 10 QTLs distributed on A03, A07, A08, A10, C06, C07, and C08 chromosomes were common in both methods. Among them, four QTLs including FT.A07.1, FT.A08, FT.C06, and FT.C07 were detected as novel. A total of 197 genes targeting FT were localized in these four regions. Finally, their mRNA-Seq of early and late flowering data showed that 14 genes (of above 197 genes) were DEGs and orthologs to 13 Arabidopsis genes controlling FT. Above all findings may facilitate the breeding efforts for developing early flowering cultivars as it has been remained an important breeding objective in *B. napus* [65].

3.4 Reproduction and Silique-Related Traits

Pollen viability (PV) and pollen fertility (PF), and silique-related traits such as silique length (SL), silique breadth (SB), silique thickness (ST), NSPS, silique volume (SV), and 1000-seed weight (TSW) are very crucial for determining the seed yield and quality [66, 67]. In case of understanding the genetic mechanisms of PV and PF, from both used parents, their 146 DH progenies were produced from microspore culture of an allohexaploid F_1 (a cross of two recently synthesized allohexaploid *Brassica* lines) [66]. The authors used DNA RAD-Seq for resequencing and then construct the linkage map of the genotypic material. From a total of 290,422 SNPs, the high-quality SNP (7950) markers were developed that segregated normally (1:1) in the mapping population. Further, the linkage map was shown to contain all 27 chromosomes from three parental genomes (A, B and C) with total distance of 5725.19 cM and an average genetic distance between two markers was 0.75 cM. Results also showed that among 146 DH lines, 91 had a complete set of all 27 chromosomes, and 21/27 chromosomes exhibited high collinearity between the linkage and physical maps. Authors also suggested that the loss of chromosomal segment or whole chromosomes in some DH progenies was linked with the reduction in PV and PF. A total of 25 additive QTLs were detected associated with PV and PF-related traits such as NSPS, SL, TSW, and seed yield. Lastly, 44 intra-genomic and 18 inter-genomic epistatic QTL pairs were also identified targeting above-mentioned phenotypic traits. Similarly, a QTL mapping approach was performed for insight into the genetic mechanisms of five silique-related traits such as SL, SB, ST, NSPS, and SV using 189 RIL population [68]. The data showed that a total of 120 QTLs were identified including 23 for SL, 25 for SB, 29 for ST, 22 for NSPS and 21 for SV, which were distributed on all chromosomes except C05. Additionally, there were 13 consensus QTLs, and one, five, two, four and one for SL, SB, ST, NSPS and SV. These QTLs were also identified in multiple environments and explained 4.38–13.0% of the total phenotypic variation. Furthermore, the candidate genes were filtered in these genetic regions and 12 genes were found as orthologs with *Arabidopsis* genome, related to silique traits. Generally, these findings suggested that the genes interact to each other in a sophisticated network for controlling silique-related traits in *B. napus*.

Apparently, NSPS and TSW are two main components in determining the seed yield. Specifically, to dissect the genetic mechanism of the above two traits, two examples are provided. First, the genetic mechanism of TSW was dissected and earlier reported QTL "‘cqSW.A03-2" was further investigated through linkage and association mapping of population (derived from a cross between elite line ZY50 and a pol cytoplasmic male sterility restorer line 7-5) [69]. The results showed that six major QTLs were targeting TSW and among, one major QTLs cqSW.A03-2, which alone explained 8.46–13.70% of the total phenotypic variation. Moreover, this QTL was consistent in multiple environments. To further investigate on the genetic basis of cqSW.A03-2, authors developed a set of near-isogenic lines that facilitated the test of self-pollinated progenies, which lead to the detection of a said QTL as a single Mendelian factor. Additionally, its allele localized at cqSW.A03-2 in ZY50 had showed a positive impact on TSW. Further, the authors filtered the above QTL region into a reduced 61.6 kb region through fine mapping, and 18 candidate genes localized in this region were detected. Both, the above predicted gene association and gene expression analyses identified a histidine kinase (BnaA03g37960D) is most likely to be the candidate gene for the locus "cqSW.A03-2". In a separate study, the two traits, NSPS and TSW were investigated using a natural population of *B. napus* grown for multiple years [17]. The data showed that, a total of 101 and 77 SNPs were significantly associated with the above traits and their phenotypic variances R^2 were ranging from 1.35% to 29.47% in the first trait and 0.78% to 34.58% in the second trait, respectively. Moreover, orthologs of 43 and 33 known genes in Arabidopsis for NSPS and TSW were located in the 65 and 49 HBs, respectively. Lastly, the authors detected five overlapping loci and three sets of loci with collinearity for the above mentioned both traits. Of these, four overlapping loci sheltered the haplotypes having the genetic impact in the same direction on NSPS and TSW. These above findings based on the results of both overlapping and independent traits targeting loci suggested that both traits can genetically be improved simultaneously in rapeseed. Briefly, the above studies related to reproduction and silique-related traits and insights on their molecular/genetic mechanisms will facilitate the MAS designed breeding program for seed yield improvement in *B. napus*.

3.5 Seed Quality-Related Traits

Oilseed rape breeding programs have long been driven by the need to increase seed oil content and nutritional value for humans, animals, and other non-food purposes. This included adjusting the oil's fatty acid (FA) balance and enhancing the nutritional value of the meals [70]. The distribution of erucic acid (EA) and seed oil content (SOC) was analyzed in a F_1 microspore derived DH population through developing a RFLP linkage map [71]. The results showed no segregation of SOC in both parents, however, in the progeny a transgressive segregation was observed, which distributed normally in the linkage map. Moreover, the authors detected three QTLs for SOC were mapped on three different linkage groups. These QTLs were considered the major QTLs as their additive effects explained 51% of total phenotypic variation for SOC in progeny. In case of EA, a noticeable three-class segregation was recorded, and two genes were mapped to two different linkage groups. Interestingly, these two genes were closely associated with two of SOC targeting QTLs (co-localized), suggesting a direct impact of EA genes on SOC. Exploring the variation of gene/s localized on locus E1 controlling EA that can be useful for breeding program aimed low EA content. To address this concern, two sequences homologous of *FAE1* gene were isolated from an immature embryo in cDNA library of *B. napus* [72]. The FAE1 gene encodes a β-ketoacyl-CoA synthase enzyme, which is a key component of very long chain FAs present in the cytoplasm. Both clones, CE7 and CE8 encoding protein of 506 and 505 amino acids, respectively, had inserts of 1647 bp and 1654 bp, with a molecular mass of ~56 kDa. Although, these two cDNA clones were highly homologous as sharing 97% nucleotide and 98% amino acid identity, however yet distinct. Southern blot analysis showed that β-ketoacyl-CoA synthase enzymes was encoded by a small multigene family in *B. napus*. On the other hand, Northern blot analysis showed the expressions of *FAE1* genes were restricted to the immature embryos. These findings further indicated that at least one the *FAE1* genes is tightly associated with E1 locus, which is one of two loci controlling the EA content in *B. napus*. In another study, degenerate PCR was designed for *FAE1* and amplified two copies of the gene, i.e., *BN-FAE1.1* and *BN-FAE1.2* in *B. napus*, *B. rapa* and *B. oleracea* [73]. Results from acrylamide gel electrophoresis revealed a clear polymorphism of these two *FAE1* genes. Further, authors mapped these genes in linkage groups and found a co-segregation with E1 and E2 loci controlling EA content. A mutation occurred in one of these two genes has also been suggested that could partially explain the low EA content. Here, a question arises how zero EA trait appeared in *B. napus*? The answer was given in an earlier published report [74]. First, the complete coding sequence of *FAE1* gene was isolated from eight zero and high EA content containing cultivars. The results showed that a four-nucleotide deletion was present between T1366 and G1369 in above-described genes in several cultivars. This mutation in *FAE1* gene resulted into a frameshift mutation causing a premature stop-codon of the translation after the 466 amino acid residue. Additionally, this deletion was predominantly occurred in the C subgenome compared to A subgenome. Further, authors confirmed the impact of deletion to expression level of *FAE1* in a yeast system in the form of truncated proteins with no enzymatic functionality and failed to generate very long chain FAs compared to control. However, in *B. napus* system, the transcription rate was normal, but it failed to translate protein with normal functionality. These findings confirmed the role of four-nucleotide deletion in the production of zero-EA cultivars of *B. napus*. To expand the scope, the genetic mechanism of EA content was dissected through the GWAS method using a panel of 60K SNP array produced from 472 *B. napus* accessions [50]. The results showed that two SNPs on A08 and C03 were significantly associated with *BnaA.FAE1* and *BnaC.FAE1*. Moreover, the authors also found that locus-harbored "BnaA.FAE1" was found to be significantly associated with SOC. These findings indicated the complexity of *E1* locus and suggest that the fine mapping of complex traits is suitable for the future breeding program. More recently, integrated approaches of GWAS and transcriptome data were used to dissect the genetic mechanism of SOC in *B. napus* [75]. The authors used 385,692 high-quality SNPs produced from high-throughput resequencing data with a minor allele frequency >0.05. In total, 17 loci were detected as significantly associated with SOC, and among them, 12 loci were distributed in A subgenome (11 in A03 and one in A01) and five loci were in C subgenome (one in C05 and four in C07). The comparative DEGs data of seeds and silique pericarp (main inflorescence and primary branches) from high and low SOC showed a total of 64 DEGs responsible for lipid metabolism. Among them, 14 genes were found to be involved in triacylglycerols biosynthesis pathway and assembly. The integrated results showed that seven DEGs were also present in the GWAS candidate gene data indicating the potential genes for future study.

Another crucial seed quality trait is a GSL content, which are sulfur-containing glycosides occurred both in vegetative and reproductive tissues. The GSLs play a significant role in plant defense mechanism against pests and pathogens. So, to produce cultivars with high yield and quality, breeders are targeting genotypes of 00 canola cultivars with high content of GSL in leaves and low GSL content in seeds [76]. The reason behind this demand was in the absence of a correlation found between GSL content in leaf and seed of *B. napus* breeding lines [77]. To dissect the genetic mechanism behind the GSL, Harper et al. [78] identified two genomic deletions in two QTLs detected controlling GSL content in seeds. In detail, the deletions in genetic regions contained genes (orthologs of *A. thaliana* HAG1 transcription factor) that were responsible for aliphatic glucosinolate biosynthesis in *A. thaliana*. These findings supported the role of associative transcriptomics underlying the complex traits in crops having polyploid background. In another study, the diverse set of germplasm with 307 accessions (SOR, WOR and SWOR, three ecotypes) were re-sequenced and GWAS was performed to associate significant SNPs and HBs with GSL content in seeds of *B. napus* [8]. The results showed a total of eight markers (four common and four specific to HBs), that were significantly associated with GSL content in seeds. To investigate further, the authors performed a transcriptome analysis of 36 accessions with extremely low and high GSL content. The integrated data (genomics variation, HBs and DEGs) suggested that five candidate genes were found common in three ecotypes and three other genes were highly expressed only in SOR-HBs. These findings highlighted the importance of multiple techniques for the identification of additional three genes in SOR-HBs, which potentially play an important role in the genetic mechanism of GSL content in *B. napus.* More details about the QTL mapping and GWAS targeting agronomic traits are present in Table 1.

4 Conclusion and Future Perspectives

Brassica napus L. is a naturally occurring polyploid crop species, produced by hybridization between *B. rapa* and *B. oleracea* ~7500 years ago [1]. Subsequently, the whole genome duplication event greatly contributed to the evolution of multiple gene families and resulted in the enhanced genome size and complexity. However, polyploidy also enhanced the genome diversity that led to the evolution of new crop types such as WOR, SOW, and SWOR, having better adaptive traits in response to dynamic environments including abiotic and biotic stresses [3–5, 83]. Molecular markers such as SNPs are the promising as they are molecular markers and have been extensively utilized in multiple crop species for the identification of genes targeting agronomic traits related to both environmental and developmental cues. To speed up the SNP identification in plant species including *B. napus*, numerous improvements have been made in the advancements of NGS high-throughput technologies that enabled researchers to call multi-million SNPs and linked with multi-thousands of homoeologous exchanges (HE). However, output from NGS techniques is in multi-Gbs, so a detailed QC check prior to the downstream analyses is considered as a pre-requisite. The purpose of SNP filtering is to remove the false positives that can be directly proportional to the false associations, eliminate the low-quality SNPs and gaining of true associations. Filtering of SNPs also includes the other challenges such as (1) comparative filtering of SNPs called from treated and untreated data sets; (2) filtering of SNPs called from plants with a polyploid genome having highly similar HE events occurred on two subgenomes and ideally polymorphic SNPs should be different than HE between subgenomes; (3) artifacts produced due to incomplete reference genome, and during library preparation, data processing and calling of SNPs. After SNP calling, the next crucial step is to validate the SNPs and that is carried out using a selected proportion of SNPs with false positive rate <20%. Currently, there are several SNP validation methods available for the use. The promising methods included AS-PCR assay-based methods including rhPCR, KASP and TaqMan due to their efficiency and cost effectiveness.

Table 1
SNP-based linkage and association mapping for different agronomic traits of *Brassica napus*

Trait category	Traits studied	Mapping population	SNP numbers	Method used (linkage or association mapping)	References
Root system architecture-related traits	10 root-related and shootbiomass traits	236 RIL lines	28,491 SNP	QTL	[7]
	12 mature root traits	338 B napus accessions	20,131 SNP	GWAS	[6]
	7 root-related traits	224 B. napus accessions	37,500 SNP	GWAS	[56]
	4 root-related traits	405 B. napus accessions	19,397 SNP	GWAS	[79]
Plant architecture-related traits	Plant height (PH)	520 B napus accessions	19,167 SNP	GWAS	[59]
	Plant height (PH), Branch initiation height (BIH), and Branch number (BN)	333 B napus accessions	32,297 SNP	GWAS	[16]
	Stem breaking resistance (SBR), stem diameter (SD), SS = SBR/[π(SD/2)2], and lodging coefficient (LC)	472 B napus accessions	19,945 SNP	GWAS	[62]
	Branch angle (BA)	208 DH lines	52,157 SNP	QTL	[61]
	Plant height (PH)	200 F2 line	3,93,214 SNP	QTL	[60]
Flowering time-related traits	DIF (Days to initial flowering)	523 B napus accessions	26,024 SNP	GWAS	[63]
	DIF (Days to initial flowering)	172 RIL lines	2795 SNP	QTL	[64]
	DIF (Days to initial flowering), DFF (Days to final flowering), FP (Flowering period), MT (Maturity time), and GP (Growth period)	172 RIL lines – QTL; 588 B napus Accessions (GWAS)	2795 SNP – RIL (QTL); 3,85,691 SNP – GWAS	QTL and GWAS	[10]
	Budding, Bolting, DTF, and IBD (Interval between bolting and DIF)	373 B napus accessions	21,856 SNP	GWAS	[65]
Silique-related traits	Silique length (SL)	331 B napus accessions	24,508 SNP	GWAS	[80]
	1000-seed weight (TSW)	189 DH lines	2237 SNP	QTL	[69]
	Number of seed per silique (NSPP) and 1000-seed weight (TSW)	157 B napus Inbred line	6,90,953 SNP	GWAS	[17]
	Silique length (SL), silique breadth (SB), silique thickness (ST), number of seed per silique (NSPP), and silique volume (SV)	189 RIL lines	11,458 SNP	QTL	[68]

(continued)

Table 1 (continued)

Trait category	Traits studied	Mapping population	SNP numbers	Method used (linkage or association mapping)	References
Seed quality-related traits	8 seed quality traits (oil and protein content, palmitic, stearic, oleic, linoleic, eicosenoic, and erucic acids content)	370 B napus accessions	21,856 SNP	GWAS	[81]
	Seed glucosinolate content (GSL)	307 B. napus accessions	5,791,961 SNP	GWAS	[8]
	Fatty acid compositions	520 B. napus accessions	11,368 SNPs	GWAS	[82]
	Oleic acid content	375 B. napus accessions	23,168 SNP	GWAS	[9]

Acknowledgments

The funding for RAG was supported by the National Natural Science Foundation of China (31950410560).

References

1. Chalhoub B, Denoeud F, Liu S, Parkin IA, Tang H, Wang X et al (2014) Early allopolyploid evolution in the post-Neolithic *Brassica napus* oilseed genome. Science 345(6199): 950–953. https://doi.org/10.1126/science.1253435
2. Mason AS, Snowdon RJ (2016) Oilseed rape: learning about ancient and recent polyploid evolution from a recent crop species. Plant Biol 18:883–892. https://doi.org/10.1111/plb.12462
3. Gill RA, Scossa F, King GJ, Golicz AA, Tong C, Snowdon RJ et al (2021) On the role of transposable elements in the regulation of gene expression and subgenomic interactions in crop genomes. Crit Rev Plant Sci 40:157–189. https://doi.org/10.1080/07352689.2021.1920731
4. Zou J, Mao L, Qiu J, Wang M, Jia L, Wu D et al (2019) Genome-wide selection footprints and deleterious variations in young Asian allotetraploid rapeseed. Plant Biotechnol J 17:1998–2010. https://doi.org/10.1111/pbi.13115
5. Song JM, Guan Z, Hu J, Guo C, Yang Z, Wang S et al (2020) Eight high-quality genomes reveal pan-genome architecture and ecotype differentiation of *Brassica napus*. Nat Plants 6: 34–45. https://doi.org/10.1038/s41477-019-0577-7
6. Ibrahim S, Li K, Ahmad N, Kuang L, Sadau SB, Tian Z et al (2021) Genetic dissection of mature root characteristics by genome-wide association studies in rapeseed (*Brassica napus* L.). Plants 10:2569. https://doi.org/10.3390/plants10122569
7. Kuang L, Ahmad N, Su B, Huang L, Li K, Wang H et al (2022) Discovery of genomic regions and candidate genes controlling root development using a recombinant inbred line population in rapeseed (*Brassica napus* L.). Int J Mol Sci 23:4781. https://doi.org/10.3390/ijms23094781
8. Wei D, Cui Y, Mei J, Qian L, Lu K, Wang ZM et al (2019) Genome-wide identification of loci affecting seed glucosinolate contents in *Brassica napus* L. J Integr Plant Biol 61:611–623. https://doi.org/10.1111/jipb.12717
9. Zhao Q, Wu J, Cai G, Yang Q, Shahid M, Fan C et al (2019) A novel quantitative trait locus on chromosome A9 controlling oleic acid content in *Brassica napus*. Plant Biotechnol J 17: 2313–2324. https://doi.org/10.1111/pbi.13142
10. Wang T, Wei L, Wang J, Xie L, Li YY, Ran S et al (2020) Integrating GWAS, linkage mapping and gene expression analyses reveals the genetic control of growth period traits in rapeseed (*Brassica napus* L.). Biotechnol Biofuels 13:134. https://doi.org/10.1186/s13068-020-01774-0
11. Bin Y, Wei C, Chaozhi M, Tingdong F, Jinxing T (2006) Mapping of quantitative trait loci for yield and yield components in *Brassica napus* L. Acta Agron Sin 32:676–682
12. Bennett EJ, Roberts JA, Wagstaff C (2011) The role of the pod in seed development: strategies for manipulating yield. New Phytol 190: 838–853. https://doi.org/10.1111/j.1469-8137.2011.03714.x
13. Clarke J, Simpson G (1978) Influence of irrigation and seeding rates on yield and yield components of *Brassica napus* cv. Tower. Can J Plant Sci 58:731–737
14. Tuncturk M, Çiftçi V (2007) Relationships between yield and some yield components in rapeseed (*Brassica napus* ssp. *oleifera* L.) cultivars by using correlation and path analysis. Pak J Bot 39:81–84
15. Chen B, Xu K, Li J, Li F, Qiao J, Li H et al (2014) Evaluation of yield and agronomic traits and their genetic variation in 488 global collections of *Brassica napus* L. Genet Resour Crop Evol 61:979–999. https://doi.org/10.1007/s10722-014-0091-8
16. Zheng M, Peng C, Liu H, Tang M, Yang H, Li X et al (2017) Genome-wide association study reveals candidate genes for control of plant height, branch initiation height and branch number in rapeseed (*Brassica napus* L.). Front Plant Sci 8:1246. https://doi.org/10.3389/fpls.2017.01246
17. Xing S, Dong H, Yang L, Huang D, Zheng F, Cui Y et al (2021) Both overlapping and independent loci underlie seed number per pod and seed weight in *Brassica napus* by comparative quantitative trait loci analysis. Mol Breed 41: 41. https://doi.org/10.1007/s11032-021-01232-1

18. Tang S, Liu DX, Lu S, Yu L, Li Y, Lin S et al (2020) Development and screening of EMS mutants with altered seed oil content or fatty acid composition in *Brassica napus*. Plant J 104:1410–1422. https://doi.org/10.1111/tpj.15003
19. Kuznetsova A, Brockhoff P, Christensen R (2017) LmerTest package: tests in linear mixed effects models. J Stat Softw 82:1–26. https://doi.org/10.18637/jss.v082.i13
20. Piquemal J, Cinquin E, Couton F, Rondeau C, Seignoret E, Perret D et al (2005) Construction of an oilseed rape (*Brassica napus* L.) genetic map with SSR markers. Theor Appl Genet 111:1514–1523. https://doi.org/10.1007/s00122-005-0080-6
21. Cai G, Yang Q, Yang Q, Zhao Z, Chen H, Wu J et al (2012) Identification of candidate genes of QTLs for seed weight in *Brassica napus* through comparative mapping among *Arabidopsis* and *Brassica* species. BMC Genet 13:105. https://doi.org/10.1186/1471-2156-13-105
22. Xu J, Qian X, Wang X, Li R, Cheng X, Yang Y et al (2010) Construction of an integrated genetic linkage map for the A genome of *Brassica napus* using SSR markers derived from sequenced BACs in *B. rapa*. BMC Genomics 11:594. https://doi.org/10.1186/1471-2164-11-594
23. Cheng X, Xu J, Xia S, Gu J, Yang Y, Fu J et al (2009) Development and genetic mapping of microsatellite markers from genome survey sequences in *Brassica napus*. Theor Appl Genet 118:1121–1131. https://doi.org/10.1007/s00122-009-0967-8
24. Fan C, Cai G, Qin J, Li Q, Yang M, Wu J et al (2010) Mapping of quantitative trait loci and development of allele-specific markers for seed weight in *Brassica napus*. Theor Appl Genet 121:1289–1301. https://doi.org/10.1007/s00122-010-1388-4
25. Wei D, Mei J, Fu Y, Disi JO, Li J, Qian W (2014) Quantitative trait loci analyses for resistance to *Sclerotinia sclerotiorum* and flowering time in *Brassica napus*. Mol Breed 34:1797–1804. https://doi.org/10.1007/s11032-014-0139-7
26. Zurneng T, Yunchang L, Oiong H (2007) Heterosis prediction based on genetic distance estimated by molecular markers in rapeseed. Chin J Oil Crop Sci 29:20–26
27. Panjabi P, Jagannath A, Bisht NC, Padmaja KL, Sharma S, Gupta V et al (2008) Comparative mapping of *Brassica juncea* and *Arabidopsis thaliana* using Intron Polymorphism (IP) markers: Homoeologous relationships, diversification and evolution of the A, B and C Brassica genomes. BMC Genomics 9:113. https://doi.org/10.1186/1471-2164-9-113
28. Vos P, Hogers R, Bleeker M, Reijans M, Lee T, Hornes M et al (1995) AFLP: a new technique for DNA fingerprinting. Nucleic Acids Res 23:4407–4414. https://doi.org/10.1093/nar/23.21.4407
29. Liu H, Du D, Guo S, Xiao L, Zhao Z, Zhao Z et al (2016) QTL analysis and the development of closely linked markers for days to flowering in spring oilseed rape (*Brassica napus* L.). Mol Breed 36:52. https://doi.org/10.1007/s11032-016-0477-8
30. Trick M, Long Y, Meng J, Bancroft I (2009) Single nucleotide polymorphism (SNP) discovery in the polyploid *Brassica napus* using Solexa transcriptome sequencing. Plant Biotechnol J 7:334–346. https://doi.org/10.1111/j.1467-7652.2008.00396.x
31. Fletcher RS, Mullen JL, Heiliger A, McKay JK (2015) QTL analysis of root morphology, flowering time, and yield reveals trade-offs in response to drought in *Brassica napus*. J Exp Bot 66:245–256. https://doi.org/10.1093/jxb/eru423
32. Kosambi DD (2016) The estimation of map distances from recombination values. In: Ramaswamy R, Kosambi DD (eds) Selected works in mathematics and statistics. Springer, New Delhi, pp 125–130. https://doi.org/10.1007/978-81-322-3676-4_16
33. Voorrips R (2002) MapChart: software for the graphical presentation of linkage maps and QTLs. J Hered 93:77–78. https://doi.org/10.1093/jhered/93.1.77
34. Broman KW, Wu H, Sen Ś, Churchill GA (2003) R/qtl: QTL mapping in experimental crosses. Bioinformatics 19:889–890. https://doi.org/10.1093/bioinformatics/btg112
35. Broman KW, Sen S (2009) A guide to QTL mapping with R/qtl. Springer, New York, 400 pp. https://doi.org/10.1007/978-0-387-92125-9
36. Wang S (2005) Windows QTL cartographer 2.5. https://brcwebportal.cos.ncsu.edu/qtlcart
37. Lander ES, Botstein D (1989) Mapping mendelian factors underlying quantitative traits using RFLP linkage maps. Genetics 121:185–199. https://doi.org/10.1093/genetics/121.1.185
38. Doerge RW, Churchill G (1996) Permutation tests for multiple loci affecting a quantitative character. Genetics 142:285–294. https://doi.org/10.1093/genetics/142.1.285
39. Chen G, Geng J, Rahman M, Liu X, Tu J, Fu T et al (2010) Identification of QTL for oil

content, seed yield, and flowering time in oilseed rape (*Brassica napus*). Euphytica 175: 161–174. https://doi.org/10.1007/s10681-010-0144-9

40. Yang J, Zhu J, Williams RW (2007) Mapping the genetic architecture of complex traits in experimental populations. Bioinformatics 23: 1527–1536. https://doi.org/10.1093/bioinformatics/btm143
41. Goffinet B, Gerber S (2000) Quantitative trait loci: a meta-analysis. Genetics 155:463–473. https://doi.org/10.1093/genetics/155.1.463
42. Sun YN, Pan JB, Shi X, Du XL, Wu Q, Qi ZM et al (2012) Multi-environment mapping and meta-analysis of 100-seed weight in soybean. Mol Biol Rep 39:9435–9443. https://doi.org/10.1007/s11033-012-1808-4
43. Lu K, Wei L, Li X, Wang Y, Wu J, Liu M et al (2019) Whole-genome resequencing reveals *Brassica napus* origin and genetic loci involved in its improvement. Nat Commun 10:1154. https://doi.org/10.1038/s41467-019-09134-9
44. Lenth RV (2016) Least-squares means: the R package lsmeans. J Stat Software 69:1–33. https://doi.org/10.18637/jss.v069.i01
45. Chen Y, Chen Y, Shi C, Huang Z, Zhang Y, Li S et al (2018) SOAPnuke: a MapReduce acceleration-supported software for integrated quality control and preprocessing of high-throughput sequencing data. Gigascience 7: gix120. https://doi.org/10.1093/gigascience/gix120
46. Kim D, Paggi JM, Park C, Bennett C, Salzberg SL (2019) Graph-based genome alignment and genotyping with HISAT2 and HISAT-genotype. Nat Biotechnol 37:907–915. https://doi.org/10.1038/s41587-019-0201-4; Li H, Durbin R (2009) Fast and accurate short read alignment with Burrows–Wheeler transform. Bioinformatics 25(14): 1754–1760
47. McKenna A, Hanna M, Banks E, Sivachenko A, Cibulskis K, Kernytsky A et al (2010) The genome analysis toolkit: a mapreduce framework for analyzing next-generation DNA sequencing data. Genome Res 20:1297–1303. https://doi.org/10.1101/gr.107524.110
48. Kang HM, Sul JH, Service SK, Zaitlen NA, Kong SY, Freimer NB et al (2010) Variance component model to account for sample structure in genome-wide association studies. Nat Genet 42:348–354. https://doi.org/10.1038/ng.548
49. Cai D, Xiao Y, Yang W, Ye W, Wang B, Younas M et al (2014) Association mapping of six yield-related traits in rapeseed (*Brassica napus* L.). Theor Appl Genet 127:85–96. https://doi.org/10.1007/s00122-013-2203-9
50. Li F, Chen B, Xu K, Wu J, Song W, Bancroft I et al (2014) Genome-wide association study dissects the genetic architecture of seed weight and seed quality in rapeseed (*Brassica napus* L.). DNA Res 21:355–367. https://doi.org/10.1093/dnares/dsu002
51. Uga Y, Sugimoto K, Ogawa S, Rane J, Ishitani M, Hara N et al (2013) Control of root system architecture by DEEPER ROOTING 1 increases rice yield under drought conditions. Nat Genet 45:1097–1102. https://doi.org/10.1038/ng.2725
52. Giehl RF, Gruber BD, von Wirén N (2014) It's time to make changes: modulation of root system architecture by nutrient signals. J Exp Bot 65:769–778. https://doi.org/10.1093/jxb/ert421
53. Roychoudhry S, Kepinski S (2015) Shoot and root branch growth angle control – wonderfulness of lateralness. Curr Opin Plant Biol 23: 124–131. https://doi.org/10.1016/j.pbi.2014.12.004
54. Chen YS, Lo SF, Sun PK, Lu CA, Ho THD, Yu SM (2015) A late embryogenesis abundant protein HVA1 regulated by an inducible promoter enhances root growth and abiotic stress tolerance in rice without yield penalty. Plant Biotechnol J 13:105–116. https://doi.org/10.1111/pbi.12241
55. Arifuzzaman M, Oladzadabbasabadi A, McClean P, Rahman M (2019) Shovelomics for phenotyping root architectural traits of rapeseed/canola (*Brassica napus* L.) and genome-wide association mapping. Mol Genet Genomics 294:985–1000. https://doi.org/10.1007/s00438-019-01563-x
56. Arifuzzaman M, Rahman M (2020) Genome wide association mapping and candidate gene mining for root architectural traits in rapeseed/canola (*Brassica napus* L.) at late growth stage. Euphytica 216:164. https://doi.org/10.1007/s10681-020-02700-z
57. Kiran A, Wakeel A, Snowdon R, Friedt W (2019) Genetic dissection of root architectural traits by QTL and genome-wide association mapping in rapeseed (*Brassica napus*). Plant Breed 138:184–192. https://doi.org/10.1111/pbr.12665
58. Li K, Wang J, Kuang L, Tian Z, Wang X, Dun X et al (2021) Genome-wide association study and transcriptome analysis reveal key genes affecting root growth dynamics in rapeseed.

Biotechnol Biofuels 14:178. https://doi.org/10.1186/s13068-021-02032-7

59. Sun C, Wang B, Yan L, Hu K, Liu S, Zhou Y et al (2016) Genome-wide association study provides insight into the genetic control of plant height in rapeseed (*Brassica napus* L.). Front Plant Sci 7:1102. https://doi.org/10.3389/fpls.2016.01102
60. Dong Z, Alam MK, Xie M, Yang L, Liu J, Helal M et al (2021) Mapping of a major QTL controlling plant height using a high-density genetic map and QTL-seq methods based on whole-genome resequencing in *Brassica napus*. G3 11:jkab118. https://doi.org/10.1093/g3journal/jkab118
61. Shen Y, Yang Y, Xu E, Ge X, Xiang Y, Li Z (2018) Novel and major QTL for branch angle detected by using DH population from an exotic introgression in rapeseed (*Brassica napus* L.). Theor Appl Genet 131:67–78. https://doi.org/10.1007/s00122-017-2986-1
62. Li H, Cheng X, Zhang L, Hu J, Zhang F, Chen B et al (2018) An integration of genome-wide association study and gene co-expression network analysis identifies candidate genes of stem lodging-related traits in *Brassica napus*. Front Plant Sci 9:796. https://doi.org/10.3389/fpls.2018.00796
63. Xu L, Hu K, Zhang Z, Guan C, Chen S, Hua W et al (2016) Genome-wide association study reveals the genetic architecture of flowering time in rapeseed (*Brassica napus* L.). DNA Res 23:43–52. https://doi.org/10.1093/dnares/dsv035
64. Jian H, Zhang A, Ma J, Wang T, Yang B, Shuang LS et al (2019) Joint QTL mapping and transcriptome sequencing analysis reveal candidate flowering time genes in *Brassica napus* L. BMC Genomics 20:21. https://doi.org/10.1186/s12864-018-5356-8
65. Helal M, Gill RA, Tang M, Yang L, Hu M, Yang L et al (2021) SNP- and haplotype-based GWAS of flowering-related traits in *Brassica napus*. Plan Theory 10:2475. https://doi.org/10.3390/plants10112475
66. Yang S, Chen S, Zhang K, Li L, Yin Y, Gill RA et al (2018) A high-density genetic map of an allohexaploid *Brassica* doubled haploid population reveals quantitative trait loci for pollen viability and fertility. Front Plant Sci 9:1161. https://doi.org/10.3389/fpls.2018.01161
67. Shi J, Zhan J, Yang Y, Ye J, Huang S, Li R et al (2015) Linkage and regional association analysis reveal two new tightly-linked major-QTLs for pod number and seed number per pod in rapeseed (*Brassica napus* L.). Sci Rep 5:14481. https://doi.org/10.1038/srep14481
68. Zhao X, Yu K, Pang C, Wu X, Shi R, Sun C et al (2021) QTL analysis of five silique-related traits in *Brassica napus* L. across multiple environments. Front Plant Sci 12:766271. https://doi.org/10.3389/fpls.2021.766271
69. Wang H, Yan M, Xiong M, Wang P, Liu Y, Xin Q et al (2020) Genetic dissection of thousand-seed weight and fine mapping of cqSW. A03-2 via linkage and association analysis in rapeseed (*Brassica napus* L.). Theor Appl Genet 133:1321–1335. https://doi.org/10.1007/s00122-020-03553-9
70. Nesi N, Delourme R, Brégeon M, Falentin C, Renard M (2008) Genetic and molecular approaches to improve nutritional value of *Brassica napus* L. seed. C R Biol 331:763–771. https://doi.org/10.1016/j.crvi.2008.07.018
71. Ecke W, Uzunova M, Weissleder K (1995) Mapping the genome of rapeseed (*Brassica napus* L.). II. Localization of genes controlling erucic acid synthesis and seed oil content. Theor Appl Genet 91:972–977. https://doi.org/10.1007/BF00223908
72. Barret P, Delourme R, Renard M, Domergue F, Lessire R, Delseny M et al (1998) A rapeseed FAE1 gene is linked to the E1 locus associated with variation in the content of erucic acid. Theor Appl Genet 96:177–186. https://doi.org/10.1007/s001220050725
73. Fourmann M, Barret P, Renard M, Pelletier G, Delourme R, Brunel D (1998) The two genes homologous to *Arabidopsis* FAE1 co-segregate with the two loci governing erucic acid content in *Brassica napus*. Theor Appl Genet 96:852–858. https://doi.org/10.1007/s001220050812
74. Wu G, Wu Y, Xiao L, Li X, Lu C (2008) Zero erucic acid trait of rapeseed (*Brassica napus* L.) results from a deletion of four base pairs in the *fatty acid elongase 1* gene. Theor Appl Genet 116:491–499. https://doi.org/10.1007/s00122-007-0685-z
75. Xiao Z, Zhang C, Tang F, Yang B, Zhang L, Liu J et al (2019) Identification of candidate genes controlling oil content by combination of genome-wide association and transcriptome analysis in the oilseed crop *Brassica napus*. Biotechnol Biofuels 12:216. https://doi.org/10.1186/s13068-019-1557-x
76. Mithen R (1992) Leaf glucosinolate profiles and their relationship to pest and disease resistance in oilseed rape. Euphytica 63:71–83. https://doi.org/10.1007/BF00023913
77. Li Y, Kiddle G, Bennett R, Doughty K, Wallsgrove R (1999) Variation in the glucosinolate content of vegetative tissues of Chinese lines of

Brassica napus L. Ann Appl Biol 134:131–136. https://doi.org/10.1111/j.1744-7348.1999.tb05245.x

78. Harper AL, Trick M, Higgins J, Fraser F, Clissold L, Wells R et al (2012) Associative transcriptomics of traits in the polyploid crop species *Brassica napus*. Nat Biotechnol 30: 798–802. https://doi.org/10.1038/nbt.2302

79. Wang X, Long Y, Wang N, Zou J, Ding G, Broadley MR et al (2017) Breeding histories and selection criteria for oilseed rape in Europe and China identified by genome wide pedigree dissection. Sci Rep 7:1916. https://doi.org/10.1038/s41598-017-02188-z

80. Hussain Q, Zhan J, Liang H, Wang X, Liu G, Shi J et al (2022) Key genes and mechanisms underlying natural variation of silique length in oilseed rape (*Brassica napus* L.) germplasm. Crop J 10:617–626. https://doi.org/10.1016/j.cj.2021.08.010

81. Tang M, Zhang Y, Liu Y, Tong C, Cheng X, Zhu W et al (2019) Mapping loci controlling fatty acid profiles, oil and protein content by genome-wide association study in *Brassica napus*. Crop J 7:217–226. https://doi.org/10.1016/j.cj.2018.10.007

82. Qu C, Jia L, Fu F, Zhao H, Lu K, Wei L et al (2017) Genome-wide association mapping and identification of candidate genes for fatty acid composition in *Brassica napus* L. using SNP markers. BMC Genomics 18:232. https://doi.org/10.1186/s12864-017-3607-8

83. Panchy N, Lehti-Shiu M, Shiu SH (2016) Evolution of gene duplication in plants. Plant Physiol 171:2294–2316. https://doi.org/10.1104/pp.16.00523

Chapter 7

Polyploid SNP Genotyping Using the MassARRAY System

Aline da Costa Lima Moraes, Danilo Augusto Sforça, Melina Cristina Mancini, Bianca Baccili Zanotto Vigna, and Anete Pereira de Souza

Abstract

Molecular marker discovery and genotyping are major challenges in polyploid breeding programs incorporating molecular biology tools. In this context, this work describes a method for single nucleotide polymorphism (SNP) genotyping in polyploid crops using matrix-assisted laser desorption ionization (MALDI) time-of-flight (TOF) mass spectrometry, the MassARRAY System.

Key words MassARRAY, Single nucleotide polymorphism (SNP), Genotyping, Polyploid, Allele dosage

1 Introduction

Polyploids are organisms with more than one pair of homologous chromosomes. They are established through evolutionary processes such as whole-genome duplication of the same genome (autopolyploid) or hybridization between two different genomes (allopolyploid) [1] and are quite common in many high-economic-value crops [2]. Despite the importance of polyploids, improving their phenotypes is challenging when using traditional breeding methods.

In these circumstances, marker-assisted selection in the genetic breeding context can be a promising approach. However, because of the complex and highly heterozygous genomes of polyploids, the generation of accurate genotypic data for them is still a challenge [3]. The vast majority of molecular and statistical methods are available for diploid or allopolyploid species, limiting the full benefit of using molecular markers in autopolyploids or even in segmental allopolyploid species, in which both auto- and allopolyploidy appear to be present across the polyploid genome [3, 4].

Yuri Shavrukov (ed.), *Plant Genotyping: Methods and Protocols*, Methods in Molecular Biology, vol. 2638,
https://doi.org/10.1007/978-1-0716-3024-2_7,

Alternatively, studies in autopolyploids or segmental allopolyploids (hereafter called polyploids) have been performed over the years using single-dose markers (SDMs) [5]. However, the limitations of this method hinder a complete understanding of genetic bases in these organisms, as SDMs represent only part of the loci available in the genome [6, 7]. Recent studies on allele dosage estimation revealed the presence of a significant quantity of multiple-dose markers, corroborating the assumptions of loss of information when using only SDMs [6–9]. Recently, following "the genomic revolution," some methods were developed to estimate allelic doses, allowing the use of multiple-dose markers in polyploids [10–15].

Almost all these currently available genotyping technologies aim to identify single nucleotide polymorphisms (SNPs) [16]. A SNP marker is a one-base variation that occurs at a specific position in the genome, and it is ideal for genetic analysis due to its high abundance throughout the genome and its high-throughput capacity [17]. Additionally, SNP markers reveal polymorphism at a given locus and the abundance of allelic variations in different genotypes and within the same genotype [3]. Thus, genotyping SNPs involves the identification of polymorphisms and the quantification of allele copies (allele dosage information), which makes this marker a powerful tool for genetic studies in polyploids [15].

Dosage is estimated when performing SNP genotyping quantitatively, with technologies such as genotyping by sequencing (GBS) [18, 19] or Sequenom iPLEX MassARRAY [20, 21], each of which can provide the relative abundance of alleles and has been largely used to genotype important crops [13, 22–25]. The MassARRAY platform (Sequenom Inc.), based on matrix-assisted laser desorption/ionization time-of-flight (MALDI-TOF) spectrometry, is widely recognized for its ability to perform accurate SNP genotyping analyses [26]. It generates spectra with two signals for each heterozygous locus, and the ratio between these signals can be used to provide the relative abundance of alleles.

The assay consists of an initial PCR that amplifies the DNA near the polymorphic site, followed by treatment to eliminate the remaining free dNTPs. Then, another PCR follows, known as the iPLEX reaction, which aims to perform extension of just the target SNP site. The product of this reaction is subjected to MALDI-TOF spectrometry, in which different amplicons with different masses will present different TOFs, and alleles can be easily identified based on their molecular weight difference [26]. In this way, since the primers are designed for bi-allelic SNPs, the equipment will generate two signals corresponding to the intensities recorded for each of the possible alleles. Therefore, the expected intensity of each signal is proportional to the dosage of the allele [21].

MassARRAY has been widely used for SNP genotyping in plants [27–30]. The assay's design relies on prior knowledge of the sequences where the SNPs of interest are located and the ability to find an adequate set of primers in this region. Because of its high throughput, sensitivity and specificity, MassARRAY is a very flexible and rapid tool for analyzing genetic variation in many plant breeding contexts. However, the "SNP genotype call" function (TYPER software), distributed in conjunction with the Sequenom iPLEX MassARRAY platform, cannot be applied to polyploids, excluding economically important crops.

Fortunately, this restriction was circumvented by applying Bayesian graphical models to signal intensity data. Serang et al. [10] converted these complex models into an online platform called SuperMASSA, which can be freely used to infer the genotype at each SNP locus in each individual from an F_1 population (among other possibilities), making it possible to estimate ploidy level and the allele dosage of SNPs [6, 22, 24, 31–33]. Later, this tool was adapted to GBS data by Pereira et al. [12] providing even more possibilities for studying complex autopolyploid organisms [7–9, 25, 34–36].

The following chapter introduces the practical aspects of the MassARRAY System for SNP genotyping in polyploids. First, an optimization of the original protocol was illustrated. Then, the obtained results were shown using the optimized protocol to genotype a bi-parental F_1 population of the hexaploid *Urochloa humidicola*, a forage grass commonly used in pastures of tropical regions [37–39]. It is expected that this information will be helpful and applicable for anyone interested in genotyping autotetraploids or even higher-ploidy species.

2 Materials

2.1 Plasticware and Consumables

1. 384-well PCR plate (*see* **Note 1**).
2. Standard microtubes (to prepare PCR cocktail).
3. Plate seals.
4. Filter pipette tips.
5. Tube racks.
6. Sealing roller tool and paddle.

2.2 Kits and Reagents

1. Nuclease-free water.
2. Genomic DNA at 5 ng/μL (*see* **Notes 2–4**).
3. 100% and 50% Ethanol
4. iPLEX Gold Reagent and Chip & Resin Kit (*see* **Note 5**).
5. PCR oligos and extension oligos.
6. Complete genotyping set – buffer, dNTPs, $MgCl_2$, and polymerase (*see* **Note 6**).

2.3 Equipment and Software

1. Micropipettes and multichannel pipettes.
2. Plate centrifuge.
3. Vortex.
4. Thermal cycler.
5. MassARRAY® Nanodispenser instrument and software.
6. MassARRAY® mass spectrometer instrument.
7. MassARRAY® TYPER software.
8. SuperMASSA software.

3 Methods

3.1 Workflow Overview

Every work must be in accordance with "Good Laboratory Practices" (*see* **Note** 7). The process for polyploid SNP genotyping using the MassARRAY System includes the following steps:

1. Design multiplex genotyping assays, in which we recommended a maximum plex of for polyploid (*see* **Notes 8** and **9**).
2. Amplify many specific fragments of genomic DNA for each individual (multiplex PCR or capture reaction).
3. Dephosphorylate unincorporated dNTPs using shrimp alkaline phosphatase (SAP reaction), making them unavailable for future reactions.
4. Create the iPLEX Gold Reaction (extension reaction), which involves the enzymatic addition of one terminator nucleotide (A, T, C, or G) to the SNP locus, producing allele-specific extension products of different masses.
5. Desalt the iPLEX reaction products using an appropriate resin.
6. Transfer the cleaned iPLEX reaction products onto a SpectroCHIP Array.
7. Analyze the SpectroCHIP Array in the MassARRAY Analyzer.
8. Perform data analysis in SuperMASSA software, which can be freely used to infer the genotype at each SNP locus, as well as estimate the ploidy level and the allele dosage of each SNP.

All these steps are illustrated in Figs. 1 and 2 and detailed below.

3.2 Designing Genotyping Assays

The MassARRAY System is a multiplex genotyping platform based on the difference in molecular weights between alternative products generated for each possible locus variant. In this context, the design of capture and extension primers is a critical step for a successful experiment (*see* Fig. 1). It is crucial to prevent primer dimer formation, which often occurs in multiplexed reactions. Additionally, to avoid confusion in the mass spectrum and improve

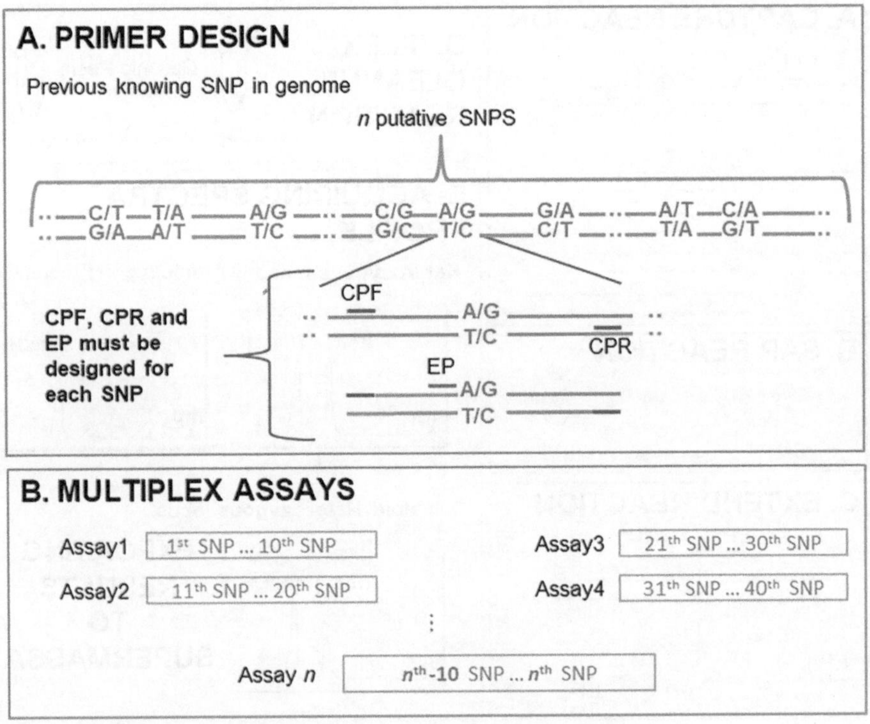

Fig. 1 Primer design and multiplex. (**a**) From a previous knowing sequence, "n" putative SNPs can be identified. The Capture Primer Forward (CPF), Capture Primer Reverse (CPR) and the Extension Primer (EP) must be designed for each SNP that will be genotyped. (**b**) The multiplex assays must be performed grouping a maximum of ten SNPs, which will be genotyped simultaneously

PCR performance, make sure that the mass of the capture primer is different from that of the extension primer and its extension products. The extension products should be 4500 and 9000 Da, differing by at least 30 Da for different SNP assays and by 5 Da within each assay (depending on the resolution of the equipment). All these parameters are controlled and adjusted by SEQUENOM Assay Design software as follows.

1. The input file for assay design must be a tab-delimited text file (.txt) containing two columns: "SNPs_ID," with the SNP names, and "Sequences," including the target sequences. The SNPs should be in the format "[C/G]," and for a one-base insertion/deletion, in the form "[C/−]." All known SNPs in the surrounding region should be replaced with "N." Sequences should be approximately 400 bp in length and contain no spaces or characters other than A, G, C, T, and N.
2. Launch Assay Design Suite (ADS) software (*see* **Note 8**). In the Assay Design window, after selecting "New Genotyping Design" and uploading the input file containing all the sequences of interest, confirm the following available options: "Preset, Low Multiplexing iPLEX Preset"; "Organism,

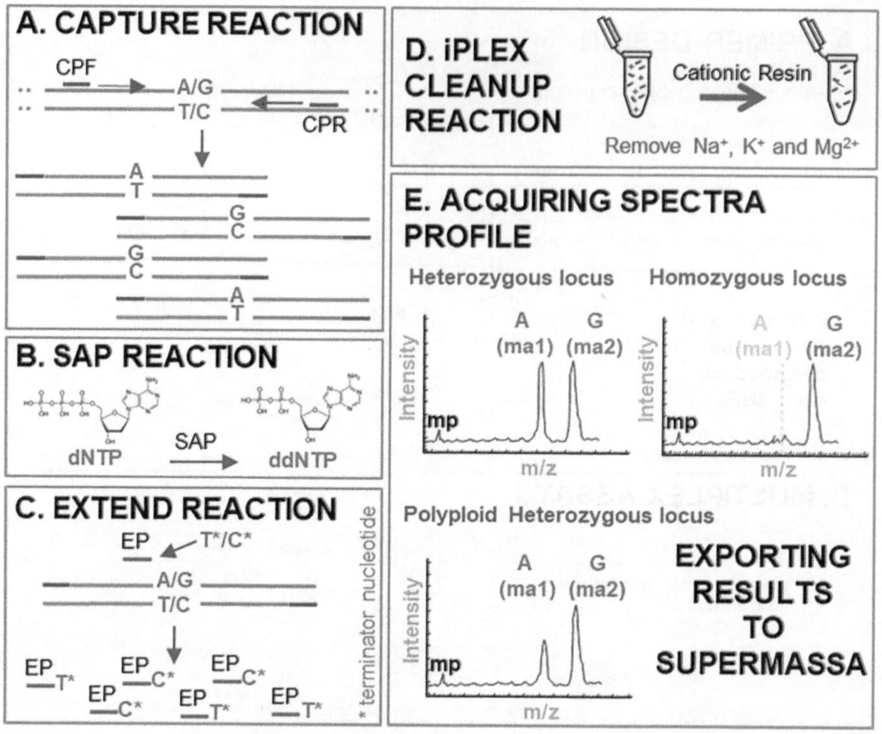

Fig. 2 SNP genotyping workflow. (**a**) The Capture Reaction delimits each SNP location in the genome and uses the CPF and CPR to perform the amplification. (**b**) The SAP Reaction inactivates all free dNTP, converting them into ddNTP (the molecule's reaction exemplified is a dATP being converted in a ddATP). (**c**) The Extend Reaction adds to EP only the SNP nucleotide. (**d**) The iPLEX Cleanup removes cations from the reaction. (**e**) The spectra profile generated by Typer Analyzer Software reveals the diploid genotype pattern, composed by the primer mass (mp), allele 1 mass (*ma1*) and allele 2 mass (*ma2*) when heterozygous locus, or mp and ma when homozygous locus. For a polyploid locus, data must be exported to be analyzed by SuperMASSA software

Other"; "Chemistry, iPLEX"; and "Multiplex Level, 10" (*see* **Notes 9** and **10**).

3. Click "Begin Run." After all the steps are complete, click "Export All" to generate the assay design file containing all the essential information for genotyping analysis and a spreadsheet of primer order (*see* **Note 11**).

3.3 Amplifying DNA for Genotyping in Multiplex (Capture Reaction)

This is the first bench stage, in which the capture reaction enriches the specific region of the DNA template containing the SNP to be genotyped (*see* Fig. 2). As already observed, we recommended a maximum of ten different regions to be simultaneously amplified in polyploids (multiplex level), as follows.

1. Normalize the genomic DNA to 5 ng/μL and add 2 μL to each well of the 384-well PCR plate, corresponding to 10 ng of DNA per reaction (*see* **Note 12**).
2. Dilute the capture primers to a stock concentration of 100 μM.

Table 1
Capture reaction primer mix preparation

Reagent	Stock concentration	Working concentration	Volume/primer	× 10 primers
Each forward primer	100 μM	1 μM	4.608 μL	46.08 μL
Each reverse primer	100 μM	1 μM	4.608 μL	46.08 μL
Nuclease-free water	–	–	–	368.64 μL
Total				**460.8 μL**

An example for a plex with ten primer pairs for 384 samples. Considering pipetting loss, the final volume should be created for 460.80 samples

Table 2
Capture reaction mix preparation

Reagent	Concentration in 5 μL	Volume	× 460.8 reactions
Nuclease-free water	NA	0.85 μL	391.68 μL
Buffer (10×)	1×	0.63 μL	290.30 μL
$MgCl_2$ (25 mM)	1.65 mM	0.33 μL	152.06 μL
dNTP (25 mM)	0.5 mM	0.1 μL	46.08 μL
Capture primer mix (1 μM each)	200 nM	1.0 μL	460.8 μL
Hot Start Taq (5 U/μL)	0.1 U/μL	0.1 μL	46.08 μL
Total volume		**3.0 μL**	**1.387 μL**

An example for a plex with ten primer pairs for 384 samples. Considering pipetting loss, the final volume should be created for 460.80 samples

3. To prepare the capture primer multiplex, mix and dilute each forward and reverse primer to a working concentration of 1.0 μM. Volumes should include 20% excess (based on the sample numbers) to account for pipetting loss. For example, considering an experiment for 384 samples, the total should be sufficient for 460.8 samples to account for pipetting loss (*see* **Note 13**).
4. Table 1 shows an example with ten assays.
5. In a microtube, prepare a PCR cocktail as described in Table 2.
6. Lightly vortex the PCR cocktail microtube and spin it down. Dispense 3 μL of the PCR cocktail into each well of the microplate with the genomic DNA.
7. Seal the microplate, gently mix or vortex it, centrifuge it at 1000 RPM, and run the following PCR profile in a thermocycler: Initial denaturation at 94 °C – 15 min; following 45 cycles including 94 °C – 20 s, AT°C – 30 s, 72 °C – 1 min, and 72 °C – 3 min; with final hold at 4 °C.
8. Proceed to the SAP reaction (Subheading 3.5).

Table 3
SAP mix preparation

Reagent	Concentration in 7 μL	Volume	× 460.8 reactions
Nuclease-free water	NA	1.53 μL	705.02 μL
SAP buffer (10×)	0.24×	0.17 μL	78.34 μL
SAP enzyme (1.7 U/μL)	0.073 U/μL	0.30 μL	138.24 μL
Total volume		**2.00 μL**	**921.6 μL**

An example for 384 samples. Considering pipetting loss, the final volume should be created for 460.80 samples

3.4 Neutralizing Unincorporated dNTPs (SAP Reaction)

In this step, SAP enzyme is used to neutralize unincorporated dNTPs from PCR amplification products by cleavage of the 5′ phosphate group (*see* Fig. 2).

1. In a microtube, prepare the SAP mix solution according to Table 3.
2. Lightly vortex the microtube containing the SAP mix solution and spin it down. Add 2 μL of SAP solution to each well of the plate.
3. Seal the microplate, spin it down, and run the following PCR profile in a thermocycler: 37 °C – 40 min; 85 °C – 5 min; and 4 °C – hold.
4. Proceed to the extension reaction (Subheading 3.6).

3.5 Creating the iPLEX Gold Reaction (Extension Reaction)

This step involves the enzymatic addition of a terminator nucleotide (ddNTPs) to the SNP locus to be genotyped (*see* Fig. 2). During the iPLEX Gold reaction, the primer is extended by one of the four terminator nucleotides, A, T, C, or G, which terminates the extension of the primer. Using DNA polymerase, the iPLEX Gold reaction produces allele-specific extension products of different masses depending on the locus variant.

1. Dilute the capture primers to a stock concentration of 200 μM.
2. To prepare the iPLEX primer multiplex, each forward and reverse primer must be mixed and diluted to a specific concentration based on its mass. When genotyping in low plex, i.e., between 2 and 18 different loci, the *2-tier* concentration adjustment method is recommended. This method divides the primers into two groups, one with low-mass and one with high-mass, and all primers in the high-mass group are doubled in concentration compared with those in the low-mass group (*see* **Notes 14** and **15** for more information).
3. Table 4 shows an example with ten assays (*see* **Notes 14** and **15** for more information). An excess of 38% (529.92 reactions out of 384 reactions) will be used.

Table 4
Extension primer mix preparation

	Group I (low mass)	Group II (high mass)	Total
Number of primers	5	5	10
Stock concentration	200 μM	200 μM	–
Volume/primer 1 reaction	0.035 μL	0.07 μL	–
Primer mix concentration[a]	7 μM	14 μM	–
iPLEX reaction concentration[b]	0.78 μM	1.56 μM	
Primer volume/plate 529.92 reactions	18.55 μL	37.09 μL	–
Group volume/plate	18.55 × 5 = 92.74 μL	37.09 × 5 = 185.45 μL	278.19 μL
H_2O	–	–	251.73 μL
Total volume			**529.92 μL**

An example for a plex with ten primer pairs for 384 samples. Considering pipetting loss, the final volume should be created for 529.92 samples
[a]Primer mix volume of 1 μL
[b]iPLEX reaction volume of 9 μL

Table 5
iPLEX mix preparation

Reagent	Concentration in 9 μL	Volume/1 reaction	× 529.92 reactions
Nuclease-free water	NA	0.56 μL	296.76
Buffer (10×)	0.222×	0.2 μL	105.98
iPLEX termination mix (10×)	0.222×	0.2 μL	105.98
Primer mix (7 μM:14 μM	0.7778 μM:1.5556 μM	1.0 μL	529.92
iPLEX enzyme (32 U/μL)	0.142 U/μL	0.04 μL	21.20
Total volume		**2.0 μL**	**1059.84 μL**

An example for 384 samples. Considering pipetting loss, the final volume should be created for 529.92 samples

4. In a microtube, prepare PCR cocktail as described in Table 5.
5. Add 2 μL of the primer extension mix to each well of the 384-well plate.
6. Seal the microplate, gently mix or vortex it, spin down and run the following PCR profile in a thermocycler: Initial denaturation, 94 °C – 30 s; with following 5 cycles including AT°C – 5 s, 80 °C – 5 s; with new 40 cycles including 94 °C – 5 s, AT°C – 5 s, and 80 °C – 5 s; and in the end final extension at 72 °C – 3 min and hold at 4 °C.
7. Proceed to iPLEX reaction cleanup (Subheading 3.7).

3.6 iPLEX Reaction Cleanup with Resin

Before the mass spectrometry analysis, iPLEX reaction products must be treated to remove salts ions such as Na^+, K^+, and Mg^{2+} from previous reactions, which may increase background noise in the mass spectra (*see* Fig. 2). Therefore, a cationic resin (clean resin) pretreated with acid reagents is used, as follows.

1. Transfer an excess amount of resin to the dimple plate. Use a plastic scraper to spread it out and remove the excess. Ensure that there is enough resin in each well of the dimple plate; precisely 6 mg of resin is necessary to clean up the iPLEX reaction.
2. Let the resin stand in the dimple plate for 10 min. Then, carefully place the iPLEX plate upside down on the dimple plate such that the wells of the iPLEX plate line up with the wells of the dimple plate.
3. Invert the plates so that the dimple plate is on top of the iPLEX plate. Gently tap the dimple plate to drop the resin into the wells with the iPLEX reaction products (*see* **Note 16**).
4. Add 16 μL of ultrapure water to each well of the iPLEX plate. Rotate the plate for 15 min on a plate rotator.
5. Centrifuge the 384-well plate for 5 min at 3220 × *g*.

3.7 Dispensing onto SpectroCHIP Arrays

After the iPLEX reaction is complete, the extension products must be transferred to the SpectroCHIP. The transfer is performed using the MassARRAY Nanodispenser, and a small volume (10–25 nL) of the reaction is dispensed onto the matrix spots on the SpectroChip as follows.

1. Before proceeding to the dispensing step, check the supply and waste tank status and confirm that the daily maintenance procedures have been completed.
2. Place the 384-well plate and a test chip (which can be an old chip) on the deck of the MassARRAY® Nanodispenser. Run the volume check to adjust the dispensing speed and change the speed to that with the best results in the volume check (15–20 μL).
3. Add 70 μL of calibrant to reservoirs on the MassARRAY® Nanodispenser.
4. Replace the test chip with a new one. Select "Transfer tab, Methods and Load the 384 MTP to 384 SpectroCHIP Array method" on the Main Menu. Dismiss the Volume Check option. On the "Aspirate/Dispense" tab, select the option "Analyte & Calibrant."
5. Run the application.

3.8 Designing the Plate Input File and Acquiring the Spectrum Profile

After transfer to the chip, product masses resulting from previous iPLEX reactions can be analyzed using the MassARRAY Mass Spectrometer. In this step, the chip matrix absorbs energy from the incident laser, leading to desorption/ionization of the DNA, and the ionized molecules are accelerated through a flight tube. The flight time is proportional to the mass-to-charge ratio (m/z). Consequently, lower-mass molecules have faster acceleration and will reach the detector at the end of the tube sooner [40]. The entire process from laser firing to signal detection takes less than 50 min to analyze 384 samples. All these steps are controlled by Sequenom-provided software, as follows:

1. Ensure that the FLEXControl, ServerControl, and MassARRAY CALLER programs are running.
2. Open the MassARRAY Typer Home window, choose "Assay Editor," and create a new project with the assay design file (generated in Subheading 3.1, **step 3**). This file needs to be imported only once, prior to running the assay for the first time.
3. Go back to the MassARRAY Typer Home window and choose "Plate" in "Plate Editor." Create a new project and add a new plate to the project.
4. Move to "Sample" and in "sample group," choose "new sample group." In the dialog box, assign the name as the group ID, press the open folder button, and select your sample list file.
5. Choose the created plate and select all the wells in the plate editor. Choose the sample group that you have created for that plate and select "apply samples."
6. Apply your assays by selecting the wells for each assay as used in the experimental procedure.
7. Check that all the samples and assays are correctly assigned to the plate by moving the cursor over the plate wells.
8. Open ChipLinker, select the plate from the navigation tree, and designate iPLEX for the Terminator Chemistry option. Select "Genotype + Area" and type "Experiment Name and Chip Barcode" for the Process Method. Add and create the application.
9. Introduce the target into the MassARRAY READER with the In/Out function on the Autoflex instrument.
10. In SpectroAQUIRE, in the Auto Run Set Up tab, enable high voltage. Click the Auto Run tab, and under Run, select Start Auto Run. The results will be automatically loaded into the database and may be accessed using the Typer Analyzer when the run is finished.

11. The SNP genotyping results for each locus are presented by Typer Analyzer as a three-peak spectrum for a heterozygous locus (*see* Fig. 2). The first peak (mp) corresponds to an unextended primer and is expected to be short in an optimized reaction where the primer is almost fully consumed. Since SNPs are mostly bi-allelic, the second and third signals correspond to the intensity recorded for one of the two possible alleles: the low-mass allele (*ma1*) and the high-mass allele (*ma2*). Alternatively, the locus may be homozygous, in which case only one allele peak will be observed (*see* Fig. 2).

3.9 Exporting Results on the MassARRAY® TYPER 4.0

MassARRAY® TYPER 4.0 – Typer Analyzer is appropriate software for diploid species, as it allows the user to analyze data generated with the MassARRAY System containing up to two signals for each SNP locus, which have the same intensity for heterozygous loci. However, for polyploid species, the expected signal intensities for a heterozygous locus are proportional to the corresponding allele dosage. Therefore, in this case, the user must export the results to be analyzed by other software as follows (see Subheading 3.10).

1. Open the MassARRAY® TYPER 4.0 program and navigate to the Typer Analyzer section. Load data into the Typer Analyzer and select the genotyping experiment for analysis from the Project Explorer menu.
2. To export the information, choose "Save all wells to file," which will create an .xml file. Each given multiplex assay will be an .xml file with the information for all 384 samples.

3.10 Dosage Estimation Using SuperMASSA Software

As a quantitative genotyping technology, the MassARRAY System generates spectra with two signals for each SNP locus, corresponding to the intensities recorded for the two possible alleles (see Subheading 3.8, **step 11**). In polyploids, the intensity of each signal is proportional to the corresponding allele dosage [10]. Therefore, the observed proportions between peaks in the spectra can be used to infer genotypes in all possible genotypic classes for a given ploidy level. For example, in a spectrum for a hexaploid locus where the signal intensity of the high-mass allele (*A*) is twice that of the low-mass allele (*B*), it is possible to infer twice as many copies of the *A* allele than of the *B* allele, suggesting the *AAAABB* configuration. On the other hand, if the two alleles have the same signal intensity, the inferred genotype will be *AAABBB* (with the two alleles presenting the same number of copies). From the observation of other proportions, one can identify all other possible configurations in the hexaploid population, *AAAAAA*, *AAAAAB*, *AABBBB*, *ABBBBB* and *BBBBBB*, totaling seven genotypic classes for a hexaploid locus. However, the "SNP genotype call" function of TYPER software, distributed together with the Sequenom iPLEX MassARRAY platform, cannot apply

Table 6
Example of a tab-delimited text file (.txt) with the individual genotype information for a hypothetical marker

Ind1 (3,1)
Ind2 (2,2)
Ind3 (0,4)
Ind4 (1,3)

this information to estimate allele dosage in polyploids. Fortunately, this restriction can be circumvented by applying Bayesian graphical models to the data. Serang et al. [10] converted these complex models into an online platform called SuperMASSA, which can be freely used to infer the genotype at all SNP loci in parents and each individual in an F_1 population, in a population in Hardy-Weinberg equilibrium, or even for population with no level information. Thus, the model makes it possible to estimate the ploidy level and the allelic dosage of SNPs (*see* **Note 17**). Analyses are performed on one marker at a time, as follows.

1. Convert the .xml file to a .txt file with information on *individual names, x-intensity, and y-intensity*, as described in Table 6.
2. Open SuperMASSA software, available at https://statgen.esalq.usp.br/SuperMASSA/.
3. Select the model, i.e., F_1 for a controlled cross (biparental population), *Hardy-Weinberg* for a diversity panel, or *Do not use population-level information.*
4. Choose the option *Draw genotype scatterplot* to generate an annotated scatterplot of allele intensities.
5. Choose the *naive posterior reporting threshold*, which is the posterior probability for the classification of each individual and may be used to verify the accuracy of genotyping calling. For example, for a given ploidy level, if the posterior threshold is set to 0.0 (default), SuperMASSA will classify all individuals, regardless of their posterior probability for the given configuration. However, with an increase in the posterior threshold level, individuals that do not reach that level are not assigned to any genotypic classes and are discarded from the analysis. Ideally, for better genotyping, as many individuals as possible must be classified, with the posterior threshold close to 1.00. A minimum value of 0.8 is recommended to retain only samples with an associated probability of 0.80 or higher in polyploids, but tests must be performed in each specific experiment.
6. Choose the *ploidy range*. A value can be specified if the ploidy is known, or an interval can be used in cases of unknown ploidy,

and SuperMASSA will use population models (e.g., F_1 or Hardy-Weinberg models) to test all even ploidies in that interval. Markers that do not fit the expected ploidy may be discarded.

7. Load the *Data file* in .txt format with genotype information (generated in Subheading 3.9, **step 1**).
8. For populations with parental information (usually bi-parental populations), load the *F_1 parent data* (optional; only for the F_1 model).
9. Run the application, and look at the results on the loaded page. The posterior probability parameter (Pr(G = g*, parameter | D)) indicates the reliability of the results, with a value close to 1 being expected. The predicted ploidy for the locus is presented by the "ploidy" parameter, and under "parameter," one will find the marker dosage for the parents (F_1 populations). Next is the list of marker dosages for each individual.

4 A Practical Example: Ploidy and Dosage Estimation in the Hexaploid *Urochloa humidicola*

The previously presented methodology was used to genotype an F_1 bi-parental population of *U. humidicola* (hexaploid, $2n = 6x = 36$), a forage grass commonly used in pastures of tropical regions [7]. The *U. humidicola* progeny consisted of 279 hybrids obtained from a cross between the sexual accession H031 (CIAT 26146) and the apomictic cultivar *U. humidicola* cv. BRS Tupi, as described by Vigna et al. [37]. SNPs were discovered in a transcriptome data set of H31 and Tupi parents, as described by Vigna et al. [38], and only SNPs shared by the two genotypes were used.

For the progeny and parents, DNA was extracted and genotyped by MassARRAY using 17 different multiplex assays, with a plex-level of 10. Primer adjustment was performed using the two-tier method. Then, SuperMASSA was used to estimate ploidy and dosage, with the following parameters: model F_1, naive posterior reporting threshold 0.7, and ploidy range 2–6. Other parameters were set to the default.

Figure 3 shows, as an example, the scatterplot output of a locus named Bhum_205, classified by SuperMASSA as hexaploid with a high posterior probability ($\approx$1.00). In Fig. 3a, SuperMASSA inferred the parental doses (both with the configuration *AAABBB*). For a cross of these two parents, there are seven possible genotypic classes for the offspring: *AAAAAA, AAAAAB, AAAABB, AAABBB, AABBBB, ABBBBB* and *BBBBBB*. Six of these classes were identified, corresponding to alleles with 0, 1, 2, 3, 4, and 5 copies (*see* Fig. 3b).

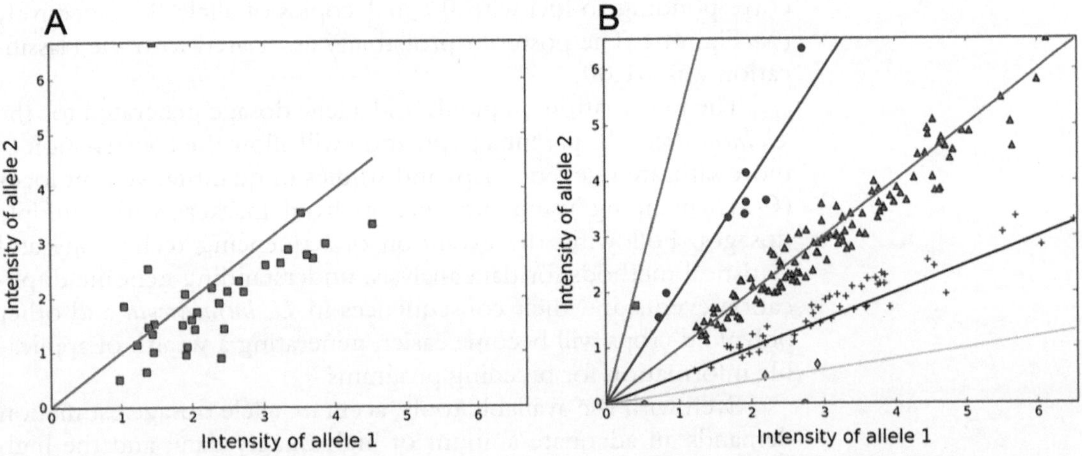

Fig. 3 Annotated scatterplot of allele intensities for the SNP Marker Bhum_205 of *Urochloa humidicola*. (**a**) Red squares represent both parents (16 replicates for each parent) with the inferred genotype *AAABBB*. (**b**) Red squares represent offspring with one copy of allele *B*, i.e., *AAAAAB*. Blue circles represent offspring with *AAAABB*. Green triangles represent offspring with *AAABBB*. Black stars represent offspring with *AABBBB*. Pink diamonds represent offspring with *ABBBBB*. Orange hexagons represent offspring with *BBBBBB*

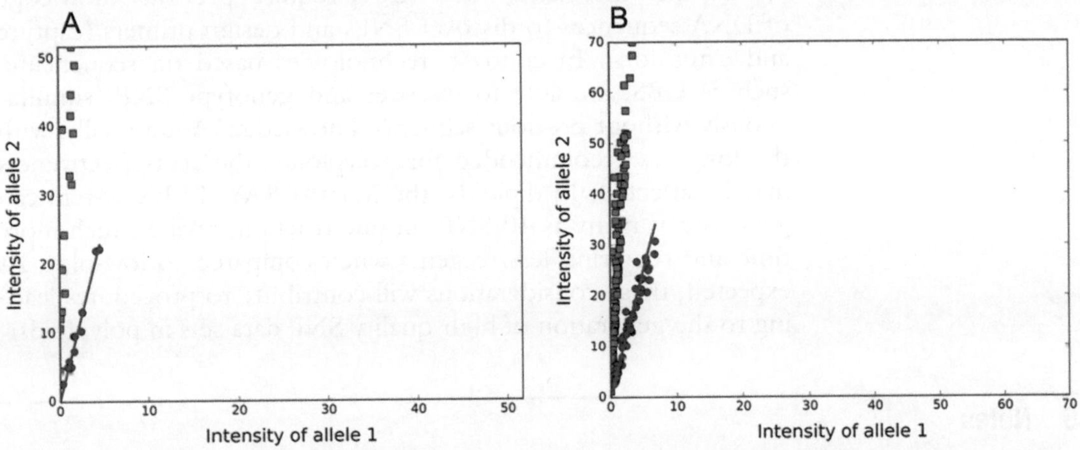

Fig. 4 Annotated scatterplot of allele intensities for the SNP Marker Bhum_063 of *Urochloa humidicola*. (**a**) Red squares represent the parent Tupi with the inferred genotype *AAAAAA*, and blue circles represent the parent H31 with the inferred genotype *AAAAAB*. For both parental genotypes, 16 replicates were used. (**b**) Red squares represent the homozygous offspring, i.e., those with *AAAAAA*. Blue circles represent offspring with *AAAAAB*

Figure 4 shows another example for *U. humidicola*, the locus Bhum_063. SuperMASSA inferred of ploidy of 6 and parental doses with the configuration *AAAAAA* for sexual accession H31 (red squares) and *AAAAAB* for the apomictic cv. Tupi (blue circles) (*see* Fig. 4b). From the combination of these parents, there are two possible genotypic classes for the offspring: *AAAAAA* and *AAAAAB*. Both were identified in the mapping population, corresponding to loci with 0 and 1 copies of allele B, respectively (*see* Fig. 4b). The posterior probability associated with the classification was ≈1.00.

The information on ploidy and allelic dosage generated for the *U. humidicola* bi-parental population will allow the construction of more saturated genetic maps and studies of quantitative trait locus (QTL) mapping using information from markers with multiple dosages. Following the evolution of sequencing technology and statistical methods for data analysis, understanding genome duplication events and their consequences in *U. humidicola* and other polyploid crops will become easier, generating a wealth of applicable information for breeding programs.

Even with the available tools, accurate allele dosage estimation demands an adequate amount of high-quality data, and the high throughput, sensitivity and specificity of the Sequenom MassARRAY platform make it a robust technology for generating this kind of data in polyploids. In addition, MassARRAY offers a targeted genotyping approach, generating a smaller, more manageable and less computationally demanding data set and allowing researchers to focus on specific regions of interest.

On the other hand, MassARRAY requires previous knowledge of DNA sequences to discover SNPs and design primers (capture and extension). In contrast, technologies based on sequencing, such as GBS, are able to discover and genotype SNPs simultaneously without previous sequence knowledge. Additionally, with the low-plex recommended for polyploids, the cost-effectiveness may be affected. In diploids, the MassARRAY iPLEX system can genotype as many as 40 SNPs in one reaction, saving much more time and requiring less reagents when compared to low-plex. As expected, these considerations will contribute to procedures leading to the generation of high-quality SNP data sets in polyploids.

5 Notes

1. Depending on the equipment configuration, reactions can be performed in 384-well plates or 96-well plates. Reagent sets are available in 24-, 96-, and 384-well formats, making the system scalable for low- to high-throughput applications.
2. DNA should be prepared to a stock concentration (for example, 50 ng/μL) and diluted to 5 ng/μL immediately before genotyping studies.
3. iPLEX genotyping can be performed on genomic DNA isolated from a variety of tissue samples with any extraction method, but quality and quantity are of considerable importance. It is strongly suggested the use of an ultraviolet (UV) spectrophotometer for assessing DNA quality (A260/

A280 = 1.7–2.0) and a fluorometric method for quantification that precisely measures only double-stranded DNA (dsDNA).

4. Only genomic DNA is discussed in this section. However, in most cases, this information is also applicable to cDNA. If cDNA templates have to be used, it is assumed that the method has to be optimized. More attention must be paid for primer design, as intronic regions must be avoided.

5. Two iPLEX reagent options are commercially available: iPLEX Pro, suitable for diagnostic and clinical applications, and iPLEX Gold, which is intended for standard genotyping and is the most commonly used in agriculture assays.

6. Although PCR reagents for the capture reaction (buffer, dNTPs, $MgCl_2$, and polymerase) are supplied in Sequenom's "Complete Genotyping Set," it is possible to use other similar solutions, but tests need to be carried out.

7. Codes of "Good Laboratory Practices."
   - Wear gloves, a lab coat and safety glasses when preparing any reaction and handling all equipment.
   - Immediately store the reagents according to label instructions after use. Reagents may be aliquoted to avoid numerous freezing and thawing cycles.
   - Be careful to not contaminate the samples, mixtures and reagents. To prevent cross-contamination or "carry over" from one sample to another, make sure to use filter tips.
   - Gently vortex reagents until the liquid is well dispersed, and spin down tubes and plates with reagents before use or cycling.
   - Make sure to use a highly sticky seal with plate PCR to avoid evaporation during cycling or when not in use. Ensure proper sealing to avoid evaporation.
   - Purified DNA samples can be stored at 4 °C for up to 24 h. To store samples for longer than 24 h, storage at −20 °C. Avoid repeated freezing and thawing of DNA.
   - After thermocycling (capture reaction, SAP and iPLEX reaction), there are stopping points in which plates can be stored at 4 °C for up to 24 h or at −20 °C for up to 2 weeks.
   - The dispensing of solutions into the 384-well plate should be performed with automated workstations, such as the MassARRAY® Liquid Handler, when possible. Alternatively, multichannel pipettes can also be used. In either case, consider an appropriate overhang, which must be higher when using multichannel pipettes.
   - Carefully check the iPLEX Gold Application Guide and MassARRAY Nanodispenser Guide.

8. MassARRAY Assay Design software can be installed or used on Sequenom's Web site https://www.mysequenom.com/default.aspx by Sequenom customers.
9. One of the greatest advantages of the MassARRAY System is its multiplexing ability; that is, it allows multiple loci to be analyzed simultaneously. This reduces the total number of DNA samples required for the analysis, making this technology cost-effective and high-throughput when compared to others.
10. Although a high multiplex level (36+) is allowed, we recommend a maximum plex of 10 for polyploid species to avoid interactions between primers and an increase in the noise generated by loci with four or more alleles. Even with increases in the multiplex level, we do not recommend decreasing reagent quantities because the number of alleles that should be amplified is more significant in polyploids (multiple series of chromosomes and consequently multiple alleles).
11. Usually, HPLC purification of extension primers is advised in the manuals of the MassARRAY System. However, in our experience, primers with standard desalted purity work well.
12. It is recommended to reserve 1–2 wells as blank controls (no DNA), and when possible, a few samples should be analyzed as replicates. When working with controlled crosses (mapping populations), we recommend replicating as many of the parents as possible (we used a minimum of 16 replicates for each parent). SuperMASSA will use the information from the parents to infer the genotypes of their hybrids, so the information must be highly reliable.
13. The protocol was present under the assumption that the reactions will be performed using the same multiplex of primers on 384 different samples. As a result, all reactions are optimized for 384 + pipetting error. Furthermore, depending on the design, the experiment can be planned with the same sample and different multiplex primers by modifying the amounts of each reagent in each reaction.
14. The concentration of extension primers in iPLEX Primer Mix must be adjusted due to the inverse relationship between peak intensity and analyte mass, ensuring primer signal intensities that are as similar as possible [26]. The primers may be sorted into groups of two (two-tier method), three (three-tier method) or four (four-tier method) based on their mass or adjusted by the regression method, which involves mixing primer at specific concentrations according to their masses.
15. Ideally, it is necessary to inspect the relative peak heights of the extension primer mix before performing the iPLEX reaction and decide if any extended oligos need to be spiked in. To do so, mix 5 μL of the extension primer mix with 95 μL of water

and add an aliquot of 20 μL to a well of a 384-well PCR plate. Clean up the mix with resin as described in Subheading 3.6 and transfer the cleaned product to a SpectroCHIP Array as described in Subheading 3.7. Set up the plate in the Sequenom Typer Plate Editor database, run the chip on the mass spectrometer as performed for sample plates, and use Typer Analyzer to visualize the resulting histogram. Any oligos with much lower intensity than other peaks need to be increased in volume. For example, if a given peak shows approximately half the intensity of the other, double the volume to balance all peaks' intensities.

16. Sometimes pouring all the resin into the 384-well plate can be challenging. To simplify this step, make sure that the resin has enough time to dry in the dimple plate. If necessary, one may increase this drying time (15 min) or tap the dimple plate more vigorously.

17. Explaining the theory underlying the analysis of MassARRAY data by SuperMASSA is beyond the scope of this guide. References Serang et al. [10] and Mollinari and Serang [31] are highly recommended for more information.

References

1. Soltis DE, Visger CJ, Soltis PS (2014) The polyploidy revolution then...and now: Stebbins revisited. Am J Bot 101:1057–1078. https://doi.org/10.3732/ajb.1400178
2. Kyriakidou M, Tai HH, Anglin NL, Ellis D, Strömvik MV (2018) Current strategies of polyploid plant genome sequence assembly. Front Plant Sci 9:1660. https://doi.org/10.3389/fpls.2018.01660
3. Bourke PM, Voorrips RE, Visser R, Maliepaard C (2018) Tools for genetic studies in experimental populations of polyploids. Front Plant Sci 9:513. https://doi.org/10.3389/fpls.2018.00513
4. Mason AS, Wendel JF (2020) Homoeologous exchanges, segmental allopolyploidy, and polyploid genome evolution. Front Genet 11:1014. https://doi.org/10.3389/fgene.2020.01014
5. Wu KK, Burnquist W, Sorrells ME, Tew TL, Moore PH, Tanksley SD (1992) The detection and estimation of linkage in polyploids using single-dose restriction fragments. Theor Appl Genet 83:294–300. https://doi.org/10.1007/BF00224274
6. Garcia AA, Mollinari M, Marconi TG, Serang OR, Silva RR, Vieira ML et al (2013) SNP genotyping allows an in-depth characterisation of the genome of sugarcane and other complex autopolyploids. Sci Rep 3:3399. https://doi.org/10.1038/srep03399
7. Ferreira R, Lara L, Chiari L, Barrios S, do Valle CB, Valério JR et al (2019) Genetic mapping with allele dosage information in tetraploid *Urochloa decumbens* (Stapf) R.D. Webster reveals insights into spittlebug (*Notozulia entreriana* Berg) resistance. Front Plant Sci 10:92. https://doi.org/10.3389/fpls.2019.00092
8. de Lara LAC, Santos MF, Jank L, Chiari L, Vilela MM, Amadeu RR et al (2019) Genomic selection with allele dosage in *Panicum maximum* Jacq. G3: Genes Genomes Genet 9:2463–2475. https://doi.org/10.1534/g3.118.200986
9. Mollinari M, Olukolu BA, Pereira G, Khan A, Gemenet D, Yencho GC et al (2020) Unraveling the hexaploid sweetpotato inheritance using ultra-dense multilocus mapping. G3: Genes Genomes Genet 10:281–292. https://doi.org/10.1534/g3.119.400620
10. Serang O, Mollinari M, Garcia AA (2012) Efficient exact maximum a posteriori computation for Bayesian SNP genotyping in polyploids. PLoS One 7:e30906. https://doi.org/10.1371/journal.pone.0030906
11. Gerard D, Ferrão L, Garcia A, Stephens M (2018) Genotyping polyploids from messy

sequencing data. Genetics 210:789–807. https://doi.org/10.1534/genetics.118.301468

12. Pereira GS, Garcia A, Margarido G (2018) A fully automated pipeline for quantitative genotype calling from next generation sequencing data in autopolyploids. BMC Bioinform 19:398. https://doi.org/10.1186/s12859-018-2433-6
13. Wadl PA, Olukolu BA, Branham SE, Jarret RL, Yencho GC, Jackson DM (2018) Genetic diversity and population structure of the USDA sweetpotato (*Ipomoea batatas*) germplasm collections using GBSpoly. Front Plant Sci 9:1166. https://doi.org/10.3389/fpls.2018.01166
14. Clark LV, Lipka AE, Sacks EJ (2019) polyRAD: genotype calling with uncertainty from sequencing data in polyploids and diploids. G3: Genes Genomes Genet 9:663–673. https://doi.org/10.1534/g3.118.200913
15. Gerard D, Ferrão L (2020) Priors for genotyping polyploids. Bioinformatics 36:1795–1800. https://doi.org/10.1093/bioinformatics/btz852
16. Clevenger J, Chavarro C, Pearl SA, Ozias-Akins P, Jackson SA (2015) Single nucleotide polymorphism identification in polyploids: a review, example, and recommendations. Mol Plant 8:831–846. https://doi.org/10.1016/j.molp.2015.02.002
17. You Q, Yang X, Peng Z, Xu L, Wang J (2018) Development and applications of a high throughput genotyping tool for polyploid crops: single nucleotide polymorphism (SNP) array. Front Plant Sci 9:104. https://doi.org/10.3389/fpls.2018.00104
18. Poland JA, Brown PJ, Sorrells ME, Jannink JL (2012) Development of high-density genetic maps for barley and wheat using a novel two-enzyme genotyping-by-sequencing approach. PLoS One 7:e32253. https://doi.org/10.1371/journal.pone.0032253
19. Elshire RJ, Glaubitz JC, Sun Q, Poland JA, Kawamoto K, Buckler ES et al (2011) A robust, simple genotyping-by-sequencing (GBS) approach for high diversity species. PLoS One 6:e19379. https://doi.org/10.1371/journal.pone.0019379
20. Oeth P, Beaulieu M, Park C, Kosman D, de Mistro G, van Den Boom D et al (2007) iPLEX assay: increased plexing efficiency and flexibility for MassARRAY system through single base primer extension with mass-modified terminators. Sequenom Application Note. Sequenom, San Diego
21. Oeth P, del Mistro G, Marnellos G, Shi T, van den Boom D (2009) Qualitative and quantitative genotyping using single base primer extension coupled with matrix-assisted laser desorption/ionization time-of-flight mass spectrometry (MassARRAY). In: Komar A (ed) Single nucleotide polymorphisms, Methods in molecular biology, vol 578. Humana Press, Totowa, pp 307–343. https://doi.org/10.1007/978-1-60327-411-1_20
22. Costa EA, Anoni CO, Mancini MC, Santos FRC, Marconi TG, Gazaffi R et al (2016) QTL mapping including codominant SNP markers with ploidy level information in a sugarcane progeny. Euphytica 211:1–16. https://doi.org/10.1007/s10681-016-1746-7
23. Campa A, Ferreira JJ (2018) Genetic diversity assessed by genotyping by sequencing (GBS) and for phenological traits in blueberry cultivars. PLoS One 13:e0206361. https://doi.org/10.1371/journal.pone.0206361
24. Sforça DA, Vautrin S, Cardoso-Silva CB, Mancini MC, Romero-da Cruz MV, Pereira G et al (2019) Gene duplication in the sugarcane genome: a case study of allele interactions and evolutionary patterns in two genic regions. Front Plant Sci 10:553. https://doi.org/10.3389/fpls.2019.00553
25. Martins FB, Moraes A, Aono AH, Ferreira R, Chiari L, Simeão RM et al (2021) A semi-automated SNP-based approach for contaminant identification in biparental polyploid populations of tropical forage grasses. Front Plant Sci 12:737919. https://doi.org/10.3389/fpls.2021.737919
26. Gabriel S, Ziaugra L, Tabbaa D (2009) SNP genotyping using the Sequenom MassARRAY iPLEX platform. Curr Protoc Hum Genet 60:2.12.1–2.12.18. https://doi.org/10.1002/0471142905.hg0212s60
27. Buggs RJ, Chamala S, Wu W, Gao L, May GD, Schnable PS et al (2010) Characterization of duplicate gene evolution in the recent natural allopolyploid *Tragopogon miscellus* by next-generation sequencing and Sequenom iPLEX MassARRAY genotyping. Mol Ecol 19:132–146. https://doi.org/10.1111/j.1365-294X.2009.04469.x
28. Godbout J, Tremblay L, Levasseur C, Lavigne P, Rainville A, Mackay J et al (2017) Development of a traceability system based on a SNP array for large-scale production of high-value white spruce (*Picea glauca*). Front Plant Sci 8:1264. https://doi.org/10.3389/fpls.2017.01264
29. Smith HM, Smith BP, Morales NB, Moskwa S, Clingeleffer PR, Thomas MR (2018) SNP markers tightly linked to root knot nematode resistance in grapevine (*Vitis cinerea*) identified by a genotyping-by-sequencing approach

followed by Sequenom MassARRAY validation. PLoS One 13:e0193121. https://doi.org/10.1371/journal.pone.0193121

30. Choudhury DR, Kumar R, Devi SV, Singh K, Singh NK, Singh R (2021) Identification of a diverse core set panel of rice from the east coast region of India using SNP markers. Front Genet 12:726152. https://doi.org/10.3389/fgene.2021.726152
31. Mollinari M, Serang O (2015) Quantitative SNP genotyping of polyploids with MassARRAY and other platforms. In: Batley J (ed) Plant genotyping, Methods in molecular biology, vol 1245. Humana Press, New York, pp 215–241. https://doi.org/10.1007/978-1-4939-1966-6_17
32. Grandke F, Singh P, Heuven HC, de Haan JR, Metzler D (2016) Advantages of continuous genotype values over genotype classes for GWAS in higher polyploids: a comparative study in hexaploid chrysanthemum. BMC Genomics 17:672. https://doi.org/10.1186/s12864-016-2926-5
33. Balsalobre TW, da Silva Pereira G, Margarido GR, Gazaffi R, Barreto FZ, Anoni CO et al (2017) GBS-based single dosage markers for linkage and QTL mapping allow gene mining for yield-related traits in sugarcane. BMC Genomics 18:72. https://doi.org/10.1186/s12864-016-3383-x
34. da Silva Pereira G, Gemenet DC, Mollinari M, Olukolu BA, Wood JC, Diaz F et al (2020) Multiple QTL mapping in autopolyploids: a random-effect model approach with application in a hexaploid sweetpotato full-sib population. Genetics 215:579–595. https://doi.org/10.1534/genetics.120.303080
35. Aono AH, Costa EA, Rody H, Nagai JS, Pimenta R, Mancini MC et al (2020) Machine learning approaches reveal genomic regions associated with sugarcane brown rust resistance. Sci Rep 10:20057. https://doi.org/10.1038/s41598-020-77063-5
36. Batista LG, Mello VH, Souza AP, Margarido G (2022) Genomic prediction with allele dosage information in highly polyploid species. Theor Appl Genet 135:723–739. https://doi.org/10.1007/s00122-021-03994-w4
37. Ferreira R, Moraes ACL, Chiari L, Simeão RM, Vigna B, Souza AP (2021) An overview of the genetics and genomics of the *Urochloa* species most commonly used in pastures. Front Plant Sci 12:770461. https://doi.org/10.3389/fpls.2021.770461
38. Vigna BB, Santos JC, Jungmann L, do Valle CB, Mollinari M, Pastina MM et al (2016) Evidence of allopolyploidy in *Urochloa humidicola* based on cytological analysis and genetic linkage mapping. PLoS One 11:e0153764. https://doi.org/10.1371/journal.pone.0153764
39. Vigna BB, de Oliveira FA, de Toledo-Silva G, da Silva CC, do Valle CB, de Souza AP (2016) Leaf transcriptome of two highly divergent genotypes of *Urochloa humidicola* (Poaceae), a tropical polyploid forage grass adapted to acidic soils and temporary flooding areas. BMC Genomics 17:910. https://doi.org/10.1186/s12864-016-3270-5
40. Irwin D (2008) The MassARRAY system for plant genomics. In: Henry R (ed) Plant genotyping II. SNP Technology. CSIRO Publishing, Collingwood, pp 98–113. https://doi.org/10.1079/9781845933821.0098

Chapter 8

qPCR Genotyping of Polyploid Species

Haiyan Wang, Jiangbo Dang, Qigao Guo, and Guolu Liang

Abstract

A simple and cost-effective method for genotyping polyploid plants using quantitative PCR (qPCR) is described in this chapter. There is no additional operation, only simultaneous amplification of alleles and reference sequences with constant copy number in the genome. The qPCR genotyping can detect the genotypes of important traits in polyploid plants without whole genome sequencing data.

Key words Quantitative PCR (qPCR), Genotyping, Polyploid, Allele

1 Introduction

Polyploidy is an important driving force for plant evolution, and they have higher selection advantages than diploids [1, 2]. The complex nature of polyploidy, especially the diversity of plant genotypes caused by multiple copies of chromosomes, makes molecular genetics research more difficult than diploid [3]. In addition, the alleles may have additive effects or more complex interactions that lead to phenotypic changes, requiring accurate identification of heterozygous genotypes [4–6].

Quantitative real time polymerase chain reaction (qPCR) is one of the most common techniques for quantification of nucleic acid molecules, widely applied to detect gene expression [7], copy number variation [8] and aneuploidy [9]. Its characteristics can reflect the quantification of a specific sequence in the total nucleic acid, which has a high application potential in quantitative genotyping. Only need a stable reference sequence and allele-specific primers, we can easily obtain the relative ratios of allele. Several quantitative genotyping techniques based on other theories have also been developed, such as quantitative fluorescent polymerase chain reaction (QF-PCR) [10, 11], competitive allele-specific PCR [12], high-resolution digestion (HRM) [13], genotyping by sequencing (GBS) [14, 15]. However, the application scope of

Yuri Shavrukov (ed.), *Plant Genotyping: Methods and Protocols*, Methods in Molecular Biology, vol. 2638,
https://doi.org/10.1007/978-1-0716-3024-2_8,

these methods is greatly limited due to the high cost and complex operation. For polyploid genotyping, qPCR technology, a more economical and convenient technique, may be a more extensive and excellent choice.

2 Materials

2.1 Genomic DNA Extraction

1. 2% CTAB extraction buffer: 100 mM Tris–HCl pH 8.0, 20 mM EDTA (disodium) pH 8.0, 1.4 M NaCl, 4% (w/v) PVP, 2% (w/v) CTAB. Store the buffer at room temperature.
2. β-mercaptoethanol stored at room temperature.
3. Proteinase K (10 mg/mL) stored at −20 °C.
4. 3 M sodium acetate solution (3 M NaAc, pH 5.2), stored at 4 °C.
5. A mixture of chloroform and isoamylol (v/v, 24:1), stored at room temperature (*see* **Note 1**).
6. Non-diluted ethanol stored at −20 °C.
7. 75% ethanol stored at −20 °C.
8. Ribonuclease A stored at −20 °C.
9. DNase & RNase free water.
10. Water baths or heat blocks (65 °C).
11. Refrigerated centrifuge.
12. Nanodrop-2000 (or similar instrument).

2.2 gDNA Quantitative PCR Amplification

1. 2×NovoStart SYBR qPCR Super Mix Plus (Novoprotein Scientific Inc., Suzhou, JS or similar SYBR qPCR Super Mix).
2. Reference sequence primers.
3. Allele-specific primers.
4. Genomic DNA diluted to 50 μL.
5. DNase & RNase free water.
6. PCR strip tubes with caps.
7. Routine real-time thermal cycler.

3 Methods

3.1 Genomic DNA Extraction

See **Notes 1–3** before beginning. Plant samples are crushed by tissue mill (at high speed) in a 2 mL centrifuge tube. Store in −80 °C freezer (*see* **Note 2**). Genomic DNA (gDNA) extraction method referred to Wen Guo's improved CTAB method [16]. But other similar extraction methods can be used, as long as the DNA purity is guaranteed, and RNA contamination is removed.

DNA is stored at 4 °C before use or −20 °C for longer term. The use of young leaves ensures high quality DNA, which is essential for reliable and reproducible (*see* **Note 3**). The isolated DNA should be RNA free.

1. Add 700 μL of 2×CTAB extraction buffer, 4.5 μL of proteinase K, 20 μL of β-mercaptoethanol in the tube with sample, thoroughly mixed, incubate 60 min in 65 °C water bath, gently invert every 10 min.
2. Remove samples from water bath and allow it to cool slightly. Add 150 μL 3 M NaAc, incubate 20 min in ice.
3. Spin for 10 min at 8000 rpm in microcentrifuge. Transfer supernatant to a new 2.0 mL centrifuge tube, discard residue in hazardous waste.
4. Add chloroform and isoamyl alcohol mixture and shake vigorously for 5 min (*see* **Note 2**).
5. Spin for 10 min at 8000 rpm in microcentrifuge. Transfer supernatant to a new 2.0 mL centrifuge tube, discard residue in hazardous waste.
6. Repeat **steps 4** and **5** (*see* **Note 4**).
7. Add 800–1000 μL non-diluted ethanol (−20 °C) and incubate 30 min in −20 °C freezer to precipitate gDNA. Spin for 5 min at 12000 rpm in microcentrifuge (4 °C). Pipette off supernatant and discard in hazardous waste.
8. Add 800–1000 μL 75% ethanol (−20 °C) for washing precipitate. Spin for 1 min at 12000 rpm in microcentrifuge (4 °C). Pipette off supernatant and discard in hazardous waste.
9. Repeat **step 8** but with non-diluted ethanol instead 75% ethanol.
10. Dry gDNA precipitate in speed-vac for 10–15 min (shorter or longer if necessary).
11. Add 120 μL DNase & RNase free water and dissolve gDNA precipitation 12 h approximately, 4 °C (*see* **Note 5**).
12. Add 1.5 μL Ribonuclease A (10 mg/mL), incubate 1 h in 37 °C.
13. Add 4 μL 3 M NaAc, incubate 20 min in ice.
14. Repeat **steps 4–10**.
15. Add 40 μL DNase & RNase free water and dissolve gDNA precipitation.
16. gDNA is quantified using Nanodrop-2000 (or similar instrument) according to the manufacturer's instructions.

3.2 Reference Sequence and Allele-Specific Primer Design

Primers in experiments must be verified in advance to ensure the best conditions and repeatability of amplification.

1. Reference sequence is a DNA sequence with a constant copy number in the genome, which was used as an internal control. The Reference sequence shows no polymorphism in plants, and its copy number only changes with ploidy (*see* **Note 6**).
2. Allele-specific primers are designed according to the sequence of the candidate alleles being targeted. Make sure of the specificity of primer amplification for the targeted genes.

3.3 Quantitative PCR Amplification

See **Notes 7–9** before beginning. Reproducibility can be improved by using very clean template gDNA, high-fidelity polymerase, the same thermal cycler and program for all replicates. Negative controls must be performed to identify primer dimers and assess reagent contamination.

1. Prepare template gDNA for qPCR amplification with a concentration of 50 ng/μL (*see* **Note 7**).
2. Prepare a qPCR mix for all samples on ice. Each reaction consists of 5 μL 2×SYBR qPCR Super Mix Plus, 0.2 μL of each primer (10 mM), 1 μL gDNA, adjust to 10 μL with DNase & RNase free water (*see* **Note 8**).
3. Perform a two-step qPCR program as follows (*see* **Note 9**): Pre-denaturation at 5 °C for 1 min, 40 cycles of reaction (95 °C for 20s, 60 °C for 1 min), followed by 60–95 °C to prepare the melting curve.

3.4 Analyzing qPCR Genotyping Date

1. According to Ct values, the relative copy number of each allele in diploid and polyploid materials is calculated according to $2^{-\Delta\Delta Ct}$ method. If the control is a heterozygous diploid, the relative copy number of allele ***A*** was $\mathbf{a_1}$, the relative copy number of allele ***A*** and ***a*** is $\mathbf{a_3}$, and the relative copy number of allele ***a*** was $\mathbf{a_2}$ ($\mathbf{a_2} = \mathbf{a_3} - \mathbf{a_1}$) (*see* Fig. 1).
2. The relative allele ***A*** signal is calculated according to the formula $\mathbf{b} = \mathbf{a_1}/(\mathbf{a_1} + \mathbf{a_2})$ ($0 \leq \mathbf{b} \leq 1$). Genotypes can be determined by the relative allele ***A*** signals of different ploidy materials, as shown in Table 1.

4 Notes

1. Chloroform precipitates protein after denaturation by squeezing out water molecules between protein molecules. This process requires vigorous shaking of the fully mixed solution, which is prone to bubble formation. Adding a small amount of isoamyl alcohol (reducing surface tension) can effectively reduce bubble formation.

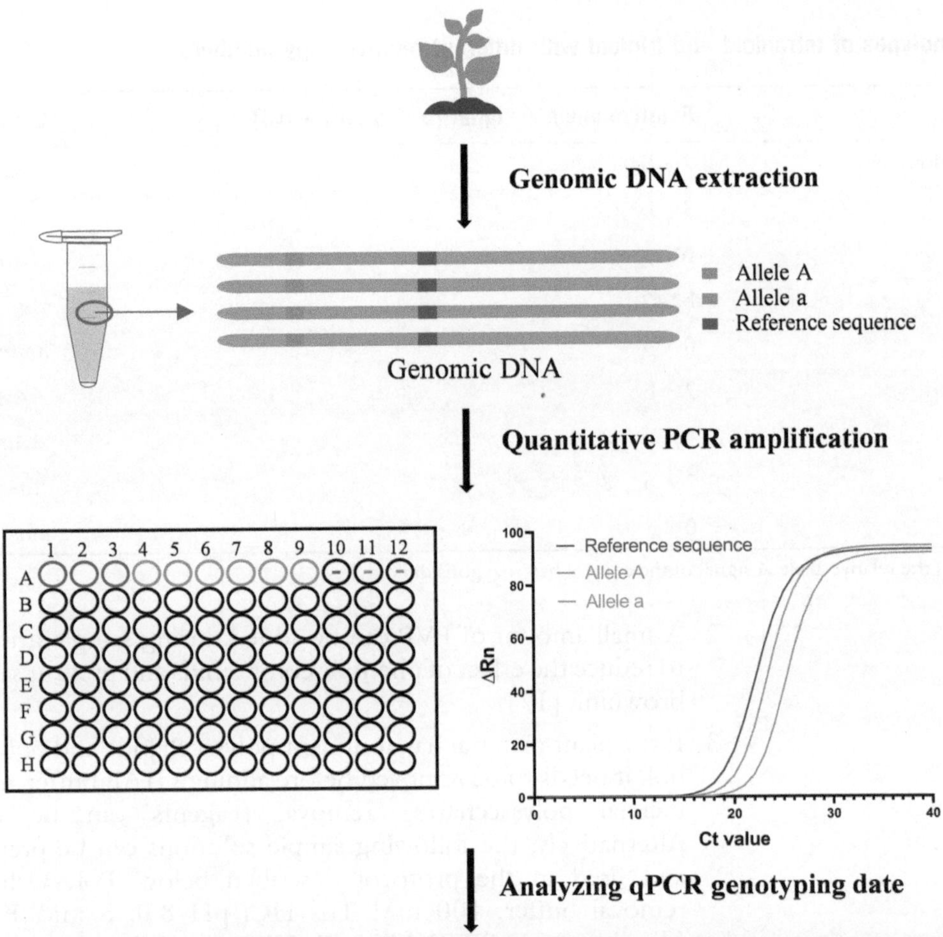

Sample	Allele A (a1) ($2^{-\Delta\Delta Ct}$)	Allele a (a2) ($2^{-\Delta\Delta Ct}$)	Genotype [$b=a_1/(a_1+a_2)$]
Sample 1	0	2	0 (aaaa)
Sample 2	0.5	1.5	0.25 (Aaaa)
Sample 3	1	1	0.5 (AAaa)
Sample 4	1.5	0.5	0.75 (AAAa)
Sample 5	2	0	1 (AAAa)

Fig. 1 Schematic flew diagram of qPCR genotyping. gDNA of young leaves is extracted for qPCR genotyping. qPCR genotyping is carried out with amplifying by allele-specific primers and reference sequence primers, and the relative copy number of each allele is calculated by $2^{-\Delta\Delta Ct}$ method, and then parameter **b** [**b** $= \mathbf{a_1}/(\mathbf{a_1} + \mathbf{a_2})$; $0 \leq \mathbf{b} \leq 1$] is calculated to determine the genotype of the tested material

Table 1
The genotypes of tetraploid and triploid with different relative copy numbers

Ploidy	Relative allele *A* signal [$b = a_1/(a_1 + a_2)$]	Genotype
Tetraploid	2 × **b***	*AAAA*
	3 × **b***/2	*AAAa*
	b*	*AAaa*
	b*/2	*Aaaa*
	0	*aaaa*
Triploid	2 × **b***	*AAA*
	4 × **b***/3	*AAa*
	2 × **b***/3	*Aaa*
	0	*aaa*

The **b*** is the relative allele ***A*** signal of the control heterozygous diploid

2. A small amount of PVP can be added during sample grinding to reduce the effect of phenolic compounds and prevent sample browning [17].
3. If the plant material has abundant polysaccharide and polyphenol, it needs to be removed before adding CTAB buffer. Commercial polysaccharide removal reagents can be used. Alternatively, the following simple solutions can be prepared and used in the protocol described below. Polysaccharide removal buffer: 100 mM Tris–HCl pH 8.0; 5 mM EDTA (disodium) pH 8.0; 1 M NaCl; 2% (w/v) PVP; 400 mM glucose; and 4% (v/v) β-mercaptoethanol. Add 500–700 μL Polysaccharide removal buffer and mix thoroughly, incubate 10 min in ice. Spin for 10 min at 12000 rpm in microcentrifuge (4 °C). Pipette off supernatant and discard in hazardous waste. Repeat this **step 1** more time.
4. After centrifugation, the supernatant in the tube was separated into chloroform and isoamyl alcohol in the lower layer, white protein precipitation in the middle, and DNA extraction liquid above. The repeat times are determined according to the amount of intermediate protein precipitation, until there is only a little precipitation in the middle.
5. The extracted DNA is dissolved and then precipitated to further purify DNA in the **steps 11–15**. If the quality requirement for DNA extraction is not high, do not proceed further, and use Ribonuclease A treatment before **step 6**.
6. Candidate reference sequences are used for the screening of plants with different ploidy levels to identify primers with high consistency for generating the reference DNA sequence. There

are many options for reference sequences but only those that can prove their stability can be used. In addition to the common *Actin* gene, some molecular markers, such as SSR markers, are a good choice.

7. Depending on the abundance of reference sequence and alleles, increasing or decreasing the concentration of gDNA results in an appropriate range of Ct values.
8. 2×SYBR qPCR Super Mix Plus contains pre-mixed DNA polymerase, qPCR Buffer, dNTPs, SYBR Green. Any qPCR kits can be used, as long as the corresponding adjustments are made beforehand. The primer amount can influence the PCR product, and has to be adapted, if required. Within the allowable range, the use of lower concentration of primers will result in better qPCR results.
9. If qPCR is unsuccessful, optimization can be used to improve the qPCR conditions. Compared with the three-step qPCR program, the two-step program has higher specificity and should be preferred. The annealing temperature depends on primer Tm value.

Acknowledgments

This qPCR genotyping research was funded by the National Key Research and Development Program of China (contract No. 2019YFD1000900), the Innovation Research Group Funds for Chongqing Universities (contract No. CXQT19005).

References

1. Blanc G, Wolfe KH (2004) Widespread paleopolyploidy in model plant species inferred from age distributions of duplicate genes. Plant Cell 16:1667–1678. https://doi.org/10.1105/tpc.021345
2. Van de Peer Y, Maere S, Meyer A (2009) The evolutionary significance of ancient genome duplications. Nat Rev Genet 10:725–732. https://doi.org/10.1038/nrg2600
3. Dufresne F, Stift M, Vergilino R, Mable BK (2014) Recent progress and challenges in population genetics of polyploid organisms: an overview of current state-of-the-art molecular and statistical tools. Mol Ecol 23:40–69. https://doi.org/10.1111/mec.12581
4. Sjöling A, Walentinsson A, Nordlander C, Karlsson A, Behboudi A, Samuelson E et al (2005) Assessment of allele dosage at polymorphic microsatellite loci displaying allelic imbalance in tumors by means of quantitative competitive-polymerase chain reaction. Cancer Genet Cytogenet 157:97–103. https://doi.org/10.1016/j.cancergencyto.2004.06.013
5. Osborn TC, Pires JC, Birchler JA, Auger DL, Chen ZJ, Lee H et al (2003) Understanding mechanisms of novel gene expression in polyploids. Trends Genet 19:141–147. https://doi.org/10.1016/s0168-9525(03)00015-5
6. Veitia RA, Bottani S, Birchler JA (2008) Cellular reactions to gene dosage imbalance: genomic, transcriptomic and proteomic effects. Trends Genet 24:390–397. https://doi.org/10.1016/j.tig.2008.05.005
7. Jing D, Chen W, Shi M, Wang D, Xia Y, He Q et al (2020) Ectopic expression of an *Eriobotrya japonica APETALA3* ortholog rescues the petal and stamen identities in Arabidopsis *ap3-3* mutant. Biochem Biophys Res Commun 523:33–38. https://doi.org/10.1016/j.bbrc.2019.11.177

8. Li J, Fan Z, Shen F, Pendleton AL, Song Y, Xing J et al (2020) Genomic copy number variation study of nine macaca species provides new insights into their genetic divergence, adaptation, and biomedical application. Genome Biol Evol 12:2211–2230. https://doi.org/10.1093/gbe/evaa200
9. Wen G, Dang J, Xie Z, Wang J, Jiang P, Guo Q et al (2020) Molecular karyotypes of loquat (*Eriobotrya japonica*) aneuploids can be detected by using SSR markers combined with quantitative PCR irrespective of heterozygosity. Plant Methods 16:1231–1245. https://doi.org/10.1186/s13007-020-00568-7
10. Masaru T, Tetsufumi S, Yasuhiro T (2019) Allele dosage–dependent selection of recessive *F3' H* allele homozygote altered anthocyanin composition in sweetpotato. Mol Breed 39: 152. https://doi.org/10.1007/s11032-019-1062-8
11. Garavello M, Cuenca J, Dreissig S, Fuchs J, Navarro L, Houben A et al (2020) Analysis of crossover events and allele segregation distortion in interspecific citrus hybrids by single pollen genotyping. Front Plant Sci 11:615. https://doi.org/10.3389/fpls.2020.00615
12. Cuenca J, Aleza P, Navarro L, Ollitrault P (2013) Assignment of SNP allelic configuration in polyploids using competitive allele-specific PCR: application to citrus triploid progeny. Ann Bot 111:731–742. https://doi.org/10.1093/aob/mct032
13. Guo Q, Xiao L, Zhou Y (2012) Rapid diagnosis of aneuploidy by high-resolution melting analysis of segmental duplications. Clin Chem 58:1019–1025. https://doi.org/10.1373/clinchem.2011.178475
14. Tränkner C, Krüger J, Wanke S, Naumann J, Wenke T, Engel F (2019) Rapid identification of inflorescence type markers by genotyping-by-sequencing of diploid and triploid F_1 plants of *Hydrangea macrophylla*. BMC Genet 20:60. https://doi.org/10.1186/s12863-019-0764-6
15. Vasumathy SK, Peringottillam M, Sundaram KT, Kumar SHK, Alagu M (2020) Genome-wide structural and functional variant discovery of rice landraces using genotyping by sequencing. Mol Biol Rep 47:7391–7402. https://doi.org/10.1007/s11033-020-05794-9
16. Wen G (2020) Construction of a molecular karyotype analysis method for detecting loquat (*Eriobotrya japonica*) aneuploids based on SSR markers and quantitative PCR. PhD Dissertation, Southwest University. https://doi.org/10.27684/d.cnki.gxndx.2020.002571
17. Lee BJ, Kim S, Lee JW, Lee HM, Eo SH (2021) Technical note: Polyvinylpyrrolidone (PVP) and proteinase-K improve the efficiency of DNA extraction from Japanese larch wood and PCR success rate. Forensic Sci Int 328: 111005. https://doi.org/10.1016/j.forsciint.2021.111005

Chapter 9

Genome-Wide Association Studies (GWAS)

Jianbo He and Junyi Gai

Abstract

Most of the breeding targets are quantitative traits. In exploring the quantitative trait locus (QTL) system of a trait, linkage mapping was established using sparse polymerase chain reaction (PCR) markers. With the genome-wide sequencing technology advanced, genome-wide association study (GWAS) was developed for natural (germplasm) populations using dense genomic markers, which facilitates the identification of the complete QTL system with their multiple alleles on genomic locations. GWAS makes use of the linkage disequilibrium (LD) due to historical saturate recombination and high-density genomic markers to detect QTLs through statistical test for the association between molecular markers and phenotypes. However, due to inbreeding and mixture of source populations, the germplasm population often has complex and unknown structure, which leads to false positives/negatives in GWAS. Various GWAS methods have been proposed to reduce false positives/negatives, including those of the general linear model and the mixed linear model, which focused mainly on finding a handful of major QTLs under single-locus model for major gene cloning and could not detect directly the multiple alleles using bi-allelic single-nucleotide polymorphism (SNP) marker. As a relatively thorough detection of QTLs with their multiple alleles is required for germplasm population, the restricted two-stage multi-locus multi-allele GWAS (RTM-GWAS) procedure was proposed for identifying the QTL system with varying multiple alleles. From the RTM-GWAS results, a QTL-allele matrix is constructed as a compact form of the population genetic constitution, which can be further used for crop genetic and breeding studies, including major gene mining, population evolution, and breeding by genetic design.

Key words Quantitative trait, Genome-wide association study (GWAS), Population structure, General linear model, Mixed linear model, Multi-locus multi allele model, Restricted two-stage multi-locus multi-allele GWAS (RTM-GWAS)

1 Introduction

Mendelian genetics started inheritance study from segregation analysis of bi-parental crosses, from which the concept of gene was established. A trait may be conferred by a single gene (oligogene), performed qualitative inheritance, or may be conferred by multiple genes and performed quantitative inheritance. Most of the traits involved in crop production are quantitative traits. Unraveling the genetic architecture of quantitative traits is of great

Yuri Shavrukov (ed.), *Plant Genotyping: Methods and Protocols*, Methods in Molecular Biology, vol. 2638,
https://doi.org/10.1007/978-1-0716-3024-2_9,

importance in crop breeding. Quantitative traits are typically controlled by a number of genes, each with relatively small and varied genetic effects, which makes challenges to identify the quantitative trait locus (QTL) underlying the phenotypic diversity.

The molecular technology has provided a technique to mark the genes. Using molecular markers to establish a genetic linkage map, then to locate the involved genes on the map is called gene mapping. A trait with its genes mapped is usually understood as the trait is genotyped. Especially, when the genomic segments are used as molecular markers, the sequence with its location is known and therefore, the trait is in fact genotyped at genomic sequence level. For quantitative traits, the mapped segment is not necessarily a precise gene but a QTL. In genotyping or mapping a gene or a QTL, linkage mapping procedure was developed at first. This is done usually by preparing a bi-parental segregating population, establishing a genetic linkage map, then using phenotype data to link the gene(s) to the genetic map [1]. Linkage mapping limits the genes to be studied within the involved pair of parents, especially limits the studied alleles within the two parents. But in natural situation, many genes and multiple alleles per gene/QTL happened. Then to understand the complete genetic system of a quantitative trait involved should be natural population(s). Therefore, the genome-wide association study (GWAS) was developed based on natural or germplasm population, which provides a potential approach for detecting the whole-genome QTLs and their corresponding alleles. GWAS detects QTLs through statistical test for the association between molecular markers and phenotypes in natural populations, known as association mapping. It makes use of the linkage disequilibrium (LD) combined with dense genomic single-nucleotide polymorphism (SNP) markers in natural populations. In details, the principle of association analysis is based on the tight LD between two neighboring segments, represented by the marker and the causal QTL/gene hidden in phenotype data, respectively, after historical saturate recombination in long-term random mating in a large-enough population. Under the conditions of non-random mating (inbreeding, especially in small size population) and mixture of source populations, the detected LD might be false ones. In this case, both false positives and false negatives could occur, resulted in a disturbed association. GWAS is assumed to have a high resolution and power for QTL detection and has also become an important approach for genetic dissection of quantitative traits. At present, GWAS has been widely used in humans, animals and plants.

Various statistical methods have been proposed to address the issue for avoiding high false positive/negative rate in GWAS caused by unknown population structure [2]. The most commonly used methods in avoiding the population structure problem mainly include the structured association approach [3], the principal component analysis approach [4] and the mixed linear model (MLM) approach [5]. The structured association method reduces false positives/negatives by incorporating inferred population structure as model covariate. The population structure is usually inferred through model-based clustering algorithms, such as the software STRUCTURE [6]. The principal component analysis method reduces false positives/negatives by taking the eigenvectors of genetic relationship matrix as model covariate. The MLM method incorporates the genetic background effect as random effect and takes the kinship matrix as covariance structure of the genetic background effect to correct the population structure bias from both admixture and relatedness [7].

The above methods are all based on a single-locus model and perform association test for each marker independently. As linkage may exist among neighboring loci, the genetic contribution of a detected QTL may be influenced by its neighboring loci, leading to an inflated contribution, in a total, causing the overflowing heritability problem of GWAS under single-locus model. That is the major reason to cause false positives more often than false negatives. Therefore, it is more appropriate to use multi-locus model in GWAS, such that multiple QTLs can be tested simultaneously in one model [8]. The MLMM [9] and LMM-lasso [10] methods under the MLM framework have been proposed for multi-locus mapping. However, the bi-allele SNP marker cannot fit directly the multi-allelic property of a natural population and the multi-locus model generally cannot be readily implemented for large-scale calculation requirement due to hundreds of thousands or even millions of SNPs involved in GWAS.

As indicated above, the previous GWAS was usually conducted based on SNP markers, but the bi-allelic SNP marker does not match QTLs with multiple alleles, which is an important property of natural or germplasm population in plants. Thus, the detection power of GWAS was reduced when using SNP to detect QTLs with multiple alleles in a natural population, because a single SNP marker could not fully capture the multi-allelic variation. In addition, plant breeding is actually a genetic operation of reassembling alleles at different loci. Thus, a thorough detection of QTLs as well as their multiple alleles is an essential requirement for GWAS as well as for plant breeding [11]. In the previous GWAS, researchers more likely interested in finding and cloning only a handful of major QTLs or genes with the others neglected. Thus, only a part of QTLs was usually detected in individual GWAS, which in fact largely limited the application of GWAS in studies on complete QTL-allele and gene-allele systems. To overcome these limitations, a restricted two-stage multi-locus multi-allele GWAS method, called RTM-GWAS, was proposed and has been applied to a series of genetic studies [8, 12]. The RTM-GWAS method constructs SNP linkage disequilibrium block (SNPLDB) markers with multi-allelic variation to fit plant germplasm population and then utilizes a two-stage multi-locus multi-allele model to detect genome-wide QTLs. From the RTM-GWAS results, a QTL-allele matrix can be constructed to represent the genetic constitution of a quantitative trait of the population. The QTL-allele matrix can be further used for gene discovery and population genetic differentiation studies. According to the QTL-allele matrix, the optimal crosses among the accessions as well as their progenies can also be predicted. Therefore, breeding by genetic design can be performed.

2 Simple Association Test Approach

GWAS makes use of the LD between an observed molecular marker and a hidden causal locus to detect QTL or gene (target association) in a natural population through statistical test for association between marker and phenotype (inferred association). LD refers to the non-random association between alleles at neighboring loci in a natural population. In a theoretical random mating population, the linkage between loci is gradually broken due to intercrossing, and with the advance of generations, the loci gradually reach a linkage equilibrium state. However, due to the tight linkage between neighboring loci, there is still a certain remnant correlation between the two loci, or linkage disequilibrium (LD). The degree of LD is often high for tightly linked neighboring loci, and decays with increase of genomic distance between loci. If a marker is in a strong LD with a causal locus, then the phenotypes will also show difference between/among the haplotypes of the marker. The simple association test is considered as a statistical test for phenotype difference between/among different haplotypes at a marker. The linear model can be written as

$$y_{ij} = \mu + g_j + \varepsilon_{ij} \tag{1}$$

where, y_{ij} is the i-th observation of phenotype for the j-th haplotype/genotype at a marker, and μ is the overall mean. g_j is the effect of j-th haplotype/genotype, and ε_{ij} is the residual effect following $\varepsilon_i \sim N(0, \sigma^2)$.

The analysis of variance method can be used to test the phenotype difference between/among haplotypes/genotypes. The null and alternative hypotheses are

$$H_0 : g_1 = g_2 = \ldots = g_j$$
$$H_A : \text{The effects are not all equal.}$$

If the phenotype difference is statistically significant, we can infer that there may be a causal locus in strong LD with the marker under testing. As showed by Weir [13], the correlation between phenotype and marker is a function of the heritability of causal locus and the LD between marker and causal locus. Therefore, the association between the phenotype and the marker is actually an inferred association caused by the LD between the marker and the causal locus, or the target association.

Model (1) can also be written in matrix form,

$$\mathbf{y} = \mathbf{1}\mu + \mathbf{Xb} + \mathbf{e} \tag{2}$$

where, **y** is a $n \times 1$ vector of phenotype values with n as the sample size; **b** is a vector of genotype effects $(g_1, g_2, \ldots, g_j)$; **X** is a design matrix of marker genotype with elements of 0 or 1; and **e** is a vector residual effects with $\mathbf{e} \sim N(0, \mathbf{I}\sigma^2)$. For example, there are three possible genotypes for a SNP marker (*AA*, *Aa* and *aa*), and the matrix **X** will have three columns corresponding to the three genotypes. The element of **X** for *i*-th individual and *j*-th genotype is determined under the genotype model

	x_{i1}	x_{i2}	x_{i3}
AA	1	0	0
Aa	0	1	0
AA	0	0	1

The additive-dominance model can also be used to form the design matrix **X**, in which the genotypic effect is partitioned into additive (a) and dominance (d) effect. For the two homozygous genotypes, *AA* and *aa*, the genotypic effects are expressed as a and $-a$, respectively, while the genotypic effect of heterozygous genotype *Aa* is d. In this case, the design matrix **X** for marker genotype is formed as

	$x_{i(a)}$	$x_{i(d)}$
AA	1	0
Aa	0	1
aa	−1	0

The natural population or germplasm population in self-pollinated plants is typically homozygous without heterozygous genotype. In such population, the design matrix of marker genotype can be further simplified to contain only additive effect.

Model (2) can be solved efficiently using the least squares method and hypothesis testing can be performed using standard statistical method such as *F*-test. In GWAS, association test is performed for each marker independently, and the *p*-value of statistical hypothesis test is obtained and a significance level, such as $\alpha = 0.05$, is used to confirm a QTL. A variety of computer programs have been developed for GWAS, including the TASSEL software [14] and the PLINK software [15, 16]. They were designed to conveniently handle the genotype and phenotype data file, and perform association test for genome-wide SNP markers.

For example, a soybean germplasm population comprising 446 released cultivars was used to perform GWAS of soybean plant height. The part of genotype data was extracted from He et al. [8]. The population was genotyped using restriction site-associated DNA sequencing (RAD-seq) and 68,050 high-quality SNPs were finally obtained using the reference genome Williams 82 (Wm82.a1.v1.1) [17]. The phenotype data of plant height was obtained through filed experiment (unpublished data). The genotype and phenotype data file should be prepared in appropriate format for GWAS depending on the software used.

Take the widely used TASSEL software for example, the following command line of the TASSEL software can be used to perform the simple association test (Please also check the users' guide of TASSEL for more details):

```
run_pipeline.pl -fork1 -importGuess genotype.vcf -fork2 -im-
portGuess phenotype.txt -combine3 -input1 -input2 -intersect
-FixedEffectLMPlugin -endPlugin -export NaiveResult
```

Figure 1 shows the scatter plot of *p*-values (on $-\log_{10}$ scale) against genomic positions for each SNP marker in GWAS of plant height in the soybean germplasm population, and this scatter plot is also called Manhattan plot in GWAS. The genome-wide threshold is usually determined using the Bonferroni correction for the collective of markers, in which the normal significance level of 0.01 or 0.05 is used for individual markers and then it is divided by the number of SNP markers for correction of multiple testing. In this example, the horizontal line in Fig. 1 indicates a genome-wide threshold of $-\log_{10}(0.01/68050) \approx 6.8$. Using the Bonferroni-adjusted threshold, a total of 817 QTLs for soybean plant height were detected (*see* Table 1). The phenotypic variation explained (in term of R^2) by each QTL ranged from 6.0% to 21.8% with a total of 7924.7%. Obviously, the simple association test may have a high false positive rate.

Under the null hypothesis, the *p*-values of association test follows a uniform distribution, thus the quantile-quantile (Q-Q) plot can be used to approximately assess the false rate of GWAS [18]. As shown in Fig. 2, the observed *p*-values of the simple linear model (Naïve) deviate significantly from the expected values, indicating that this method may have a high false positive rate. The inflation factor (λ_{GC}) was used to approximately assess the false rate of GWAS for qualitative trait in human beings. λ_{GC} is derived from the genomic control method [19] for GWAS and is defined as the

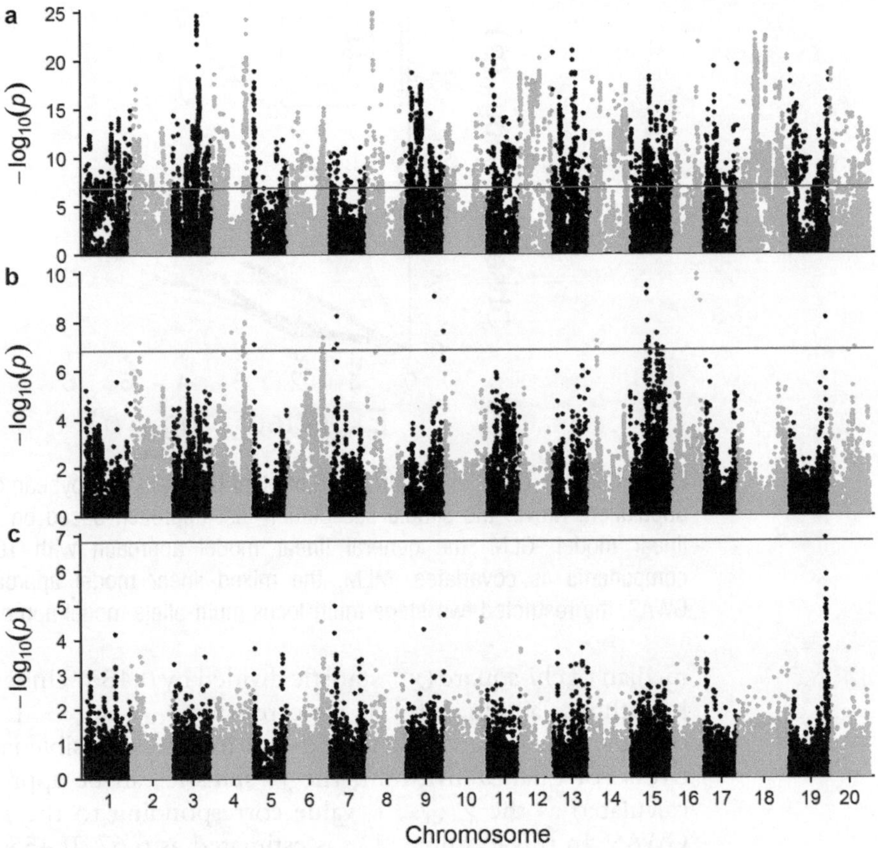

Fig. 1 Manhattan plot of GWAS of plant height in the soybean germplasm population. GWAS was conducted in a soybean germplasm population comprising 446 released cultivars, which were genotyped with 68,050 SNPs. (Part of data extracted from He et al. [8]). Association test was performed using (**a**) the simple linear model approach (Naïve), (**b**) the general linear model approach with 10 principal components as covariates (GLM), and (**c**) the mixed linear model approach (MLM)

Table 1
Comparison of GWAS methods in QTL detection for plant height of the soybean germplasm population

Method	QTLs	Individual R^2 (%)	Total R^2 (%)	λ_{GC}
Naïve	817	6.0–21.8	7924.7	14.66
GLM	26	4.0–5.9	117.8	2.83
MLM	1	6.5	6.5	0.80
RTM-GWAS	50	0.2–23.6	92.2	1.38

Note: GWAS was conducted in a soybean germplasm population comprising 446 released cultivars, which were genotyped with 68,050 SNPs. (Part of data extracted from He et al. [8]). Naïve, the simple association test approach based on the simple linear model. GLM, the general linear model approach with 10 principal components as covariates. MLM, the mixed linear model approach. RTM-GWAS, the restricted two-stage multi-locus multi-allele model approach. R^2 is the phenotypic variation explained by detected QTLs. λ_{GC} is the inflation factor and defined as the median of chi-square (χ^2) statistic divided by 0.455, where the χ^2-statistic is calculated as the $\chi^2(1)$ quantile corresponding to the p-value in GWAS

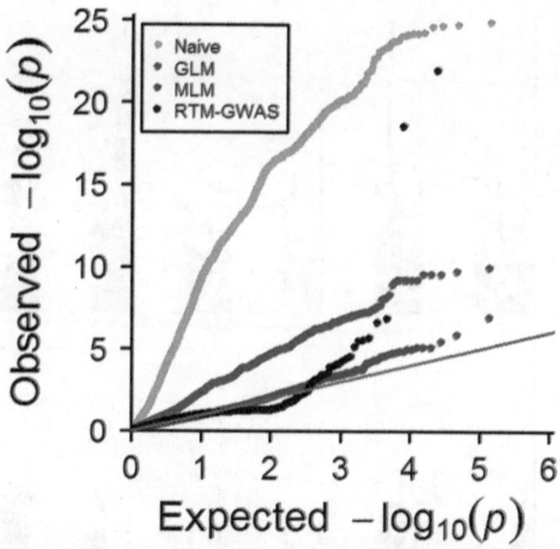

Fig. 2 Quantile-quantile plot of GWAS of plant height in the soybean germplasm population. Naïve, the simple association test approach based on the simple linear model. GLM, the general linear model approach with 10 principal components as covariates. MLM, the mixed linear model approach. RTM-GWAS, the restricted two-stage multi-locus multi-allele model approach

median of chi-square (χ^2) statistic divided by 0.455. Under the null hypothesis, a value of 1 is expected for λ_{GC}. $\lambda_{GC} > 1$ indicates possible false positives, while $\lambda_{GC} < 1$ indicates possible false negatives. For quantitative trait, the χ^2-statistic can be approximately calculated as the $\chi^2_{(df = 1)}$ value corresponding to the p-value in GWAS. In this example, λ_{GC} is estimated as $6.67/0.455 = 14.66$ (6.67 is the median of $\chi^2_{(df = 1)}$ values corresponding to p-values of GWAS), and also indicates that the GWAS results have a high false positive rate.

3 General Linear Model Approach

In general linear model (GLM) approach, the major point is to minimize the disturbance of population structure bias in identification of associations. To reduce false positives/negatives caused by unknown population structure in GWAS, the population structure is pre-estimated firstly and then incorporated as model covariates. Thus, model (2) is extended to the GLM:

$$\mathbf{y} = \mathbf{1}\mu + \mathbf{W}\mathbf{a} + \mathbf{X}\mathbf{b} + \mathbf{e} \tag{3}$$

where, **a** is a vector of population structure effects, and **W** is a matrix corresponding to population structure. Statistical analysis of model (3) is the same as model (2) and can also be performed using the GLM function in the TASSEL software.

In practice, the population structure is usually inferred using the STRUCTURE software or the EIGENSOFT software (https://github.com/DReichLab/EIG). The association test based on GLM for soybean plant height was performed using the GLM function in the TASSEL software:

```
run_pipeline.pl -fork1 -importGuess genotype.vcf -fork2 -im-
portGuess phenotype.txt -fork3 -importGuess pca.txt -combine4
-input1 -input2 -input3 -intersect -FixedEffectLMPlugin -end-
Plugin -export GLMResult
```

A total of 26 QTL were detected for soybean plant height (*see* Table 1). The phenotypic variation (R^2) explained by each marker ranged from 4.0% to 5.9% with a total of 117.8%. The p-values of the GLM method are shown in Fig. 1b, and the Q-Q plot (*see* Fig. 2) indicated that the false rate is reduced in comparison with the simple linear model. The λ_{GC} for GLM is estimated as 2.83, which indicates that false positives existed, but much smaller than that of the simple linear model (14.66).

The key of the GLM method is using an estimated population structure to correct the population structure bias. One approach is the STRUCTURE method. The population is assumed to be an admixture of several sub-populations, and then each individual is estimated for its genetic source proportions corresponding to sub-populations. However, the genetic structure of plant germplasm population varies greatly, which can be caused from both admixture and inbreeding. Therefore, separation of the whole population into subpopulations does not necessarily match the real population structure. Another approach is the principal component analysis method. The genetic relationship matrix (GRM) is firstly estimated based on genome-wide SNPs, and then the principal components are extracted and modeled as covariates to correct the population structure bias. Despite the population structure varies uncertainly due to variable admixture components, inbreeding schemes, and even both, the GRM matrix may be used as a general approach to estimate the varied population structure biases without any pre-set assumptions.

4 Mixed Linear Model Approach

In mixed linear model approach, it is supposed that a quantitative trait may involve some major genes and a group of minor genes. The former is to be identified individually, while the latter cannot be identified individually and is to be excluded as a whole to minimize its disturbance to the former. Assume that the effects of minor genes of a quantitative trait follows normal distribution $a_k \sim N(0, \tau^2)$, where a_k is the effects of k-th minor gene, τ^2 is the genetic variance of the minor gene. Thus, the total effect of minor genes for i-th individual is

$$u_i = \sum\nolimits_k w_{ik} a_k$$

where, w_{ik} is the genotype indicator taking a value of 0 or 1 for i-th individual and k-th minor gene.

Thus, the variance-covariance matrix of the random minor gene effects is

$$\mathrm{Var}(\mathbf{u}) = \tau^2 \mathbf{W}\mathbf{W}' = \sigma_g^2 \mathbf{K}$$

where, $\sigma_g^2 = m \cdot \tau^2$, and $\mathbf{K} = \mathbf{W}\mathbf{W}'/m$, with m as the number of minor genes.

The minor gene effects can be incorporated into model (2) to account for the background effect caused by multiple levels of relatedness. Thus, model (2) is extended to a mixed linear model (MLM),

$$y = \mathbf{1}\mu + \mathbf{X}\mathbf{b} + \mathbf{u} + \mathbf{e} \tag{4}$$

where, $\mathbf{u}$ is a vector of total effect of minor genes with $\mathbf{u} \sim N\left(0, \sigma_g^2 \mathbf{K}\right)$.

The variance-covariance matrix of phenotype vector is $\mathbf{V} = \sigma_g^2 \mathbf{K} + \sigma_e^2 \mathbf{I}$, and multivariate normal distribution for $\mathbf{y}$ is

$$f(\mathbf{y}) = \frac{1}{(2\pi)^{n/2}} \exp\left[-\frac{1}{2}(\mathbf{y} - \mathbf{X}\mathbf{b})'\mathbf{V}^{-1}(\mathbf{y} - \mathbf{X}\mathbf{b})\right] \tag{5}$$

where, $\mathbf{X} = \{1\ \mathbf{X}\}$, and $\mathbf{b}' = \{\mu\ \mathbf{b}'\}$.

The unknown parameters in model (4) can be estimated by using the maximum likelihood estimation method, as well as various methods optimized for GWAS [20–22]. Once the parameters are estimated, the association test can be performed using F-test [23]. The association test based on MLM can be performed using the MLM function of TASSEL software:

```
run_pipeline.pl -fork1 -importGuess genotype.vcf -fork2 -importGuess phenotype.txt -fork3 -importGuess kinship.txt -combine4 -input1 -input2 -intersect -combine5 -input3 -input4 -mlm -export MLMResult
```

For the above soybean example, only one marker with an R^2 of 6.5% was detected to be significantly associated with soybean plant height (*see* Table 1). The p-values of the MLM method are shown in Fig. 1c, and the Q-Q plot (*see* Fig. 2) indicated that the false positive rate is largely reduced in comparison with the simple linear model. The λ_{GC} for MLM is estimated as 0.80, which is even smaller than the expected value 1.0, indicating false negatives existed.

Generally, the mixed linear model has lower false positive rate than the general linear model method, and the mixed model method has also become the preferable method for GWAS in plants. The key of the MLM method is to specify the appropriate variance-covariance structure of random minor gene effects [24]. In practice, the genetic relationship matrix calculated from genome-wide SNP markers is used as variance-covariance structure. In the above example of soybean, the detection power of the MLM GWAS approach was very low with only one QTL detected. Therefore, the mixed model method may have high false negative rate and how to obtain an appropriate variance-covariance structure is to be further investigated.

5 Multi-Locus Multi-Allele Model (RTM-GWAS) Approach

The basic genetic assumption of RTM-GWAS is that a quantitative trait is conferred by a group of genes with varied multiple alleles and varied allele effects, a part of which can be detected individually while another part may not be identified individually depending on genotyping and phenotyping precision. The RTM-GWAS method [8] provides a novel approach to identifying the QTL system with multiple alleles in natural or germplasm populations. Two main innovations were proposed in the RTM-GWAS method. The first one is to construct multi-allelic SNPLDB markers based on high-density SNP markers for QTL detection. SNPLDB marker can have multiple haplotypes/alleles at a locus, so it fits the abundant multiple allelic variation in plant natural or germplasm population. In addition, the genetic similarity coefficient matrix calculated from SNPLDB is used to correct the population structure bias based on large size of the population. The second one is to use a restricted two-stage multi-locus multi-allele model to detect whole genome QTL. As multiple QTLs are fitted simultaneously in one multi-locus model, unbiased estimation of genetic effects can be obtained in comparison to the single-locus model. GWAS usually involves a large number of markers, it can be time-consuming to directly solve the multi-locus model. RTM-GWAS adopts an efficient two-stage analysis strategy to reduce the computational cost. In the first stage, a large number of markers unrelated to the quantitative trait are eliminated, and in the second stage, the multi-locus model is fitted based on the reduced markers.

5.1 Constructing SNPLDB Marker for Multiple Alleles Detection

The tight and loose linkage between SNPs leads to a block structure of genome, and the haplotype sequences within a block may transmit without recombination to the offspring. The combinations of SNPs within the same block can form multiple haplotypes. Thus, a genomic block can be considered as a genetic locus, while its haplotypes can be considered as its alleles. Linkage disequilibrium (LD) is a standard measurement of the recombination history in natural population, so genome-wide blocks can be defined

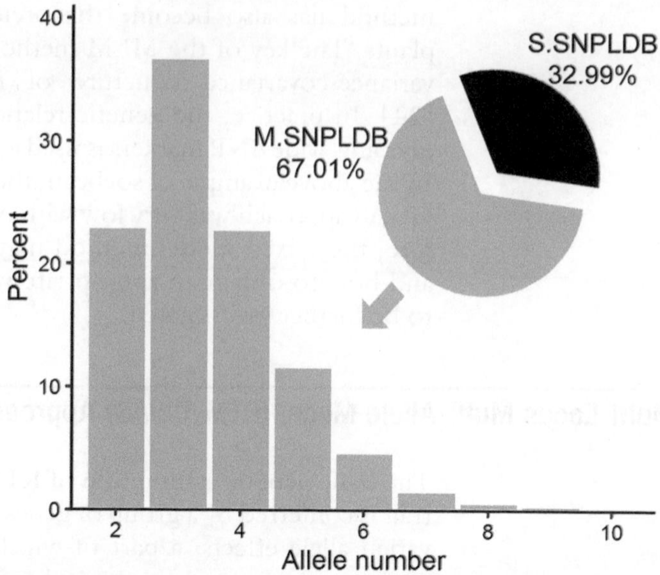

Fig. 3 Distribution of allele number of SNPLDBs in the soybean germplasm population. M.SNPLDB means SNPLDB with multiple SNPs, while S.SNPLDB means SNPLDB with a single SNP

according to the degree of LD among SNPs. In the RTM-GWAS method, the block partitioning approach based on the confidence interval of LD is used to find genomic blocks [25]. Then, the SNPs within a genomic block can be grouped into an SNPLDB marker with multiple haplotypes as its alleles. The SNPLDB genotype for each individual is determined by its corresponding SNP haplotypes.

Using the RTM-GWAS software (https://gitee.com/njau-sri/rtm-gwas), a total of 11,771 SNPLDBs was constructed based on 68,050 SNPs in the soybean germplasm population consisting of 446 accessions in the above example. The number of alleles ranges from 2 to 10, and 6092 SNPLDBs have more than two alleles. There are 5679 bi-allelic SNPLDBs, in which 3883 SNPLDBs are actually SNPs and 1796 SNPLDBs contain multiple SNPs but with only two alleles (*see* Fig. 3). SNPLDB provides more information about multi-allelic variation than SNP and can theoretically fit QTLs with different allele numbers. Therefore, GWAS based on SNPLDB is more reasonable than SNP in plant germplasm population.

5.2 Detecting QTL-Allele System Using Efficient Multi-Locus Model

GWAS usually involves hundreds of thousands or even millions of markers, but actually most of them are not related to the target trait. To effectively reduce the model space of the multi-locus model, RTM-GWAS adopts an efficient restricted two-stage analysis strategy. The GLM method is performed in the first stage for preliminary selection of candidate markers using a normal significance level (e.g., 0.05), thus the markers unrelated to the target trait are eliminated. In the second stage, QTL detection using multi-locus model is performed based on candidate markers selected from the first stage. The multi-locus model is constructed based on the GLM model,

$$\mathbf{y} = \mathbf{1}\mu + \mathbf{W}\mathbf{a} + \sum_k \mathbf{X}_k \mathbf{b}_k + \mathbf{e} \quad (6)$$

The stepwise regression featured with forward selection and backward elimination is implemented in RTM-GWAS to solve model (6). As multiple QTLs are fitted in one model, the phenotypic variation explained by detected QTLs is restricted within the trait heritability value. A normal significance level of 0.01 or 0.05 is suggested for QTL detection in RTM-GWAS as the built-in experiment-wise error control of multi-locus model. This is different from the single-locus model which often requires an additional correction for all-marker joint testing based on single-locus model testing for individual markers. As all QTL are fitted simultaneously for joint statistical hypothesis test, the error rate can be controlled by using the normal significance level, and multiple test corrections are not needed.

Using the RTM-GWAS method, 3729 out of 11,771 SNPLDBs were preselected in the first stage using a threshold of 0.05, and 50 QTLs were detected in the second stage. The phenotypic variation explained (R^2) by each marker ranged from 0.2% to 23.6% with a total of 92.2% (*see* Table 1). The Q-Q plot (*see* Fig. 2) showed that the false rate is well controlled in RTM-GWAS with λ_{GC} estimated as 1.38. RTM-GWAS detects the genome-wide QTL and multiple alleles through the multi-locus multi-allele model and uses the trait heritability value as the upper limit of the total phenotypic contribution of the detected QTLs. Thus, false positives were well-controlled. The two-stage strategy reduced the computational complexity of the multi-locus model, and the GSC matrix based on SNPLDB markers further controlled false rate caused by population structure bias. Under the multi-locus model, the model test included all the markers, no requirement for adjusting the significance level based on $\alpha = 0.05$ or 0.01, the same level as the corrected one in single-locus model methods. Finally, all the innovations made more reasonable QTLs detected (*see* Table 1).

The effects of multiple alleles can be also estimated from the RTM-GWAS software (*see* Fig. 4), and all the QTLs and their multiple alleles along with the estimated allele effects can be organized into a QTL-allele matrix as a compact form of the population genetic constitution. In the above soybean example, there are a total of 167 alleles on the 50 QTLs, with a range of 2~7 alleles for a single QTL.

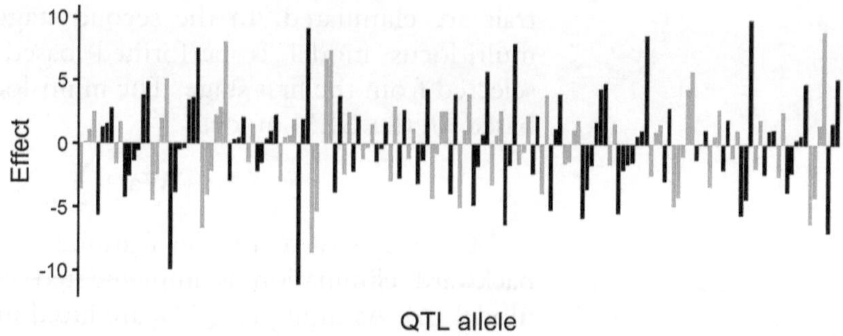

Fig. 4 Allele effect of plant height QTLs detected in the soybean germplasm population. One vertical bar represents one allele, and different QTLs are distinguished alternately in black and grey color. QTLs are arranged by physical position

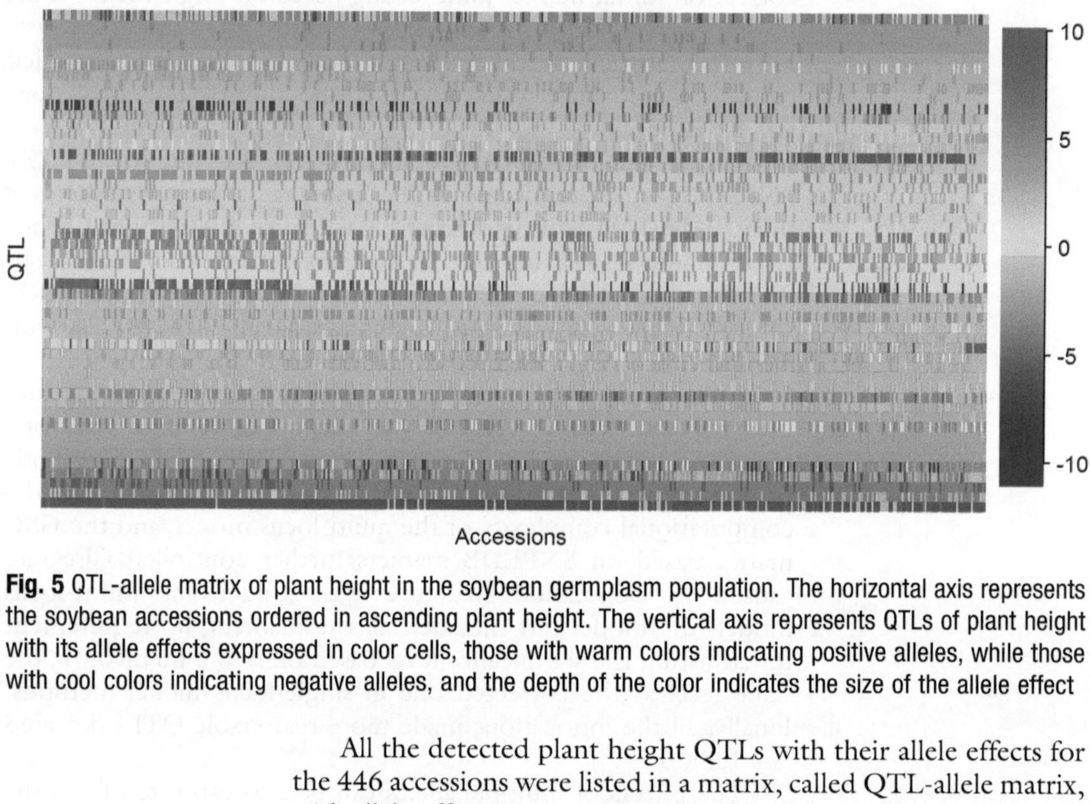

Fig. 5 QTL-allele matrix of plant height in the soybean germplasm population. The horizontal axis represents the soybean accessions ordered in ascending plant height. The vertical axis represents QTLs of plant height with its allele effects expressed in color cells, those with warm colors indicating positive alleles, while those with cool colors indicating negative alleles, and the depth of the color indicates the size of the allele effect

All the detected plant height QTLs with their allele effects for the 446 accessions were listed in a matrix, called QTL-allele matrix, with allele effect expressed in color darkness as indicated in Fig. 5. The QTL-allele matrix provides the complete QTL-allele information of a quantitative trait in the population, and thus can be further used for gene discovery and plant breeding by design, as well as evolutionary studies [12]. For example, to study the genes of soybean plant height, large effect QTLs can be selected from the detected QTLs, the QTL of largest effect is on Chromosome 19, which is also detected by the MLM method.

5.3 Potential Applications of RTM-GWAS

The results of the 50 QTLs with 167 alleles of the above soybean plant height indicates the powerfulness of the RTM-GWAS. In fact, more QTLs and alleles were reported for other traits and populations. Some people doubted about too loose of the significance threshold. As indicated above, the significance threshold in RTM-GWAS and the other three approaches were the same as commonly used 0.05 or 0.01, the difference in fact lies in that under single-locus model, Bonferroni correction should be used for multiple tests of the complete marker set, while no correction is required for RTM-GWAS because under multi-locus model, the model test covered the complete marker set. Especially, in RTM-GWAS, the identified QTLs are controlled with their total contribution to phenotypic variance within the trait heritability value to avoid the false positives. In addition, as the quantitative traits are controlled by numerous QTLs/genes, the genetic operation of plant breeding involves in fact a great number of QTLs with their alleles rather than a few QTLs/genes in some qualitative traits. Therefore, for a relatively thorough detection of QTL/gene system with their alleles, the germplasm population should be studied. The RTM-GWAS can meet the requirement.

The major application of RTM-GWAS is to detect the QTL-allele system of a plant germplasm population. The RTM-GWAS method has been applied to soybean for QTL-allele detection [26–34], and has also been applied to cotton [35, 36], rice [37], and wheat [38].

The RTM-GWAS is demonstrated to be also effective in bi-parent population [39–41] and multi-parent population [42–46] because these bi-parent and multi-parent populations usually are similar to a random-mating population if without selection acted, therefore, less possibility of population structure bias problem.

The QTL allele matrix established by RTM-GWAS can be used for studies on the dynamic changes of QTLs and their multiple alleles, such as genetic differentiation and population-specific allele emergence and exclusion [47–50]. For example, based on the 52 and 59 QTLs with 241 and 246 alleles detected by RTM-GWAS for days to flowering and maturity in soybean [48], QTL-allele matrix was established and subsequently separated into geographic submatrices. Comparisons for allele changes among them revealed that the genetic adaptation from the origin to geographic subpopulations was characterized mainly by emergence of new alleles and new loci but little allele exclusion, while the forming of extreme early maturity groups was mainly due to allele exclusion but not new allele emergence.

The QTL-allele matrix provides the required genetic information for parental cross design. Based on the QTL-allele matrix, the phenotype of progeny populations derived from parental crosses can be predicted for optimal cross selection. He et al. [8] proposed a method for prediction of the breeding potentials of parental crosses and was implemented in the *Cross* software (https://gitee.com/njau-sri/cross). For example, based on 73 QTLs with 273 alleles detected by RTM-GWAS, the prediction of recombination potentials in the seed protein contents (SPC) in Northeast China soybean germplasms indicated that the mean of SPC in overall crosses was 43.29% (+2.52% improvement) and the maximum was 50.00% (+9.23% improvement) in the SPC, and the maximum transgressive potential was 3.93% [51].

Compared with other GWAS methods, RTM-GWAS usually detect more QTLs, and the total contribution of QTL to phenotypic variation is also closer to the trait heritability. Thus, RTM-GWAS provides possible way to study the complete genetic system of quantitative traits. Therefore, in addition to examining the entire QTL-allele system, candidate gene annotation for understanding the gene network can be also inferred from the relatively thorough identification of QTLs.

6 Example: QTL-Allele System of Seed Protein Content in Northeast China Soybeans

This example is based on the experiment data in Feng et al. [51]. The study aimed at firstly exploring the seed protein content (SPC) variation and QTL-allele system of Northeast China soybeans, and then exploring the QTL-allele recombination potential for optimal cross design. The steps of the RTM-GWAS procedure were described below.

6.1 Plant Materials and Field Experiment

The Northeast China soybean germplasm population (NECSGP) consists of 361 representative accessions and were planted with a blocks-in-replication design for 2 years at Tieling, Northeast China. According to soybean maturity group, these accessions were grouped into six blocks and four replications were implemented each year. At the maturity stage, the plants in each plot were threshed and dried after harvest, and then the seed protein content (SPC) was measured by using the FOSS NearInfared grain analyzer Infratec 1241. The SPC of the NECSGP ranged from 36.60% to 46.07%, with an average of 40.77% (*see* Fig. 6a). The heritability of SPC over two environments was estimated as 0.83.

6.2 SNP Genotyping

All accessions were sequenced with RAD-seq technology at BGI Tech, Shenzhen, China. Sequence reads were aligned against the genome Williams 82 (Wm82.a1.v1.1) [17]. SNP calling was performed at population level, and 82,966 high-quality SNPs were obtained after quality control and imputation of missing genotype calls.

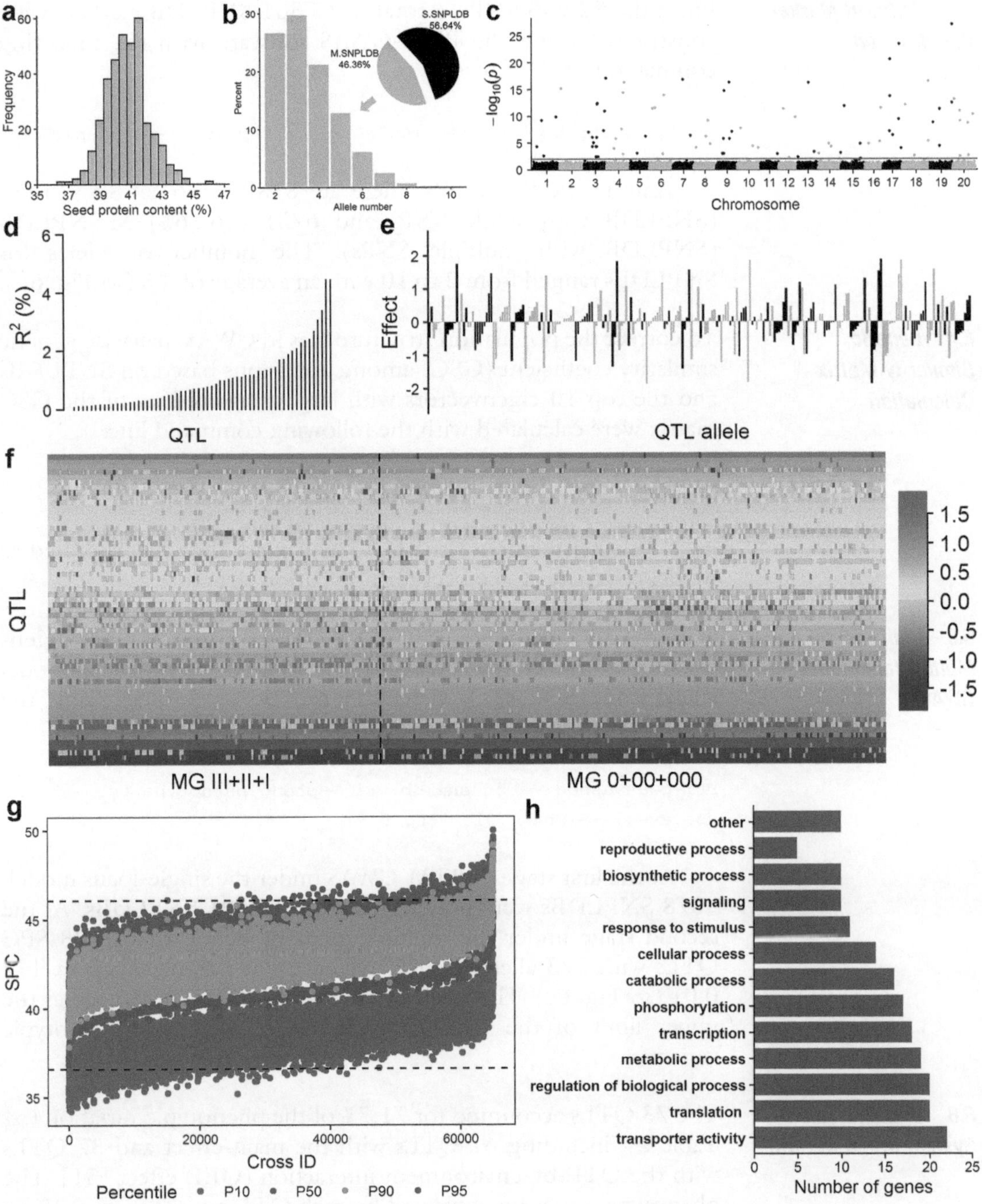

Fig. 6 Genome-wide association study of seed protein content in Northeast China soybeans. (**a**) Histogram of seed protein content (SPC); (**b**) Distribution of allele number of SNPLDB markers; (**c**) Manhattan plot; (**d**) The phenotypic variation explained by the detected SPC QTLs; (**e**) Allele effects of the detected SPC QTLs; (**f**) QTL-allele matrix of SPC. MG III + II + I mean soybean accessions from maturity group III, II or I, while MG 0 + 00 + 000 mean soybean accessions from maturity group 0, 00 or 000; (**g**) Predicted SPC in progeny populations derived from all possible crosses among the 361 accessions; (**h**) Gene Ontology (GO) annotations of the candidate genes for SPC QTLs

6.3 SNPLDB Marker Construction

From the 82,966 SNPs, a total of 15,501 SNPLDB markers were constructed using the RTM-GWAS software with the following command line:

```
rtm-gwas-snpldb --maxlen 100000 --vcf snp.vcf --out snpldb
```

Among the SNPLDBs, there are 8780 (56.64%) S.SNPLDB (SNPLDB with single SNP) and 6721 (46.36%) M.SNPLDB (SNPLDB with multiple SNPs). The number of alleles for SNPLDBs ranged from 2 to 10 with an average of 3.5 (*see* Fig. 6b).

6.4 Genetic Similarity Matrix Calculation

To correct the population structure bias in GWAS, pairwise genetic similarity coefficient (GSC) among accessions based on SNPLDBs and the top 10 eigenvectors with largest eigenvalues of the GSC matrix were calculated with the following command line:

```
rtm-gwas-gsc --vcf snpldb.vcf --out gsc
```

The eigenvectors of the GSC matrix are then incorporated as model covariates to correct for population structure bias.

6.5 Multi-Locus Multi-Allele Model GWAS

With the SPC phenotype data, SNPLDB genotype data and eigenvectors of the GSC matrix, the multi-locus multi-allele model analysis for SPC QTL detection can be performed using the RTM-GWAS software with the following command line:

```
rtm-gwas-assoc --vcf snpldb.vcf --pheno phenotype.txt --covar
gsc.evec --alpha 0.01 --rsq 0.83
```

At the first stage of RTM-GWAS under the single-locus model, 9078 SNPLDBs were preselected from 15,501 SNPLDBs. At the second stage under the multiple-locus model, a total of 73 SPC QTLs with 273 alleles were detected using a significance level of 0.01 (*see* Fig. 6c). The trait heritability ($h^2 = 0.83$) was used as the upper limit of the total contribution of QTLs to phenotypic variation.

6.6 SPC QTL-Allele System of the NECSGP

The 73 QTLs accounted for 71.7% of the phenotypic variation (*see* Table 2), including 61 QTLs with the main effect and 37 QTLs with the QTL-by-environment interaction (QEI) effect [51]. The phenotypic variation explained by each QTL ranged from 0.1% to 4.43% (*see* Fig. 6d), and the 61 main effect QTLs were further classified into 25 large contribution QTLs ($R^2 \geq 1\%$) and 36 small contribution QTLs ($R^2 < 1\%$).

The main effects (additive effect) of QTLs ranged from −1.89 to 1.88 (*see* Fig. 6e). The 61 SPC main-effect QTLs and their allele effects for each of the 361 accessions were organized as a QTL-allele matrix (*see* Fig. 6f), which is a compact form of the genetic constitution of the NECSGP.

Table 2
Twenty-five large contribution ($R^2 \geq 1\%$) QTLs detected for seed protein content in Northeast China soybeans (36 small contribution QTLs with $R^2 < 1\%$ not listed)

SNPLDB	Model −lg*P*	Main effect QTL		QTL × Env QTL	
		−lg*P*	R^2 (%)	−lg*P*	R^2 (%)
2_3224504:3386591	15.17	41.01	2.24		
3_27140827:27314739	12.22	23.15	1.36	9.18	0.61
3_28211045	12.33	26.81	1.29		
3_43342844	10.88	21.46	1.02		
3_45585122:45784092	17.08	55.56	3.03		
4_34306302	14.34	34.21	1.67		
4_36804595:36956694	16.27	44.42	2.32		
6_6871018:6911128	11.37	20.11	1.05	5.41	0.30
6_12122955	11.48	24.00	1.15		
6_31465819:31481467	13.90	33.44	1.69		
7_21000729	16.82	41.48	2.05		
9_17559197:17575587	14.66	35.57	1.85		
9_28009070:28068638	16.28	50.34	2.64		
12_10419180:10443940	11.33	20.84	1.09	2.00	0.12
13_22819454	16.17	40.01	1.97	2.81	0.11
15_7483684:7680382	11.82	23.17	1.29	5.46	0.35
16_33388199:33538361	11.42	22.41	1.17		
17_15717972:15782305	20.59	63.37	3.36		
17_35769952:35842671	23.46	80.02	4.42		
18_7317014:7504444	12.02	23.54	1.42		
18_58197364:58341700	16.45	44.25	2.37	6.33	0.37
19_21347881:21386285	11.33	23.02	1.32	2.36	0.20
19_47660663:47715521	27.19	82.27	4.43	8.09	0.43
20_2795994	13.07	24.11	1.15	5.75	0.24
20_14150016:14150305	10.94	21.74	1.09		
LC-QTL		25	48.42		
SC-QTL		36	14.30	37	8.98
Total	73	61	62.72	37	8.98

Note: SNPLDB was named with a prefix of chromosome number and a suffix of genomic interval (2_3224504:3386591) or position (3_28211045). −lg*P* means the *p*-value on the −log10 scale of statistical test. Model −lg*P* means the *p*-value of joint statistical test for the main effect plus QEI effect model expressed in −lg*P*. LC-QTL: large contribution QTL ($R^2 \geq 1\%$). SC-QTL: small contribution QTL ($R^2 < 1\%$)

6.7 SPC QTL-Allele Changes in the Evolution from Late to Early Maturity Groups

The QTL-allele matrix from RTM-GWAS also provides a tool for population genetic differentiation and evolutionary analysis. For example, from the SPC QTL-allele matrix (*see* Fig. 6f), genetic dynamic changes can be inferred. The accessions in the NECSGP covered six maturity groups (MGs), including three late MGs (I, II, III) and three early MGs (0, 00, 000). The number of alleles of SPC QTLs changed during the evolution from late to early MGs. Overall, 265 alleles in early MGs were inherited from the late MGs, while six alleles (two negatives and four positives) emerged, and two alleles of positive effect excluded (*see* Table 3). Different patterns of allele changes were also observed, the number of emerged alleles were much less than that of excluded alleles from the late MGs to the early MG 00 and 000.

6.8 Prediction of Recombination Potential for Optimal Cross Design

The recombination potential of progeny populations derived from parental combinations can be predicted based on the QTL-allele matrix, and then optimal parental crosses can be selected [8]. For example, all possible 64,980 single crosses among the 361 accessions in the NECSGP was generated in silico based on the SPC QTL-allele matrix, and 2000 homozygous progenies were simulated for each cross. The 95th percentile of SPC was used to predict the recombination potential of each cross. A total of 1803 crosses showed higher SPC than the maximum SPC (46.07%) in the NECSGP (*see* Fig. 6g). The five best crosses were listed in Table 4. The cross between L54 (MG 000) and L5 (MG 0) exhibited the highest 95th percentile of the predicted SPC (50.00%), with an 8.53% increase in SPC compared with the maximum SPC in the NECSGP.

Table 3
The QTL-allele changes of seed protein content among maturity groups

	Total allele		Inherent allele		Emerged allele		Excluded allele	
MG	Allele	QTL	Allele	QTL	Allele	QTL	Allele	QTL
I + II + III	267 (142,125)	73						
I + II + III vs 0	268 (143,125)	73	262 (141,121)	73	6 (2,4)	6	5 (1,4)	5
I + II + III vs 00	222 (120,102)	73	217 (118,99)	73	5 (2,3)	5	50 (24,26)	35
I + II + III vs 000	180 (95,85)	73	179 (95,84)	73	1 (0,1)	1	88(47,41)	46
I + II + III vs 0 + 00 + 000	271 (144,127)	73	265 (142,123)	73	6 (2,4)	6	2 (0,2)	2

Note: The number outside parentheses is the total of alleles; the numbers in parentheses are the number of negative effect alleles and positive effect alleles, respectively; Inherent allele means alleles passed from the compared MG; Emerged allele means the alleles in MG 0 and 00 new to those in the partner of MG I + II + III; Excluded allele means the alleles excluded in the compared MG

Table 4
Optimal crosses for high seed protein content in the Northeast China soybean germplasm population

P_1	P_2	Y_1	Y_2	Mean	SD	P90	P95	P99
L54	L5	46.07	44.71	45.37	2.86	49.16	50.00	51.54
L381	L54	43.93	46.07	45.04	2.82	48.71	49.63	51.14
L326	L54	42.34	46.07	44.15	3.07	48.13	49.15	51.12
L177	L54	44.34	46.07	44.66	2.76	48.34	49.31	50.75
L37	L54	41.70	46.07	43.88	3.27	48.24	49.32	50.73

Note: P_1 and P_2 are the parents of a simple cross. Υ_1 and Υ_2 are the average SPC of parents P_1 and P_2, respectively. MEAN and SD indicate the mean and standard deviation of homozygous progeny population; P90, P95, and P99 indicate 90th, 95th, and 99th percentile of the homozygous progeny value in the cross

6.9 Annotation of Candidate Gene System of SPC

The QTL-allele matrix from RTM-GWAS can be used for candidate gene discovery. In this example, a total of 120 candidate genes on 34 SPC QTLs were annotated and functionally classified into 13 Gene Ontology biological process categories, including transporter activity, translation, regulation of the biological process, metabolic process, transcription, phosphorylation, catabolic process, cellular process, response to stimulus, signaling, biosynthetic process, reproductive process, and others (*see* Fig. 6h).

Acknowledgments

This work was financially supported through the grant from the National Key Research and Development Program of China (2021YFF1001204).

References

1. Wang J, Crossa J, Gai J (2020) Quantitative genetic studies with applications in plant breeding in the omics era. Crop J 8:683–687. https://doi.org/10.1016/j.cj.2020.09.001
2. Price AL, Zaitlen NA, Reich D, Patterson N (2010) New approaches to population stratification in genome-wide association studies. Nat Rev Genet 11:459–463. https://doi.org/10.1038/nrg2813
3. Pritchard JK, Stephens M, Rosenberg NA, Donnelly P (2000) Association mapping in structured populations. Am J Hum Genet 67:170–181. https://doi.org/10.1086/302959
4. Price AL, Patterson NJ, Plenge RM, Weinblatt ME, Shadick NA, Reich D (2006) Principal components analysis corrects for stratification in genome-wide association studies. Nat Genet 38:904–909. https://doi.org/10.1038/ng1847
5. Yu J, Pressoir G, Briggs WH, Vroh Bi I, Yamasaki M, Doebley JF et al (2005) A unified mixed-model method for association mapping that accounts for multiple levels of relatedness. Nat Genet 38:203–208. https://doi.org/10.1038/ng1702
6. Pritchard JK, Stephens M, Donnelly P (2000) Inference of population structure using multilocus genotype data. Genetics 155:945–959. https://doi.org/10.1093/genetics/155.2.945
7. Sul JH, Martin LS, Eskin E (2018) Population structure in genetic studies: confounding factors and mixed models. PLoS Genet 14:e1007309. https://doi.org/10.1371/journal.pgen.1007309
8. He J, Meng S, Zhao T, Xing G, Yang S, Li Y et al (2017) An innovative procedure of genome-wide association analysis fits studies

on germplasm population and plant breeding. Theor Appl Genet 130:2327–2343. https://doi.org/10.1007/s00122-017-2962-9

9. Segura V, Vilhjálmsson BJ, Platt A, Korte A, Seren Ü, Long Q et al (2012) An efficient multi-locus mixed-model approach for genome-wide association studies in structured populations. Nat Genet 44:825–830. https://doi.org/10.1038/ng.2314
10. Rakitsch B, Lippert C, Stegle O, Borgwardt K (2013) A Lasso multi-marker mixed model for association mapping with population structure correction. Bioinformatics 29:206–214. https://doi.org/10.1093/bioinformatics/bts669
11. Gai J, Chen L, Zhang Y, Zhao T, Xing G, Xing H (2012) Genome-wide genetic dissection of germplasm resources and implications for breeding by design in soybean. Breed Sci 61: 495–510. https://doi.org/10.1270/jsbbs.61.495
12. He J, Gai J (2020) QTL-allele matrix detected from RTM-GWAS is a powerful tool for studies in genetics, evolution, and breeding by design of crops. J Integr Agric 19:1407–1410. https://doi.org/10.1016/S2095-3119(20)63199
13. Weir BS (2008) Linkage disequilibrium and association mapping. Annu Rev Genomics Hum Genet 9:129–142. https://doi.org/10.1146/annurev.genom.9.081307.164347
14. Bradbury PJ, Zhang Z, Kroon DE, Casstevens TM, Ramdoss Y, Buckler ES (2007) TASSEL: software for association mapping of complex traits in diverse samples. Bioinformatics 23: 2633–2635. https://doi.org/10.1093/bioinformatics/btm308
15. Purcell S, Neale B, Todd-Brown K, Thomas L, Ferreira MAR, Bender D et al (2007) PLINK: a tool set for whole-genome association and population-based linkage analyses. Am J Hum Genet 81:559–575. https://doi.org/10.1086/519795
16. Chang CC, Chow CC, Tellier LC, Vattikuti S, Purcell SM, Lee JJ (2015) Second-generation PLINK: rising to the challenge of larger and richer datasets. GigaSci 4:7. https://doi.org/10.1186/s13742-015-0047-8
17. Schmutz J, Cannon SB, Schlueter J, Ma J, Mitros T, Nelson W et al (2010) Genome sequence of the paleopolyploid soybean. Nature 463(7278):178–183. https://doi.org/10.1038/nature08670
18. Saxena R, Voight BF, Lyssenko V, Burtt NP, de Bakker PIW, Chen H et al (2007) Genome-wide association analysis identifies loci for type 2 diabetes and triglyceride levels. Science 316(5829):1331–1336. https://doi.org/10.1126/science.1142358
19. Devlin B, Roeder K (1999) Genomic control for association studies. Biometrics 55:997–1004. https://doi.org/10.1111/j.0006-341X.1999.00997.x
20. Lippert C, Listgarten J, Liu Y, Kadie CM, Davidson RI, Heckerman D (2011) FaST linear mixed models for genome-wide association studies. Nat Methods 8:833–835. https://doi.org/10.1038/nmeth.1681
21. Zhou X, Stephens M (2012) Genome-wide efficient mixed-model analysis for association studies. Nat Genet 44:821–824. https://doi.org/10.1038/ng.2310
22. Jiang L, Zheng Z, Qi T, Kemper KE, Wray NR, Visscher PM et al (2019) A resource-efficient tool for mixed model association analysis of large-scale data. Nat Genet 51:1749–1755. https://doi.org/10.1038/s41588-019-0530-8
23. Kang HM, Zaitlen NA, Wade CM, Kirby A, Heckerman D, Daly MJ et al (2008) Efficient control of population structure in model organism association mapping. Genetics 178: 1709–1723. https://doi.org/10.1534/genetics.107.080101
24. Yang J, Zaitlen NA, Goddard ME, Visscher PM, Price AL (2014) Advantages and pitfalls in the application of mixed-model association methods. Nat Genet 46:100–106. https://doi.org/10.1038/ng.2876
25. Gabriel SB, Schaffner SF, Nguyen H, Moore JM, Roy J, Blumenstiel B et al (2002) The structure of haplotype blocks in the human genome. Science 296(5576):2225–2229. https://doi.org/10.1126/science.1069424
26. Zhang Y, He J, Wang Y, Xing G, Zhao J, Li Y et al (2015) Establishment of a 100-seed weight quantitative trait locus–allele matrix of the germplasm population for optimal recombination design in soybean breeding programmes. J Exp Bot 66:6311–6325. https://doi.org/10.1093/jxb/erv342
27. Meng S, He J, Zhao T, Xing G, Li Y, Yang S et al (2016) Detecting the QTL-allele system of seed isoflavone content in Chinese soybean landrace population for optimal cross design and gene system exploration. Theor Appl Genet 129:1557–1576. https://doi.org/10.1007/s00122-016-2724-0
28. Zhang Y, He J, Wang H, Meng S, Xing G, Li Y et al (2018) Detecting the QTL-allele system of seed oil traits using multi-locus genome-wide association analysis for population characterization and optimal cross prediction in

soybean. Front Plant Sci 9:1793. https://doi.org/10.3389/fpls.2018.01793

29. Zhang Y, He J, Meng S, Liu M, Xing G, Li Y et al (2018) Identifying QTL-allele system of seed protein content in Chinese soybean landraces for population differentiation studies and optimal cross predictions. Euphytica 214:157. https://doi.org/10.1007/s10681-018-2235-y
30. Li S, Xu H, Yang J, Zhao T (2019) Dissecting the genetic architecture of seed protein and oil content in soybean from the Yangtze and Huaihe river valleys using multi-locus genome-wide association studies. Int J Mol Sci 20:3041. https://doi.org/10.3390/ijms20123041
31. Fu M, Wang Y, Ren H, Du W, Yang X, Wang D et al (2020) Exploring the QTL–allele constitution of main stem node number and its differentiation among maturity groups in a Northeast China soybean population. Crop Sci 60:1223–1238. https://doi.org/10.1002/csc2.20024
32. Wang W, Zhou B, He J, Zhao J, Liu C, Chen X et al (2020) Comprehensive identification of drought tolerance QTL-allele and candidate gene systems in Chinese cultivated soybean population. Int J Mol Sci 21:4830. https://doi.org/10.3390/ijms21144830
33. Wang L, Liu F, Hao X, Wang W, Xing G, Luo J et al (2021) Identification of the QTL-allele system underlying two high-throughput physiological traits in the Chinese soybean germplasm population. Front Genet 12:600444. https://doi.org/10.3389/fgene.2021.600444
34. Fahim AM, Liu F, He J, Wang W, Xing G, Gai J (2021) Evolutionary QTL-allele changes in main stem node number among geographic and seasonal subpopulations of Chinese cultivated soybeans. Mol Genet Genomics 296:313–330. https://doi.org/10.1007/s00438-020-01748-9
35. Su J, Wang C, Ma Q, Zhang A, Shi C, Liu J et al (2020) An RTM-GWAS procedure reveals the QTL alleles and candidate genes for three yield-related traits in upland cotton. BMC Plant Biol 20:416. https://doi.org/10.1186/s12870-020-02613-y
36. Wang C, Ma Q, Xie X, Zhang X, Yang D, Su J et al (2022) Identification of favorable haplotypes/alleles and candidate genes for three plant architecture-related traits via a restricted two-stage multilocus genome-wide association study in upland cotton. Ind Crop Prod 177:114458. https://doi.org/10.1016/j.indcrop.2021.114458
37. Kong W, Zhang C, Zhang S, Qiang Y, Zhang Y, Zhong H et al (2021) Uncovering the novel QTLs and candidate genes of salt tolerance in rice with linkage mapping, RTM-GWAS, and RNA-seq. Rice 14:93. https://doi.org/10.1186/s12284-021-00535-3
38. Chidzanga C, Fleury D, Baumann U, Mullan D, Watanabe S, Kalambettu P et al (2021) Development of an Australian bread wheat nested association mapping population, a new genetic diversity resource for breeding under dry and hot climates. Int J Mol Sci 22:4348. https://doi.org/10.3390/ijms22094348
39. Pan L, He J, Zhao T, Xing G, Wang Y, Yu D et al (2018) Efficient QTL detection of flowering date in a soybean RIL population using the novel restricted two-stage multi-locus GWAS procedure. Theor Appl Genet 131:2581–2599. https://doi.org/10.1007/s00122-018-3174-7
40. Liu F, He J, Wang W, Xing G, Gai J (2020) Bi-phenotypic trait may be conferred by multiple alleles in a germplasm population. Front Genet 11:559. https://doi.org/10.3389/fgene.2020.00559
41. Fahim AM, Pan L, Li C, He J, Xing G, Wang W et al (2021) QTL-allele system of main stem node number in recombinant inbred lines of soybean (*Glycine max*) using association versus linkage mapping. Plant Breed 140:870–883. https://doi.org/10.1111/pbr.12956
42. Li S, Cao Y, He J, Zhao T, Gai J (2017) Detecting the QTL-allele system conferring flowering date in a nested association mapping population of soybean using a novel procedure. Theor Appl Genet 130:2297–2314. https://doi.org/10.1007/s00122-017-2960-y
43. Khan MA, Tong F, Wang W, He J, Zhao T, Gai J (2018) Analysis of QTL–allele system conferring drought tolerance at seedling stage in a nested association mapping population of soybean [*Glycine max* (L.) Merr.] using a novel GWAS procedure. Planta 248:947–962. https://doi.org/10.1007/s00425-019-03143-0
44. Khan MA, Tong F, Wang W, He J, Zhao T, Gai J (2019) Using the RTM-GWAS procedure to detect the drought tolerance QTL-allele system at the seedling stage under sand culture in a half-sib population of soybean [*Glycine max* (L.) Merr.]. Can J Plant Sci 99:801–814. https://doi.org/10.1139/cjps-2018-0309
45. Khan MA, Tong F, Wang W, He J, Zhao T, Gai J (2020) Molecular characterization of QTL-allele system for drought tolerance at seedling stage and optimal genotype design

using multi-locus multi-allele genome-wide association analysis in a half-sib population of soybean (*Glycine max* (L.) Merr.). Plant Genet Res Crop Evol 18:295–306. https://doi.org/10.1017/S1479262120000313
46. Ali MJ, Xing G, He J, Zhao T, Gai J (2020) Detecting the QTL-allele system controlling seed-flooding tolerance in a nested association mapping population of soybean. Crop J 8:781–792. https://doi.org/10.1016/j.cj.2020.06.008
47. Liu X, He J, Wang Y, Xing G, Li Y, Yang S et al (2020) Geographic differentiation and phylogeographic relationships among world soybean populations. Crop J 8:260–272. https://doi.org/10.1016/j.cj.2019.09.010
48. Liu F, He J, Wang W, Xing G, Zhao J, Li Y et al (2021) Genetic dynamics of flowering date evolved from later to earlier in annual wild and cultivated soybean in China. Crop Sci 61:2336–2354. https://doi.org/10.1002/csc2.20462
49. Liu X, Li C, Cao J, Zhang X, Wang C, He J et al (2021) Growth period QTL-allele constitution of global soybeans and its differential evolution changes in geographic adaptation versus maturity group extension. Plant J 108:1624–1643. https://doi.org/10.1111/tpj.15531
50. Fu M, Wang Y, Ren H, Du W, Wang D, Bao R et al (2020) Genetic dynamics of earlier maturity group emergence in south-to-north extension of Northeast China soybeans. Theor Appl Genet 133:1839–1857. https://doi.org/10.1007/s00122-020-03558-4
51. Feng W, Fu L, Fu M, Sang Z, Wang Y, Wang L et al (2022) Transgressive potential prediction and optimal cross design of seed protein content in the northeast China soybean population based on full exploration of the QTL-allele system. Front Plant Sci 13:896549. https://doi.org/10.3389/fpls.2022.896549

Chapter 10

Transcriptomic Approach for Global Distribution of SNP/Indel and Plant Genotyping

Claudia Muñoz-Espinoza, Marco Meneses, and Patricio Hinrichsen

Abstract

Single Nucleotide Polymorphisms (SNPs) are the most common structural variants found in any genome. They have been used for different genetic studies, from the understanding of genetic structure of populations to the development of breeding selection markers. In this chapter we present the use of transcriptomic data obtained from contrasting phenotypes for a target trait, in searching of SNPs and insertions/deletions (InDels). This approach has the advantage that the identified markers are in or close to differentially expressed genes, and so they have higher chances to tag the genes underlying the phenotypic expression of a particular trait.

Key words RNA-Seq, Molecular markers, Reference genome, Structural variants, Differentially expressed genes, Population

1 Introduction

Single Nucleotide Polymorphisms (SNPs) and Insertion/Deletions (InDels) are among the most popular members of a large family of genomic structural variants (SVs), also called DNA polymorphisms, that have been used for different genetic studies including genetic diversity, fingerprinting, paternity studies, linkage mapping, assisted selection in breeding, and gene tagging, among others [1, 2]. During the last decade, SNPs have become the most used genotyping tool, explained by their ubiquity and simple molecular nature, most of them biallelic, affording an exhaustive coverage of the genomes [3, 4]. At the same time, it is possible to simultaneously analyze thousands of loci at a relatively low cost per marker compared to other types of SVs, being optimal to perform genome-wide association studies (GWAS) [5, 6]. Furthermore, it has been reported that SNPs exhibit a high potential to perform phenotype-genotype association analysis [7, 8], even being the causal variants of phenotypic differences [9]. The increasing availability of

Yuri Shavrukov (ed.), *Plant Genotyping: Methods and Protocols*, Methods in Molecular Biology, vol. 2638,
https://doi.org/10.1007/978-1-0716-3024-2_10,

thoroughly sequenced genomes has also contributed to the preference of SNPs over other markers since variant discovery has been facilitated with the alignment among whole genomes [10].

In this framework, the feasibility of RNA-Seq experiments based on a high-throughput transcriptome sequencing has paved the way to the massive identification of polymorphisms located in coding regions, eventually excluding highly repetitive genome regions [9, 11]. An additional advantage of this approach is that the molecular markers obtained are more likely transferable, considering their location in more conserved genomic regions [12]. Using transcribed genes (initially called ESTs) as a substrate to identify SNPs, remains a promissory strategy for the search on SVs more directly associated to genomic regions that are actively expressed under certain circumstances, and hence, to track both synonymous and non-synonymous changes associated to target genes [13], that are responsible of or are associated to the phenotypic expression of a particular trait.

Next Generation Sequencing platforms have been instrumental in the SNPs discovery, not just because of an increase in the speed at which sequence data is generated, but also because of the quality and length of fragments sequenced, both improved dramatically [14]. In plant species, the discovery of novel molecular markers represents a key task for breeding programs, especially for the study of economically relevant and complex traits [15, 16]. For most non-model crops species, the lack of a reference genome affects the genetic marker availability [17]; interestingly, in these cases the transcriptomic data has proved to be suitable for molecular markers discovery, even when a reference genome is not available [16–19].

The feasibility of RNA-Seq data for the detection of SVs have been reported in different contexts, including human beings [20], animal models [21], and plants [9, 11, 17–19, 22–25]. Several bioinformatic pipelines have been described for marker discovery [12, 17, 21, 26]. Interestingly, the development of transcriptomic data in plants has included many different tissues and organs such as apical meristem, roots, peduncle, fruits, seeds, leaves (juvenile/mature), buds, and shoots [17, 18, 23, 24]. Also, expression data has been collected under diverse physiological conditions [18, 27], such as control/treatment salinity response in bread wheat [23], and non-infected/infected leaves of *Platanus acerifolia* with sycamore lace bug pest (*Corythucha ciliate*) [19], underscoring its versatility.

In this chapter, the development of SNPs, and by extension of other SVs such as SSRs and InDels, based on transcriptomics data will be approached.

2 Materials

2.1 Consumables

1. XT extraction buffer: 0.2 M sodium borate decahydrate (borax), 30 mM EGTA, 1% p/v deoxycholate, 1% p/v sodium dodecyl sulfate (SDS), balanced at pH 8.2 with NaOH (*see* **Note 1**).
2. β-mercaptoethanol.
3. Solution of 4 M LiCl.
4. Solution of 3 M sodium acetate.
5. Polyvinylpyrrolidone (*see* **Note 2**).
6. Surfactant NP-40 (look for "NP-40 Alternative" if the product is discontinued and therefore cannot be provided by supply chain).
7. RNAse-free H_2O. Alternatively, DEPC-treated water can be used for any solution used after extraction step with XT buffer.
8. Chloroform:Isoamyl alcohol, 1:1.
9. Isopropyl alcohol.
10. 80% v/v ethanol.
11. TruSeq RNA Sample Prep Kit v2: several buffers and reagents, including adapters. Make sure to check the full list of contents supplied in the selected kits, as the same for needed consumables and equipment (*see* **Note 3**).
12. High fidelity reverse transcriptase, available from different providers.
13. AMPure XP beads.
14. 80% v/v ethanol.
15. 96-well microplate 0.3 mL and microseal adhesive.

2.2 Equipment

1. Centrifuge tubes Oak Ridge type (~50 mL volume capacity).
2. Thermoregulated bath.
3. Centrifuge (for example, Sorvall RC-6 Plus).
4. Refrigerated microcentrifuge (for example, Sorvall Biofuge PrimoR).
5. Fume hood.
6. Mortar and pestle.
7. Anti RNAse spray to clean bench and facilities.
8. Magnetic stand for pelleting beads in the PCR plate.
9. Thermocycler.
10. Microplate shaker (for example, Bioshake XP high-speed lab shaker).

11. Standard plate microfuge to spin-down samples.
12. Agilent Technologies 2100 Bioanalyzer (recommended to quantify the integrity of RNA extractions and prepared libraries).

3 Methods

A general scheme of the methodology is shown in Fig. 1 with following description of different steps in detail.

3.1 RNA Extraction (Modified "Hot Borate" Method)

1. Sample plant material in liquid nitrogen. Store at −80 °C until further extraction (*see* **Note 4**).
2. Ground frozen tissue in liquid nitrogen with mortar and pestle. For woody species, such as trees and fruit crops, add 0.2 g of polyvinylpyrrolidone (PVP) per gram of processed tissue at the moment of grinding (*see* **Note 5**).
3. Add 50 μL of β-mercaptoethanol to preheated 5 mL of XT extraction buffer at 80 °C and transfer the homogenized samples to the tubes. Vortex during 45 s and incubate during 1 h at 45 °C, mixing gently by immersion every 15 min (*see* **Note 6**).
4. Centrifuge at 9000 rpm (12000 × *g*) for 30 min at 4 °C. Transfer supernatant to a new tube and add 1 volume of 4 M LiCl. Mix gently and incubate at 4 °C overnight (*see* **Note 7**).
5. Centrifuge at 11500 rpm (20000 × *g*) for 40 min at 4 °C. Discard supernatant and gently resuspend the pellet with 500 μL RNAse-free H_2O and 50 μL 3 M sodium acetate. Transfer mixture to 2 mL microtube and add 500 μL of choloroform:isoamyl alcohol (1:1). Vortex for 30 s.
6. Centrifuge at maximum speed during 15 min at 4 °C (Reference: 13000 rpm corresponding to 17000 × *g*). Transfer aqueous phase to a new tube and add 1 volume of isopropyl alcohol and centrifuge during at maximum speed during 1 h at 4 °C (*see* **Note 8**).
7. Discard supernatant and wash with ethanol 80% v/v. Centrifuge at maximum speed during 15 min at 4 °C and discard the supernatant. Ethanol washing and centrifugation can be repeated multiple times if the samples require to do so. Dry to room temperature for 30 min or in a dry bath at 37 °C for 10–15 min.
8. Resuspend the pellet in 50 μL of RNAse-free H_2O (*see* **Note 9**).
9. Verify that total RNA meets the quality and quantity standards to perform downstream analysis (*see* **Note 10**).

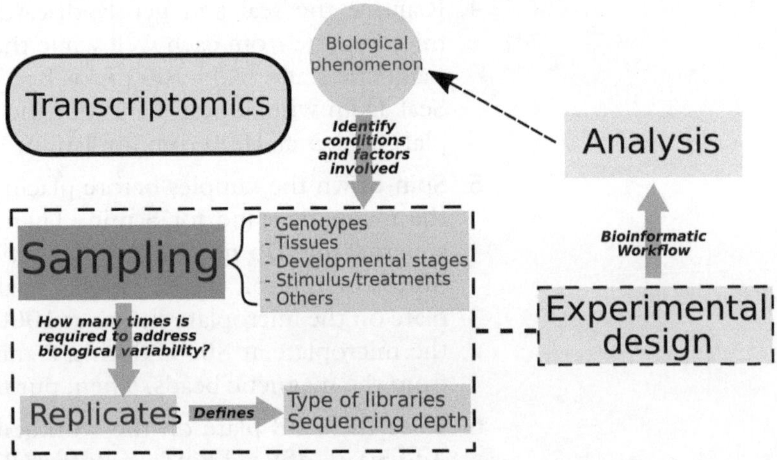

Fig. 1 Simplified scheme of the experimental design to perform a transcriptomic experiment

3.2 Library Synthesis and Sequencing

Most of the steps here were transcribed from manufacturer's instructions on the respective manual. So, it is recommended to check it to see the full protocol. Following the steps thoroughly is encouraged since they are standardized for the sequencer to be used (*see* **Note 11**).

3.2.1 RNA Purification and Fragmentation

1. From RNA extractions, 2.5 μg aliquots are used to isolate mRNA from it through selective binding with poly-A tails using the RNA Purification Beads (RPB) included in the TruSeq RNA Sample Prep Kit v2. As the manufacturer recommends, the aliquot of 2.5 μg RNA should be diluted in 50 μL of RNAse-free ultrapure water. Vortex the RPB solution before adding it to ensure full resuspension of the content and add 50 μL to each well of the plate to bind the poly-A RNA to the oligo-dT beads. Gently pipette the entire volume up and down around six times to mix completely. Since the plate contains the RNA Purification Beads it will be addressed as RPB plate onwards.
2. Apply seal adhesive to the RPB plate and put it on the microplate shaker at 1000 rpm for 1 min to mix thoroughly. Spin down on a standard microcentrifuge. Incubate at 65 °C for 5 min to fully denature RNA. Immediately after that, place the RPB plate in ice during 1 min.
3. Stand the RPB plate in the workbench for 5 min at room temperature (20–24 °C) to allow binding of mRNA to the oligo-dT tails of the magnetic beads. Then, place the RPB plate on the magnetic stand for another 5 min to pellet the beads from the solution.

4. Remove the seal and gently discard the supernatant with a micropipette from each well while the RPB plate is still on the magnetic stand. Add 200 μL of Bead Washing Buffer (BWB). Seal again with adhesive and put the RPB plate on the microplate shaker at 1000 rpm for 1 min.
5. Spin-down the samples before placing the RPB plate again on the magnetic stand for 5 min. Then, remove and discard the supernatant. Do not let the magnetic beads dry out and add 50 μL of Elution Buffer (ELB). Seal again and put the RPB plate on the microplate shaker at 1000 rpm for 1 min. Incubate the microplate at 80 °C during 2 min to detach RNA strings from the magnetic beads. Then, put the plate on ice for 1 min.
6. Place the RPB plate on the workbench and remove the seal. Add 50 μL of Bead Binding Buffer (BLB) and seal again to mix by using the microplate shaker at 1000 rpm for 1 min. Then, let it at room temperature for 5 min.
7. After incubation at room temperature, place the RPB plate on the magnetic stand to pellet the magnetic beads. Discard the supernatant. Remove the plate from the magnetic stand and add 200 μL of BWB to each well. Seal the plate and mix by using the microplate shaker at 1000 rpm for 1 min.
8. After a brief spin down, place the RPB plate on the magnetic stand for 5 min. Remove the seal and discard the supernatant. Add 19.5 μL of "Elute, Prime, Fragment Mix" (EPF) to each well of the plate. Seal and set the microplate shaker at 1000 rpm for 1 min.
9. Remove the seal and transfer the entire content of each well to a new plate. Since this new plate will contain the reagents for the RNA fragmentation, it will be labeled and referred as an RNA Fragmentation Plate (RFP). Label the plate and seal it again.
10. Place the sealed RFP plate on a pre-programmed thermocycler (94 °C for 8 min, 4 °C hold).
11. Remove the RFP plate from thermocycler when it reaches 4 °C and centrifuge briefly.

3.2.2 RNA First Strand Synthesis

1. Place the RFP plate on the magnetic stand at room temperature for 5 min. While still on the stand, remove the seal and transfer 17 μL of the supernatant (fragmented and primed RNA) to its corresponding well on a new hard-shell plate.
2. Spin-down the First Strand Master mix (FSM) tube. Add 1 μL of SuperScript II reverse transcriptase for each 9 μL of FSM mix to be used. This will constitute the "FSM mix" cited below (*see* **Note 12**).
3. Add 8 μL of the FSM mix (with the already added SuperScript II reverse transcriptase) to each well of the hard-shell plate with

the transferred supernatant from **step 1**. Since this plate contains now all the reagents for the cDNA first strand synthesis it will be named "CDP" (cDNA plate). Seal the CDP plate and put it on the microplate shaker at 1600 rpm during 20 s to mix thoroughly.

4. Incubate the CDP plate in the thermocycler following the next temperatures: 25 °C for 10 min, 42 °C for 50 min, 70 °C for 15 min and hold at 4 °C.

3.2.3 RNA Second Strand Synthesis

1. Remove the adhesive seal and add 25 μL of Second Strand Master mix (SSM). Seal again and put the CDP plate on a shaker at 1600 rpm during 20 s.
2. Place the sealed CDP plate on a thermocycler and incubate at 16 °C for 1 h. Remove the CDP plate from the thermocycler and let it stand to room temperature until further step.

3.2.4 RNA Purification

1. Vortex the AMPure XP beads and add 90 μL to each well of a new MIDI plate labeled with the corresponding identifier/barcode. Transfer the entire content from each well of the previous CDP plate to the corresponding ones of the new MIDI plate. Now, since this will constitute a cDNA clean up plate it will be denominated "CCP plate" onwards. Seal this new CCP plate and set the microplate shaker at 1800 rpm for 2 min to mix thoroughly.
2. Let the CCP plate to incubate at room temperature for 15 min. After incubation, spin-down the samples briefly. Remove the seal and put the CCP plate on a magnetic stand to pellet the beads during 5 min.
3. Remove and discard 135 μL of supernatant. With the plate still on the magnetic stand, add 200 μL of freshly prepared ethanol 80% v/v without disturbing the beads. Incubate at room temperature for 30 s. Remove and discard all the supernatant from each well.
4. Add 200 μL of ethanol 80% v/v. Incubate at room temperature for 30 s. Remove and discard all the supernatant from each well, repeating this washing twice.
5. Let the plate stand at room temperature without removing it from the magnetic stand for 15 min to dry. Then, remove the CCP plate from the stand and add 52.5 μL of Resuspension Buffer (centrifuge the buffer before its use). Seal the CCP plate and mix by putting it on the microplate shaker at 1800 rpm for 2 min.
6. Incubate the CPP plate during 2 min at room temperature. Then, briefly spin-down the samples. Remove the seal and place the CPP plate during 5 min on the magnetic stand. Now, transfer 50 μL of supernatant to a new MIDI plate to

perform the next step. Now that this is a double stranded DNA, it is safe for pausing the protocol for up to 7 days, sealing the plate and storing it at −15 to −25 °C.

3.2.5 RNA End Repair

1. Add 10 μL of Resuspension Buffer and add 40 μL of End Repair Mix (ERM) to each well. Since the content of the plate is prepared for the insertion of subsequent modifications, in the form of adenylation and ligation, this will be denominated "Insert Modification Plate" (IMP). Seal the IMP plate and shake it at 1800 rpm for 2 min.
2. Spin-down the samples and put the IMP plate to incubate at 30 °C for 30 min. Place the samples on ice until further step.
3. Add 160 μL of well-mixed AMPure XP beads to each well of the IMP plate containing 100 μL of solution. Seal the plate and mix it thoroughly by shaking at 1800 rpm for 2 min. Then, incubate at room temperature for 15 min.
4. Put the IMP plate on the magnetic stand for 5 min to pellet the AMPure XP beads at room temperature. Remove the adhesive seal.
5. Then, gently remove and discard 127.5 μL of supernatant of each well from IMP plate; repeat this step twice.
6. With the IMP still on the stand, add 200 μL of fresh prepared ethanol 80% v/v to wash. Wait 30 s at room temperature and then remove and discard the supernatant; repeat this step twice.
7. Let IMP plate stand at room temperature for 15 min to dry, and remove the plate from the magnetic stand. Resuspend the pellet with 17.5 μL of Resuspension Buffer. Seal the IMP plate and put it on a shaker at 1800 rpm for 2 min.
8. Spin-down the plate, remove the seal and incubate for 2 min at room temperature.
9. Place the IMP plate on the magnetic stand again to pellet the beads. Then, transfer 15 μL of the supernatant of the corresponding wells to a new MIDI plate labeled accordingly. This is a safe stopping point, the plate can be sealed and stored at −25 to −15 °C for up to 7 days.

3.2.6 Adenylate 3′-Ends

1. Add 2.5 μL of Resuspension Buffer to each well of the plate. Then, add 12.5 μL A-Tailing Mix. Since the reagent will prepare the samples for adapter ligation, the plate is denominated as "ALP." Shake the ALP plate at 1800 rpm for 2 min to mix thoroughly.
2. Spin-down the samples. Then, place the plate to incubate at 37 °C for 30 min and subsequently at 70 °C for 5 min.
3. Put the ALP plate on ice during 1 min. Proceed immediately to ligate adaptors.

3.2.7 Ligation of Adaptors

1. Make sure to have chosen adaptors to index correctly all your samples before any step. Add 2.5 μL Resuspension Buffer. Add 2.5 μL of Ligation Mix to each well of the ALP plate.
2. Return the Ligation Mix at −25 to −15 °C storage immediately after usage.
3. Add 2.5 μL of the corresponding RNA Adaptor Index to each well. Mix the content of the ALP plate by shaking it at 1800 rpm for 2 min.
4. Spin the samples and incubate the ALP plate at 30 °C for 10 min.
5. Add 5 μL of Stop Ligation buffer to each well to inactivate the mix. Seal the ALP plate and mix thoroughly by shaking at 1800 rpm for 2 min to ensure the reaction to stop.
6. Spin-down the samples. Remove the seal and add 42 μL mixed AMPure XP beads to each well of the ALP plate. Seal the plate and mix at 1800 rpm for 2 min in the microplate shaker.
7. Incubate at room temperature for 15 min.
8. Spin-down the ALP plate and put it on the magnetic stand for 5 min. Once the liquid is clear, remove and discard 79.5 μL of supernatant from each well.
9. With the ALP plate still on the magnetic stand, add 200 μL ethanol 80% v/v to each well without perturbing the beads. Incubate at room temperature for 30 s. Then, remove and discard the supernatant without removing it from the magnetic standard; repeat this step twice.
10. Keep the ALP plate on the magnetic stand for 15 min to let the beads to dry.
11. Remove the ALP plate from the magnetic stand. Then, add 52.5 μL of Resuspension Buffer to each well of the ALP plate. Seal the plate and mix by shaking at 1800 rpm for 2 min. Let it rest at room temperature for 2 min more and centrifuge to spin-down the samples.
12. Place the ALP plate on the magnetic stand for 5 min. Transfer 50 μL from the supernatant to another MIDI plate labeled correspondingly. Add another 50 μL well-mixed AMPure XP Beads solution to the recently transferred solution, to perform another round of DNA clean up. Since this plate is Clean up ALP Plate, it will be denominated "CAP" plate. Seal the CAP plate and shake it at 1800 rpm for 2 min to mix thoroughly.
13. Incubate the CAP plate at room temperature for 15 min. Put the CAP plate on the magnetic stand to pellet the beads for 5 min. After unsealing, remove and discard the 95 μL of supernatant from each well of the CAP plate.

14. While the plate is still on the magnetic stand, add 200 μL of freshly prepared ethanol 80% v/v to wash. Incubate the CAP plate at room temperature for 30 s. Then, remove and discard all the supernatant from each well. Repeat this step once more for a total of two washing cycles.
15. With the CAP plate still on the stand let it dry for 15 min at room temperature. Remove the CAP plate from the magnetic stand and add 22.5 μL of Resuspension Buffer to each well. Seal the CAP plate and put it on the microplate shaker at 1800 rpm for 2 min. Incubate at room temperature for another 2 min. Centrifuge briefly to spin down the samples.
16. Place the CAP plate on the magnetic stand during 5 min. Remove the seal and transfer 20 μL of the supernatant to a new hardshell plate with the corresponding label for each well.

3.2.8 Enrichment of DNA Sequences with Ligated Adaptors

1. Add 5 μL of PCR primer cocktail to each well of the plate.
2. Add 25 μL of PCR Master Mix to each well. Seal the plate and shake it at 1800 rpm for 20 s. Then, spin-down the samples.
3. Transfer the plate to the thermocycler to initiate the enrichment of fragments with the ligated adaptors. The thermocycler program should be as follows: (1) 98 °C for 30 s, (2) 15 cycles of [98 °C for 10 s; 60 °C for 30 s; 72 °C for 30 s], (3) 72 °C for 5 min and (4) hold at 10 °C.
4. Add 50 μL of well-mixed AMPure Beads XP to each well of a new MIDI plate with the corresponding labels. Remove the seal from the plate with the amplified fragments and transfer the entire contents to the new MIDI plate which already contains the 50 μL of AMPure Beads XP on each well (make sure the labels correspond to each other). Seal the new plate and shake it at 1800 rpm for 20 s to mix thoroughly.
5. Incubate the plate for 15 min at room temperature. Then, transfer it to the magnetic stand for another 5 min more. With the plate still on the magnetic stand, remove and discard 95 μL of supernatant from each well.
6. Add 200 μL of ethanol 80% v/v with the plate still on the magnetic stand. Incubate for 30 s. Remove and discard all the supernatant; repeat this step twice.
7. With the plate still on the magnetic stand, let it dry at room temperature for 15 min.
8. Resuspend the pellet with 32.5 μL of Resuspension Buffer. Seal again the plate and shake it at 1800 rpm for 20 s. Let it incubate at room temperature for 2 min.
9. Place the plate on the magnetic stand for 5 min. Remove the seal and transfer the supernatant to a new plate with the corresponding labels/barcodes for each well.

3.2.9 Final Steps for Library Sequencing

1. Validate the libraries by fragment analysis with Agilent Technologies Bioanalyzer or an equivalent device.
2. Normalize and pool libraries. This will depend greatly on the experimental design along with the availability of the lanes to be used for sequencing. Therefore, it is strongly recommended to coordinate such task with the provider of sequencing service.

3.3 SNP and InDel Calling and Annotation

1. The first step in the sequence data analysis corresponds to the quality control of initial reads obtained from RNA-seq libraries, which is implemented using FASTQC software (*see* Fig. 2).
2. Subsequently, the trimming of the reads for low quality is applied and Q20 reads are selected. In our experience, good quality reads are aligned to the reference genome (if it is available), using Tophat software [25, 28].
3. A maximum of two mismatches per read could be accepted, and multiple reads (hits >20) are discarded.
4. Then, PCR-duplicates are removed by using remove_duplicates tool from Picard Tools [29] (*see* Fig. 2).
5. The next step will be the SNP/InDel calling performed with the Structural Variant Calling of Genome Analysis Toolkit (GATK) [30] and the Haplotype caller tool.
6. In the case of InDels, it is highly recommended to carry out re-alignments in the analyzed regions (*see* Fig. 2).
7. Subsequently, SNP and InDel polymorphisms are catalogued using the type of polymorphism description and consecutively numbered as "TSRNASNP" or "TSRNAINDEL," respectively.
8. Then, the output data file is transformed to the variant call format file (*.vcf), which corresponds to a standardized generic format to lay up DNA polymorphism data, that includes variant positions, reference and alternative bases, and genotypes per sample [31].
9. To proceed with the selection of polymorphisms associated with the trait of interest, the total of SNPs and short InDels identified, several filtering criteria can be applied using VCFtools options (*see* Fig. 2) [31].
10. In our work based on the analysis of individuals from a bi-parental segregating population exhibiting contrasting phenotypes for a target trait [25], the following options were considered:
    (a) the selection of biallelic polymorphisms, using --min-alleles 2 --max-alleles 2.
    (b) the missing data filter frequently used considers a value lower than 5% (i.e., a certain polymorphism is supported

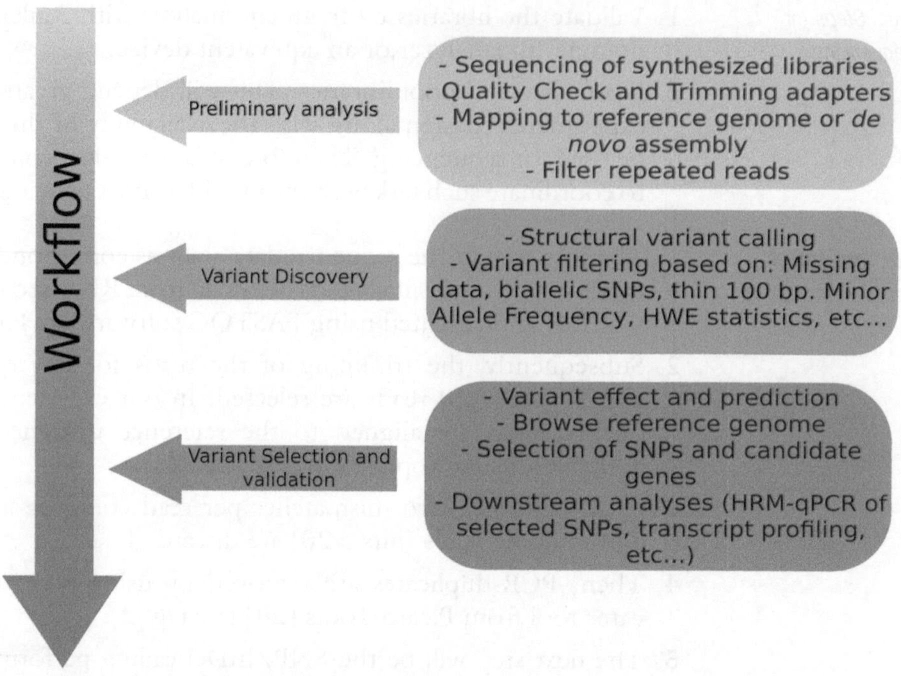

Fig. 2 Bioinformatics workflow based on RNA-Seq data, implemented to identify SNP and InDel molecular markers associated with berry weight trait in *Vitis vinifera*. (Modified from Ref. [25])

in the 95% of the available samples), especially in the case of the analysis of varieties. However, in our previous work no missing data was allowed, applying --max-missing <1>.

(c) this criterion will depend on the experimental validation platform considered in the genotyping step. In our experience using High-Resolution Melting analysis (qPCR-HRM), in case of the presence of two or more SNPs in the resulting amplicon could produce truncated curves, which must be avoided. Therefore, a thin or minimal separation between two SNPs of 100 bp, as --thin <100> was applied.

(d) in the case of Minor Allele Frequency (MAF), a common criterion analyzing varieties is higher than 0.05 (5%). However, in our experience working with individuals from a bi-parental crossing exhibiting contrasting phenotypes for a target trait, four levels were evaluated (0.1, 0.2, 0.3, and 0.4) and a MAF of 0.2 was selected, using --maf <0.2>. In fact, working with a bi-parental crossing, the segregation patterns observed on both parents for SNP/InDel were "abxaa" and "abxab"; therefore, MAF values ranged between 21% and 28.6%.

(e) a fixation index (Fst) of 1 was used to calculate populations differentiation, comparing the two groups of segregants with contrasting phenotypes for the trait of interest, using --weir-fst-pop. Therefore, in the case of selected SNP/InDel, the genotype observed in one group of segregants should be the same, representing a different genotypic class for the predicted in the second segregant group. In addition, it could be evaluated to use values between 0.7 and 1.0, depending on the target trait.

11. Subsequently, the Variant Annotation and Effect prediction tool SnpEff software [32] is recommended to achieve the functional annotation of selected putative SNPs/InDel, based on the gene model of the reference genome of the target species. Considering these in silico evidence, intragenic and non-synonymous coding SNPs can be selected for further analysis.
12. In the next step, the coverage of the putative SNPs and InDels is analyzed using the Integrative Genomic Viewer (IGV) [33], and polymorphisms exhibiting more than 20 supporting reads can be selected for primers design.

3.4 Primer Design

1. The validation of SNPs and InDels putative markers can be performed using several platforms. In our experience, validation and later genotyping step was done by High-Resolution Melting analysis (qPCR-HRM).
2. Then, design specific primers using PRIMER 3 software [34] according to Muñoz-Espinoza et al. [25], and the following parameters: amplicon length among 100 to 160 bp, primers length between 18 and 23 bp, and annealing temperature (Tm) range 58–62 °C.
3. Check primer dimer formation as well as primer complementarity in silico using Operon software or similar utilities [34]. Synthesize primers at a provider that guaranties oligos quality and quantity.

4 Notes

1. Take the proper measures before adding deoxycholate when preparing XT buffer. Make sure to wear gloves and mask since this reagent is highly toxic when the dust is breathed accidentally. Once all reagents are dissolved in water, adjust pH with NaOH and meet the required volume.
2. Polyvinylpyrrolidone (PVP) is a water-soluble high molecular weight polymer which can effectively prevent undesired interaction of phenolic compounds with nucleic acids in

downstream steps, after tissue disruption. Adjusting the ratio of sample/PVP for tissues with high phenolic content such as fruits and woody tissues could improve quality and quantity of RNA extraction.

3. Depending on the research scope, the type of library could vary according to the sequencing platform used to obtain long-reads or short-reads. As reviewed by Stark et al. [35], 95% of the published RNA-seq data from the Short Read Archive (SRA) is based on Illumina technologies and most transcriptomic studies are based on this type of reads. On the other hand, longer-read data can be obtained using Pacific Biosciences and Oxford Nanopore. Therefore, it is recommended to assess first what kind of transcript data is desired and opting for a methodology accordingly. For short-reads data, the common goal is to capture transcript abundance of mRNA species; for this, isolation of coding DNA, cDNA synthesis, adaptor ligation, and PCR amplification are accomplished by using available kits. The protocol presented here is oriented to short-reads technologies. The reagent kit used in our experiments (TruSeq RNA Sample Prep Kit v2) accomplishes library preparation in several steps. (1) For RNA purification and fragmentation remarkable items are Bead Binding Buffer (BBB), Bead Washing Buffer (BWB), Elute-Prime-Fragment Mix (EPF), Elution Buffer (ELB), Resuspension Buffer (RSB), RNA Purification Beads (RPB). (2) For cDNA first strand synthesis a remarkable item is First Strand Master mix (FSM). (3) For cDNA second strand synthesis remarkable items are Second Strand Master mix (SSM) and Resuspension Buffer (RSB). (4) For end repair remarkable items are End Repair Mix (ERM), Resuspension Buffer (RSB). (5) For adenylation of 3′-End remarkable items are A-Tailing mix (ATL) and Resuspension Buffer (RSB). (6) For ligation of adapters remarkable items are Ligation Mix (LIG), RNA Adapters Indexers (choose from the ones supplied in the kit accordingly to your samples), Stop Ligation Buffer (STL). (7) For enrichment of DNA fragments with ligated adapters remarkable items are PCR Master Mix (PMM), PCR Primer Cocktail (PPC).

4. A proper experimental design (*see* Fig. 1) includes identifying major factors and considering an appropriate number of replicates. However, the number of replicates is limiting, particularly when the interest is revealing the changes in gene expression level in a detailed way. Then, before deciding the sampling scheme, it is mandatory to characterize as well as possible the biological phenomenon of interest to choose key developmental stages or physiological-sanitary conditions.

5. As other authors have noted [36], fine grinding facilitates all the following steps in RNA isolation protocols. Previously

weighted PVP-containing microtubes could be prepared in advance to not delay this step since timing is critical.

6. Make sure of adding β-mercaptoethanol under a fume hood and bring the grounded samples to it just after being processed with mortar and pestle. Minimizing the interval from the grinding of the tissue to adding it to the tubes containing XT buffer is critical to have good yield and quality of extracted RNA.

7. The presented protocol [37, 38] considers two steps with selective precipitation of RNA (using lithium salts and isopropanol, respectively). This is a two-days protocol yielding RNA in high quantity and quality (in our hands, applied on grapevine berries). However, in our experience the heavy-salt protocol could affect 260/230 ratio. Depending on the tissue, faster protocols such as column-based ones could be a good choice, working properly. It is highly recommended to store an aliquot of the RNA used for library synthesis, since it would be needed to perform other analyses such as qPCR to validate genes of interest, or to repeat the transcriptomic assay itself.

8. Isopropyl alcohol precipitation could be extended overnight if a pause is required.

9. Several approaches exist to measure RNA quantity and quality. Spectrophotometric methods serve to determine concentration and purity through usage of calibration curve and absorbance ratios, respectively (260/280 for RNA purity and 260/230 for contaminants in stock solution; ratios of around 1.8–2.0 are acceptable in both cases). Alternatively, many commercial kits or devices can be used to estimate quantity accurately (e.g., Qubit). On RNA quality, denaturing agarose gel electrophoresis is one of the most common methods to assess integrity of samples. In this case, a clear band near the loading point of the sample must be seen to reflect good quality. Otherwise, (partially) degraded RNA will show smearing and not a clear band. Moreover, the intensity of signal will also correlate with the quantity of loaded sample, so it is a good indicator for stock solution concentration. An alternative to measure integrity in a more quantitative fashion is to measure the RNA integrity number (RIN), which can be accomplished by the usage of dedicated devices and softwares such as Fragment Analyzer (e.g., PROSize® 2.0 version 1.3.1.1 – Advanced Analytical Technologies, Inc., Ames, IA, USA). RIN values go from 0 to 10, with 10 being an intact RNA sample with no degradation; RIN value around 7.0 is the minimum to be considered for RNA sequencing. As could be expected, samples from woody tissues tend to have lower RIN values, and then a

compromise must be adopted, occasionally accepting samples with lower RIN values.

10. Using commercial elution buffers for preparing stock solutions of isolated RNA is recommended for the long-term storage of RNA samples. Providing a more stable environment for the labile RNA molecule could be optimal, since gene expression validation will require an aliquot of the same RNA from which cDNA library was prepared.
11. The preferred method is entirely based on the manufacturer's suggestions for the usage of TruSeq RNA Sample Preparation v2 Protocol. It is strongly recommended to previously define the sequencer to be used to harmonize it with the type of libraries to be prepared, while also considering the sequencing depth desired and quality controls needed [26].
12. Any other high-fidelity enzymes (not prone to errors and with a high processivity) can just work optimally for this step.

Acknowledgments

The preparation of this chapter was supported by grants FONDECYT-ANID 11190936 (CM-E) and 1221410 (PH).

References

1. Mammadov J, Aggarwal R, Buyyarapu R, Kumpatla S (2012) SNP markers and their impact on plant breeding. Int J Plant Genomics 2012:728398. https://doi.org/10.1155/2012/728398
2. Adhikari S, Saha S, Biswas A, Rana TS, Bandyopadhyay TK, Ghosh P (2017) Application of molecular markers in plant genome analysis: a review. Nucleus 60:283–297. https://doi.org/10.1007/s13237-017-0214-7
3. Garrido-Cardenas JA, Mesa-Valle C, Manzano-Agugliaro F (2018) Trends in plant research using molecular markers. Planta 247:543–557. https://doi.org/10.1007/s00425-017-2829-y
4. Rasheed A, Hao Y, Xia X, Khan A, Xu Y, Varshney RK et al (2017) Crop breeding chips and genotyping platforms: Progress, challenges, and perspectives. Mol Plant 10:1047–1064. https://doi.org/10.1016/j.molp.2017.06.008
5. De Donato M, Peters SO, Mitchell SE, Hussain T, Imumorin IG (2013) Genotyping-by-sequencing (GBS): a novel, efficient and cost-effective genotyping method for cattle using next-generation sequencing. PLoS One 8:e62137. https://doi.org/10.1371/journal.pone.0062137
6. He J, Zhao X, Laroche A, Lu ZX, Liu HK, Li Z (2014) Genotyping-by-sequencing (GBS), an ultimate marker-assisted selection (MAS) tool to accelerate plant breeding. Front Plant Sci 5:484. https://doi.org/10.3389/fpls.2014.00484
7. Collins FS, Guyer MS, Chakravarti A (1997) Variations on a theme: cataloging human DNA sequence variation. Science 278(5343):1580–1581. https://doi.org/10.1126/science.278.5343.1580
8. Cooper DN, Smith BA, Cooke HJ, Niemann S, Schmidtke J (1985) An estimate of unique DNA sequence heterozygosity in the human genome. Hum Genet 69:201–205. https://doi.org/10.1007/BF00293024
9. Ashrafi H, Hill T, Stoffel K, Kozik A, Yao J, Chin-Wo S et al (2012) De novo assembly of the pepper transcriptome (*Capsicum annuum*): a benchmark for in silico discovery of SNPs, SSRs and candidate genes. BMC Genomics 13:571. https://doi.org/10.1186/1471-2164-13-571

10. Sun Y, Shang L, Zhu QH, Fan L, Guo L (2022) Twenty years of plant genome sequencing: achievements and challenges. Trends Plant Sci 27:391–401. https://doi.org/10.1016/j.tplants.2021.10.006
11. Torre S, Tattini M, Brunetti C, Fineschi S, Fini A, Ferrini F et al (2014) RNA-Seq analysis of *Quercus pubescens* leaves: *De novo* transcriptome assembly, annotation and functional markers development. PLoS One 9:e112487. https://doi.org/10.1371/journal.pone.0112487
12. Cordeiro G, Casu R, McIntyre C, Manners J, Henry R (2001) Microsatellite markers from sugarcane (*Saccharum* spp.) ESTs cross transferable to erianthus and sorghum. Plant Sci 160:1115–1123. https://doi.org/10.1016/S0168-9452(01)00365-X
13. Picoult-Newberg L, Ideker TE, Pohl MG, Taylor SL, Donaldson MA, Nickerson DA et al (1999) Mining SNPs from EST databases. Genome Res 9:167–174. https://doi.org/10.1101/gr.9.2.167
14. Goodwin S, McPherson JD, McCombie WR (2016) Coming of age: ten years of next-generation sequencing technologies. Nat Rev Genet 17:333–351. https://doi.org/10.1038/nrg.2016.49
15. Van Damme V, Gómez-Paniagua H, de Vicente MC (2011) The GCP molecular marker toolkit, an instrument for use in breeding food security crops. Mol Breed 28:597–610. https://doi.org/10.1007/s11032-010-9512-3
16. Badenes ML, Fernández i Martí A, Ríos G, Rubio-Cabetas MJ (2016) Application of genomic technologies to the breeding of trees. Front Genet 7:198. https://doi.org/10.3389/fgene.2016.00198
17. Thakur O, Randhawa GS (2018) Identification and characterization of SSR, SNP and InDel molecular markers from RNA-Seq data of guar (*Cyamopsis tetragonoloba,* L. Taub.) roots. BMC Genomics 19:951. https://doi.org/10.1186/s12864-018-5205-9
18. Salgado LR, Koop DM, Pinheiro DG, Rivallan R, Le Guen V, Nicolás MF et al (2014) De novo transcriptome analysis of *Hevea brasiliensis* tissues by RNA-seq and screening for molecular markers. BMC Genomics 15:236. https://doi.org/10.1186/1471-2164-15-236
19. Li F, Wu C, Gao M, Jiao M, Qu C, Gonzalez-Uriarte A et al (2019) Transcriptome sequencing, molecular markers, and transcription factor discovery of *Platanus acerifolia* in the presence of *Corythucha ciliata*. Sci Data 6:128. https://doi.org/10.1038/s41597-019-0111-9
20. Shukla N, Levine MF, Gundem G, Domenico D, Spitzer B, Bouvier N et al (2022) Feasibility of whole genome and transcriptome profiling in pediatric and young adult cancers. Nat Commun 13:2485. https://doi.org/10.1038/s41467-022-30233-7
21. Adetunji MO, Lamont SJ, Abasht B, Schmidt CJ (2019) Variant analysis pipeline for accurate detection of genomic variants from transcriptome sequencing data. PLoS One 14:e0216838. https://doi.org/10.1371/journal.pone.0216838
22. Tian W, Paudel D, Vendrame W, Wang J (2017) Enriching genomic resources and marker development from transcript sequences of *Jatropha curcas* for microgravity studies. Int J Genomics 2017:8614160. https://doi.org/10.1155/2017/8614160
23. Karam A, El-assal SES, Hussein BA (2022) Transcriptome data mining towards characterization of single nucleotide polymorphisms (SNPs) controlling salinity tolerance in bread wheat. Biotechnol Biotechnol Equip 36:389–400. https://doi.org/10.1080/13102818.2022.2081516
24. Vatanparast M, Shetty P, Chopra R, Doyle JJ, Sathyanarayana N, Egan AN (2016) Transcriptome sequencing and marker development in winged bean (*Psophocarpus tetragonolobus*; Leguminosae). Sci Rep 6:29070. https://doi.org/10.1038/srep29070
25. Muñoz-Espinoza C, Di Genova A, Sánchez A, Correa J, Espinoza A, Meneses C et al (2020) Identification of SNPs and InDels associated with berry size in table grapes integrating genetic and transcriptomic approaches. BMC Plant Biol 20:365. https://doi.org/10.1186/s12870-020-02564-4
26. Conesa A, Madrigal P, Tarazona S, Gomez-Cabrero D, Cervera A, McPherson A et al (2016) A survey of best practices for RNA-seq data analysis. Genome Biol 17:13. https://doi.org/10.1186/s13059-016-0881-8
27. Wang Z, Gerstein M, Snyder M (2009) RNA-Seq: a revolutionary tool for transcriptomics. Nat Rev Genet 10:57–63. https://doi.org/10.1038/nrg2484
28. Trapnell C, Pachter L, Salzberg SL (2009) TopHat: discovering splice junctions with RNA-Seq. Bioinformatics 25:1105–1111. https://doi.org/10.1093/bioinformatics/btp120
29. Li H, Handsaker B, Wysoker A, Fennell T, Ruan J, Homer N et al (2009) The sequence

alignment/map format and SAMtools. Bioinformatics 25:2078–2079. https://doi.org/10.1093/bioinformatics/btp352

30. Auwera GA, Carneiro MO, Hartl C, Poplin R, Del Angel G, Levy-Moonshine A et al (2013) From FastQ data to high-confidence variant calls: the genome analysis toolkit best practices pipeline. Curr Protoc Bioinform 43:483–492. https://doi.org/10.1002/0471250953.bi1110s43

31. Danecek P, Auton A, Abecasis G, Albers CA, Banks E, DePristo MA et al (2011) The variant call format and VCFtools. Bioinformatics 27: 2156–2158. https://doi.org/10.1093/bioinformatics/btr330

32. Cingolani P, Platts A, Wang LL, Coon M, Nguyen T, Wang L et al (2012) A program for annotating and predicting the effects of single nucleotide polymorphisms, SnpEff. Fly 6:80–92. https://doi.org/10.4161/fly.19695

33. Thorvaldsdóttir H, Robinson JT, Mesirov JP (2013) Integrative Genomics Viewer (IGV): high-performance genomics data visualization and exploration. Brief Bioinform 14:178–192. https://doi.org/10.1093/bib/bbs017

34. Untergasser A, Nijveen H, Rao X, Bisseling T, Geurts R, Leunissen JAM (2007) Primer3Plus, an enhanced web interface to Primer3. Nucleic Acids Res 35:W71–W74. https://doi.org/10.1093/nar/gkm306

35. Stark R, Grzelak M, Hadfield J (2019) RNA sequencing: the teenage years. Nat Rev Genet 20:631–656. https://doi.org/10.1038/s41576-019-0150-2

36. Allen GC, Flores-Vergara M, Krasynanski S, Kumar S, Thompson WF (2006) A modified protocol for rapid DNA isolation from plant tissues using cetyltrimethylammonium bromide. Nat Protoc 1:2320–2325. https://doi.org/10.1038/nprot.2006.384

37. Gudenschwager O, González-Agüero M, Defilippi BG (2012) A general method for high-quality RNA isolation from metabolite-rich fruits. S Afr J Bot 83:186–192. https://doi.org/10.1016/j.sajb.2012.08.004

38. González-Agüero M, García-Rojas M, Di Genova A, Correa J, Maass A, Orellana A et al (2013) Identification of two putative reference genes from grapevine suitable for gene expression analysis in berry and related tissues derived from RNA-Seq data. BMC Genomics 14:878. https://doi.org/10.1186/1471-2164-14-878

Chapter 11

Specific-Locus Amplified Fragment Sequencing (SLAF-Seq)

Yang Zhou and Huitang Pan

Abstract

Specific length amplified fragment sequencing (SLAF-seq) technology is a simplified genome sequencing technology based on next-generation sequencing. SLAF-seq technology has several distinguishing characteristics: 1. Deep sequencing to ensure accuracy of genotyping; 2. Effectively reduce sequencing costs; 3. Pre-designed simplified representation scheme to optimize marker efficiency; 4. Doubled barcode system for large populations. The advantages and technical process of SLAF-seq are described briefly with summarized results for the application of SLAF-seq in development of molecular markers, construction of high-density genetic map and gene mapping in ornamental plants. Finally, the difficulties and prospects of this method are discussed in application.

Key words Genotyping, SLAF, Molecular markers, SNP

1 Introduction

Sequencing technology has been developed from the first-generation Sanger sequencing to the next-generation high-throughput sequencing to the third-generation single molecule sequencing [1–3]. The next-generation and third-generation sequencing technologies have solved the practical problem that the first-generation sequencing technology cannot carry out on a large scale, and greatly accelerated the research in the fields of whole genome sequencing, metagenomics sequencing, DNA methylation sequencing, RNA-seq (transcriptome) and mutation identification (SNP detection) [2, 3]. Although the third-generation sequencing technology has the advantages of ultra-long reading length, fast operation, no template amplification and direct detection of apparent modification sites, the sequencing error rate and sequencing cost are higher than those of the first-generation and the next-generation, which have not been widely used at present [3–5]. Therefore, the next-generation of high-throughput sequencing technology is more conducive for researchers to quickly and comprehensively apply it to biological genome research at a lower price.

Yuri Shavrukov (ed.), *Plant Genotyping: Methods and Protocols*, Methods in Molecular Biology, vol. 2638,
https://doi.org/10.1007/978-1-0716-3024-2_11,

This technology can determine hundreds of thousands to millions of nucleotide sequences at one time, and has a large amount of data output in a single run, so it is called high-throughput sequencing technology. This technology mainly includes metagenome sequencing, epigenomics, whole genome de novo sequencing, whole genome sequencing and simplified genome sequencing (SLAF-seq).

SLAF-seq technology is a simplified genome sequencing based on the next-generation high-throughput sequencing and it has sprung up rapidly in recent years. SLAF-seq technology interrupts the genome by restriction endonuclease and performs high-throughput sequencing on specific regions of the genome to obtain a large number of genetic polymorphism tag sequences to show the characteristics of the whole genome sequence. SLAF-seq combines simplified library, specific fragment amplification methods with high-throughput, high efficiency and accuracy of polymorphism labels, automatic repeat avoidance, complex genomes (*see* **Note 1**), no reference genome, groups and species of different sizes and types, and high-cost performance (*see* **Note 2**) [6]. SLAF-seq technology has been successfully applied to plant molecular markers development, high-density genetic mapping construction, QTL mapping of important trait genes, population genetics and assisted whole genome sequencing [7, 8]. Based on the brief introduction of the advantages and technical process of SLAF-seq, this chapter also introduces its progress in the field of ornamental horticulture, discusses the application difficulties, and looks forward to its development prospect to provide scientific reference for relevant application research in the future.

2 Application of SLAF-Seq in Ornamental Plants

Most of the genetic backgrounds of ornamental plants are complex and lack of genomic data support. Zhang et al. [9] crossed *Prunus mume* "Liu-Ban" and "Fentai Chuizhi" to obtain a hybrid population, and were used as segregating mapping population, to construct a high-density genetic linkage map containing 8007 SLAF markers, with an average distance of 0.195 cM between markers. QTLs related to weeping traits were located in the region from 69.63 to 75.52 cM of group 7. Nine structural genes related to lignification and nine transcriptional regulatory genes were predicted to be closely related to weeping traits of *P. mume*, which provided an important theoretical basis for accelerating of molecular breeding in *P. mume*. Cai et al. [10] used SLAF makers to construct the first high-density genetic map of tree peony with a total map distance of 920.67 cM and an average distance of 0.774 cM between markers. Based on this genetic map, 27 quantitative traits were analyzed, 49 QTLs were identified, which can explain 8.3~71.9% of phenotypic variation. Ye et al. [11] developed SLAF molecular markers to identify the traits of plant height, internode length and primary branch point height in *Lagerstroemia*, which can be used as marker-assisted selection for further breeding programs. Song et al. [12] constructed a high-density genetic linkage map of *Chrysanthemum* used SLAF technology. The total map distance was 3693.23 cM and the average distance was 0.76 cM between markers, and 123 QTLs related to flower type traits were found, including three major QTLs control corolla tube merged degree (CTMD) and four major QTLs control relative number of ray florets (RNRF).

At present, high-density genetic maps through simplified genome sequencing technology were constructed in *Salix matsudana*, *Osmanthus fragrans* and *Dendrobium nobile*, which has laid a scientific foundation for characteristics mapping and genetic breeding of ornamental plants [13–15].

3 SLAF Method

3.1 Experimental Scheme Design Based on Bioinformatics Information

First, we performed SLAF pre-designed experiments. The size of DNA fragments digested with restriction enzymes was evaluated by bioinformatics, as training data. Three criteria are considered [6]: (1) The number of SLAFs must be suitable for the specific requirements of the research project. The number of SLAFs to be developed depends on many factors such as the research purpose, the size of the genome of the species, etc. (*see* **Note 3**); (2) The SLAFs must be evenly distributed throughout the sequence for the accuracy, trying to ensure a uniform distribution. Nevertheless, uniform distribution can often be difficult to achieve in the results. (3) Repeated SLAFs must be avoided. These considerations can improve the efficiency of SLAF-seq. To maintain the sequence depth consistency of different fragments, select a narrow length range (about 30–50 bp) and carry out preliminary PCR amplification to check the RRL (reduced representation library) features in the target length range. The target length range usually includes fragments with similar amplification features on the gel. When non-specific amplification bands appear on the gel, then the pre-design step has to be repeated from the beginning to produce a new batch of DNA digested fragments.

3.2 Construct SLAF Library According to the Scheme of Preliminary Experiment

The SLAF library is constructed in accordance using the preliminary experiment scheme, the genomic DNA of each qualified sample have to be digested respectively. After digestion with a restriction enzyme, the obtained fragments (SLAF tags) are connected as described below, sequenced, amplified, purified, mixed, separated by electrophoresis and cut from gel. The target fragments are selected (*see* Fig. 1).

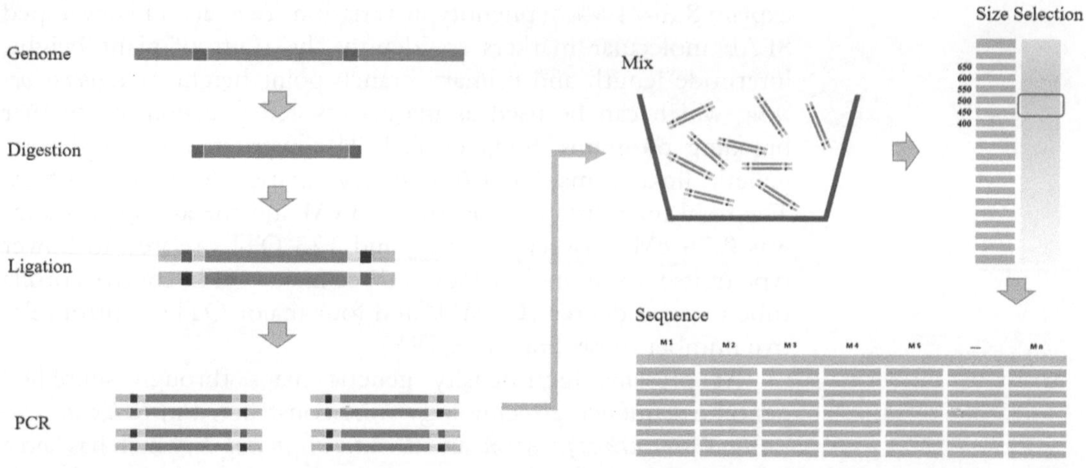

Fig. 1 SLAF-seq flowchart. Genomic DNA was digested by groups of enzymes designed for individuals. Double barcodes were added to two round PCR reactions to discriminate each individual and to facilitate the pooling of samples for size selection, which maintained consistent fragment size among individuals. Deep sequencing for the pooled RRLs with the Illumina paired-end sequencing protocol

The SLAF library can be constructed according to the pre-designed scheme such as it was shown on the example of *Lagerstroemia*. Genomic DNA was digested into 450–500 bp fragments (*see* **Note 4**) using suitable restriction enzyme combinations, including *Eco*RI + *Nla*III + *Mse*I. The restriction-ligation reactions were heat inactivated at 65 °C and then digested with the additional restriction enzyme *Nla*III at 37 °C. These digested products were diluted in 30 μL elution buffer and mixed with dNTPs, Taq DNA polymerase, and *Mse*I-primer containing barcode 1 for a PCR. The PCR products have to be purified and mixed with Cycle Pure Kit. The reaction mixtures with *Mse*I, T4 DNA ligase, ATP and adapter are incubated at 37 °C. The appropriate fragments with indexes and adaptors were isolated using any suitable Gel extraction kit. The purified fragments were subjected to PCR amplification and barcode 2 was added. The samples were gel-purified, and 450–500 bp of DNA fragments were excised with following dilution prior Illumina sequencing.

3.3 High-Throughput Sequencing

After testing of the produced library and quality control, the fragments have to be sequenced typically with Illumina HiSeq 2500 or Illumina GAII instruments. Each sequencing cycle has to be monitored accurately with calculated ratio between the original and high-quality readings of the GC content. The mass fraction greater than Q20 is considered for quality control because this level of the mass fraction (= 20) indicates for 1% of error probability and 99% of the confidence. Sequencing depth is determined based on research goals and population type (*see* **Note 5**). In molecular marker development program, the average sequencing depths were more than 20-fold in the parents and 100-fold in the progeny pools in average. Sequence similarity was detected by BLAT [16], and sequences with over 90% identity were defined as a SLAF locus. In each SLAF locus cite, genetic polymorphism can be identified especially between parents. All polymorphic SLAFs have to be genotyped in offspring, and any siblings in progenies with more than 80% SLAF in parents are showing the corresponding integrity of SLAF markers in the identified genotypes. Potential SLAFs can be identified as markers during comparison between offspring and parent genotypes with clear and significant differences.

3.4 Data Processing and Analysis

Leverage SLAF-poly.pl software (Beijing Biomarker company. www.biomarker.com.cn) is basically used for the data processes, evaluates the data obtained by sequencing. The computer software can provide confirmed results regarding obtained SLAF fragments after their sequencing in-depth and passed quality control. After that, research can be absolutely confident that their SLAF markers can fully represent the whole genome information of the studying species and to carry out their relevant research and analysis. However, other computer programs and software can be used for SLAF analysis and processing like those described in [6].

3.5 Conclusion

As a next-generation high-throughput simplified genome sequencing technology, SLAF-seq method is widely used in various fields of life science research with the advantages of simple operation, high-throughput, high efficiency, short experimental cycle, lower cost and no reference genome. It has strong advantages in the development of a large number of polymorphic molecular markers, the construction of high-density genetic linkage map, QTL gene mapping, genetic diversity analysis and assisted whole genome assembly.

However, the sequence generated by SLAF-seq technology has short reading length. It is, therefore, difficult to span genetic regions with high repeats and base preference. The filtering of repeated sequences can also lead to the reducing or complete lack of research on genome replication events. This might be the main limitation of SLAF-seq method for the research of complex genomic regions. When compared with the reference genome, the short read sequences may be aligned and show similarity in multiple locations in the genome. In the process of genetic map construction and analysis using database, data deviation will be introduced to filter out some results. Based on this occurring problem, the development of methods and analysis software for detecting and correcting of the deviations should be strengthening. Further upgrade and development of the analysis tools in various plant models and visual operation with the received data is required and welcome.

With the further improvement of sequencing technology and the reduction of price, SLAF-seq method will be more widely used in the field of plant science. In future research, SLAF-seq technology can be combined with transcriptome, metabolome, proteome, transgenic manipulation, CRISPR-Cas9 gene editing technology and other genetic and molecular methods to study "in-depth" the phenotypic, physiological, biochemical and genetic mechanisms related to important agronomic traits in crops. In this regard, SLAF-seq method is perfect for the realization of efficient and accurate molecular marker-assisted breeding, key trait mapping, genetic evolution relationship exploration and germplasm resources protection and utilization in various pant species.

4 Notes

1. Simplified genome technologies, such as SLAF, are not only suitable for species without reference genome, but also a very reasonable choice for economy reasons of the genotyping method in species with existing but very big or huge reference genome.
2. SLAF and other simplified genomic technologies represent only sequenced fragments after digestion with restriction enzymes, and this strategy greatly reduced the complexity of the studying genome.
3. According to different research purposes, the requirements for the number of molecular markers are also varied. If it is necessary to study functional interval scanning and gene mining in the whole genome, such as genome-wide association analysis (GWAS) and selection pressure analysis, tens of thousands of molecular markers with high-density are required. In case of phylogenetic relationship, geographical population structure, pedigree detection and other research, the molecular marker density does not need to be too high. Generally, only hundreds to thousands of molecular markers are needed to complete the analysis.
4. The selection of the length range of the restriction fragment varies according to the species and restriction combination.
5. Different linkage groups have various requirements for the number of markers during plant genotyping and molecular mapping. The recombination events are occurring with higher frequencies in more developed populations, like recombinant inbred lines (RIL) after several generations and with larger number of genotypes. Theoretically, increasing the marker density can effectively improve the quality of genetic map, so the more markers are better for molecular mapping and further plant genotyping.

References

1. Sanger F, Nicklen S, Coulson AR (1977) DNA sequencing with chain-terminating inhibitors. Proc Natl Acad Sci U S A 74:5463–5467. https://doi.org/10.1073/pnas.74.12.5463
2. Shendure J, Ji H (2008) Next-generation DNA sequencing. Nat Biotechnol 26:1135–4115. https://doi.org/10.1038/nbt1486
3. van Dijk EL, Jaszczyszyn Y, Naquin D, Thermes C (2018) The third revolution in sequencing technology. Trends Genet 34: 666–681. https://doi.org/10.1016/j.tig.2018.05.008
4. Metzker ML (2010) Sequencing technologies – the next generation. Nat Rev Genet 11:31–46. https://doi.org/10.1038/nrg2626
5. Niedringhaus TP, Milanova D, Kerby MB, Snyder MP, Barron AE (2011) Landscape of next-generation sequencing technologies. Anal Chem 83:4327–4341. https://doi.org/10.1021/ac2010857
6. Sun X, Liu D, Zhang X, Li W, Liu H, Hong W et al (2013) SLAF-seq: an efficient method of large-scale de novo SNP discovery and genotyping using high-throughput sequencing. PLoS One 8:e58700. https://doi.org/10.1371/journal.pone.0058700
7. Barchi L, Lanteri S, Portis E, Vale G, Volante A, Pulcini L et al (2012) A RAD tag derived marker based eggplant linkage map and the location of QTLs determining anthocyanin pigmentation. PLoS One 7:e43740. https://doi.org/10.1371/journal.pone.0043740
8. Jia J, Zhao S, Kong X, Li Y, Zhao G, He W et al (2013) *Aegilops tauschii* draft genome sequence reveals a gene repertoire for wheat adaptation. Nature 496(7443):91–95. https://doi.org/10.1038/nature12028
9. Zhang J, Zhang Q, Cheng T, Yang W, Pan H, Zhong J et al (2015) High-density genetic map construction and identification of a locus controlling weeping trait in an ornamental woody plant (*Prunus mume* Sieb. et Zucc). DNA Res 22:183–191. https://doi.org/10.1093/dnares/dsv003
10. Cai C, Cheng FY, Wu J, Zhong Y, Liu G (2015) The first high-density genetic map construction in tree peony (*Paeonia* sect. *Moutan*) using genotyping by specific-locus amplified fragment sequencing. PLoS One 10: e0128584. https://doi.org/10.1371/journal.pone.0128584
11. Ye Y, Cai M, Ju Y, Jiao Y, Feng L, Pan H et al (2016) Identification and validation of SNP markers linked to dwarf traits using SLAF-seq technology in *lagerstroemia*. PLoS One 11(7): e0158970. https://doi.org/10.1371/journal.pone.0158970
12. Song X, Xu Y, Gao K, Fan G, Zhang F, Deng C et al (2020) High-density genetic map construction and identification of loci controlling flower-type traits in chrysanthemum (*chrysanthemum × morifolium* Ramat.). Hortic Res 7: 108. https://doi.org/10.1038/s41438-020-0333-1
13. He Y, Yuan W, Dong M, Han Y, Shang F (2017) The first genetic map in sweet osmanthus (*Osmanthus fragrans* Lour.) using specific locus amplified fragment sequencing. Front Plant Sci 8:1621. https://doi.org/10.3389/fpls.2017.01621
14. Lu J, Liu Y, Xu J, Mei Z, Shi Y, Liu P et al (2018) High-density genetic map construction and stem total polysaccharide content-related QTL exploration for chinese endemic *Dendrobium* (Orchidaceae). Front Plant Sci 9:398. https://doi.org/10.3389/fpls.2018.00398
15. Ma JQ, Huang L, Ma CL, Jin JQ, Li CF, Wang RK et al (2015) Large-scale SNP Discovery and genotyping for constructing a high-density genetic map of tea plant using specific-locus amplified fragment sequencing (SLAF-seq). PLoS One 10:e0128798. https://doi.org/10.1371/journal.pone.0128798
16. Kent WJ (2002) BLAT - the BLAST-like alignment tool. Genome Res 12:656–664. https://doi.org/10.1101/gr.229202

Chapter 12

Modifications of Kompetitive Allele-Specific PCR (KASP) Genotyping for Detection of Rare Alleles

Anthony Brusa, Eric Patterson, and Margaret Fleming

Abstract

KASP is commonly used to genotype bi-allelic SNPs and In/Dels, and the standard protocol works well when both alleles are nearly equally prevalent in the DNA template. To detect rare alleles in bulked samples or to distinguish more than three genotypes, such as tri-allelic loci or mutations across orthologous genes in polyploids, adjustments to the protocol and/or data analysis are required. In this chapter, we present modified protocols for these non-traditional applications, including reaction conditions that enhance the fluorophore signal from rare alleles, resulting in increased KASP assay sensitivity. We also describe alternative KASP data analysis approaches that increase statistical certainty of genotyping calls. Furthermore, this increased assay sensitivity enables high-throughput genotyping using KASP, as samples can be pooled and tested in a single reaction. For example, rare alleles can be detected in mixed seed pools when present in ratios as low as 1 in 200. The assay modifications presented here expand the options available for complex genotyping, and retain KASP's advantages of being cheap, fast, and accurate.

Key words Data transformation, In/Del, KASP, Genotyping, Rare alleles, SNP, Species identification, Primer ratios, FRET cassette, Pooled DNA

1 Introduction

1.1 General Description of Assay

The problem of determining which single nucleotide polymorphism (SNP) is present at a locus has been addressed with a variety of techniques, including many that begin with locus amplification using PCR. Kompetitive Allele-Specific PCR (KASP) assays are a rapid and robust PCR-based approach for genotyping SNPs, which are currently used for breeding and inheritance studies, species identification, and for diagnostics such as resistance to herbicides or pathogens. There are several key advantages of KASP assay design. First is the inclusion of two forward primers in a single SNP genotyping reaction. Both forward primers will bind the same locus; however, the final (3′) base of each forward primer is complementary to one of the two SNP alleles. Optimized assay conditions ensure that the 3′base of each primer only anneals to its exact

Yuri Shavrukov (ed.), *Plant Genotyping: Methods and Protocols*, Methods in Molecular Biology, vol. 2638,
https://doi.org/10.1007/978-1-0716-3024-2_12,

complement, and thus only the SNP allele(s) present in the template DNA are amplified. Second, during primer design, each forward primer is given a unique oligonucleotide "tail" at the 5′-end. The assay reaction mix includes quenched, FAM- and HEX-labeled oligonucleotides; the 5′-tails of the forward primers are complementary to either the FAM- or HEX-labeled oligonucleotides. As PCR proceeds, the fluorophore-labeled sequence anneals to the amplicon's 5′-tail and is no longer quenched. Ultimately, this means that if one allele is present, either FAM or HEX fluorescence is detected, while both FAM and HEX fluorescence are identified in heterozygotes. Fluorescence is reported in arbitrary "fluorescence units," and the relative fluorescence (ratio of FAM to HEX fluorescence) can be used for a wide range of applications including standard diploid genotyping, rare allele detection, and polyploid genotyping.

Rare alleles are alleles which are present at a level well below that expected from normal segregation. Rare allele detection assays are most often used when working with bulked samples, such as detecting low level contamination of weed seed in a sample of agronomic seeds. In these cases, the allele of concern may be present in levels as low as 1–2%. Polyploid genotyping is an important application for many crops (e.g., wheat, quinoa, strawberries, etc.). If the SNP being targeted is only found on one set of species-derived chromosomes then the KASP assay will function as normal. However, if the target is found on multiple species-derived chromosomes then segregation will not follow the expected pattern. KASP can still be used in this case, but analysis and interpretation needs to be adjusted to account for differences in expected allele frequencies.

1.2 Primer Design

KASP primer design uses slightly different guidelines than traditional PCR primers. The effects of the FAM or HEX tail sequence on the forward primers (such as primer or amplicon length, or T_m) are not considered, as they do not interfere with normal amplification of the target. Recommended starting parameters include:

1. primer length = 21–30 bp
2. amplicon length = 42–100 bp (optimally 50 bp)
3. T_m = 62–66 °C
4. GC content = 30–55% (optimally 50%)
5. no more than four repeated single or dinucleotides
6. no more than three Gs and/or Cs in the final (3′-end) five bases
7. predicted secondary structure ΔG more positive than –9 kcal/mol
8. high binding specificity for the annealing site, which ideally is unique in the genome (*see* **Note 1**).

Table 1
Sequences of the standard FAM and HEX tails used in KASP

Fluorophore	5′-Tail sequence (5′ → 3′)
FAM: allele A	GAA GGT GAC CAA GTT CAT GCT
HEX: allele B	GAA GGT CGG AGT CAA CGG ATT

These sequences should be present on the 5′-end of the allele specific primer

Table 2
Annealing regions of a KASP marker for genotyping a 2 bp InDel

Annealing sequence	Sequence (5′ → 3′)
Allele 1 (insertion)	AGT GCT GTC GTG TCG TAG CTAG
Allele 2 (deletion)	AGT GCT GTC GTG TCG TAG AG

These parameters apply to both forward primers and the single reverse primer. By convention, the SNP-distinguishing primers are always called forward primers, regardless of their orientation with respect to the template.

SNPs neighboring the target SNP (or of interest) can be addressed in several ways. First, while the forward primers are restricted to ~20 bases adjacent to the SNP of interest, they can be placed either upstream or downstream of the SNP, so check both placements in case one includes fewer "ancillary" SNPs than the other. Second, if the SNPs are inherited together and form a diagnostic haplotype, the co-inherited alleles can be included in the forward primers to enhance diagnostic specificity. Third, if the additional SNPs are not diagnostic but cannot be avoided, "degenerate" bases can be used to encompass the allele possibilities at each additional SNP locus (*see* **Note 2**). Once primer sequences have been finalized, the appropriate tail sequence is added to the 5′-end of each forward primer. As FAM fluorescence is generally brighter and more consistent than HEX fluorescence, using the FAM tail for the rare allele (allele A) is recommended (and it was consistently used in this way in our research described below). Tail sequences are specific to the master-mix. Currently, LGC Biosearch Technologies uses the following sequences for their primer tails (*see* Table 1).

KASP assays are not limited to bi-allelic SNPs. For instance, insertion/deletion events (In/Dels) can be distinguished with KASP assays as long as the final base pair of the In/Del differs in the two genotypes. If a diagnostic In/Del sequence is as shown in Table 2, where the underlined CT in allele 1 is deleted in allele 2, in Bold C in allele 1 and A in allele 2 can be treated as a bi-allelic SNP for KASP assay design. In this way, almost any small polymorphism can be turned into a KASP assay [1].

A tri-allelic SNP locus may also be suitable for genotyping with a single KASP assay as well. For example, consider SNPs that result in a target-site herbicide resistance trait: the wild-type sequence (herbicide-susceptible) is CCC (Proline), but either of two mutations, UCC (Serine) or ACC (Threonine), confer resistance. An effective resistance diagnostic assay only needs to separate plants with wild-type C from plants with either U or A in the codon's first position, as either mutation will cause resistance. In this case, the KASP assay would use three forward primers, with the HEX tail added to the primer for the wild-type allele, and the FAM tail added to primers for each mutant allele. FAM fluorescence would signal herbicide resistance. The exact genotype for all resistant plants could then be determined—if needed—with a follow-up, bi-allelic KASP assay.

1.3 Template DNA

Template DNA can be obtained with a variety of DNA extraction methods. As with many PCR-based methods, KASP assays are highly sensitive and can amplify even small numbers of starting DNA template molecules (and details on using this to the researcher's advantage are given below). Practically speaking, this means crude DNA extractions that still contain polyphenolics, ethanol, sugars, or other PCR-inhibitory compounds can be substantially diluted so that PCR is not inhibited; although the DNA is also diluted, it can now be amplified. KASP is particularly insensitive to these contaminants compared to other fluorescence-based PCR reactions, such as expression analysis with qPCR or even TaqMan. Even relatively crude CTAB extractions can be used if the goal is high-throughput or low input genotyping [2, 3]. Theoretically, even cDNA could be amplified with KASP to quantify relative expression of two alleles or very similar orthologous genes.

1.4 PCR Conditions

KASP assays are set up as two-steps, touchdown PCRs. Beginning with ten cycles touchdown that slowly decreases the annealing/extension temperature (e.g., from 61 to 55 °C at −0.6 °C every cycle) provides stringent annealing conditions that enrich amplification with the correct forward primer. A further 20–40 cycles using the final touchdown temperature (e.g., 55 °C) for annealing/extension are sufficient to discriminate between the four outcomes (no amplification, homozygous-FAM, homozygous-HEX, heterozygous), which is usually done as an endpoint assay. If the initial fluorescence data are uninformative, reaction plates can go through more rounds of PCR ("recycling"), using three cycles at a time before re-reading and up to four rounds of recycling (12 PCR cycles total). However, over-amplification makes the fluorophores difficult to distinguish, as well as producing non-specific amplification (e.g., primer dimers) in the no-template controls (*see* Fig. 1).

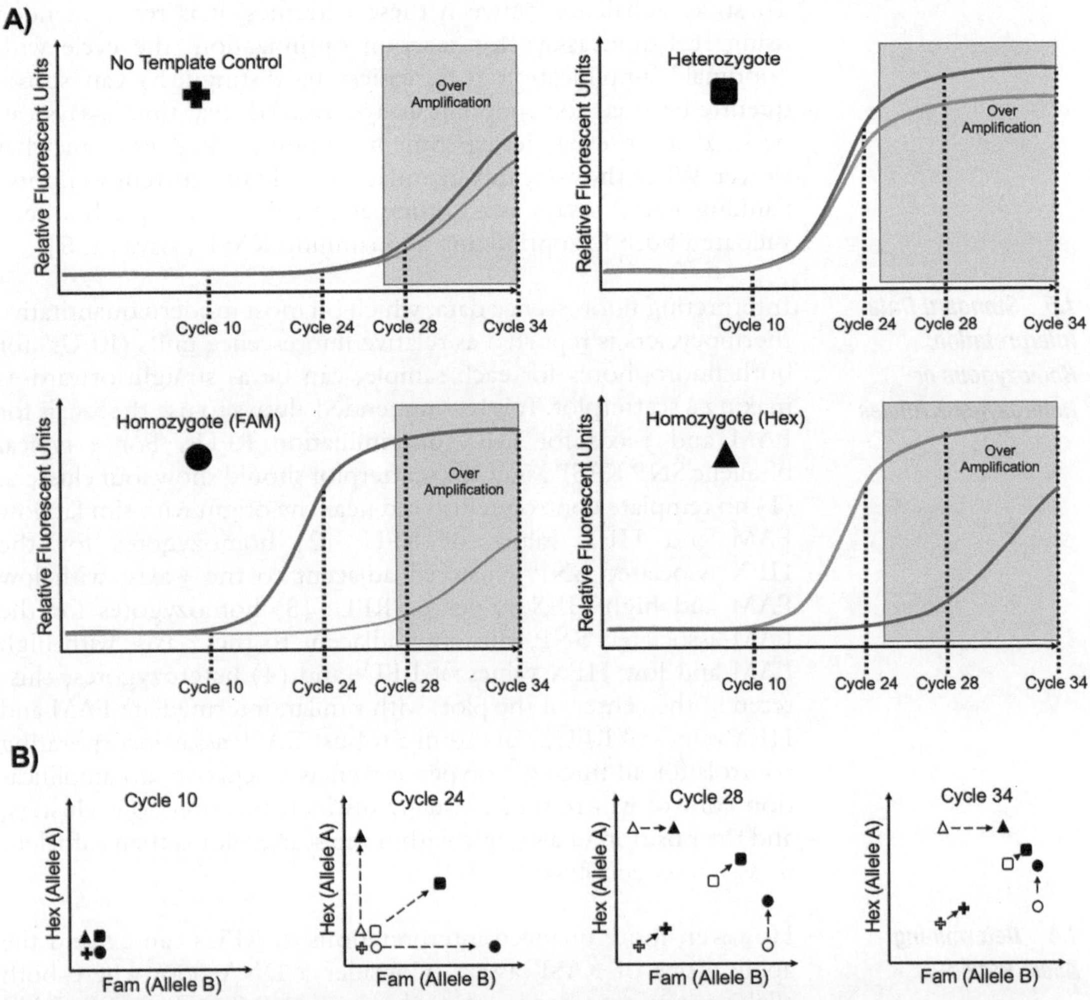

Fig. 1 (**a**) Four examples of real-time amplification curves for samples that are no-template controls (+), heterozygous (■), homozygous for FAM (●), and homozygous for HEX (▲). (**b**) Scatterplots of FAM and HEX fluorescence for the four sample types at successive cycles, indicated in panel (**a**). If over-amplification occurs (gray region, panel **a**), sample genotypes become less distinct as background and non-specific amplification obfuscate the results. Additionally, no-template controls begin to accumulate non-specific products which greatly alters the normalization of fluorescence. Collecting real-time data helps identify the optimal cycle for maximum differentiation between genotypes

To strike a balance between these extremes, it is recommended using real-time assays for reaction optimization; the cycle with "optimal" amplification (i.e., easiest to distinguish) can subsequently be used for endpoint assays. Indeed, real-time assays can be used for the whole experiment, although they are somewhat slower. While the assay mix manufacturers do not currently endorse running KASP assays as real-time assays, this technique has been validated both for optimizing and running KASP assays [4, 5].

1.5 Standard Data Interpretation: Homozygous or Heterozygous Alleles

Interpreting fluorescence data, which on most modern quantitative thermocyclers is reported as relative fluorescence units (RFUs) for both fluorophores for each sample, can be as straightforward as making a scatterplot. It is recommended always to use the *x*-axis for FAM and *y*-axis for HEX discrimination RFUs. For a typical bi-allelic SNP KASP assay, the scatterplot should show four clusters: (1) no template controls, clustered near the origin with similar, low FAM and HEX values of RFU; (2) homozygotes for the HEX-associated SNP, clustered adjacent to the *y*-axis, with low FAM and high HEX values of RFU; (3) homozygotes for the FAM-associated SNP, clustered adjacent to the *x*-axis, with high FAM and low HEX values of RFU; and (4) heterozygotes, clustered in the center of the plot, with similar, intermediate FAM and HEX values of RFU. Data from a robust KASP assay incorporating controls for all three genotypes as well as a negative, no amplification control will result in visually distinct, unambiguous clusters, and the position of a sample within the scatterplot is then sufficient to assign its genotype.

1.6 Determining Allele Ratios

However, more nuanced interpretations of RFUs can expand the applications of KASP assays. Consider a DNA pool where both alleles are present but in ratios that are not 0:0 (NTC), 1:0 (FAM allele), 0:1 (HEX allele), or 1:1 (Heterozygous). This situation could arise, for example, when quantifying mutations in homologous genes across the subgenomes of a polyploid (*see* Fig. 2) [6], or when the presence of a relatively rare allele needs to be detected in samples composed of pooled tissues from multiple individuals [4]. In both examples, FAM and HEX fluorescence would be expected in most samples, but the ratio of FAM:HEX fluorescence would deviate from those described above.

For the first example, consider hexaploid breadwheat, *Triticum aestivum*. Plants of bread wheat have three homologous genes of Acetyl CoA Carboxylase (ACCase), with one homolog in each genome (A, B, and D). A SNP was identified that could occur in any of three genomes of ACCase locus and caused resistance to an ACCase-inhibiting herbicide [7]. After breeding and selection for homozygosity at this locus, resistance was found to be dose-dependent: breadwheat lines with the herbicide-resistant ACCase allele in all three genomes were more resistant than lines with the

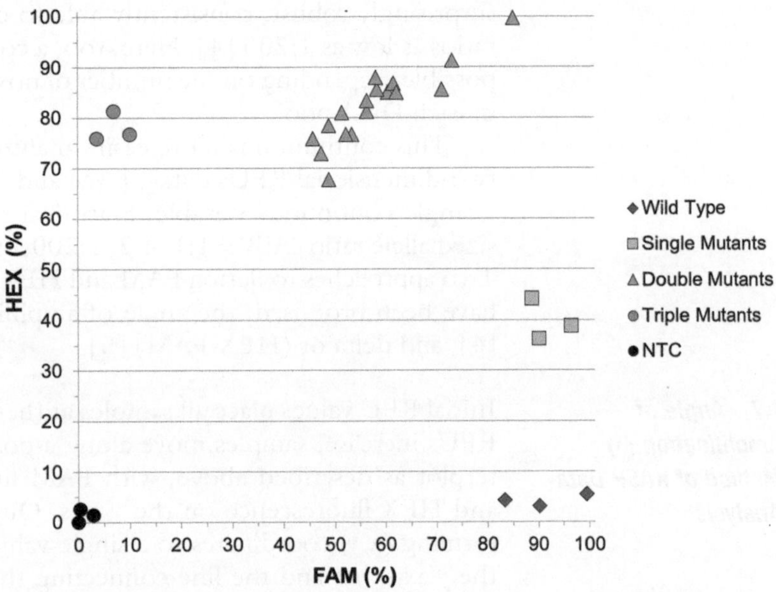

Fig. 2 A KASP scatterplot showing the power of KASP for analyzing novel allele ratios in hexaploid breadwheat. Samples amplify and cluster in a dose-dependent manner: the presence of the mutant (FAM-tagged) allele on multiple genomes shifts samples closer to the FAM axis

allele in only one genome. A traditional three-primer KASP assay amplified all three ACCase homologues, which showed not only presence or absence of the resistant allele but also dosage. FAM: HEX ratios of 0:6, 2:4, 4:2, and 6:0 were easily distinguished in the scatterplot: while individuals cluster in the interior of the plot, they are not all in the center, as expected for a diploid heterozygote (*see* Fig. 2). This type of assessment switches the focus of data analysis from three fixed states (homozygous for the FAM-tagged allele [*AA* genotype], homozygous for the HEX-tagged allele [*BB* genotype], or heterozygous with *AB* genotype) to a continuum of allele ratios.

This way of thinking about KASP data leads to an interesting question: What is the greatest difference in allele ratios that KASP can detect? 1:50? 1:100? For example, the absence of prohibited noxious weed species, such as *Amaranthus palmeri*, from seed lots must be validated before shipping them across state lines. When both the seed lot and the possible weeds are small-seeded and morphologically identical, this can be a slow and difficult task. However, if a sequence can be found that is common among the species and has a suitable polymorphism, KASP assay can greatly increase throughput. The ratio of alleles present in DNA extracted from a pool of seeds can reveal how many individuals from that pool have the rare (weedy) allele. It was found and reported that, when the DNA is clean and the assay is properly replicated, it is surprisingly robust, consistently able to detect an allele present at ratios as low as 1:200 [4]. Here, too, a continuum of allele ratios is possible, depending on the number of noxious weed seeds included in each DNA pool.

This continuum is more easily analyzed after transforming the two-dimensional RFUs data—FAM and HEX for each sample—to a single continuous variable. Statistical support for the hypothesized allele ratio (A:B = 1:1, 4:2, 1:200, etc.) can then be provided. Two approaches to flatten FAM and HEX RFUs to a single variable have been proposed, the angle of amplification or arctan method [4], and delta or (HEX-FAM) [5].

1.7 Angle of Amplification (θ) Method of KASP Data Analysis

Initial RFU values place all samples at the origin at PCR cycle 0. As RFUs increase, samples move along a positive trajectory in a scatterplot as described above, with FAM fluorescence on the *x*-axis and HEX fluorescence on the *y*-axis. Our first approach to transforming (*x*,*y*) coordinates to a single value uses the angle between the *y*-axis line and the line connecting the sample's starting point (0,0) with the sample's coordinates after amplification. This is the angle of amplification, θ. The value of θ indicates the proportion of allele A (FAM-tagged) to allele B (HEX-tagged). Values of 0°, 45°, and 90° correspond to ratios of B only, A:B = 1:1, and A only, respectively (θ_1, θ_2, θ_3 in Fig. 3). When $0° < \theta < 45°$, more B is present than A: e.g., if $\theta \approx 23°$, A:B = 1:3. Conversely, when $45° < \theta < 90°$, more A is present than B: e.g., if $\theta \approx 67°$, A:B = 3:1. Rare alleles will have angle values close to, but statistically distinct from, zero. Figure 4 shows an example of a 1-in-200 allele before and after transformation.

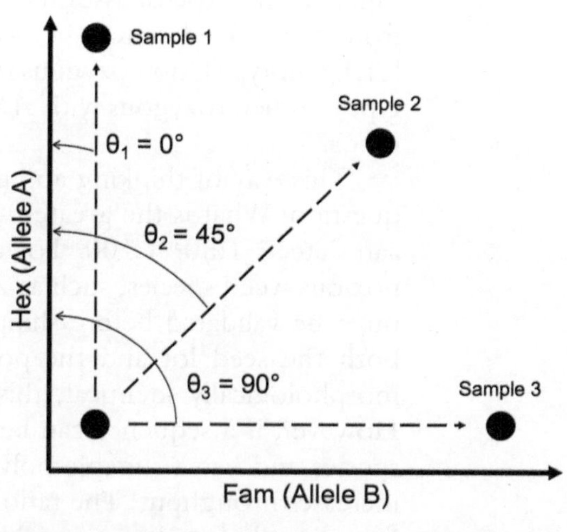

Fig. 3 An example KASP scatter plot demonstrating how the angle of amplification can be used as a single value that simplifies genotyping calls by translating (*x*,*y*) coordinates into a single value, θ

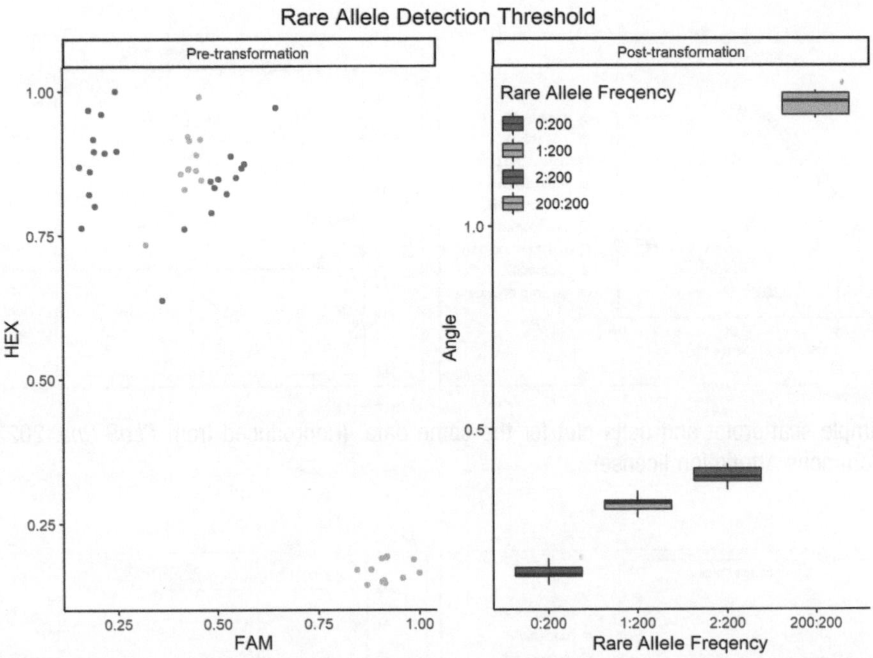

Fig. 4 An example of arctan transformation and calling for rare alleles (*Amaranthus palmeri* contamination in pooled seeds). (Figure is based on data published in *Pest Management Science* [4])

1.8 Delta Method of KASP Data Analysis

In a perfectly executed KASP assay, homozygotes show no fluorescence for the opposite fluorophore, and heterozygotes have exactly equal fluorescence from both fluorophores. Using this model, our second method simply finds the difference of HEX and FAM fluorescence, or delta—our convention is to define delta as HEX – FAM. However, data interpretation, especially across multiple assays, is greatly improved by first rescaling HEX and FAM values as percentages, then calculating delta. Then, for heterozygotes, equal HEX and FAM RFUs translate to delta = 0; for allele A (FAM-tagged), delta <0; for allele B (HEX-tagged), delta >0 (*see* Fig. 5). While a continuous range of delta values is possible, based on allele dosage from either gene copy number variation or DNA concentration, heterozygotes reliably have delta values near 0 regardless of DNA concentration. The delta values above or below which a sample is considered homozygous can be set independently for each assay (96-well plate) and for each allele. The delta method is particularly useful for situations where genotypes are poorly resolved on the scatterplot, which can occur from overamplification, primer dimerization, or other technical limitations (*see* Fig. 6).

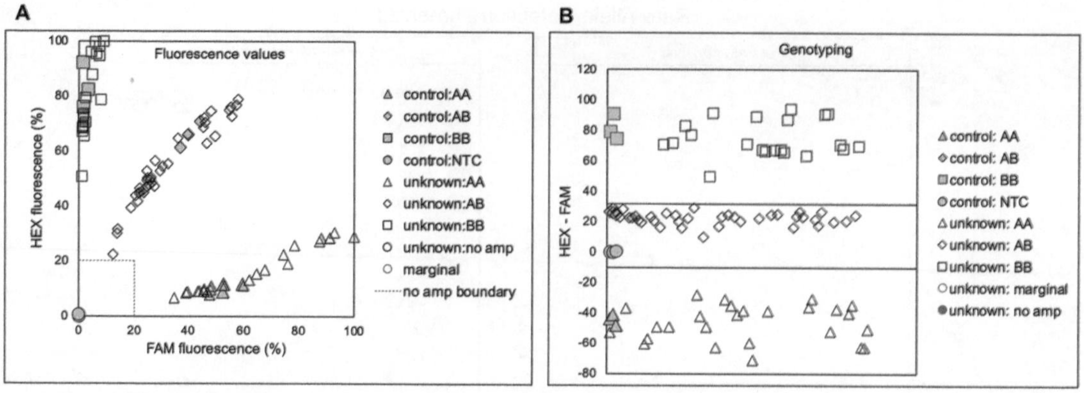

Fig. 5 Example scatterplot and delta plot for the same data. (Reproduced from *PLoS One*, 2022 [5] under Creative Commons Attribution license)

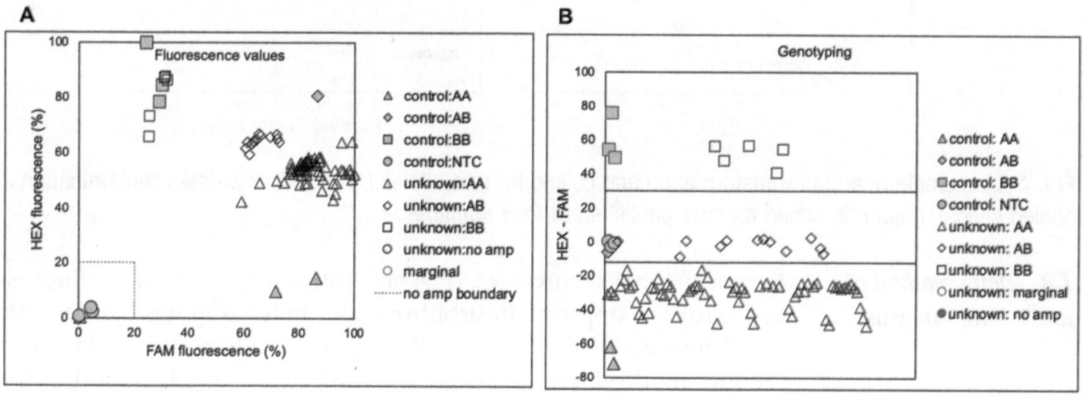

Fig. 6 Example scatterplot and delta plot for a KASP assay with sub-optimal amplification. This assay was intentionally over-amplified to see signal from *AA* controls, which were at a much lower DNA concentration than the unknowns. This led to samples with the *BB* genotype developing FAM fluorescence, and unknowns with the *AA* genotype developing HEX fluorescence. Nevertheless, the boundaries between the three genotypes are clear, especially when both plots are used together to interpret the data. (Reproduced from *PLoS One*, 2022 [5] under Creative Commons Attribution license)

2 Materials

1. KASP 2 × Master-mix.
2. For each sample, 5 μL of DNA at an approximate concentration of 10–20 ng/μL (*see* **Note 3**).
3. 1.5 mL microcentrifuge tubes.
4. One forward primer for each allele, one with a HEX tail and the other with a FAM tail (*see* **Note 4**).
5. Reverse primer.
6. Nuclease-free water.

7. Standard 96-well PCR plate.
8. Optically clear plate sealing tape.
9. Touchdown-capable thermal cycler (*see* **Note 5**).
10. Refrigerator (optional, *see* **Note 6**).
11. FRET-capable plate reader (*see* **Note 7**).
12. Method for transferring data from the plate reader (e.g., a USB drive).
13. Access to R, SPSS, SAS, Mathematica, Excel or other statistical software.

3 Methods

3.1 Preparation of Reaction Mixture for Rare Allele Detection

1. Create a 10 μM solution of each of the three KASP primers in nuclease-free water (*see* **Note 8** on alternative primer concentrations).
2. Create 150 μL of the following primer mix. The mix is specific for the alleles to be detected (*see* **Note 9**):
   - 3 μL of 10 μM forward primer (common allele)
   - 20 μL of 10 μM forward primer (rare allele)
   - 45 μL of 10 μM reverse primer
   - 82 μL of nuclease-free water
3. Create the reaction mixture by combining 432 μL of KASP 2 × Master-mix with 12 μL of the primer mix from **step 2**. This volume (444 μL) is sufficient for one 96-well plate (*see* **Note 10**).

3.2 Plate Preparation

1. Pipette 4 μL of reaction solution into each well of a standard 96-well PCR plate.
2. Add 4 μL of DNA (at approximately 10–20 ng/μL) to each well. Include no template controls (NTCs, using 4 μL of water instead of DNA) and controls for both common and rare alleles. Controls can be technical replicates. It is recommended allotting eight wells for common allele controls to assist in data interpretation and at least three wells each for NTC and common allele controls; having three wells of heterozygote controls is helpful but not required (*see* **Note 11**).

3.3 Running KASP Assay

Using a Lightcycler or other RT-PCR machine, run the following program:

15 m @ 94 °C

Touchdown program for ten cycles at (*see* **Note 5**):

20 s @ 94 °C

60 s @ 61 °C to 55 °C (increment with each cycle) (*see* **Note 12**).

Followed by 30 cycles of:

20 s @ 94 °C

60 s @ 55 °C (*see* **Note 12**)

30 s @ 28 °C (*see* **Note 13**)

Read fluorescence at 465–510 and 533–580 nm (*see* **Note 14**).

3.4 Data Interpretation

To interpret the results of a KASP assay for rare alleles we recommend one of the following methods (*see* **Note 15**).

3.4.1 Arctan Transformation

1. Export raw FAM (465–510 nm) and HEX (533–580 nm) RFUs from light cycler software as a .csv file. Import into your statistical software of choice (R, SPSS, SAS, Mathematica, etc.).
2. For each sample, perform an arctan transformation on the ratio of FAM to HEX RFUs to calculate the angle of amplification (θ) (*see* **Note 16**).

$$\theta = \arctan \frac{\text{FAM}}{\text{HEX}}. \quad (1)$$

3. Discard any reads that fall in the bottom 20% of your plot (*see* **Note 17**).
4. Generate a box-and-whisker plot (*see* Fig. 4 and **Note 18**).
5. A statistical comparison between the common allele control and test sample will validate if a given sample contains at least one individual of the rare genotype. The control replicates are used to construct a null distribution, from which *p*-values of the test samples can be determined.
6. As a service to the reader we have provided annotated code written in R using ggplot2:

   library(ggplot2) #loads the ggplot2 package (*see* **Note 19**).

   Data$angle = atan(Data$‘465-510‘/Data$‘533-580‘) #use arctan to find angle based on X-Y coordinates (*see* **Note 20**).

   Ggplot(Data, aes(x = Treatment, y = angle, fill = Treatment)) +

   geom_boxplot() #generate a box plot

   wilcox.test(angle ~ sample, Data) #determine p-values

3.4.2 Delta Transformation

1. Export raw FAM (465–510 nm) and HEX (533–580 nm) RFUs from lightcycler software as a .csv file. Import into your statistical software of choice (Excel, R, SPSS, SAS, Mathematica, etc.).
2. For each sample from a single 96-well plate, rescale fluorescence values from 0–100%:

$$\text{FAM\%} = 100 \times \frac{x - \min_{\text{FAM}}}{\max_{\text{FAM}} - \min_{\text{FAM}}} \quad (2)$$

$$\text{HEX\%} = 100 \times \frac{y - \min_{\text{HEX}}}{\max_{\text{HEX}} - \min_{\text{HEX}}} \quad (3)$$

where FAM% and HEX% are the rescaled values, x is the sample's raw FAM RFU value, $\min_{\text{FAM}}$ is the plate's smallest FAM RFU value, $\max_{\text{FAM}}$ is the plate's largest FAM RFU value, y is the sample's raw HEX RFU value, $\min_{\text{HEX}}$ is the plate's smallest HEX RFU value, and $\max_{\text{HEX}}$ is the plate's largest HEX RFU value.

3. Make a scatterplot with FAM% and HEX% on the x- and y-axes, respectively.
4. Remove any samples where FAM% < 20 and HEX% < 20, and ensure all NTCs and no other controls fall in this range.
5. For each sample, calculate delta:

$$\Delta = \text{HEX–FAM}$$

6. Plot delta on one axis (here, the y-axis), and arrange values along the other (x-) axis by sample order or another arbitrary scheme.
7. Compare the delta plot with the scatterplot to decide on cutoff values between genotypes. For heterozygotes delta ≈ 0, for homozygous HEX (allele B) delta > 0, and for homozygous-FAM (allele A) delta < 0. However, delta values from samples with the same genotype will vary with template DNA concentrations.

4 Notes

1. If a KASP primer has multiple binding sites within the genome it will interfere with the results of the assay. If a non-specific primer must be used, one solution is pre-amplification, using standard PCR to enrich for the unique region around the SNP of interest, followed by using that product in the standard KASP protocol described above. This will increase the signal strength of the target of interest by providing an excess of target annealing sites for KASP.

Table 3
Degenerate bases and their corresponding codes

Degenerate base	Nucleotides indicated
R	A or G
Y	C or T
S	G or C
W	A or T
K	G or T
M	A or C
B	C or G or T
D	A or G or T
H	A or C or T
V	A or C or G
N	Any base

Primers may be designed using these degenerate bases if the annealing region for a primer is polymorphic

2. Degenerate bases may be used in KASP primers when the annealing site for the primer contains a polymorphic base that is uninformative. Different letters used in the primer sequence indicate different polymorphism combinations (*see* Table 3) [8]. For example, the best reverse primer sequence may include an A/T SNP. In this case the primer would be ordered with the degenerate base W at that position, and the "primer" shipped to you would actually be an equal mixture of two primers with either A or T at that position.
3. A minimum concentration of 5 ng/μL of genomic DNA is recommended. Higher concentrations will also work, although concentrations above 50 ng/μL may see reduced amplification. Contrary to intuition, reducing the amount of DNA (or diluting the sample) often results in a stronger fluorescence signal. Any successful DNA isolation should be compatible with this approach, including crude leaf extracts.
4. We have chosen to use the FAM dye for rare alleles and the HEX dye for common alleles. The dye for the rare allele was chosen to correspond to the color channel with the most consistent readings (FAM), but dyes may be swapped during primer design if necessary.
5. A touchdown-capable thermal cycler is strongly recommended. The touchdown protocol improves the specificity of binding for the PCR reaction.

6. Depending on equipment availability, it may be useful to store plates after amplification and read them in a batch. Completed KASP reactions can be stored up to a week at room temperature before being read and recycled if needed.

7. A combination thermal cycler and plate reader should be used if possible. This will allow fluorescence to be read after each PCR cycle, which provides useful information for troubleshooting. Fluorescence signal amplification curves can be used to assess the rate of PCR activity to determine if reactions are over- or under-amplified (*see* Fig. 1) or if the primer fails to perform.

   Any machine being used for these protocols should be checked for two important specifications, ROX calibration and HEX/FAM read capability. Most modern thermocyclers do not require ROX, but older machines may require "high ROX" or "low ROX" KASP master-mix. In older machines, the detector array is at a fixed position; wells further from the detector will fluoresce differently than nearby wells, and the ROX signal is used to normalize HEX and FAM fluorescence readings by well position. Newer machines do not require ROX calibration because they have a mobile detector array that reads each well at the same distance and angle.

   Most machines are capable of reading HEX and FAM fluorescence wavelengths, or compounds with similar spectra. HEX fluorescence may be read by machines capable of reading JOE or VIC, while FAM can be read by machines capable of reading SYBR Green.

8. Effective primer concentrations can range from 10 to 100 μM, as long as all primers use the same base concentration. Adjusting this value will alter the overall rate of the reaction, which may help improve data-point clustering. Lower concentrations will require more PCR cycles and total fluorescence may be reduced if primer availability becomes limiting.

9. These volumes are the recommended amounts for rare allele detection. For assays where rare alleles are not expected, the following volumes for the primer mix should be used instead:

   18 μL of 100 μM forward primer (FAM tail)

   18 μL of 100 μM forward primer (HEX tail)

   45 μL of 100 μM reverse primer

   69 μL of nuclease-free water.

10. The total volume of reaction solution may be scaled based on the anticipated number of reactions. Excess reaction solution may be stored in a −20 °C freezer for future use, but should only be refrozen once. Repeated freeze-thaw of KASP reagents will result in reduced assay sensitivity.

11. Positive controls are the rare allele you are attempting to detect, while negative controls are the common allele. No

Template Controls (NTCs) are reactions without template DNA. NTC data points should remain near the origin. Positive and negative controls should fall near the *x* and *y* axes, respectively.

12. The standard annealing/extension temperature for KASP reactions is a touchdown over ten cycles from 61 to 55 °C, then 55 °C for the remainder of the program. These temperatures should be adjusted based on the predicted Tm values of your primers, especially if they vary greatly. If KASP fails to produce recognizable clusters we recommend reducing the cycling temperature to improve annealing specificity and/or running a temperature gradient to find ideal annealing temperatures for the second part of the PCR.
13. KASP assays cannot be read above 40 °C. Some thermal cyclers are unable to go as low as 28 °C. For these machines use the lowest available cycling temperature for this step. While this and the following step are only required for real-time assays, we have found that real-time and endpoint results correspond better if this step (30 s @ 28 °C) is also included in endpoint assays. For a traditional endpoint assay, this step and the following step are performed only once, after the final amplification cycle.
14. This protocol is for a real-time assay with plate reads after every amplification cycle to provide real time and/or troubleshooting information. 465–510 and 533–580 nm correspond to the optimal fluorescence peaks of the FAM and HEX dyes, respectively. At the time of writing KASP master-mix is only available with these two dyes. If your master-mix uses different dyes, change the fluorescence ranges as appropriate.
15. To date, standard software on lightcyclers is ***not*** designed for rare allele calling. The clustering algorithms will call a sample with low levels of a rare allele as lacking that allele, giving you false negative results. A data transformation is required to draw a distinction between these two groups.
16. Across subsequent cycles, KASP fluorescence points move from the origin out along a trajectory determined by the relative ratios of the two alleles. FAM-heavy samples will travel along the *x*-axis, while HEX-heavy samples will travel along the *y*-axis. Samples with a mix of both alleles will move diagonally. The angle of movement indicates the relative ratio of rare to common alleles, while the distance from the origin is a function of DNA concentration.
17. Any samples near the origin, defined by the FAM and HEX RFUs of the no-template controls, have failed to amplify properly and should be discarded. Failure to remove these data points can interfere with accurate calling. We have found a

cutoff of 20% of the maximum fluorescence for each fluorophore to work well, but this value can be adjusted based on your results. NTCs should always fall within the cutoff range.

18. To generate sufficient variation for a box and whisker plot, we recommend running eight technical replicates of your negative control (common allele). If a sample falls outside of the distribution of data points for the negative control then we conclude that the sample has at least one individual with the rare allele.
19. We make use of the ggplot2 package for generating figures. Other packages such as lattice may be used, including the base R functions.
20. The arctan transformation can be used to find the angle relative to either axis by swapping numerator and denominator. As written, the angle is calculated relative to the *y*-axis (HEX RFUs). This is recommended for rare allele detection, because the higher reliability of FAM fluorescence allows us to discriminate confidently between smaller RFU values, which translates to smaller angles relative to the *y*-axis.

References

1. de Figueiredo MRA, Küpper A, Malone JM, Petrovic T, de Figueiredo ABTB, Campagnola G et al (2022) An in-frame deletion mutation in the degron tail of auxin coreceptor *IAA2* confers resistance to the herbicide 2,4-D in *Sisymbrium orientale*. Proc Natl Acad Sci U S A 119: e2105819119. https://doi.org/10.1073/pnas.2105819119
2. Doyle JJ, Doyle JL (1987) A rapid DNA isolation procedure for small quantities of fresh leaf tissue. Phytochem Bull 19:11–15
3. Patterson EL, Fleming MB, Kessler KC, Nissen SJ, Gaines TA (2017) A KASP genotyping method to identify northern watermilfoil, Eurasian watermilfoil, and their interspecific hybrids. Front Plant Sci 8:752. https://doi.org/10.3389/fpls.2017.00752
4. Brusa A, Patterson EL, Gaines TA, Dorn K, Westra P, Sparks CD et al (2021) A needle in a seedstack: an improved method for detection of rare alleles in bulk seed testing through KASP. Pest Manag Sci 77:2477–2484. https://doi.org/10.1002/ps.6278
5. Fleming MB, Miller T, Fu W, Li Z, Gasic K, Saski C (2022) Ppe. XapF: high throughput KASP assays to identify fruit response to *Xanthomonas arboricola* pv. *pruni* (*Xap*) in peach. PLoS One 17:e0264543. https://doi.org/10.1371/journal.pone.0264543
6. Cuenca J, Aleza P, Navarro L, Ollitrault P (2013) Assignment of SNP allelic configuration in polyploids using competitive allele-specific-PCR: application to citrus triploid progeny. Ann Bot 111:731–742. https://doi.org/10.1093/aob/mct032
7. Bough RA, Westra P, Gaines TA, Westra EP, Haley S, Erker B et al (2021) The CoAXium® wheat production system: a new herbicide-resistant system for annual grass weed control and integrated weed management. Outlook Pest Manag 32:151–157. https://doi.org/10.1564/v32_aug_04
8. https://www.bioinformatics.org/sms/iupac.html

Chapter 13

Amplifluor-Based SNP Genotyping

Manmode Darpan Mohanrao, Senapathy Senthilvel, Yarabapani Rushwanth Reddy, Chippa Anil Kumar, and Palchamy Kadirvel

Abstract

Amplifluor, a genotyping system used to analyze single nucleotide polymorphisms (SNPs), is supplied by Merck-Millipore. Amplifluor is based on polymerase chain reaction (PCR) with two competing allele-specific primers and a SNP specific common reverse primer. Sequence information flanking SNP of interest and fluorescent plate reader for end-point measurement or qPCR machine for real time measurement are required for the execution of the Amplifluor assay. In this chapter, the principle and working protocol of the Amplifluor assay based on end-point fluorescence detection of SNP allele is presented with an example.

Key words Amplifluor, Single nucleotide polymorphisms (SNPs), Genotypic assays

1 Introduction

Single nucleotide polymorphisms (SNPs) represent variation among different individuals of a species for single base pair at the corresponding position of their genomes. Each SNP locus is defined by the flanking sequence to the polymorphic nucleotide. SNPs are extremely abundant, about one SNP every 100–300 bp of the plant genome. Different SNP densities are observed among genomes of different species and among different genomic regions of the same species. The SNPs have emerged as the marker of choice owing to their abundance, ease in discovery and extremely high throughput with relatively low cost per data point due to developments in next generation sequencing technologies. There are different SNP genotyping methods/platforms available, which include scoring of single locus to millions of SNP loci simultaneously. The Invader assay technology (Third Wave Technology) [1], KASP (LGC Biosearch Technologies), TaqMan (Applied Biosystems), Amplifluor (Merck-Millipore), Illumina GoldenGate platform [2], molecular inversion probe (MIP) technology [3]

Yuri Shavrukov (ed.), *Plant Genotyping: Methods and Protocols*, Methods in Molecular Biology, vol. 2638,
https://doi.org/10.1007/978-1-0716-3024-2_13,

and whole genome microarray platforms are some of the popular genotyping technologies available. Genotyping by sequencing (GBS) is now becoming a popular method for simultaneous SNP discovery and genotyping.

All the SNP genotyping systems mentioned above have their own advantages and disadvantages. The selection of SNP genotyping method for a particular experiment depends upon the number of SNPs to be used for genotyping, number of individuals to be genotyped and the cost required for genotyping. Most of the breeding applications like marker-assisted selection need a few to a hundred SNPs to be genotyped. SNP genotyping methods based on allele-specific PCR, which offers single to moderate multiplexing in comparatively less cost and requires only the routine lab equipment, are the best option for breeding applications. "Amplifluor SNP genotyping system" is one such technology. It is a patented technology, which facilitates high throughput end-point as well as real-time detection of nucleic acids.

Amplifluor SNP genotyping assay utilizes the Amplifluor Universal Primer (UniPrimer) system [4]. Amplifluor Universal systems are based on energy transfer from an excited fluorophore to a quencher that results in quenching of the fluorescence. This quenching is achieved by linking the fluorophore and the quencher, 4-(dimethylamino)azo benzene sulfonic acid (DABSYL) to an oligonucleotide primer. The optimized design of the UniPrimer results in a large increase in fluorescent emission only upon its incorporation in the amplified products generated during each PCR cycle. The other essential components of this system are reaction buffer, dNTPs and *Taq* polymerase. Variable components used in this system are three unlabelled primers, which are SNP specific and the template (test) DNA.

One of the two Amplifluor Universal Primers is labelled with fluorescein (FAM) while the other Amplifluor Universal Primer is labelled with sulforhodamine (SR) or 6-carboxy-4′,5′-dichloro-2′,7′-dimethoxyfluorescein (JOE). Each UniPrimer includes four elements as follows: (i) the fluorophore, which is located at the 5′-base of the UniPrimer, (ii) the quencher, which is linked to the nucleotide complementary to the 5′-base (iii) the hairpin loop and stem, which is formed due to complementarity between few nucleotides at 5′-end and nucleotides in the vicinity of the nucleotide base which is attached to quencher and (iv) a specific tail region of 21 nucleotide base pairs at 3′-end of the UniPrimer (*see* Fig. 1). The specific tail regions at 3′-end of UniPrimer are identical to the tail region at 5′-end of allele-specific primers. Sequence of specific tail regions is discussed later under designing of SNP specific primers.

The Amplifluor SNP genotyping system works on the basis of the following principle. In a PCR reaction mixture, the very first step is the annealing of allele-specific primers to the target region of

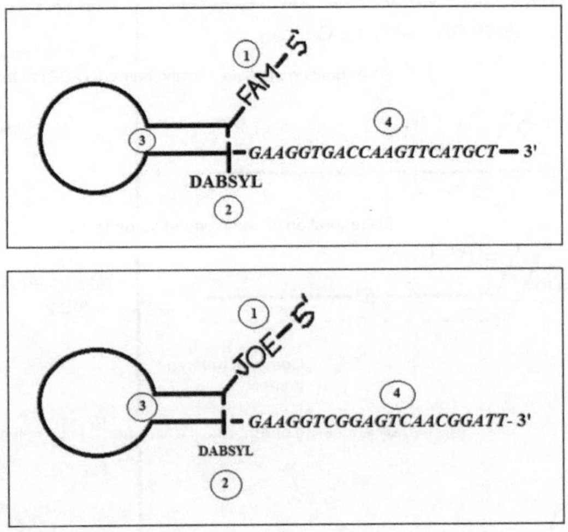

Fig. 1 Structure of Amplifluor UniPrimer with FAM and JOE. (Adapted and modified from refs. [10, 11])

DNA, and its elongation by *Taq* polymerase. This will form DNA sequence having allele-specific primer sequence at 5′-end and complementary sequence of common reverse primer. The second step is annealing of common reverse primer to its complementary sequence in sequence generated in the first step. Annealing of common reverse primer and its elongation by *Taq* polymerase synthesizes tail sequence complement. In the third step, specific annealing of Amplifluor UniPrimer to the PCR product generated in step 2 containing complementary sequence of specific tail occurs. In the fourth step, unfolding of hairpin-like structure of Amplifluor UniPrimer occurs as a result of elongation of common reverse primer by *Taq* polymerase. As soon as the hairpin structure unfolds, link between fluorophore and quencher gets disrupted and results in generation of fluorescent signal (*see* Fig. 2).

2 Materials

2.1 Primer Design

1. DNA sequence containing the SNP of interest (*see* **Note 1**).
2. Software for designing of primers (*see* **Note 2**).

2.2 Amplifluor Assay

1. Genomic DNA of the samples with a concentration of 1–10 ng/μL (*see* **Note 3**).
2. Amplifluor master mix components supplied by the manufacturer (*see* **Note 4**).
3. *Taq* polymerase (*see* **Note 5**).
4. SNP specific Assay mix including two allele-specific primers and a common reverse primer.

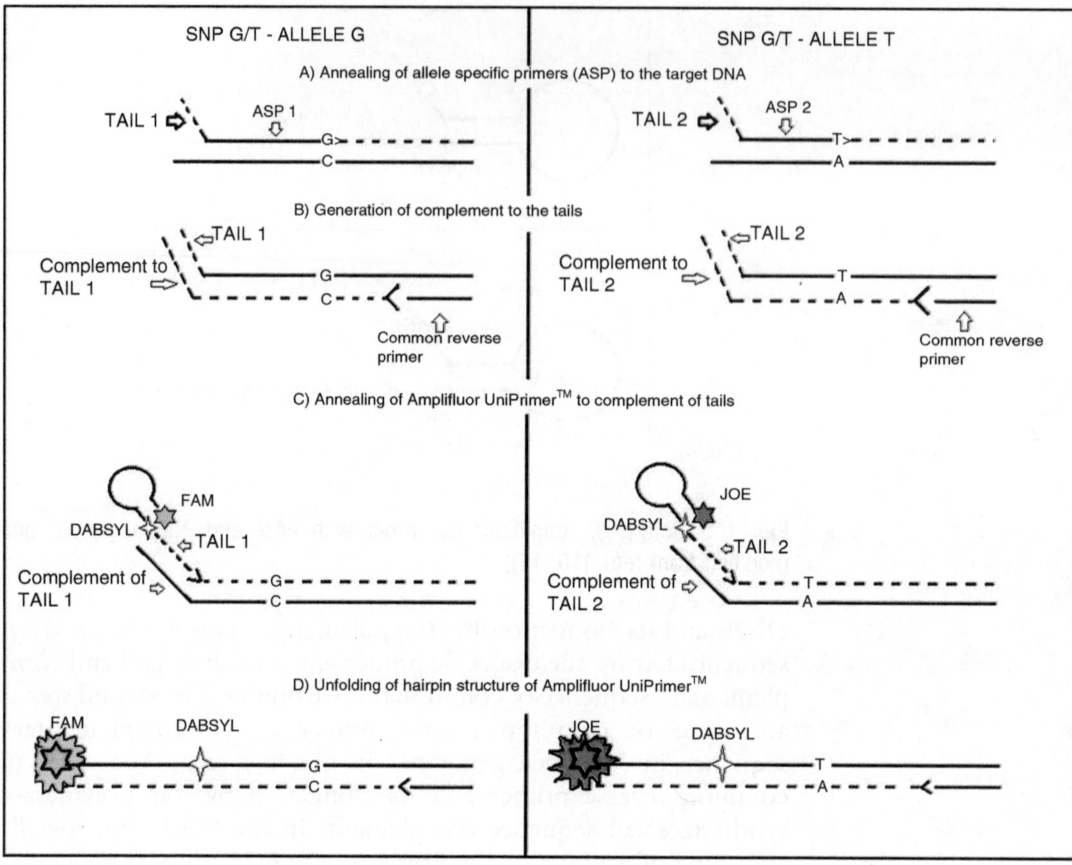

Fig. 2 Flowchart explaining working principle of Amplifluor SNP genotyping system. Arrowhead ">" or "<" represents direction of polymerization by *Taq* polymerase. (Adapted and modified from refs. [10, 11])

5. Nuclease free water.
6. PCR plate and optical clear seal (*see* **Note 6**).
7. Fluorescent plate reader or qPCR instrument (*see* **Note 7**).

3 Methods

3.1 Design of SNP Specific Primers

Amplifluor SNP genotyping system requires SNP specific set of three unmodified (non-labelled) oligonucleotides consisting of two allele-specific primers having two identified variants of SNP at 3′-end and one common reverse primer. The single common reverse primer is designed to generate amplicon with the optimum length of 70–150 bp. Each allele-specific primer has specific tail extension on their 5′-end, which is identical to the tail extension at 3′-end of Amplifluor UniPrimer. An example of SNP specific set of three unmodified oligonucleotides representing wild and mutated allele in fatty acid desaturase 2–1 gene, *CtFAD2–1*, associated with oleic acid content in safflower [5] is presented in **Note 8**.

Table 1
Preparation of 20 × Primer-mix (200 μL)

Components	Stock concentration (μM)	Volume (μL)
Allele-specific primer with FAM tail	100	1
Allele-specific primer with JOE tail	100	1
Common reverse primer	100	15
dH_2O		183

Table 2
Components of PCR cocktail

Reagent	For 1 reaction, μL (96-well plate)	For 1 reaction, μL (384-well plate)
dH_2O	4.6	1.3
dNTPs (2.5 mM each)	0.8	0.4
20 × Amplifluor UniPrimer FAM	0.5	0.25
20 × Amplifluor UniPrimer JOE	0.5	0.25
10 × Reaction mix S plus[a]	1.0	0.5
20 × SNP specific primer mix	0.5	0.25
Taq polymerase (5 U/μL)	0.1	0.05
Genomic DNA (~5 ng/μL)	2.0	2.0
Total	10.0	5.0

[a]Contains ROX which is used as passive reference dye for normalization of FAM and JOE values

3.2 Genotyping

1. Preparation of 20 × Primer mix: The 20 × SNP specific primer mix contains 0.5 μM of each allele-specific primer and 7.5 μM of the common reverse primer in dH_2O or 1 × TE. 20 × Primer mix must be prepared as shown in Table 1 (*see* **Note 9**).
2. Select the plate scale suitable for genotyping project as the subsequent use of the volume of PCR master mix components depends on the plate scale (*see* **Note 10**).
3. Pipette DNA samples into the wells of the plate.
4. Prepare a mixture of genotyping master mix and the SNP specific assay mix as shown in Table 2 (*see* **Note 11**).
5. Add the mixture of master mix and assay mix to the DNA samples.
6. Cover the PCR plate with an optical seal.
7. Briefly centrifuge the plate.

Table 3
PCR programme for Amplifluor SNP genotyping

Steps	Temp (°C)	Time	Cycle
Pre-heat block	96	–	–
Initial denaturation	96	4 min	1
Denaturation Annealing Extension	96 55–60 72	10–15 s 5 s 10 s	20
Denaturation Annealing Extension	96 55 72	10–15 s 20 s 40 s	22–24
Final extension	72	3 min	1
Hold	20	Hold	1

Table 4
The excitation/emission settings for fluorescence reading

Amplifluor label	Excitation (nm)	Emission (nm)
FAM	485 + 7	535 + 15
JOE	520	548
Sulforhodamine	585 + 10	620 + 4
ROX	575	610

8. Run the PCR plate in a Thermal cycler with the thermal cycling conditions as presented in (*see* Table 3 and **Note 12**).
9. Measure fluorescence in a plate reader pre-set to read fluorescence at a specific wavelength (*see* Table 4).
10. Interpret the fluorescence data using the clustering software (*see* **Notes 13** and **14**; Fig. 3).

4 Notes

1. DNA sequence of at least 50 bp flanking the SNP of interest must be available for designing the primers. Quality of DNA sequence determines the success rate of designed primers for the assay.
2. Set of three primers (two allele-specific forward primers and a SNP specific common reverse primer) can be designed using software such as WASP tool available at https://bioinfo.biotec.or.th/WASP [6] with the option "no mismatch" at the penultimate position of primers.

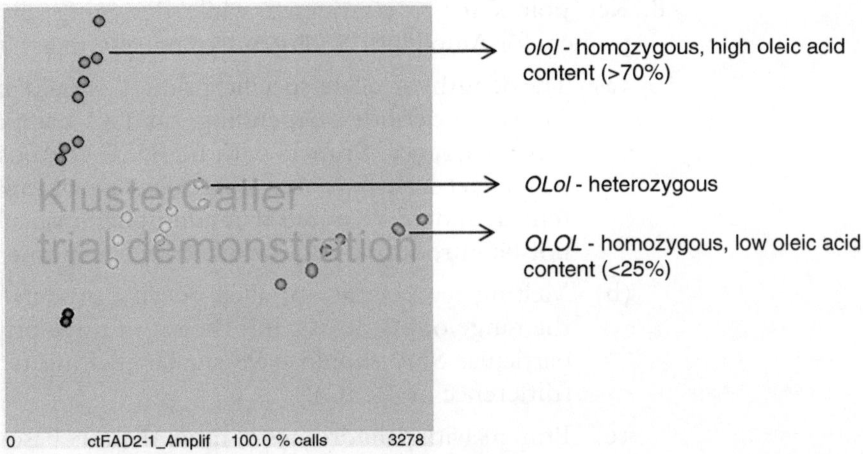

Fig. 3 Visualization of SNP alleles of *CtFAD2–1* using Amplifluor assay. Blue dots, *OLOL*-homozygous alleles for low oleic acid content (<25%); green dots, *OLol*-heterozygous; red dots, *olol*-homozygous alleles for high oleic acid content (>70%) in a set of safflower genotypes [5]

3. Working DNA concentration of 1–10 ng/μL is enough for amplification in Amplifluor genotyping assay. Higher the quality of DNA, lesser the concentration of working samples is required. Good quality DNA samples of concentration 5 ng/μL yielded a robust amplification in safflower [5].
4. Components of Amplifluor System supplied by manufacturer including 20 × Amplifluor UniPrimer FAM, 20 × Amplifluor UniPrimer JOE, 10 × Reaction mix S plus and dNTPs (2.5 mM each) must be stored properly at recommended temperature (−20 °C for short term or at −80 °C for long term storage). Frequent thawing and freezing of these components affect their activity; therefore, it should be minimized to the extent possible for realizing a robust amplification. Components of Amplifluor System can be stored in small aliquots to reduce the number of thawing and freezing cycles.
5. Optimal *Taq* polymerase: Antibody mediated "Hot start" *Taq* polymerase is considered optimal for Amplifluor SNP genotyping. Platinum *Taq* DNA polymerase of Invitrogen [5, 7], Titanium *Taq* of BD Clontech [8] and JumpStart *Taq* of Sigma [9] have been successfully used.
6. PCR plates made up of opaque material should be used. The optical clear seal which sticks very tightly to the plate should be used to prevent evaporation issues of the reaction mix.
7. Perkin-Elmer Victor X3 [5], Perkin-Elmer Wallac Victor 1420 Fluorescence microplate reader and FLUOstar Optima (BMG Labtech) [7] have been successfully used for end-point measurement of fluorescence values for Amplifluor SNP genotyping. ABI PRISM sequence detection systems (Applied Biosystems) have also been used for end-point as well as real time detection of fluorescent readings.

8. Key points to be considered while designing allele-specific primers for Amplifluor SNP genotyping system are as follows:
   (a) The length of allele-specific primers should be around 17–25 nucleotides depending on GC composition of DNA sequence. Primers with high GC composition may be shorter in length, and longer in case of low GC content. It should be noted that additional mismatches need not be introduced into the allele-specific primers.
   (b) Melting temperature of allele-specific primers can be in the range of 56–66 °C, but the set of three primers for a particular SNP should have similar melting temperature (difference of <2 °C).
   (c) Primers with palindromes of more than six bases of length should be avoided.
   (d) At times, allele-specific primers having same primer length may not work. In such case, adjustment of the length of allele-specific primers can be done accordingly to match the melting temperature of two allele-specific primers. Increasing the length of the primer having comparatively less melting temperature among two allele-specific primers may resolve this problem.

   A set of three primers (unmodified oligonucleotides) used in our study for amplifying wild and mutated allele of fatty acid desaturase 2–1 gene, *CtFAD*2–1, associated with oleic acid content in safflower [5] is given below.

   Allele specific primer 1 5′*GAAGGTGACCAAGTTCATGCT*CAGGCGAAACGGTTGTAGGG

   Allele specific primer 2 5′*GAAGGTCGGAGTCAACGGATT*CAGGCGAAACGGTTGTAGGT

   Common Reverse primer: 5′ CAGCCTGTTCGCCACTCTCA

   (The specific tail extensions in allele-specific primers are in italics and underlined).

   Tail 1—The 21 bp sequence "*GAAGGTGACCAAGTTCATGCT*" is attached to 3′-end of Amplifluor Universal primer labelled with FAM, and at 5′-end of one of the allele-specific primer.

   Tail 2—The 21 bp sequence "*GAAGGTCGGAGTCAACGGATT*" is attached to 3′-end of Amplifluor Universal primer labelled with JOE, and at 5′-end of one of the allele-specific primer.

9. The working stock of 20 × SNP specific Primer mix (200 μL) is sufficient for >700 reactions (for 384-well PCR plate format). This Primer mix, once prepared, can be stored at −20 °C for future use.

Table 5
Alternative programs used for Amplifluor SNP genotyping

	[5]			[7]		
Steps	**Temp (°C)**	**Time**	**No. of cycles**	**Temp (°C)**	**Time**	**No. of cycles**
Pseudo hot start	–	–	–	94	5–10 s	1
Initial denaturation	95	15 min	1	95	4 min	1
Denaturation Annealing Extension	95 65 (−0.6 °C/cycle)	20 s 1 min	10	94 55 72	10 s 20 s 40 s	35
Denaturation Annealing Extension	94 59	20 s 1 min	26	– –	– –	–
Final extension	–	–	–	72	3 min	1
Hold	4	Hold	1	20	Hold	1

10. PCR plate such as 96-well, 384-well and 1536-well can be used as per the requirement. When 384-well PCR plate is used, volume of the PCR master mix could be reduced to half (compared to 96-well plate).
11. All the components of PCR except 20 × SNP specific primer mix, *Taq* polymerase, nuclease free water and template DNA can be ordered as the Amplifluor SNP genotyping system kit supplied by the manufacturer (Merck-Millipore).
12. The temperature for annealing may be optimized for individual assays as annealing temperature may differ. The sequences with high GC content may require longer duration for denaturation at 96 °C. Some of the PCR programmes (Thermal cycling conditions) reported in the literature are given Table 5.
13. Interpretation of fluorescence data: The fluorescence data can be viewed graphically, using MS Excel or software such as "KlusterCaller" (LGC Genomics, UK), by plotting the FAM and JOE/HEX data on the *x*- and *y*-axes, respectively, to visualize the alleles as illustrated in Fig. 3 citing the example from [5]. At times, some samples may show higher fluorescence values, although these higher values seem reliable, they might affect the clustering of the other samples. In such incidents, excluding those samples can help in a better clustering of the remaining samples.
14. Sometimes, a heterozygous cluster may be positioned closer to a homozygous cluster due to assay biasness towards one allele. In such cases concentration of allele specific primers used in preparing 20 × SNP specific assay mix should be optimized

(by reducing the concentration of the primer towards which assay is showing biasness and increasing the concentration of the other primer). Also, having control DNA, for homozygotes and heterozygotes will help in discriminating the samples correctly.

References

1. Lyamichev V, Mast AL, Hall JG, Prudent JR, Kaiser MW, Takova T et al (1999) Polymorphism identification and quantitative detection of genomic DNA by invasive cleavage of oligonucleotide probes. Nat Biotechnol 17:292–296. https://doi.org/10.1038/7044
2. Oliphant A, Barker DL, Stuelpnagel JR, Chee MS (2002) BeadArray™ technology: enabling an accurate, cost-effective approach to high-throughput genotyping. BioTechniques 32: S56–S61
3. Hardenbol P, Yu F, Belmont J, MacKenzie J, Bruckner C, Brundage T et al (2005) Highly multiplexed molecular inversion probe genotyping: over 10,000 targeted SNPs genotyped in a single tube assay. Genome Res 15:269–275. https://doi.org/10.1101/gr.3185605
4. Nazarenko IA, Bhatnagar SK, Hohman RJ (1997) A closed tube format for amplification and detection of DNA based on energy transfer. Nucleic Acids Res 25:2516–2521. https://doi.org/10.1093/nar/25.12.2516
5. Kadirvel P, Veerraju C, Senthilvel S, Yadav P, Usha Kiran B, Shaik M et al (2020) Marker-assisted selection for fast-track breeding of high oleic lines in safflower (*Carthamus tinctorious* L.). Ind Crops Prod 158:112983. https://doi.org/10.1016/j.indcrop.2020.112983
6. Wangkumhang P, Chaichoompu K, Ngamphiw C, Ruangrit U, Chanprasert J, Assawamakin A et al (2007) WASP: a Web-based Allele-Specific PCR assay designing tool for detecting SNPs and mutations. BMC Genomics 8:275. https://doi.org/10.1186/1471-2164-8-275
7. Bengra C, Mifflin TE, Khripin Y, Manunta P, Williams SM, Jose PA et al (2002) Genotyping of essential hypertension single-nucleotide polymorphisms by a homogeneous PCR method with universal energy transfer primers. Clin Chem 48:2131–2140. https://doi.org/10.1093/clinchem/48.12.2131
8. Ochs-Balcom HM, Cicek MS, Thompson CL, Tucker TC, Elston RC, Plummer SJ et al (2008) Association of vitamin D receptor gene variants, adiposity and colon cancer. Carcinogenesis 29:1788–1793. https://doi.org/10.1093/carcin/bgn166
9. Tsai YJ, Choudhry S, Kho J, Beckman K, Tsai HJ, Navarro D et al (2006) The *PTGDR* gene is not associated with asthma in 3 ethnically diverse populations. J Allergy Clin Immunol 118:1242–1248. https://doi.org/10.1016/j.jaci.2006.07.045
10. Merck-Millipore (2013) Amplifluor SNPs HT genotyping system for assay development: Amplifluor manual. https://www.sigmaaldrich.com/deepweb/assets/sigmaaldrich/product/documents/136/578/amplifluormanual-s7909.pdf
11. Jatayev S, Kurishbayev A, Zotova L, Khasanova G, Serikbay D, Zhubatkanov A et al (2017) Advantages of Amplifluor-like SNP markers over KASP in plant genotyping. BMC Plant Biol 17:254. https://doi.org/10.1186/s12870-017-1197-x

Chapter 14

SNP Genotyping with Amplifluor-Like Method

Gulmira Khassanova, Sholpan Khalbayeva, Dauren Serikbay, Shynar Mazkirat, Kulpash Bulatova, Maral Utebayev, and Yuri Shavrukov

Abstract

For SNP genotyping, amplification of fluorescence (Amplifluor) is a popular and actively developing method in the plant sciences. The "Amplifluor-like" is a "home-made" modification of the original commercial Amplifluor method. Amplifluor-like genotyping requires two essential components: (1) two allele-specific forward primers targeting the SNP site with one common reverse primer; and (2) a universal part with two non-allele-specific molecular probes containing one of the two used fluorophores and a quencher. Allele discrimination is based on the fluorescence score, where the dominance of one dye over the other confirms the presence of each specific SNP allele. The Amplifluor-like method is similar to commercial KASP and original Amplifluor methods but is much cheaper because all components can be ordered as regular and modified oligos. The easily adaptable Amplifluor-like method can be modified by any researcher to make it suitable for available instruments, reagents and conditions in low-budget laboratories for SNP genotyping of any plant species with identified genetic polymorphism.

Key words Amplifluor-like, Genotyping, Fluorescence, FRET, SNP, Universal molecular probes

1 Introduction

There is an enormous range of methods for plant genotyping based on SNP (Single nucleotide polymorphism), where the difference in a single nucleotide can be detected and used for differentiation in genotyping (reviewed in refs. [1–5]). Fluorescence resonance energy transfer (FRET) is one of the approaches popularly used in recent times and provides the platform for several new actively developing methods for plant genotyping (reviewed in refs. [6, 7]).

The principles of the amplification with fluorescence method (Ampliflour) were described earlier [8–10]. Since that time, Amplifluor has been commercialized and manufactured by Millipore, which currently is a part of the Merck-Sigma-Aldrich Company. Amplifluor is successfully used in research and for diagnostic purposes, but the cost remains a little high [11, 12], compared to other methods [13–16].

Yuri Shavrukov (ed.), *Plant Genotyping: Methods and Protocols*, Methods in Molecular Biology, vol. 2638,
https://doi.org/10.1007/978-1-0716-3024-2_14,

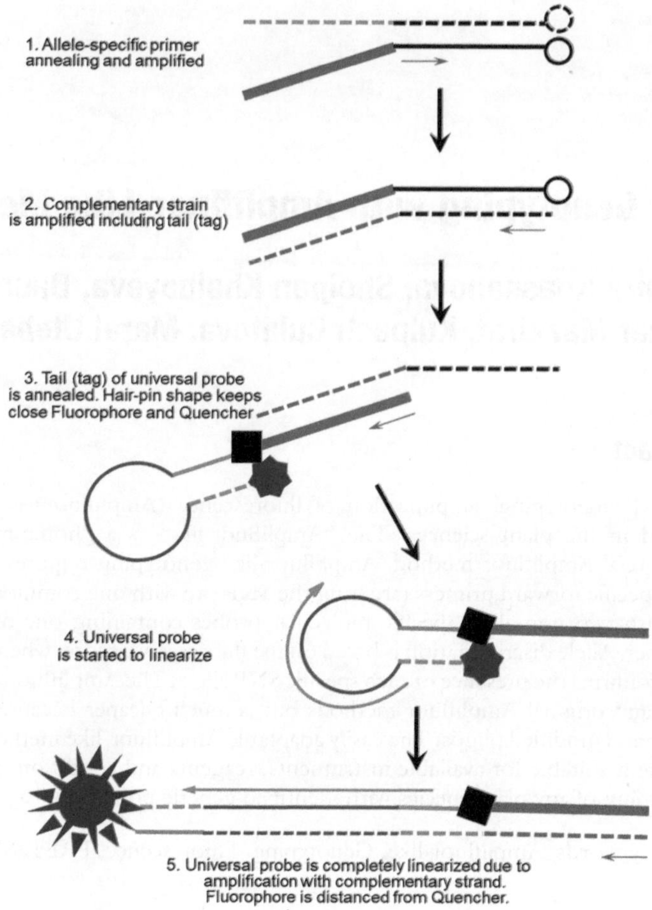

Fig. 1 The principle and main steps of the Amplifluor-like SNP genotyping method

The "Amplifluor-like" method for SNP genotyping is a "home-made" modification version of the original Amplifluor. Based on published papers and open access information on Amplifluor technology, provided particularly in Rickert et al. [17], our Amplifluor-like method was developed paying respect to the original method and avoiding commercial overlaps [18]. It was successfully applied in research with chickpea [19], sugar beet [20], barley [21, 22], wheatgrass [23], and bread wheat [24].

The principles of Amplifluor-like SNP genotyping were also published earlier [18] and are summarized in Fig. 1. This is a FRET-based SNP genotyping method involving PCR amplification with one of two used fluorophores, which correspond to two alternate SNPs and, therefore, has some similarities with other methods, particularly KASP [16, 18]. The Amplifluor-like method has two major parts: (1) An allele-specific part with three non-labelled primers including two allele-specific forward primers targeting the SNP site and one common reverse primer to make initial amplification; and (2) A universal part with two non-allele-specific molecular probes containing one of two used fluorophores and a quencher. The detection of fluorescence at a higher level from one dye compared to the other used dye confirms the presence of each particular SNP allele with the final score used for allele discrimination (*see* Fig. 1).

One of the main attractions of the Amplifluor-like method is the complete freedom to change or re-arrange any components, including the structure and design of allele-specific primers and universal probes, as well as the amplification program, such as the temperature and duration of each step.

2 Materials

1. Autoclaved milli-Q water is recommended.
2. DNA extracted by any method and diluted to a concentration of 10–50 ng/μL.
3. Two non-labelled allele-specific (AS) forward primers (about 22 bp each) targeting SNP at the 5′-end, with attached specific Tail-1 or Tail-2 (21 bp each).
4. One common reverse primer (about 22 bp).
5. Two Universal molecular probes (Uni-1 and Uni-2) with FAM or HEX/VIC on the 5′-end, a quencher in the middle, and Tails on the 3′-end, with a total length of 41–43 bp.
6. Reagents for regular PCR, including any Taq-polymerase kit (preferably, "Hot-start") with supplementary buffer and $MgCl_2$, and additional dNTP mix.
7. Consumable PCR microplates and clear seals. Alternatively, PCR microtubes or strips can be substituted.
8. Cold block or basis for setting up of the PCR cocktail and DNA sample loading to minimize evaporation.
9. Any available qPCR instrument with at least two channels for fluorophores with different wave-length absorptions. Any manufacturer, brand and year are suitable.
10. A Microplate reader capable of detecting fluorescence or a regular PCR cycler represent alternative options.

3 Methods

1. Identify a high-quality SNP by Sanger sequencing in two genotypes, which will be used as references for further genotyping (*see* **Note 1**).
2. Develop a pair of allele-specific (AS) forward primers, which are ideally 22 bp and 50% GC content but differ by only one nucleotide in the last position of SNP in the 3′-terminus (or in some cases, it is possible to have the SNP in the penultimate position at the 3′-end) (*see* **Note 2**).
3. Design and check an alternative pair of AS-forward primers in reverse-compliment orientation, where the SNP is also located in the terminal (or in penultimate) position at 3′-end, similar to that described in **step 2** and **Note 2**. Compare and choose which design orientation is better (*see* **Note 3**).
4. Single common reverse primers must be designed after and in conjunction with a pair of the selected AS-forward primers targeting a fragment of amplification with SNP for around 100 bp but not less than 50 bp and not longer than 200 bp (**Note 4**).
5. Attach 21-bp Tail-1 or Tail-2 to the 5′-end of the corresponding AS-forward primers (AS-F1 and AS-F2) making it longer prior to ordering (*see* **Note 5**).
6. Two universal probes (UPs) must be ordered separately in advance (Uni-1 and Uni-2). Each of these UPs contain one of two fluorophores, FAM or HEX/VIC, in the 5′-end, with a quencher in the middle and 21-bp Tail-1 or Tail-2 at the 3′-end, identical to the tails of the AS-forward primers (AS-F1 and AS-F2) (*see* **Note 6**).
7. For genotyping, DNA can be extracted using any method suitable and convenient for researchers. After quality control, DNA should be diluted to the working concentration (10, 20 or 50 ng/μL) and prepared in individual microtubes, strips or microplates (*see* **Note 7**).
8. The choice of suitable Taq-polymerase is very important with preference given to the more expensive Hot-start Taq-polymerase, but regular Taq polymerase may be acceptable in most cases. Preliminary testing is required to confirm the optimal conditions for amplification with fluorescence and allele discrimination (*see* **Note 8**).
9. Prepare two pre-mixes before starting SNP genotyping: (a) three allele-specific primers (AS-F1, AS-F2 and AS-R); and (b) two UPs (Uni-1 and Uni-2) (*see* **Note 9**).

Table 1
The composition of the PCR Master-mix cocktail for the Amplifluor-like method of SNP genotyping

Component	Concentration	Volume (μL)	Final concentration
5 × Buffer (without $MgCl_2$)	5×	2.0	1×
$MgCl_2$	25 mM	1.2	3.0 mM
dNTP	2 mM	1.0	0.2 mM
AS-primers pre-mix:			
AS-primer F1 AS-primer F2 AS-primer R	1.5 μM 1.5 μM 7.8 μM	1.0	0.15 μM 0.15 μM 0.78 μM
UPs pre-mix:			
Uni-1-FAM: Uni-2-HEX (or VIC)	2.5 μM 2.5 μM	1.0	0.25 μM 0.25 μM
Taq polymerase	5 units/μL	0.2	0.1 units/μL
Sterile water	–	1.2	–
Master-mix	–	**7.0**	–
DNA template	10 ng/μL	3.0	30 ng/μL
Total	–	**10.0**	–

10. Prepare the Master-mix cocktail following the list of components, their concentration and volume per single reaction (*see* Table 1), multiply according to the number of reactions in microplate (*see* **Note 10**).
11. Using a cold-block, load the prepared Master-mix cocktail in each position of a 96- or 384-well microplate and add DNA template or sterile water (for NTC), apply the plate seal and ensure that the reaction mix is at the bottom of the wells (*see* **Note 11**).
12. A no template control (NTC) is strongly recommended and should be used in all experiments with SNP genotyping regardless the methods used (*see* **Note 12**).
13. The qPCR instruments used must be prepared and set-up according to the manufacturer's recommendation. The amplification program is recommended as presented in Table 2, but can be modified if this is more convenient provides better results. Real-time fluorescence can be monitored after each cycle and recorded (*see* **Note 13**).
14. The microplate can be scanned for fluorescence at the "end-point" after all amplification cycles are complete. This is an "Optional" step, and a regular PCR cycler can also be used in

Table 2
PCR protocol for genotyping using the Amplifluor-like method of SNP genotyping

Step	Temperature (°C)	Duration	Notes
1	95	5 min	Initial denaturation. Can be variable depending on Taq-polymerase
2	95	10 s	First step denaturation
3	56	10 s	First step annealing
4	72	20 s	First step extension
5	95	10 s	Second step denaturation
6	51	20 s	Second step annealing
7	72	40 s	Second step extension and Microplate read
8			25–30 "doubled" cycles repeat from step 2

place of a qPCR instrument, but it requires a Microplate Reader with channels and filters appropriate for the fluorophores used (*see* **Note 14**).

15. SNP genotyping results are presented as an allele discrimination plot with different color dots or other shapes in the presence of NTC or normalized with that data (*see* **Note 15**).
16. The number of biological and technical replicates is important for each experiment and is very dependent on the nature of the plant materials and experimental conditions used. Generally, they can vary between one and three replicates (*see* **Note 16**).
17. Accuracy, reproducibility and significance of difference. Generally, the accuracy of the SNP genotyping is determined by comparison with the reference genotype and known (sequenced) SNP fragment. Reproducibility refers to the running of biological and technical replicates that show identical or very similar results. Significant differences between SNP alleles are determined automatically by the software included in the used qPCR instruments (*see* **Note 17**).

4 Notes

1. It is strongly recommended to sequence and verify the presence of the SNP that will be targeted for genotyping using the Amplifluor-like method. Nevertheless, in some circumstances, it is possible to make genotyping based on published information about an SNP or retrieve it from public databases. However, in such cases, there is a risk of wasting time and resources in the study of monomorphic genetic material in the absence of SNP—a situation akin to Confucius's classical aphorism "It is hard to find a black cat in a dark room, especially if there is no cat".

2. Forward primers can be designed as regular PCR primers and checked using any suitable program for GC percentage and melting temperature, but more importantly for the absence of self-annealing and potential hairpins, which must be avoided. After confirmation of no self-annealing and hairpins, the length of the forward primers can be adjusted, making it 1–2 bp longer in lower GC content and a shorter length for higher GC content. Such adjustment in the oligonucleotide length can be important for further steps since the melting temperature of AS-forward and reverse primers must be as close as possible. It is important to note also that a pair of AS-forward primers differs by only one nucleotide, and their melting temperature will be identical only in two cases designated as SNP positions "S" (C or G) and "W" (A or T). Slightly different melting temperature will result for all other four combinations of SNP positions: "K" (G or T); "M" (A or C); "R" (A or G); and "Y" (C or T). In the latter cases, average melting temperature for both AS-forward primers can be considered, but usually such a difference in melting temperature is relatively small.
3. It is strongly recommended as a matter of some importance that a pair of AS-forward primers should be designed and checked in both orientations (direct and reverse-compliment) with subsequent comparison and final selection. An example of such a comparison is presented in Fig. 2 based on the *LcELF3* (Early flowering 3) gene in lentil (*Lens culinaris* Medik.). The AS-forward primers designed in direct orientation showed very strong self-annealing (*see* Fig. 2a) and this variant has to be discarded. In contrast, the design of AS-forward primers in the reverse-compliment orientation for the same SNP in the same gene was successful with no self-annealing, dimerization or potential hairpins (*see* Fig. 2b).
4. The design of the common reverse primer has to be the same as in a pair of AS-forward primers as described in **Note 2** and shown in Fig. 2b, with similar melting temperature.
5. The sequences of Tail-1 and Tail-2 attached to the AS-forward primers are shown in Fig. 2c. Therefore, both AS-forward primers are longer by 21-bp from the 5′-end while the SNP remains in the terminal position in the 3′-end of both AS-forward primers. Generally, the length of oligonucleotides for each of the two AS-Forward primers is around 43 bp and for the common reverse primers is about 22 bp, but the oligonucleotide lengths can be slightly adjusted depending on GC content and melting temperature. It is important to note that all three oligonucleotides (two AS-forward and one common reverse) do not contain fluorophores or quenchers. The cost of AS-forward primers and the single common reverse primer is

A. ***LcEFL3 – (Early flowering 3).*** Design of forward primers in direct orientation

```
GTCAGGTTCTTTTATGATTTTAAATAGGAATGTTTGGTAGTCCGATGATTTGATTTTTGTTATT
TTGGTGGTTTATTGGTATTTTCTCCTACCTTCTTTTTTATATTTTCTTGCATTTATCATTAGCC
AGGGCTGGGTTTGAAAATATATATGAAAA[W]ATCCTTTGAAGAGAAGAAATGTAGGACCAGGT
AATGTGAATTATAGGGACACTCCTATCTAACGTTGCCATTTTTTCCCTTACCTTTTAAAGCATC
TTTTATGCATCCTTAATTAAAAACCAGGGATGTAAGTTGGATAGTTACATTTCAATTTTCAATC
AAGGAGTTGTATGTATGGCTT
[W] = [A/G]

LcEFL3-SNP3-F: GGTTTGAAAATATATATGAAAAW
```

Potential hairpin formation:

```
None !

3' Complementarity:
None !

All potential self-annealing sites are marked in red (allowing 1 mis-match):
5' GGTTTGAAAATATATATGAAAAT       3'
3'   TAAAAGTATATATAAAAGTTTGG     5'

5' GGTTTGAAAATATATATGAAAAT       3'
3'     TAAAAGTATATATAAAAGTTTGG   5'

5'   GGTTTGAAAATATATATGAAAAT     3'
3' TAAAAGTATATATAAAAGTTTGG       5'

5' GGTTTGAAAATATATATGAAAAT       3'
3'       TAAAAGTATATATAAAAGTTTGG 5'
```

B. ***LcEFL3.*** Design of forward primers in reverse-compliment orientation

```
LcEFL3-SNP3-F (RevCom): WATCCTTTGAAGAGAAGAAATGTAG

LcEFL3-SNP3-F1: CTACATTTCTTCTCTTCAAAGGATT
LcEFL3-SNP3-F2: CTACATTTCTTCTCTTCAAAGGATA

LcEFL3-SNP3-R: GGAATGTTTGGTAGTCCGATG

Amplicon size = 156 bp
```

C. Design of the ordered AS-primers

LcEFL3-F1: 5′-*GAAGGTGACCAAGTTCATGCT*-CTACATTTCTTCTCTTCAAAGGATT-3′
LcEFL3-F2: 5′-*GAAGGTCGGAGTCAACGGATT*-CTACATTTCTTCTCTTCAAAGGATA-3′
LcEFL3-R: 5′-GGAATGTTTGGTAGTCCGATG-3′

Fig. 2 Design of non-labelled allele-specific (AS) primers for the example of lentil gene *LcEFL3*, where the SNP position is indicated by the red letter. (**a**) Unsuccessful design of AS-primers (in grey) in direct orientation with highly self-annealing sites. (**b**) Good design of AS-primers in reverse orientation. (**c**) The same design primers with attached tails as ordered for synthesis

relatively cheap as they are regular oligonucleotides, and so they can be ordered through any company producing such oligonucleotides and usually require only the smallest production volume.

6. The structure and sequences of the two UPs are presented in Fig. 3. FAM and HEX represent a typical combination of the used fluorophores attached to the 5′-end of the UPs (with maximum absorption for FAM = 494 nm and for HEX = 535 nm). However, other dyes can be used if they are

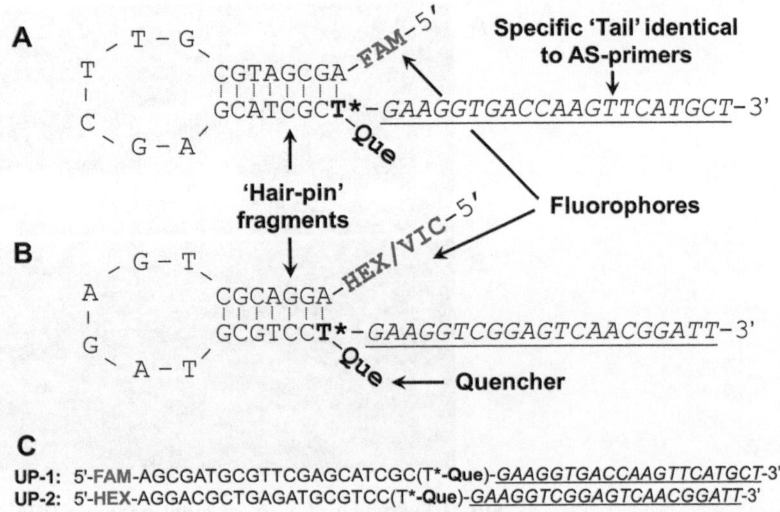

Fig. 3 Structure and design of two universal probes (UPs) with: (**a**) FAM; and (**b**) HEX or VIC fluorophores; and (**c**) their linear order

clearly distinguished by appropriate filters for absorption and emission wavelengths in the instrument used. For example, Thermo Fisher Scientific qPCR instruments are designed for their own fluorophore (VIC), and some companies can produce UPs with VIC dye (with maximum absorption for VIC = 538 nm). However, the spectra for absorption and emission of VIC and HEX are very similar (538 and 535 nm, respectively), and they can be perfectly replaceable. For other fluorophores, their characteristics and cost for attachment in the 5′-end must first be checked. The majority of quenchers are also suitable for the UPs. Historically, Dabcyl was the first quencher used, Black Hole Quencher 1 (BHQ1) is currently the most popular, while Eclipse is one of the more modern quenchers. The cost of the quencher is the most expensive component of the UPs, and for economic reasons, the cheapest available quencher can be selected and used since they all can quench the fluorescence completely. If possible, it is recommended to conduct a test to determine how two UPs work in a certain laboratory, with the available equipment, reagents, material for SNP genotyping with AS-primers and staff experience level. After that, once ordered and prepared, the stock of two UPs can be stored at −20 °C for a very long time and a dilution of the pre-mix can be used in separate experiments for SNP genotyping.

7. Generally, there is no restriction on which method is used for DNA extraction and further genotyping, as long as any contamination present does not inhibit the Taq-polymerase enzyme during amplification. For example, the regular CTAB method [25] showed very good DNA quality. A greater

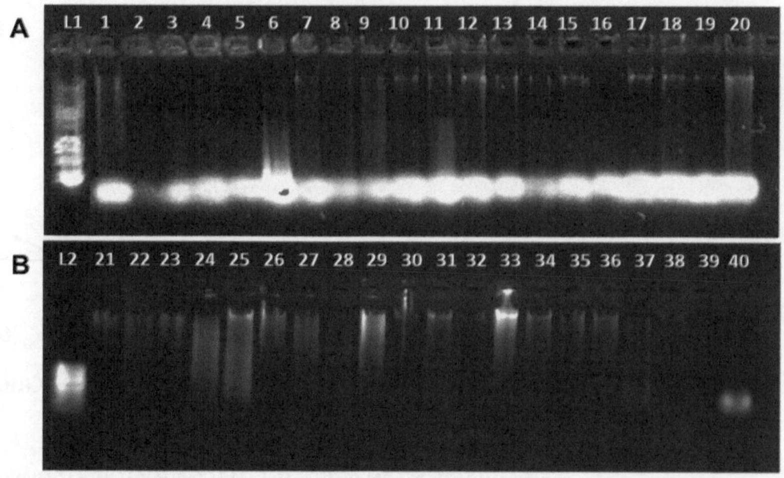

Fig. 4 Examples of 1 μL of lentil genomic DNA (Phenol-chloroform extraction method) separated in a 1% agarose gel. (**a**) DNA pellets were dissolved in TE buffer without RNase and RNA is visible as bright spots in the bottom of each lane; and (**b**) DNA samples dissolved in TE buffer with RNase, where RNA was destroyed. All DNA samples had different concentrations and majority of them are suitable for SNP genotyping after adjustment of concentrations. Samples 6 and 37 showed some degradation of DNA but still can be used. Sample 40 failed for quality control and DNA extraction was repeated. L1 and L2 are markers for 1 Kb and 100 bp fragment size on the gel

amount with excellent quality DNA is produced using the Phenol-chloroform method, but more careful attention is required in working with phenol as it is a very dangerous reagent [26]. Excellent quality DNA is received through commercial kits sold by various companies, but the volumes purified can be limited while the cost of DNA extraction kit might be relatively high. Many of the commercial kits for DNA extraction are adapted for 96-well microplates. However, it is also convenient to use the Freeze-drying method adapted for 96-collecting tubes and microplates [27]. DNA quality control is recommended but not compulsory, and good quality DNA is shown. The range of working concentration for template DNA can be quite diverse and usually is in the range of 20–100 ng. Any format of individual microtubes, strips or microplates is acceptable (*see* Fig. 4).

8. There is no strong prescription to use Taq-polymerase of any certain brand or manufacturer. This means that any Taq-polymerase can be used if they effectively amplify the relatively small PCR fragments with incorporated UPs and release fluorescence. However, the Hot-start Taq polymerase is well-known to be more effective, starting amplification that generates a visible signal at earlier cycles, with quicker development using a lower number of enzyme activity units per

reaction. At the same time, regular Taq-polymerase could also be also tested using a greater amount of enzyme units, with slower developing amplification and signal detection starting at later cycle number. This option gives greater flexibility for researchers to check and test existing regular Taq polymerase, which can be used for SNP genotyping if it shows effective amplification. It is also important to note that Taq-polymerases are very sensitive to concentrations of Mg^{2+} ions in the reaction mix and, therefore, the content of magnesium can be slightly adjusted to produce the best performance of the preferred Taq-polymerase for SNP genotyping. However, extra-care is required to check for the presence or absence of magnesium in the supplied Buffer. Because Taq-polymerase is a commercial product, each manufacturer produces Taq-polymerase with certain characteristics and specific targets. Magnesium chloride or sulphates is very often already included in the supplementary buffer, typically with 1.5 mM Mg^{2+}, but it might be variable, which must be considered. In any case, the final step will be to confirm the optimal conditions of the Taq-polymerase used, including the concentration of magnesium in the reaction mix, for the best amplification with fluorescence and allele discrimination.

9. Two pre-mixes must be prepared before starting SNP genotyping: (a) three allele-specific primers; and (b) two UPs. The pre-mixes are prepared for convenience for simpler and quicker preparation of the Master-mix. The concentration of components in both pre-mixes (A and B) is 10×, which makes it comfortable for using 1 μL of each pre-mix per 10 μL of reaction. Both pre-mixes can be kept in the fridge (+4 °C) for a few days during the SNP genotyping experiment but must be moved to the freezer (−20 °C) for longer usage or for storage. The Universal probes (UPs), both as stock and in pre-mix B are sensitive to light and so microtubes must be made of dark plastic or wrapped in foil.
10. The accuracy of all micropipettes used must be considered during Master-mix preparation. Always prepare Master-mixes to allow for a greater volume than is required for the experiment. For example, for 96 samples over an entire microplate (including all NTC), the Master-mix calculations should be for ×105 or ×110 samples. This adjusts for the small amount of mix that may adhere and be discarded on pipette tips through the process. A little bit of Master-mix should be left-over after loading all wells.
11. Microplates are usually used for genotyping with qPCR instruments with a regular 96-well format and a 10 μL reaction volume, which is convenient for manual operation. High-throughput technology requires a robotic system which

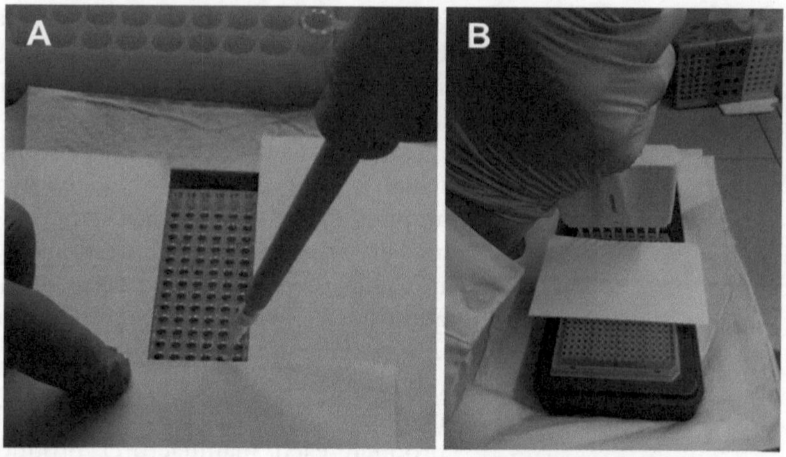

Fig. 5 Examples of manual preparation of 384-well microplates for plant genotyping with the Amplifluor-like method. (**a**) Cold-block was used as the base for microplates and sticky notepads to frame a clear area for Master-mix loading using a digital pipette. (**b**) Final loading of DNA templates using a multichannel pipette

operates very accurately and quickly with a reduced reaction volume of 5 μL in 384-well microplates, but obviously such robotic systems are very expensive. It is still possible to prepare 384-well microplates manually using digital pipettes for Master-mix loading and multichannel pipettes for addition of DNA samples, and working on a cold-block surface to minimize evaporation (*see* Fig. 5). Generally, such an approach for the manual preparation of a 384-well microplate had very good genotyping results.

12. For NTC (No template control), the Master-mix used is the same as used for regular samples for genotyping, but sterile water is added instead of template DNA. Usually, 2 NTC but sometime 3–4 NTC are used for 96-well microplates, but at least one NTC is required for fewer samples for genotyping. Background of fluorescence after amplification in the absence of template DNA is automatically identified and compared with samples but different equipment shows NTC differently (*see* Fig. 6). Users must be consistent in the type of analysis used and presentation of SNP genotyping results.
13. Each qPCR instrument requires its own set-up following the manufacturer's supplied manuals using general options and with monitoring of amplification curves after each cycle in Real-time format (*see* Fig. 7). Users should test all options in their qPCR instruments and choose the best for their experiments.

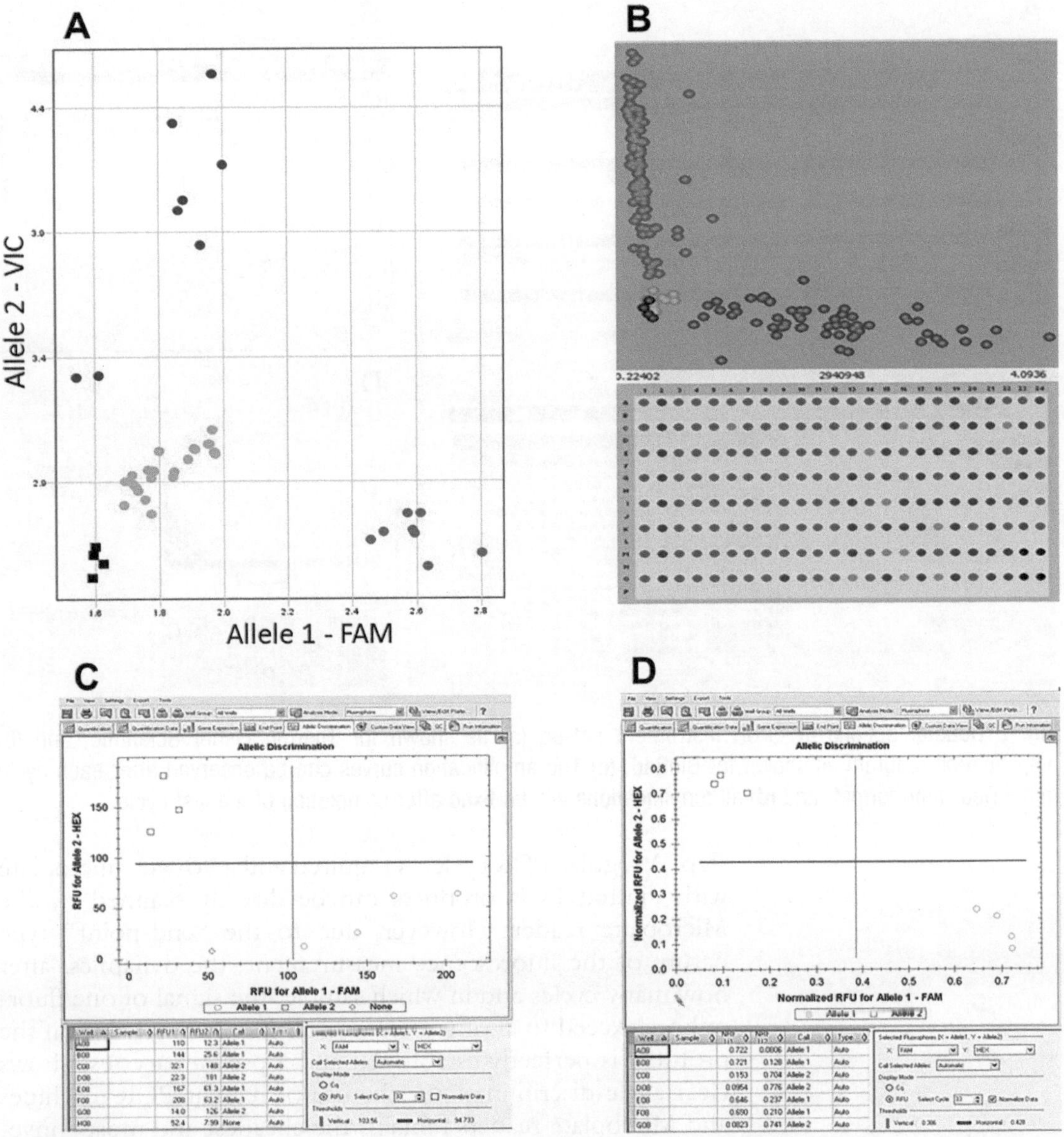

Fig. 6 Presentation of No template control (NTC) in different equipment. For example, (**a**) marked by black squares in Thermo Fisher Scientific qPCR instruments; (**b**) marked by black dots in KASP markers via LGC Genomics instruments and software. In contrast, BioRad qPCR instruments provide two options: (**c**) to use Relative expression units (RFU) and NTC as black diamonds; or (**d**) to use Normalized RFU for NTC value without NTC in the allele discrimination plot

14. A qPCR instrument is a relatively expensive piece of equipment and is very good for Real-time monitoring of fluorescence accumulation after each cycle of amplification. However, it can be replaced by a Microplate reader that measures fluorescence from the fluorophores used at the post-amplification

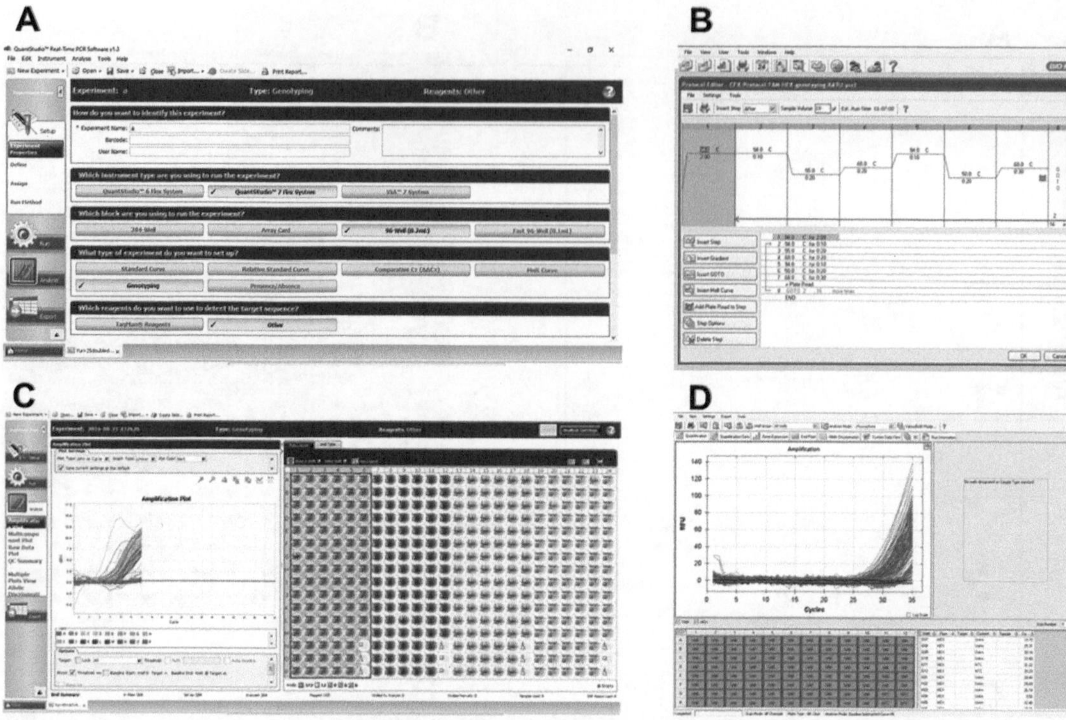

Fig. 7 General options of qPCR instrument set-up (**a**) as shown for Thermo Fisher Scientific; and (**b**) amplification program as shown for BioRad. (**c**) The amplification curves can be observed after each cycle in the Real-time format; and (**d**) all amplifications will be fixed after completion of the last cycle

step. A regular PCR cycler is required and a 96-well microplate with diluted PCR products can be directly scanned by the Microplate reader. However, due to the "end-point" type nature of the fluorescence measurement, the dynamics, after how many cycles and in which sample, the signal of one fluorophore exceed to another, are unavailable. Nevertheless, if the protocol is perfectly established and a preliminary test shows clear allele discriminations, the use of regular PCR machines and Microplate readers remains the cheapest and most convenient option for SNP genotyping. The example of the Microplate Reader is shown in Fig. 8, which is followed by the corresponding results in the different format.

15. Results for SNP genotyping in Amplifluor-like and similar methods based on FRET are simple for interpretation. Such results are presented automatically using computer software included in the qPCR instrument. Each result for a separate DNA sample is shown by the fluorescence levels of both used fluorophores, for example for FAM and HEX or FAM and VIC or another combination. This includes a graphical allele discrimination plot, where *X*- and *Y*-axes are automatically set for relative fluorescence units (RFU), for example, FAM and HEX,

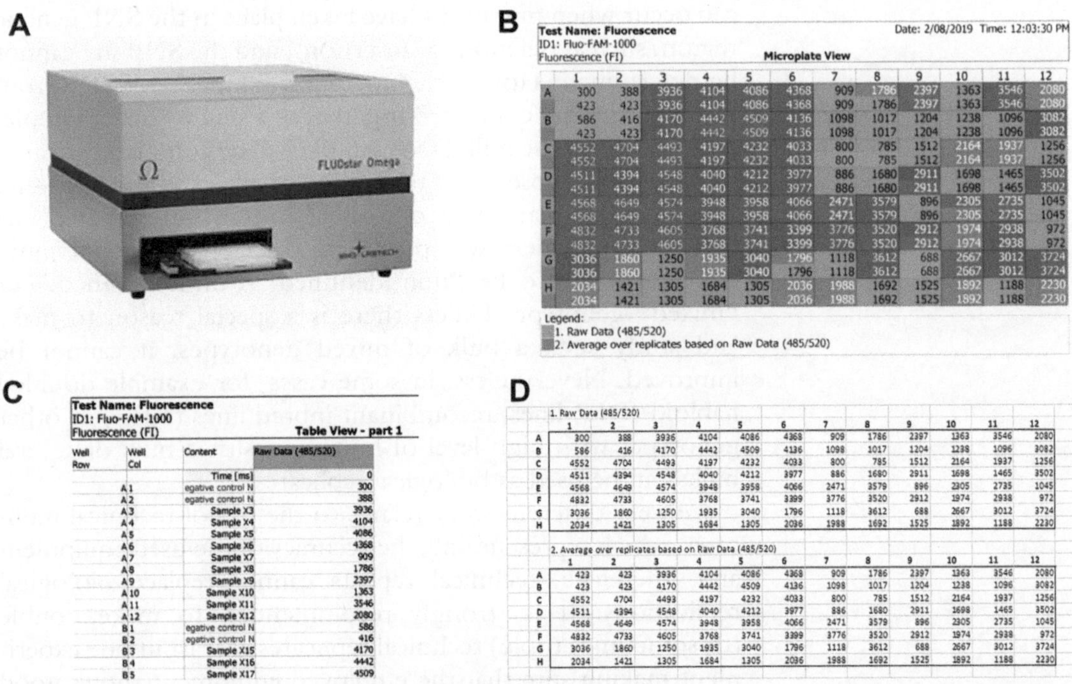

Fig. 8 (**a**) Microplate Reader FLUOstar Omega (BMG LabTech, Germany) for post-amplification measurement. (**b**) Graphical presentation of FAM fluorescence in 96-well microplate; (**c**) the same data presented in Table format; and (**d**) microplate plan in Excel file

respectively. For convenience, the same data are also presented in a Table format with numbers for RFU in both used fluorophores, respectively. In some qPCR instruments, like those produced by BioRad, USA, fluorescence data can be transformed into normalized RFU relative to NTC as shown in Fig. 6d.

In all cases, computer software programs make SNP genotyping automatic, with the identification of both types of homozygotes "*aa*" and "*bb*" for corresponding alleles 1 (FAM) and 2 (HEX), for example, respectively. Heterozygotes "*ab*" are determined as genotypes having both fluorescences present at similar levels. Additionally, genotypes showing mixed amplification of fluorophores with clear discrimination are classified as "non-identified," "undetermined," or "mixed," and such genotypes need further analyses.

16. Generally, biological replicate represents mean individual plants and SNP genotype can be strongly determined as one of two homozygotes or heterozygote. In such a case, 1 or 2 biological replicates are sufficient to make correct SNP genotyping, if the allele discrimination score was very clear. It is possible but not reasonable to make more biological replicates for each clearly scored genotype. A more complicated variant

can occur when mutations have taken place in the SNP genetic region, such as deletions or insertions, and the SNP site cannot be determined. However, sometimes, samples collected from several plants are combined together in a single "bulk" sample. The analysis of the bulk DNA sample will be completely related to presence or absence of two different alleles in the SNP site. Even a single plant with one allele among four other plants with another allele will provide a mixed SNP genotyping, which is likely to be "non-identified," "undetermined," or "mixed" genotype. Unless there is a special reason to make and study such a bulk of mixed genotypes, it cannot be improved. Nevertheless, in some cases, for example doubled haploid (DH) lines, recombinant inbred lines (RILs) or other genotypes with high level of homozygosity, a bulk or several bulks can be used as biological replicates.

A very different story relates to the use of technical replicates, which represent only the accuracy of the used equipment and, therefore, technical repeats cannot replace biological replications. It is strongly recommended to make double (or sometimes triple) technical replicates at least in one experiment making sure that the employed equipment shows good reproducibility of the produced SNP genotyping results.

17. The reference genotypes with two known alternative alleles based on sequences of the genetic fragment containing the SNP are strongly required in each round of genotyping for accuracy of allele identification. The simplest way is to use a segregating population at any generation if both parents are reference genotypes with different SNP alleles. In this case, the inheritance of maternal and paternal alleles in siblings can be very accurately scored for homo- and heterozygotes compared to reference parents. The accuracy of SNP genotyping in non-pedigree related germplasm collection is also based on reference genotypes. However, the frequency of each allele can be unpredictable and, therefore, SNP genotyping must be made with special care for accuracy with allele discrimination.

    Good results including SNP genotyping are always reproducible, showing identical or very similar results with biological and technical replicates. However, different equipment, reagents and, more importantly, differing experience levels of research staff can be an additional component affecting the reproducibility of SNP allele genotyping. Therefore, highly qualified staff, and good quality instruments and reagents will improve reproducibility because the biological material remains unchanged.

The question about significant differences between scored SNP alleles often arises during statistical analysis. However, the computer software attached to the qPCR instrument used can provide SNP genotyping automatically as qualitative but not quantitative traits. Of course, the Amplifluor-like FRET method is based on PCR amplification with gradual incorporation of one used fluorophore. Therefore, fluorescence levels are measured as a quantitative value, but the final conclusion is qualitative, for example in bi-allelic variants, one of two homozygotes, heterozygotes or undetermined genotypes. In this regard, SNP homozygotes "*aa*" and "*bb*" will be always significantly different between each other, if such results are accurate (similar to referent genotypes) and reproducible (with replicates) because only one or another nucleotide is present in the bi-allelic variant of SNP. Therefore, traditional statistical treatment of means, variability and level of significance is not usually required for SNP genotyping in such methods as Amplifluor-like.

Acknowledgments

This study was supported by Bolashak International Fellowships, Center for International Programs (GK and SK), and Research project AP14869777 (GK), Ministry of Education and Science, Kazakhstan. We want to thank Carly Schramm for critical comment of the chapter.

References

1. Chen X, Sullivan PF (2003) Single nucleotide polymorphism genotyping: biochemistry, protocol, cost and throughput. Pharmacogenomics J 3:77–96. https://doi.org/10.1038/sj.tpj.6500167
2. Kim S, Misra A (2007) SNP genotyping: technologies and biomedical applications. Annu Rev Biomed Eng 9:289–320. https://doi.org/10.1146/annurev.bioeng.9.060906.152037
3. Ragoussis J (2009) Genotyping technologies for genetic research. Annu Rev Genomics Hum Genet 10:117–133. https://doi.org/10.1146/annurev-genom-082908-150116
4. You Q, Yang X, Peng Z, Xu L, Wang J (2018) Development and applications of a high throughput genotyping tool for polyploid crops: single nucleotide polymorphism (SNP) array. Front Plant Sci 9:104. https://doi.org/10.3389/fpls.2018.00104
5. Morgil H, Gercek YC, Tulum I (2020) Single nucleotide polymorphisms (SNPs) in plant genetics and breeding. In: Çalışkan M (ed) The recent topics in genetic polymorphisms. InTech Open, London, pp 825–400. https://doi.org/10.5772/intechopen.91886
6. Schramm C, Shavrukov Y, Anderson P, Kurishbaev A, Jatayev S (2019) Development of single nucleotide polymorphism (SNP) markers for cereal breeding and crop research: current methods and future prospects. In: Ordon F, Friedt W (eds) Advances in breeding techniques for cereal crops. BD Publishing, Cambridge, pp 327–362. https://doi.org/10.19103/AS.2019.0051.16
7. Kaur A, Kaur P, Ahuja S (2020) Förster resonance energy transfer (FRET) and applications

thereof. Anal Methods 12:5532–5550. https://doi.org/10.1039/d0ay01961e

8. Nazarenko IA, Bhatnagar SK, Hohman RJ (1997) A closed tube format for amplification and detection of DNA based on energy transfer. Nucleic Acids Res 25:2516–2521. https://doi.org/10.1093/nar/25.12.2516
9. Myakishev MV, Khripin Y, Hu S, Hamer DH (2001) High-throughput SNP genotyping by allele-specific PCR with universal energy-transfer-labeled primers. Genome Res 11: 163–169. https://doi.org/10.1101/gr.157901
10. Khripin Y (2006) High-throughput genotyping with energy transfer-labeled primers. In: Didenko VV (ed) Methods in molecular biology: fluorescent energy transfer nucleic acid probes: designs and protocols, vol 335. Humana Press, Totowa, pp 215–240
11. Bengra C, Mifflin TE, Khripin Y, Manunta P, Williams SM, Jose PA et al (2002) Genotyping of essential hypertension single-nucleotide polymorphisms by a homogeneous PCR method with universal energy transfer primers. Clin Chem 48:2131–2140. https://doi.org/10.1093/clinchem/48.12.2131
12. Fuhrman LE, Shianna KV, Aballay A (2008) High-throughput isolation and mapping of *C. elegans* mutants susceptible to pathogen infection. PLoS One 3:e2882. https://doi.org/10.1371/journal.pone.0002882
13. Giancola S, McKhann HI, Bérard A, Camilleri C, Durand S, Libeau P et al (2006) Utilization of the three high-throughput SNP genotyping methods, the GOOD assay, Amplifluor and TaqMan, in diploid and polyploid plants. Theor Appl Genet 112:1115–1124. https://doi.org/10.1007/s00122-006-0213-6
14. La Paz JL, Esteve T, Pla M (2007) Comparison of real-time PCR detection chemistries and cycling modes using Mon810 event-specific assays as model. J Agric Food Chem 55: 4312–4318. https://doi.org/10.1021/jf063725g
15. Gašparič BM, Tengs T, La Paz JL, Holst-Jensen A, Pla M, Esteve T et al (2010) Comparison of nine different real-time PCR chemistries for qualitative and quantitative applications in GMO detection. Anal Bioanal Chem 396:2023–2029. https://doi.org/10.1007/s00216-009-3418-0
16. Kadirvel P, Veerraju C, Senthilvel S, Yadav P, Usha Kiran B, Shaik M et al (2020) Marker-assisted selection for fast-track breeding of high oleic lines in safflower (*Carthamus tinctorious* L.). Ind Crops Prod 158:112983. https://doi.org/10.1016/j.indcrop.2020.112983
17. Rickert AM, Borodina TA, Kuhn EJ, Lehrach H, Sperling S (2004) Refinement of single-nucleotide polymorphism genotyping methods on human genomic DNA: amplifluor allele-specific polymerase chain reaction versus ligation detection reaction-TaqMan. Anal Biochem 330:288–297. https://doi.org/10.1016/j.ab.2004.03.035
18. Jatayev S, Kurishbaev A, Zotova L, Khasanova G, Serikbay D, Zhubatkanov A et al (2017) Advantages of Amplifluor-like SNP markers over KASP in plant genotyping. BMC Plant Biol 17:254. https://doi.org/10.1186/s12870-017-1197-x
19. Khassanova G, Kurishbayev A, Jatayev S, Zhubatkanov A, Zhumalin A, Turbekova A et al (2019) Intracellular vesicle trafficking genes, *RabC*-GTP, are highly expressed under salinity and rapid dehydration but down-regulated by drought in leaves of chickpea (*Cicer arietinum* L.). Front Genet 10:40. https://doi.org/10.3389/fgene.2019.00040
20. Yerzhebayeva R, Abekova A, Konysbekov K, Bastaubayeva S, Kabdrakhmanova A, Absattarova A et al (2018) Two sugar beet chitinase genes, *BvSP2* and *BvSE2*, analysed with SNP Amplifluor-like markers, are highly expressed after Fusarium root rot inoculations and field susceptibility trial. PeerJ 6:e5127. https://doi.org/10.7717/peerj.5127
21. Baidyussen A, Aldammas M, Kurishbayev A, Myrzabaeva M, Zhubatkanov A, Sereda G et al (2020) Identification, gene expression and genetic polymorphism of zinc finger A20/AN1 stress-associated genes, *HvSAP*, in salt stressed barley from Kazakhstan. BMC Plant Biol 20:156. https://doi.org/10.1186/s12870-020-02332-4
22. Baidyussen A, Jatayev S, Khassanova G, Amantayev B, Sereda G, Sereda S et al (2021) Expression of specific alleles of zinc-finger transcription factors, *HvSAP8* and *HvSAP16*, and corresponding SNP markers, are associated with drought tolerance in barley populations. Int J Mol Sci 22:12156. https://doi.org/10.3390/ijms222212156
23. Absattar T, Absattarova A, Fillipova N, Otemissova A, Shavrukov Y (2018) Diversity array technology (DArT) 56K analysis, confirmed by SNP markers, distinguishes one crested wheatgrass *Agropyron* species from two others found in Kazakhstan. Mol Breed 38:37. https://doi.org/10.1007/s11032-018-0792-3
24. Shavrukov Y, Zhumalin A, Serikbay D, Botayeva M, Otemisova A, Absattarova A et al (2016) Expression level of the DREB2-type gene, identified with Amplifluor SNP markers,

correlates with performance and tolerance to dehydration in bread wheat cultivars from Northern Kazakhstan. Front Plant Sci 7:1736. https://doi.org/10.3389/fpls.2016.01736

25. Murray MG, Thompson WF (1980) Rapid isolation of height molecular weight plant DNA. Nucleic Acids Res 8:4321–4326. https://doi.org/10.1093/nar/8.19.4321
26. Weining S, Langridge P (1991) Identification and mapping of polymorphisms in cereals based on the polymerase chain reaction. Theor Appl Genet 82:209–216. https://doi.org/10.1007/BF00226215
27. Shavrukov Y, Gupta NK, Miyazaki J, Baho MN, Chalmers KJ, Tester M et al (2010) *HvNax3* – a locus controlling shoot sodium exclusion derived from wild barley (*Hordeum vulgare* ssp. *spontaneum*). Funct Integr Genomics 10:277–291. https://doi.org/10.1007/s10142-009-0153-8

Chapter 15

Semi-Thermal Asymmetric Reverse PCR (STARP) Genotyping

Awais Rasheed

Abstract

PCR-based individual Single nucleotide polymorphism (SNP) genotyping methods are preferred due to their flexibility, high-throughput, and improved accuracy. Semi-thermal asymmetric reverse PCR (STARP) is one of the SNP genotyping methods developed to reduce operational cost with improved platform compatibility. STARP is a unique method which can be used either as a gel-free SNP genotyping by detection of fluorescent signals or polyacrylamide gel-based size separation. SNP assay designing using sequence information and detection methods of STARP are described in detail.

Key words Single nucleotide polymorphisms (SNP), Molecular markers, Co-dominant markers, Molecular diagnostics

1 Introduction

Technological surge into Next-generation sequencing (NGS) method is making genome sequencing information cheaper and more accessible [1]. Genes controlling important phenotypes in crop plants are being discovered at a rapid pace. This in turn, providing more accurate and efficient molecular diagnostics tools to select important phenotypes in crop plants [2, 3]. Huge efforts have been made to develop genotyping platforms compatible with crop breeding programs given the needs of different scenarios of Marker-assisted selection (MAS). Such genotyping technologies should be flexible without compromising high-throughput, low-cost and compatible with multiple platforms [4].

The most commonly used platform for individual SNP genotyping is the KASP technology [5], based on detection of fluorescent signals generated by successful PCR amplification of either of the two SNP. A general limitation in using KASP markers is the higher cost because the Master-mix is available from only one company, LGC Genomics, which is a commercial proprietary. However, KASP is still relatively expensive due to the commercial Master-mix. Similarly, Amplifluor (Amplification with fluorescence)

Yuri Shavrukov (ed.), *Plant Genotyping: Methods and Protocols*, Methods in Molecular Biology, vol. 2638,
https://doi.org/10.1007/978-1-0716-3024-2_15,

SNP genotyping technology, having the same basic principles as KASP, is also a SNP genotyping method based on the AS-PCR platform and open PCR system. The Amplifluor-like system [6] has lower cost and much greater flexibility than KASP since the universal probes are designed by the researchers. Nevertheless, there are still limitations in mandatory requirements of the relatively expensive Taq DNA polymerase (GenLab, Astana, Kazakhstan) or Maxima Hot-Start Taq polymerase (Thermo Fisher Scientific, USA). RNase H2 enzyme-based amplification (rhAmp) is a reliable and cost-effective option for targeted SNP genotyping, which was released by the Integrated DNA Technologies Inc [7], but this technology is dependent on the indispensable enzymes. In addition, both KASP and Amplifluor-like systems cannot be applied in conventional polyacrylamide gel electrophoresis (PAGE) used by many crops breeding groups, particularly in developing countries.

Although several other commercial alternates are available now, however there was strong need to develop some open-source SNP genotyping method, where a wide range of commercially available Master-mix could be used for SNP genotyping. An example is the development of Semi-thermal asymmetric reverse PCR (STARP) [8] and Amplifluor-like [6] methods which can be used with wide range of commercial Master-mixes. This would have significant impact on the cost of the SNP genotyping. The STARP technology was then used to convert KASP markers into STARP format, thus further reducing the cost due to use of wide range of Master-mix [9].

STARP is a new low-cost method for flexible SNP genotyping [8]; it is a compatible platform with both conventional PAGE and high-throughput genotyping systems, combining the main advantages of TaqMan, KASP, Amplifluor, and rhAmp. Compared with KASP, STARP uses two universal priming element-adjustable primers (PEA primers) rather than the fluorescent probes used in the KASP Master-mix, greatly reducing the cost of PCR reagents. A 4-bp insertion comes into the final PCR products between two genotypes in STARP, allowing discrimination of allelic variants in conventional gel electrophoresis. In addition, the substituted nucleotides are introduced into two asymmetrically modified allele-specific primers (AMAS primers) at the third or fourth base, increasing the specificity of PCR amplification.

2 Materials

2.1 Primer Design

1. A sequence of DNA at least 100–150 bp long including the SNP/InDel of interest (*see* **Note 1**).
2. Software for primer design or alternatively using the assay design guidelines in [8] for fluorescent-based gel-free genotyping (*see* **Note 2**).

3. Primer design for PAGE-based genotyping using STARP assay (*see* **Note 3**).
4. The PEA primers (PEA1 and PEA2) for fluorescent-based gel-free genotyping (*see* **Note 4**).
5. The PEA primers (PEA1 and PEA2) for PAGE-based genotyping using STARP assay (*see* **Note 5**).

2.2 STARP Assay

1. A concentration of 50–100 ng genomic DNA per samples (*see* **Note 6**).
2. Two universal PEA primers (PEA primer 1 and PEA primer 2).
3. Genotyping Master-mix, which contains 10 × NH^{4+} Buffer (0.45 μL), 5 M of betaine (0.8 μL), 1% (w/v) bovine serum albumin (BSA, 0.2 μL), 50 mM of $MgCl_2$ (0.15 μL), 2.5 mM of each dNTP (0.1 μL), 0.05 μL of Taq DNA polymerase (5 U/μL) (*see* **Note 7**).
4. Two asymmetrically modified allele-specific primers (AMAS primer 1 and AMAS primer 2) and one common primer (100 μM).
5. Nuclease free water.
6. A multimode fluorescent plate reader (e.g., Pherastar, BMG Labtech) with a software for clustering the genotype information (e.g., KlusterCaller, Biosearch Technologies) or qPCR machine (e.g., CFX96/CFX384 real-time system, Bio-Rad) with a specialized clustering software for data interpretation (e.g., CFX Maestro software, Bio-Rad).
7. Polyacrylamide gel electrophoresis apparatus for conventional electrophoresis or DNA analyzer (e.g., IR2 4300/4200, LI-COR, USA).
8. A 96-well or 384-well PCR plate, which might be clear, white or black, and compatible with qPCR machine or plate reader, and optical clear seal (*see* **Note 8**).

3 Methods

3.1 Primer Design

1. The first step is to design the primers for STARP assay, if the pre-designed primers cannot be used.
2. After designing the primers, the sequences of PEA primers can be sent to any oligo-nucleotide synthesizing company (*see* **Note 9 and 10**).
3. The newly synthesized primers should be properly tested for both homozygous and heterozygous genotypes on a set of DNA samples containing both genotypes.

4. For heterozygous genotypes, a mixture of DNA from both homozygous types can be used.
5. Similarly, a no-template control (NTC), either buffer for DNA dissolving or nuclease free water, should be used.

3.2 STARP Assay

The STARP PCR reaction should include an NTC, and if possible then positive controls of both genotypes. DNA mixture from both genotypes can also be used a heterozygote control.

1. First step is to prepare the Master-mix containing two PEA primers, PCR consumables and nuclease free water according to Table 1. An assay mix of two AMAS primers and common reverse primer should be made and stored (*see* **Note 11**).
2. Select the suitable 96- or 384-well PCR plate according to the genotyping needs and conduct following steps in accordance with the chosen scale (*see* **Note 12**).
3. Pipette DNA samples in wells of PCR plate or optionally dried DNA (e.g., in a dry oven at 45 °C for 30 min or speed-vac).
4. Prepare the Master-mix according to Table 1, and add nuclease free water equivalent to DNA volume, if DNA was dried.

Table 1
STARP assay reagent mix for genotyping

	Hydrated DNA		Dried DNA	
Components	**96-well plate**	**384-well plate**	**96-well plate**	**384-well plate**
10 × NH^{4+} buffer	0.9	0.45	0.9	0.45
5 M betaine	1.6	0.8	1.6	0.8
1% (w/v) bovine serum albumin	0.4	0.2	0.4	0.2
50 mM of $MgCl_2$	0.3	0.15	0.3	0.15
2.5 mM of each dNTP	0.2	0.1	0.2	0.1
10 µM of common reverse primer	0.1	0.05	0.1	0.05
10 µM of PEA primer1	0.05	0.025	0.05	0.025
10 µM of PEA primer1	0.05	0.025	0.05	0.025
10 µM of AMAS primer1	0.02	0.01	0.02	0.01
10 µM of AMAS primer2	0.02	0.01	0.02	0.01
Taq DNA polymerase (5 U/µL)	0.1	0.05	0.1	0.05
DNA	5	2.5	Dried DNA	Dried DNA
Nuclease free water	1.26	0.63	6.26	3.13
Total	10	5	10	5

Table 2
PCR thermal cycling conditions for STARP genotyping

Step	Temperature	Time	Cycles
Initial denaturing	94 °C	3 min	1
Denaturation Annealing and elongation	94 °C 56 °C (−1.0 °C each cycle)	20 s 2 min	6
Denaturation Annealing and elongation	94 °C 62 °C	20 s 60 s	39 (for gel-free fluorescence signals) and 36 (for gel-based size separation)
Final extension	62 °C	2 min	1
Cooling down and plate read	35 °C	60 s	1

5. Pipette this Master-mix to the DNA samples (the volume of the assay mix can be ignored).
6. Cover the plate with an optical seal.
7. Centrifuge the PCR plate for 15–20 s.
8. Place the plate in PCR cycler or in qPCR instrument to start the PCR amplification with thermal cycling conditions as present in Table 2. After completion of the PCR cycling, the fluorescence is measured after cooling down to 35 °C (*see* **Note 13**) or the products can be resolved on PAGE or DNA analyzer (*see* **Note 14**).
9. Interpret data by using clustering software for fluorescent based genotyping (*see* **Note 15** and Fig. 1).
10. Record the data from PAGE gels for both genotypes by detecting the +4 bp size of QHEX allele/genotype.
11. Some common issues in STARP genotyping (*see* **Note 16**).

4 Notes

1. At least 75–100 bp DNA sequence is required flanking the target SNP on both upstream and downstream. In most cases, usually AMAS primers can be designed from much shorter region around the target SNP. Care must be taken to rule out the possibility of any further nucleotide variations within the amplified region.
2. Some allele-specific primer designing tools such as BatchPrimer3 (https://wheat.pw.usda.gov/demos/BatchPrimer3) or Genome-specific Primer (GSP, https://probes.pw.usda.gov/GSP) in case of polyploid species can be used. After designing

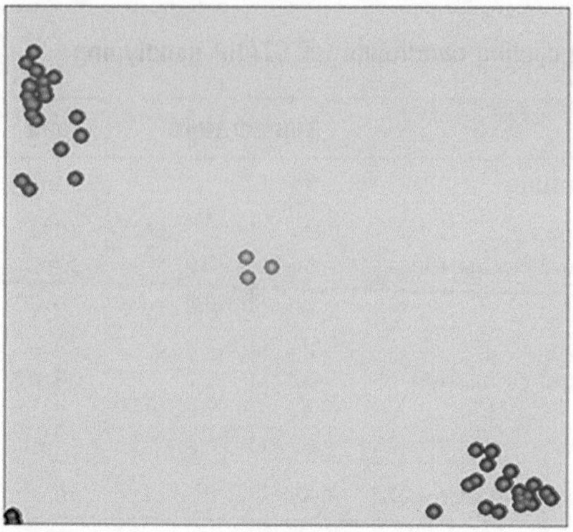

Fig. 1 The clustering of *Pinb-D1* assay amplified by primers mentioned in **Note 2** and Fig. 2 using STARP genotyping protocol on 40 diverse wheat accessions. The bi-allelic discrimination by fluorescent signals showing clustering of alleles on the *X*-axis (FAM) and *Y*-axis (HEX). Accessions colored blue have the QFAM-labelled allele (*Pinb-D1a*). Accessions colored red have the QHEX-labelled allele (*Pinb-D1b*). Green dots represent the heterozygous genotypes. Black dots represent the NTC (No-template control)

allele specific primer, an artificial mismatch at third or fourth positions from the 3′-terminus should be added. The third or fourth nucleotide in the allele-specific sequences from the 3′--terminus is substituted based on the following principle: A is changed to (→) C, T → C, G → A, and C → T. An example is given in Fig. 2 for the AMAS primer of *Pinb-D1* gene in wheat. Two tails (Tail1 and Tail2) are added to each of the allele specific primers mentioned in Table 3. The sequences of both tails are shown in color in Fig. 2.

3. Allele-specific primer design for PAGE based STARP genotyping is the same as mentioned in **Note 2**. However, a 4 bp (AGAG-3′) is inserted in PEA primer 2 to induce a size separation able to resolve on PAGE and to discriminate genotypes with both alleles. The AG motif length can be changed to adjust the difference in length between the PCR fragments of two alleles.
4. The sequences of two PEA primers (PEA1-QFAM and PEA2-QHEX) are described which include FAM (3′-TTGGTCGA-5′) and HEX (3′-ACTCGTCA-5′) sequences. The other components of the PEA-QFAM and PEA2-QHEX are explained in [8].

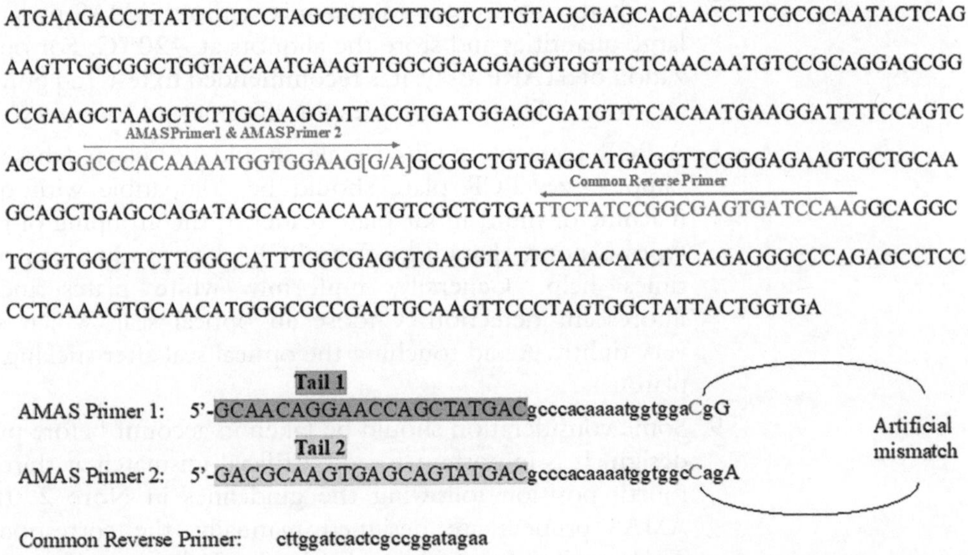

Fig. 2 The sequence of *Pinb-D1* gene in wheat showing the position of SNP and primers (red color). Two AMAS primers were designed at the site of SNP, each with unique nucleotide of SNP at 3′ terminus. An artificial mismatch is introduced at third and fourth position from 3′ terminus in AMAS1 and AMAS2, respectively. Tail1 and Tail2 sequences are added on AMAS1 and AMAS2 primers at 5′ terminus, respectively

Table 3
Sequence information of the PEA primers used in STARP assay

Primer	Sequence
PEA1-QFAM	5′-FAM-AGCTGGTT-Sp9[a]-GCAACAGGAACCAGC-T(Dabsyl)-ATGAC-3′
PEA2-QHEX	5′-HEX-ACTGCTCA-Sp9-GACGCAAGTGAGCAG-T(Dabsyl)-ATGAC-3′
PEA1-700	5′-IRDye700-AGCTGGTT-Sp9-GCAACAGGAACCAGCTATGAC-3′
PEA2-700	5′-IRDye700-ACTGCTCAAGAG-Sp9-GACGCAAGTGAGCAGTATGAC-3′
PEA1-800	5′-IRDye800-AGCTGGTT-Sp9-GCAACAGGAACCAGCTATGAC-3′
PEA2-800	5′-IRDye800-ACTGCTCAAGAG-Sp9-GACGCAAGTGAGCAGTATGAC-3′

[a]Triethylene glycol as PCR blocker

5. In case of size separation-based genotyping, either of the pair PEA1-700/PEA2-700 or PEA1-800/PEA2-800 primers can be used which are labelled with fluorophore IRDye 700 and IRDye 800, respectively.

6. An important aspect is to use the uniform DNA concentration of the samples with almost uniform DNA quality. The amount of DNA to be used in STARP genotyping depends on the genome size of the plant species. A DNA amount of 40–50 ng was successfully used for STARP genotyping in wheat [9]. Usually, the DNA extracted with CTAB method and without RNase treatment gives better results in wheat.

7. It is better to prepare Master-mix without AMAS primers in large quantities and store the aliquots at −20 °C. For optimization of STARP assay, it is recommended to test Taq polymerase from different companies for efficiency and accuracy.
8. A PCR plate of required scale should be selected based on sample size. PCR plate should be compatible with qPCR machine or multi-mode plate reader. If the grouping of genotypes is poor, changing of the PCR plate product can sometimes help. Generally, uniformly white plates increase fluorescent detection. Choose an optical seal, which sticks very tightly. Avoid touching the optical seal after sticking with plate.
9. Some consideration should be taken in account before primer design. It is important to add artificial mismatch at third and fourth position following the guidelines in **Note 2**. If the AMAS primers are designed manually, the corresponding Tail1 and Tail2 should be added at both allele specific primers. In case of polyploid species, carefully genome-specific primer should be designed. A BLASTN analysis of allele specific primers will be further helpful for undesired amplification.
10. Primer manufacturing: As inexpensive solution for small projects, ordering approximately 25 nmol desalted primer oligos can be sufficient. Yet, allele-specific primers including tail are very long (around 50 nucleotides). Thus, for allele-specific primers sometimes higher amounts of oligos (about 50 nmol) are necessary to be ordered.
11. Two allele specific primers (AMAS1 and AMAS2) and common reverse primer can be mixed, and an aliquot can be made for frequent use (*see* Table 4).
12. The reaction volume can be modified according to plate scale and to reduce further costs. In 96-well plates, usually 10 μL volume is used. However, this volume can be reduced progressively up to 5 μL, if the allele discrimination in the genotype

Table 4
Assay mix for STARP genotyping using two ASAM and common reverse primers

Primers	Starting concentration (μM)	Final concentration (μM)	Volume (μL)
ASAM1	100	12	12
ASAM2	100	12	12
Common reverse primer	100	30	30
Nuclease free water			46

separation is poor. The modifying primer concentartion is sometime needed according to the sensitivity of equipment used for fluorescence detection. STARP is equivocally compatible with robotics platforms such as SNPLine from LGC Genomics, Biosearch Technologies. The reaction volume can be reduced further in case of using 1536-well plates in the automated SNP genotyping platforms.

13. Fluorescence detection parameters of STARP are similar to those of KASP, and in qPCR machine, FAM and HEX fluorescent detection can be used.
14. In conventional PAGE, the PCR products can be resolved, and the alleles can be identified based on size difference of 4 bp between ASAM1 and ASAM2 genotypes. The size of ASAM genotype/allele is 4 bp larger than ASAM1 genotype. In case of resolution problem, AG motif can be progressively increased for better resolution.
15. Genotype calls can be automatically made using standard data interpretation software embedded in qPCR systems. In most of the cases, manual corrections should be made for good clustering. For example, some samples with very high signals can be removed from the results to get good clusters. Likewise, manual corrections can be made to identify the heterozygotes, which in some cases might cluster nearby homozygous genotypes. As long as different groups visually exist, manual correction should be acceptable.
16. Some problems may arise due to unidentified errors. Therefore, systematic troubleshooting will be helpful to sort out the problem and solve the modification in any of the steps. The guidelines provided by LGC for troubleshooting KASP genotyping (https://biosearch-cdn.azureedge.net/assetsv6/KASP-troubleshooting-guide.pdf) may be applied to STARP genotyping. In some cases, modifying cycling conditions is helpful to improve the genotype call rate. For example, a slight modification in annealing temperature improves the call rate of the separation of cluster is poor. In case of ambiguous clusters, additional cycles also improve the visualization of clustering.

References

1. Le Nguyen K, Grondin A, Courtois B, Gantet P (2019) Next-generation sequencing accelerates crop gene discovery. Trends Plant Sci 24:263–274. https://doi.org/10.1016/j.tplants.2018.11.008
2. Rasheed A, Xia X (2019) From markers to genome-based breeding in wheat. Theor Appl Genet 132:767–784. https://doi.org/10.1007/s00122-019-03286-4
3. Torkamaneh D, Boyle B, Belzile F (2018) Efficient genome-wide genotyping strategies and data integration in crop plants. Theor Appl Genet 131:499–511. https://doi.org/10.1007/s00122-018-3056-z
4. Rasheed A, Hao Y, Xia XC, Khan A, Xu Y, Varshney RK, He ZH (2017) Crop breeding chips and genotyping platforms: progress, challenges and

perspectives. Mol Plant 10:1047–1064. https://doi.org/10.1016/j.molp.2017.06.008
5. Semagn K, Babu R, Hearne S, Olsen M (2014) Single nucleotide polymorphism genotyping using Kompetitive allele specific PCR (KASP): overview of the technology and its application in crop improvement. Mol Breed 33:1–14. https://doi.org/10.1007/s11032-013-9917-x
6. Jatayev S, Kurishbaev A, Zotova L, Khasanova G, Serikbay D, Zhubatkanov A et al (2017) Advantages of Amplifluor-like SNP markers over KASP in plant genotyping. BMC Plant Biol 17:254. https://doi.org/10.1186/s12870-017-1197-x
7. Ayalew H, Tsang PW, Chu C, Wang J, Liu S, Chen C et al (2019) Comparison of TaqMan, KASP and rhAmp SNP genotyping platforms in hexaploid wheat. PLoS One 14:e0217222. https://doi.org/10.1371/journal.pone.0217222
8. Long YM, Chao WS, Ma GJ, Xu SS, Qi LL (2017) An innovative SNP genotyping method adapting to multiple platforms and throughputs. Theor Appl Genet 130:597–607. https://doi.org/10.1007/s00122-016-2838-4
9. Wu Y, Li M, He Z, Dreisigacker S, Wen W, Jin H et al (2020) Development and validation of high-throughput and low-cost STARP assays for genes underpinning economically important traits in wheat. Theor Appl Genet 133:2431–2450. https://doi.org/10.1007/s00122-020-03609-w

Chapter 16

Modified Allele-Specific qPCR (ASQ) Genotyping

Aigul Amangeldiyeva, Akmaral Baidyussen, Marzhan Kuzbakova, Raushan Yerzhebayeva, Satyvaldy Jatayev, and Yuri Shavrukov

Abstract

The allele-specific qPCR (ASQ) method for SNP (single nucleotide polymorphism) detection is based on the FRET (fluorescence resonance energy transfer) system, a system using position-dependent fluorescent dyes and quenches. The modified ASQ method requires two separate components: (1) the *allele-specific part*, two AS primers targeting the SNP with identity in the penultimate positions at the 3′-end and specific tags in the 5′-end, and (2) the *universal part*, two universal probes (UPs) with corresponding tags and different fluorescent dyes in the 5′-end and a single common universal probe with a quencher in the 3′-ends (Uni-Q), complementary to all UP tags. There are two major variations of the ASQ method, with either short 4-bp tags (variant A) or longer 6-bp tags (variant B), both of which have been successfully used for SNP genotyping in plants. The modified ASQ method is much cheaper compared to other similar FRET-based methods because the most expensive parts, the universal probes, have a short and linear structure, where fluorophores and quenchers are located in the ends but not incorporated inside of the sequences.

Key words Allele-specific primers, Fluorescent dyes and quenches, Genotyping, Single nucleotide polymorphism (SNP), Universal molecular probes

1 Introduction

Many different methods can be used for plant genotyping, and researchers can choose which among them is the most suitable and convenient (reviewed in refs. 1, 2). Moreover, most or perhaps all of these methods are undergoing continual development and modifications aimed at improving genotyping quality and availability, broadening the range of genotyping targets, and reducing costs.

For this purpose, the allele-specific qPCR (ASQ) method was used as the original "prototype" [3]. This method for single nucleotide polymorphism (SNP) visualization is based on a fluorescence resonance energy transfer (FRET) system, where a dye as the donor produces a fluorescent signal and a quencher as the acceptor quenches the fluorescence when in close proximity [4–6]. These principles are used in some very popular FRET-based methods,

Yuri Shavrukov (ed.), *Plant Genotyping: Methods and Protocols*, Methods in Molecular Biology, vol. 2638,
https://doi.org/10.1007/978-1-0716-3024-2_16,

including Amplifluor, Kompetitive allele-specific PCR (KASP), and Semi-thermal asymmetric reverse PCR (STARP), where these components are kept separate. That is, they have (1) non-labeled allele-specific primers (ASPs) and (2) universal probes (UPs) labeled with fluorophores and a quencher [7, 8]. Simple and cheap ASPs can be designed and ordered for each SNP separately, while the relatively expensive UPs with fluorophores and quenchers are ordered just once in a stock, which can be used over a very long time in many different SNP target analyses.

The modified ASQ method requires two separate components: (1) the *allele-specific part*, two AS primers targeting the SNP with identity in the penultimate positions at the 3′-end and specific tags in the 5′-end, and (2) the *universal part*, two universal probes (UPs) with corresponding tags and different fluorescent dyes in the 5′-end and a single common universal probe with a quencher in the 3′-ends (Uni-Q), complementary to all UP tags (*see* Fig. 1).

The modified ASQ method is much cheaper compared to other similar FRET-based methods because the most expensive parts, universal probes, have a short linear structure, where fluorophores and quenchers are located at the ends but not incorporated within the sequences themselves [9]. Additionally, the modified ASQ method is easily adjustable, where any component can be changed or rearranged completely, including the design and composition of allele-specific primers and universal probes, as well as the amplification program, including the temperature and duration of each step. The modified ASQ method has been successfully applied in plant genotyping [9] as well as in medical research applications [10]. Finally, this method opens 'avenue' for multiplex genotyping using different dyes and a single quencher.

2 Materials

1. Autoclaved milli-Q water is recommended.
2. DNA extracted through any method and diluted to a concentration of 10–50 ng/μL.
3. Two non-labeled allele-specific (AS) forward primers (about 22 bp each) targeting SNP in the penultimate position at the 5′-end and with attached tails and tags.
4. One common reverse primer (about 22 bp).
5. Two universal molecular probes (Uni-1 and Uni-2) with for example, 6-fluorescein amidite (FAM) or HEX/VIC, on the 5′-end and a tail with a tag at the 3′-end.
6. One common universal probe (Uni-Q) with a quencher in the 3′-end.

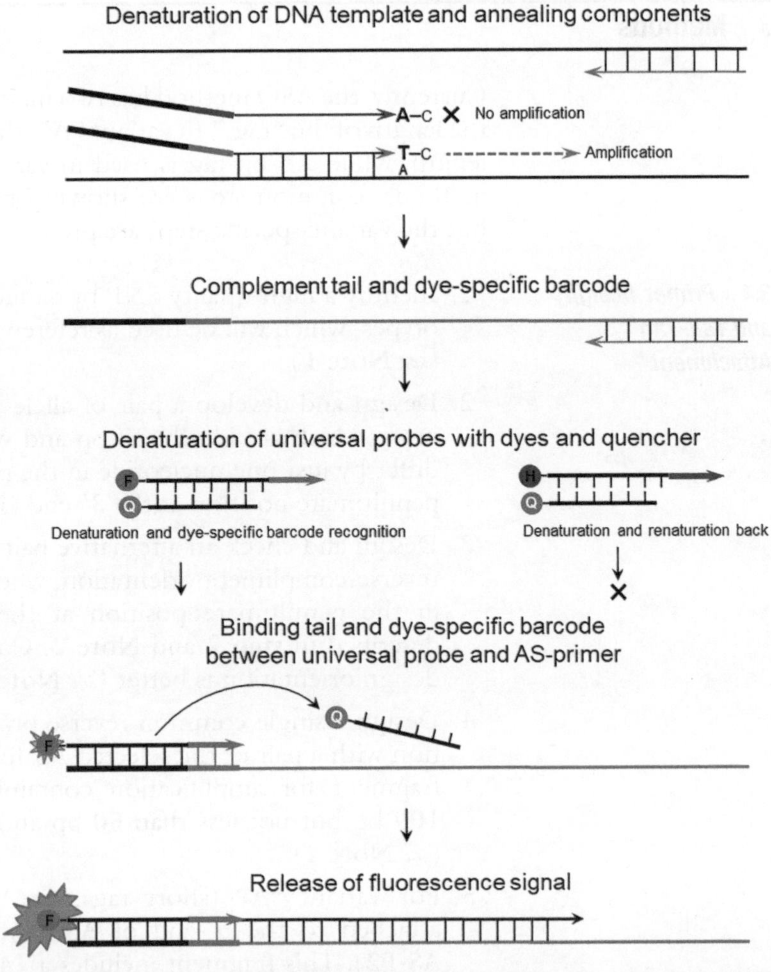

Fig. 1 Schematic presentation of the allele-specific qPCR (ASQ) genotyping method. (Modified from ref. 9)

7. Reagents for regular PCR, including any Taq polymerase kit (preferably "hot start") with supplementary buffer and $MgCl_2$, and additional dNTP mix.
8. Consumable PCR microplates and clear seals. PCR microtubes or strips can be used as alternatives.
9. Cold block for preparing the PCR cocktail and loading DNA sample to minimize evaporation.
10. Equipment required includes any qPCR instrument available with at least two channels for fluorophores with different wavelength absorptions. Any manufacturer, brand, and year are suitable.
11. Microplate reader and regular PCR cycler represent an optional alternative.

3 Methods

Currently, the ASQ method has two major variations depending on the length of the "tag." In variant "A", the tag is four nucleotides in length, while a 6-bp tag is used in variant "B". In the described methods, common steps are shown for both variations together, but the variant-specific steps are presented separately.

3.1 Primer Design and Tail-Tag Attachment

1. Identify a high-quality SNP by Sanger sequencing in two genotypes, which will be used as references for further genotyping (*see* **Note 1**).
2. Design and develop a pair of allele-specific (AS) forward primers, which are ideally 22 bp and with 50% GC content but differ by just one nucleotide in the position of the SNP in the penultimate position at the 3′-end (*see* **Note 2**).
3. Design and check an alternative pair of AS-forward primers in reverse-compliment orientation, where the SNP is also located in the penultimate position at the 3′-end, similar to that described in **step 2** and **Note 2**. Compare and choose which design orientation is better (*see* **Note 3**).
4. Design a single common reverse primer after and in conjunction with a pair of the selected AS-forward primers targeting a fragment for amplification containing the SNP of around 100 bp but not less than 50 bp and not longer than 200 bp (*see* **Note 4**).
5. For variant "A" (short tag), an 18-bp fragment must be attached to the 5′-end of AS-forward primers (AS-F1 and AS-F2). This fragment includes a 14-bp identical "tail" and a 4-bp specific "tag." The sequence of tails is shown here: `5′-GTCCTTGCGAAGGC-3′`, and they are identical in both allele-specific forward primers. In contrast, "tag1" [`5′-ATCC-3′`] is specific for AS-F1, while "tag2" [`5′-CAAC-3′`] is specific for AS-F2. The attached fragment gives the allele-specific primers a final length of 18 bp prior to ordering (*see* **Note 5**).
6. For variant "B" (long tag), the principles of the attached fragment remain the same but differ in length and sequences. The sequence of the 13-bp tails is as follows: `5′-CCAGCTGAACGGT-3′`, which is common for both allele-specific forward primers. In contrast, the 6-bp "tag1" specific for AS-F1 is `5′-TCGACC-3′`, and "tag2" specific for - AS-F2 is `5′-ACCTGC-3′`. Therefore, the attached fragment to the 5′-end of allele-specific forward primers makes them 19 bp in length prior to ordering (*see* **Note 6**).

3.2 Molecular Probes Setup for Ordering

7. Three universal probes must be ordered in advance for the ASQ method. For variant A (short tag), two universal probes (UP1

and UP2), 18 bp in length, with a sequence identical to the tag and tail attached to AS primers are as described in **step 5** above. Fluorophore FAM or HEX/VIC must be attached to the 5′-end of UP1 and UP2, respectively, where FAM is specific for the 4-bp tag of AS-F1 while HEX or VIC is specific for the 4-bp tag of AS-F2. The third universal probe is Uni-Q, 14 bp in length with a quencher. The sequence of Uni-Q is complementary to tails in both AS primers and both UPs. A quencher must be attached to the 3′-end of Uni-Q (*see* **Note 7**).

8. For variant B (long tag), the principles of three universal probes are exactly the same, and they also must be ordered in advance. Two universal probes (UP1 and UP2), 19 bp in length, with a sequence identical to the tag and tail attached to AS primers are as described in **step 6** above. Fluorophore FAM or HEX/VIC must be attached to the 5′-end of UP1 and UP2, respectively, where FAM is specific for the 6-bp tag of AS-F1 while HEX or VIC is specific for the 6-bp tag of AS-F2. The third universal probe is Uni-Q, 13 bp in length with a quencher. The sequence of Uni-Q is complementary to tails in both AS primers and both UPs. A quencher must be attached to the 3′-end of Uni-Q (*see* **Note 8**).

3.3 DNA Requirement, Master-Mix, and Reaction Setup

9. For genotyping, DNA can be extracted using any method suitable and convenient for researchers. After quality control, DNA is then diluted to a working concentration (10, 20, or 50 ng/μL) and prepared in individual microtubes, strips, or microplates (*see* **Note 9**).
10. The choice of suitable Taq polymerase is very important, with the more expensive "hot-start" Taq polymerase preferred, but regular Taq polymerase is still an option. In any case, preliminary tests are strongly required to confirm optimal conditions for amplification with fluorescence and allele discrimination (*see* **Note 10**).
11. Before starting SNP genotyping, prepare two premixes: first, the ***primer mix***, which includes the three allele-specific primers (AS-F1, AS-F2, and R). The composition and ratios of components in the primer mix are presented in **Note 11**.
12. Prepare the second premix, the ***probe mix***, including three universal molecular probes (UP1, UP2, and Uni-Q). Their composition and the ratios of the probe mixes are presented in **Note 12**.
13. Prepare the master-mix following the list of components, their concentration, and the volume per reaction (*see* Table 1) specific for each variation, A and B, but it might be quite variable depending SNP and experimental conditions. The volume of

Table 1
The composition of PCR master-mix cocktail for two variants, A and B, of the modified ASQ genotyping method

Component	Concentration	Volume (μL)	Final concentration	Concentration	Volume (μL)	Final concentration
	Variant A (short tag)			Variant B (long tag)		
Buffer (clear, without $MgCl_2$)	5×	2.0	1×	5×	2.0	1×
$MgCl_2$	25 mM	1.2	3.0 mM	25 mM	1.0	2.5 mM
dNTP	2 mM	1.0	0.2 mM	2 mM	1.0	0.2 mM
	AS-primer premix			*AS-primer premix*		
ASP-F1 ASP-F2 ASP-R	1 μM 1 μM 5 μM	1.0	0.1 μM 0.1 μM 0.5 μM	2 μM 2 μM 4 μM	1.0	0.2 μM 0.2 μM 0.4 μM
	Universal probe premix			*Universal probe premix*		
UP-FAM UP-HEX (or -VIC)	2 μM 2 μM	1.0	0.2 μM 0.2 μM	0.5 μM 0.5 μM	1.0	0.05 μM 0.05 μM
Uni-Q	6 μM		0.6 μM	2.0 μM		0.2 μM
Taq polymerase	5 units/μL	0.08	0.04 units/μL	5 units/μL	0.08	0.04 units/μL
Sterile water	–	1.72	–	–	1.92	–
Master-mix	–	*8.0*	–	–	*8.0*	–
DNA template	10 ng/μL	2.0	20 ng/μL	10 ng/μL	2.0	20 ng/μL
Total	–	*10.0*	–	–	*10.0*	–

the master-mix has to be multiplied according to the number of reactions required (*see* **Note 13**).

3.4 Microplate Preparation and Loading

14. Using a cold block or over ice, load the prepared master-mix cocktail in each position of a 96- or 384-well microplate. Add DNA samples for the template, seal with clear or ultra-clear tape for qPCR, mix, and collect the reaction mix to the bottom of wells (*see* **Note 14**).

15. A no-template control (NTC) is strongly recommended and must be used in all experiments with SNP genotyping regardless of the method used. However, in some cases, an NTC can produce fluorescence, which is referred to colloquially as the "Nightmare of NTC fluorescence" (*see* **Note 15**).

Table 2
PCR amplification protocol of the ASQ method used for genotyping

	Temperature	Duration	Temperature	Duration	
Step	**Variant A (short tag)**		**Variant B (long tag)**		**Notes**
1	94 °C	2 min	94 °C	2 min	Initial denaturation; can be variable depending on Taq polymerase
2	94 °C	10 s	94 °C	10 s	First-round denaturation
3	55 °C	20 s	58 °C	20 s	First-round annealing
4	68 °C	20 s	69 °C	20 s	First-round extension
5	10 cycles repeat for **steps 2–4**		10 cycles repeat for **steps 2–4**		
6	94 °C	10 s	94 °C	10 s	Second-round denaturation
7	62 °C	20 s	–	–	Second-round annealing
8	68 °C	20 s	70 °C	30 s	Second-round annealing-extension
9	55 °C	30 s	55 °C	45 s	Cooling, extension, and microplate read
10	26–27 cycles repeat for **steps 6–9**		30–32 cycles repeat for **steps 6–9**		

3.5 qPCR Instrument Setup, Fluorescence Amplification, SNP Genotyping, and Allele Discrimination

16. The qPCR instrument used should be prepared and set up following the supplied manual. The amplification program recommended is shown in Table 2 for both variations, A and B, but it could be modified for better results (*see* **Note 16**).
17. Real-time fluorescence of each fluorophore can be monitored after each cycle and recorded automatically (*see* **Note 17**).
18. SNP genotyping results are presented as an allele discrimination plot with different colored dots or shapes in the presence of NTC or normalized with that data (*see* **Note 18**).

3.6 Examples

This example was provided by coauthor, Ms. Aigul Amangeldiyeva.

3.6.1 Example of Plant Genotyping Using Variant A (Short-Tag) ASQ Method

1. Plant species: sugar beet (*Beta vulgaris* L.). The target gene was *BvSE2* encoding an acid chitinase and mapped to chromosome 3S, as published earlier [11].
2. The AS primer set: tail1 + F1, tail2 + F2, and R; the universal molecular probes: UP1-FAM, UP2-HEX, and Uni-OQA. All primers and probes, including Uni-OQA with Onyx Quencher A, were ordered from and manufactured by Sigma-Merck, USA.
3. The master-mix was prepared with the following final concentrations: $MgCl_2$ = 3.0 mM (including 1.5 mM from 5×

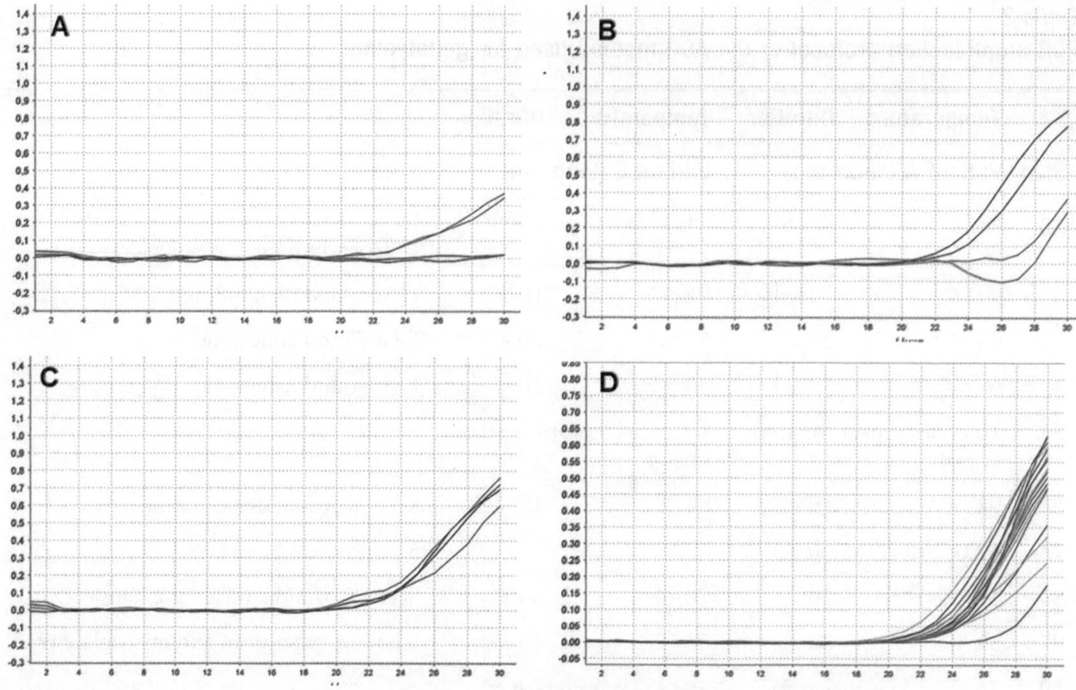

Fig. 2 Examples of the application of the ASQ method, variant A with a short tag for SNP genotyping in the sugar beet gene *BvSE2*. (**a**) Two homozygote genotypes *aa* were selected showing the amplification of FAM with a red line; (**b**) two other homozygotes genotypes *bb* showed the amplification of VIC with a blue line; (**c**) two heterozygote genotypes *ab* with the amplification of both FAM and VIC in red and blue lines, respectively. (**d**) Summary amplification of fluorescence in small-scale genotyping and in the same experiment with 16 samples (14 samples with DNA and two samples with NTC). Results and images were produced by Aigul Amangeldieva in the frame of her research project

buffer); *primer mix*: F1, F2, and R in the ratio 0.1 + 0.1 + 0.5 (mM), respectively; and *probe mix*: UP1, UP2, and Uni-Q in the ratio 0.2 + 0.2 + 0.6 (mM), respectively.

4. Regular Go-Taq polymerase (Promega, USA) was used.
5. Program: 94 °C, 2 min and ten initial cycles, including 94 °C, 10 s; 55 °C, 20 s; and 68 °C, 20 s, with following 27 cycles, including 94 °C, 10 s; 62 °C, 20 s; 68 °C, 20 s; and 55 °C, 30 s.
6. A QuantStudio-5 Real-Time PCR Cycler (Thermo Fisher Scientific, USA) was used. Amplification curves and allele discrimination were analyzed.
7. The amplification curves are presented in Fig. 2, while allele discrimination was similar to those presented later in the chapter.

3.6.2 Example of Plant Genotyping Using Variant B (Long-Tag) ASQ Method

This example was voluntarily provided by Mr. Boy Valenza Damiri, Molecular Biology Laboratory, Biotechnology Department, Research and Development Division, PT East-West Seed, Indonesia (https://www.panahmerah.id).

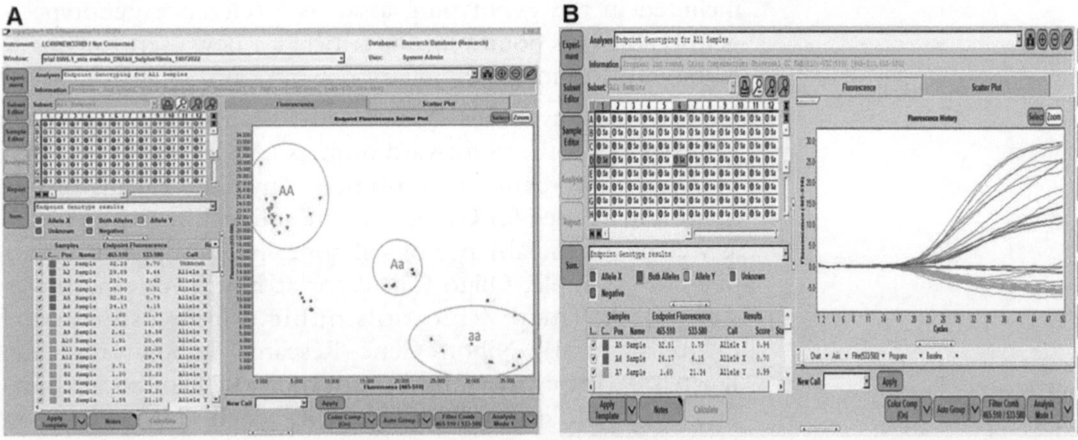

Fig. 3 Example of the application of the ASQ method, variant B with a long tag, provided by Mr. Boy Valenza Damiri, PT East-West Seed, Indonesia, for SNP genotyping in tomato (*Solanum lycopersicum*). (**a**) Allele discrimination plot and (**b**) amplification curves with FAM and HEX during sample genotyping

1. Plant species: tomato (*Solanum lycopersicum* L.). The target gene was located in chromosome 6.
2. The AS primer set: tail1 + F1, tail2 + F2, and R; the universal molecular probes: UP1-FAM, UP2-HEX, and Uni-BHQ1. All were purchased from IDT (Integrated DNA Technologies).
3. The master-mix was prepared with the following final concentrations: $MgCl_2$ = 2.0 mM (including 1.5 mM from 5× buffer); *primer mix*: F1, F2, and R in the ratio 0.1 + 0.1 + 0.3 (mM), respectively; *probe mix*: UP1, UP2, and Uni-Q in the ratio 0.2 + 0.2 + 0.5 (mM), respectively.
4. Regular PCRBIO Taq polymerase (PCR Biosystems, UK) was used.
5. PCR program: 95 °C, 20 s, and ten initial cycles, including 95 °C, 10 s; 54 °C, 20 s; and 68 °C, 20 s, with following 30 cycles, including 95 °C, 10 s; 68 °C, 30 s; and 55 °C, 30 s.
6. An LC480 Roche-demo qPCR instrument was used. Amplification curves and a scatter plot of end-point results were analyzed.
7. The results were received and provided by Mr. Boy Valenza Damiri (*see* Fig. 3).

4 Notes

1. Because the ASQ method of genotyping is relatively new, sequencing is strongly recommended. The identified SNP has to be clear in high quality sequences. It is important that one or more genotypes with the identified and confirmed SNP are

included in the genotyping assay as a reference genotype or genotypes. This point remains critical for new users starting to work with this method.

2. Similar to many other PCR-based FRET methods of plant genotyping, both AS forward primers (AS-F1 and AS-F2) can be designed as regular PCR primers. Any oligo-calculator programs can be used for GC percentage and melting temperature, as well as for the absence of self-annealing and potential hairpins, for example, Oligo Calculator at Northwestern University. USA (http://biotools.nubic.northwestern.edu/OligoCalc.html); Nihon Gene Research Laboratories Inc., Japan (http://www.ngrl.co.jp/tools/0217oligocalc.htm); Java-based Molecular Biologist Workbench (JaMBW) at the University of Pittsburgh, USA (https://www.bioinformatics.org/JaMBW/3/1/9); or OligoEvaluator from the commercial website of Sigma-Merck (http://www.oligoevaluator.com). There is a range of more specialized computer programs and software for primer design and other methods, including (https://primerdigital.com/fastpcr.html) those described in previous papers [9, 10], but users are not restricted to use only those programs published.

3. It is strongly recommended that a pair of AS forward primers be designed and checked in both orientations (direct and reverse compliment), with the best then chosen. The position of AS primers with short and long tags remains unchanged, and examples of both modifications of AS primers with short and long tags for the same SNP are presented in Fig. 4 based on the previously studied *HvSAP8* gene in barley [12].

4. The design of the common reverse primer has to be the same as in the pair of AS-forward primers described in **Note 2** and shown in Fig. 4b, with a similar melting temperature.

5. The general design of all AS primers with attached trails and tags is presented in Fig. 4b. For variant A, with a short tag, in the 5′-end of AS primers, a 14-bp common tail is attached. It is shown in lowercase letters with a following 4-bp specific short tag, as indicated in Fig. 4b in uppercase, in bold, and in italics. This 18-bp tail tag is attached to AS primers highlighted in green and in uppercase, and the SNP position matches with the penultimate nucleotide at the 3′-end of AS primers indicated in red. In fact, both forward AS primers differ only in the 4-bp tag and SNP at the penultimate position. Other fragments of both AS primers with a tail tag are identical.

6. For variant B, with a long tag, the principle of AS primers and attached tail and tag remains the same (*see* Fig. 4b). The difference is only in the 19-bp length of the attached tail tag, including a 13-bp common tail shown in lower case, with a following

(A)

Barley, *HvSAP8*

AAATTTCCCCTGCTCTGCTCCGATGCCGGTAGGTGATTAGGCCGCTGTTGGCTAATTATAAGC
CAGGCAGGTTGCATCGTTTAGTGCTGGTGCCTTCAGCTCTGTGTTTGACCAAGCAGCCACCAC
TATCCTTTTGTAAGATGCTT[Y]GTAGAGCACCTTGCTTTACCAGAGTTGGATTCAGTAGGCT
AAAACATACTGT**R**TGTTTGAAGATTATGCCATGTTCATCTCTAAATTGTTAAAGCCGCAGTGA

[Y] = [C/T]

(B)

AS primers

Variant A (short tag)

HvSAP8-F7: 5′-gtccttgcgaaggc***ATCC***GGTAAAGCAAGGTGCTCTAC**G**A-3′
HvSAP8-F8: 5′-gtccttgcgaaggc***CAAC***GGTAAAGCAAGGTGCTCTAC**A**A-3′
HvSAP8-R: 5′-TCGTTTAGTGCTGGTGCCTTCA-3′

Varian B (long tag):

HvSAP8-F9: 5′-ccagctgaacggt***TCGACC***GGTAAAGCAAGGTGCTCTAC**G**A-3′
HvSAP8-F10: 5′-ccagctgaacggt***ACCTGC***GGTAAAGCAAGGTGCTCTAC**A**A-3′
HvSAP8-R: 5′-TCGTTTAGTGCTGGTGCCTTCA-3′

Fig. 4 Design of non-labeled allele-specific (AS) primers for the modified ASQ method. (**a**) Fragment of the gene sequence where the position of two forward AS primers is highlighted in green; SNP position is indicated in red, while the position of reverse primers is shown in yellow, targeting a 91-bp amplicon size; (**b**) the sequence of AS parts in the 3′-end of two forward primers, F7 and F8 for variant A (short tag) and F9 and F10 for variant B (long tag), is also designed in green, and SNP is in red, but they are present in reverse-compliment orientation based on the original sequence. The probe-specific part of the AS primers is attached in the 5′-end and shown without color. It contains 14-bp and 13-bp common tails, shown in lower case, with a following 4-bp short tag and 6-bp long tag indicated in uppercase, in bold, and in italics, for variants A and B, respectively. The common reverse primer is highlighted in yellow

6-bp long tag indicated in uppercase, in bold, and in italics. The AS primers remain unchanged for both variants (A and B) and are highlighted in green, and the SNP positions are indicated in red (*see* Fig. 4b). In general, the length of oligonucleotides for each of the two AS forward primers is around 40–41 bp (for variants A and B, respectively), and the common reverse primers are about 22 bp. It is important to note that all three oligonucleotides (two AS-forward and one common reverse) do not contain fluorophores or quenchers. The cost of these three AS-forward and reverse primers is therefore relatively cheap. As regular oligonucleotides, they can be ordered through any company producing such regular oligonucleotides and usually require the smallest production volume.

7. The structure and sequences of the three UPs are presented in Fig. 5. For the short tag (variant A), fluorophores FAM and HEX (or VIC) are attached to the 5′-end of two UPs, but other fluorophores can be used instead. The common 14-bp tail and UP-specific 4-bp short tag remain identical to those in AS

Molecular probes

Variant A (short tag)
UP1-short: 5′-FAM-gtccttgcgaaggc***ATCC***-3′
UP2-short: 5′-HEX-gtccttgcgaaggc***CAAC***-3′
UniQ-short: 5′-gccttcgcaaggac-BHQ1-3′

Variant B (long tag)
UP1-long: 5′-FAM-ccagctgaacggt***ACGGCA***-3′
UP2-long: 5′-HEX-ccagctgaacggt***CGTTGC***-3′
UniQ-long: 5′-accgttcagctgg-BHQ1-3′

Fig. 5 Structure and design of three universal probes for two variants of the modified ASQ method with short tag (variant A) and with long tag (variant B). Two universal probes, UP1 and UP2, contain fluorophores FAM and HEX (or VIC) in the 5′-end and are indicated in purple and blue, respectively. They contain 14-bp and 13-bp common tails, shown in lower case, with a following 4-bp short tag and 6-bp long tag indicated in uppercase, in bold, and in italics, for variants A and B, respectively. One common Uni-Q with quencher BHQ1 in the 3′-end is highlighted in grey

primers (*see* **Note 5**). The majority of quenchers are also suitable for the UPs. Historically, Dabcyl was the first quencher used, but Black Hole Quencher 1 (BHQ1) is currently the most popular. The cost of the quencher is the most expensive component of the UPs, and for economic reasons, the cheapest available quencher can be selected and used since they all can quench the fluorescence completely. It is important and strongly recommended to make a test to see how the three UPs work together for the specific plant material, genetic fragment surrounding the SNP, AS primers used for genotyping, as well as the equipment, reagents, and experience level of the staff working in a given laboratory. Some adjustments might be required to arrive at the most optimal combination of all three UPs. The ordered stock of the three UPs can be stored at −20 °C for a very long time, and dilutions for the premix can be used in separate experiments for SNP genotyping.

8. For the long tag (variant B), fluorophores FAM and HEX (or VIC) are also attached to the 5′-end of two UPs, as was described for variant A (*see* **Note 7**). The difference is only with the common 13-bp tail and UP-specific 6-bp long tag, but they are also identical to the corresponding AS primers (*see* **Note 6**).
9. In general, there is no restriction as to which method can be used for DNA extraction and further genotyping if contamination does not inhibit the Taq polymerase enzyme during amplification. For example, regular cetyltrimethylammonium bromide (CTAB), phenol-chloroform, or freeze-dry method can be used for DNA extraction. The range of the working

concentration of the template DNA can be quite diverse and usually is within 20–100 ng. Any format of individual microtubes, strips, or microplates is acceptable.

10. There is no strong prescription to use Taq polymerase of a certain brand and manufacturer. This means that practically any Taq polymerase can be used if it effectively amplifies relatively small PCR fragments with incorporated UPs and releases fluorescence. Taq polymerase can be also tested with higher enzyme units if amplification starts at a later cycle number and develops slower. It is strongly recommended that researchers check and test existing regular Taq polymerases for their adjustment and application in SNP genotyping if they show effective amplification. The final step will be to confirm optimal conditions for amplification with fluorescence and allele discrimination.

11. The first premix is a 10× *primer premix* with three allele-specific primers, F1, F2, and R, where 1 μL of the 10× *primer premix* is used per 10 μL of reaction (*see* Table 1). For variants A and B with short and long tags, different combinations of all three AS primers have been found to be optimal. Empirically, in our experiments, for short tags (variant A), the combination of 1 μM + 1 μM + 5 μM of AS primers F1, F2, and R, respectively, was found as the most suitable. In contrast, such combination of AS primers was different in genotyping experiments with long tags (variant B) with the combination of 2 μM + 2 μM + 4 μM of AS primers F1, F2, and R, respectively (*see* Table 1). However, the combination and ratio of all three AS primers can be changed following the best performance with the SNP of interest and in the specific conditions studied.

12. Similar to the first *primer premix* described in **Note 11**, the second 10× *probe premix* can be also variable, and it was adjusted in our experiments as follows. For variant A with a short tag, the 10× combination of UP1-(FAM) + UP2-(HEX) + Uni-Q was 2 μM + 2 μM + 6 μM, respectively. The combination of analogous probes in the 10× *probe premix* in variant B with a long tag was different as follows: 0.5 μM + 0.5 μM + 2 μM, respectively (*see* Table 1). In rare cases, one of fluorophores can start to 'dominate' providing a false over-representation in all samples. Simple adjustment with reducing ratio of 'false-dominated' dye and slightly incresing of alternative fluorophore can make perfect and correct scoring results. Both *primer premix* and *probe premix* can be kept in the fridge (+4 °C) for SNP genotyping experiments and stored at −20 °C for longer-term storage. All three universal probes (UP1, UP2, and Uni-Q) are sensitive to light, and microtubes must be in dark plastic or wrapped in foil.

13. For cocktail-mix preparation in both variants, A and B, it is important to follow the protocol. However, a relatively wide variability is acceptable if the final amplification and allele discrimination are good enough. Importantly, it also was found that adding of dimethyl sulfoxide (DMSO) in about 5% concentration (0.5 μL per 10μL of reaction) can improve significantly the amplification and allele discimination. Additionally, researchers must pay close attention to the composition of commercial reagents like Taq polymerase, where magnesium is often included in the supplied buffer, for example, 1.5 mM or 3.0 mM of Mg^{2+}, at final concentration. For example, regular buffers for GoTaq polymerase (Promega, USA) contain 1.5 mM magnesium in the final concentration while Flexi buffers for GoTaq Flexi polymerase from the same manufacturer do not contain magnesium at all. Therefore, presence or absence of Mg^{2+} in the supplied buffers must be considered and included in the final concentration of magnesium in the reaction mix for the most optimal performance of the Taq polymerase used.
14. The preparation and loading of the master-mix and the addition of template DNA samples must all be carried out in cold conditions, for example, in a cold block or similar to minimize the risk of evaporation.
15. NTC (no-template control) is an important part of correct SNP genotyping and allele discrimination. However, for the ASQ method in both variants (A and B), sometimes NTC samples produce amplification and register fluorescent signals from one of two dyes used, FAM or HEX/VIC. The reason for this phenomenon remains unclear and is often called the "Nightmare of NTC fluorescence." This has happened only in some cases, and it was found that the addition of DMSO, the reduction of Uni-FAM or Uni-HEX/VIC in the *probe mix* and in combination with fewer cycle numbers for SNP genotyping resulted in significantly smaller NTC fluorescence and the improvement of SNP scoring, as shown in Fig. 2.
16. Amplification programs are recommended for both variants, A and B, as indicated in Table 2. It was checked and optimized in each of the variants. However, some modifications are still possible at any step, including the changing of temperature and duration or the insertion or deletion of any step.
17. The monitoring of fluorescence in each of the dyes used is very convenient using a real-time qPCR instrument, which automatically records fluorescence after each cycle of amplification. However, in general, amplification with any fluorescence begins no earlier than after 14–15 and even 20 cycles. Samples with non-typical or "errant" amplification must be discarded or

repeated one more time to avoid unpredictable or spontaneous failure of the reaction for SNP genotyping. For example, it might be considered in cases of very early detection of fluorescent signal after only a few cycles or if the amplification was not smoothly exponential, which is a typical requirement of all PCR-based methods.

18. Results for SNP genotyping using the ASQ method are similar to those with Amplifluor-like and other methods based on FRET with a simple automatic presentation of the results through computer software included with the qPCR instrument (*see* Fig. 6). The fluorescence levels of both dyes used are presented in picture and table formats with an allele discrimination plot. *X*- and *Y*-axes are automatically set for relative fluorescence units (RFU), for example, for FAM and HEX, respectively. An SNP genotyping score is produced automatically with the identification of both types of homozygotes *aa* and *bb* for corresponding alleles 1 (FAM) and 2 (HEX or VIC), respectively. Heterozygotes *ab* are also determined as the

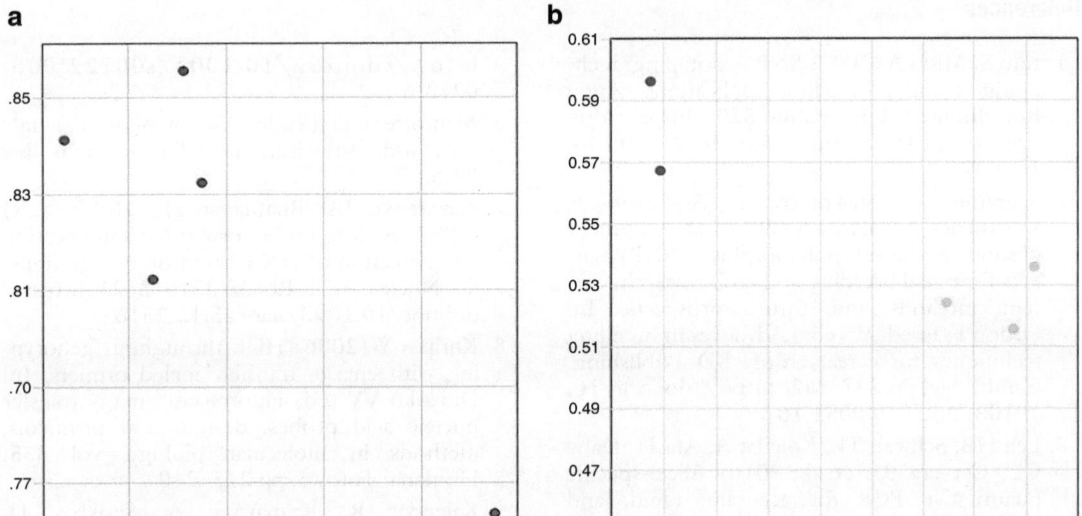

Fig. 6 Examples of allele discrimination using modified ASQ method with variant A (short tag) based on SNP of the *TaVrn-A1* gene in bread wheat, *Triticum aestivum* L. (**a**), and the *LcEFL3* gene in lentil *Lens culinaris* Medik (**b**). Homozygote genotypes *aa* and *bb* are designated with red and blue dots, respectively, while heterozygote genotypes *ab* are shown with green dots. Black squares represent NTC (no-template control) with sterile water instead of a DNA sample. Relative fluorescence units for FAM and VIC are calculated automatically by computer software installed in the qPCR instrument and present in *X*- and *Y*-axes, respectively. The results and images were produced by Dauren Serikby (**a**) and Marzhan Kuzbakova (**b**) in the frame of their master's degree and PhD-degree scholarship projects, respectively

genotype with both fluorescence signals present at about similar levels. Additionally, genotypes showing mixed amplification of fluorophores with clear discrimination are classified as "non-identified," "undetermined," or "mixed," and such genotypes need further analyses. The reference genotypes with two known alternative alleles based on sequences of the genetic fragment containing the SNP are strongly required in each round of genotyping for the accuracy of allele identification.

Acknowledgments

We want to thank Ruslan Kalendar for the strategic idea and the design of molecular probes with the corresponding website. We also appreciate the help and support of Boy Valenza Damiry and Mr. Muryanto, PT East-West Seed, Indonesia, for an example of successful genotyping results. Finally, we acknowledge Carly Schramm for the critical comments on the chapter.

References

1. Kim S, Misra A (2007) SNP genotyping: technologies and biomedical applications. Annu Rev Biomed Eng 9:289–320. https://doi.org/10.1146/annurev.bioeng.9.060906.152037
2. Schramm C, Shavrukov Y, Anderson P, Kurishbaev A, Jatayev S (2019) Development of single nucleotide polymorphism (SNP) markers for cereal breeding and crop research: current methods and future prospects. In: Ordon F, Friedt W (eds) Advances in breeding techniques for cereal crops. BD Publishing, Cambridge, pp 327–362. https://doi.org/10.19103/AS.2019.0051.16
3. Lee HB, Schwab TL, Koleilat A, Ata H, Daby CL, Cervera RL et al (2016) Allele-specific quantitative PCR for accurate, rapid, and cost-effective genotyping. Hum Gene Ther 27:425–435. https://doi.org/10.1089/hum.2016.011
4. Chen X, Sullivan PF (2003) Single nucleotide polymorphism genotyping: biochemistry, protocol, cost and throughput. Pharmacogenomics J 3:77–96. https://doi.org/10.1038/sj.tpj.6500167
5. Giancola S, McKhann HI, Bérard A, Camilleri C, Durand S, Libeau P et al (2006) Utilization of the three high-throughput SNP genotyping methods, the GOOD assay, Amplifluor and TaqMan, in diploid and polyploid plants. Theor Appl Genet 112:1115–1124. https://doi.org/10.1007/s00122-006-0213-6
6. Mamotte CD (2006) Genotyping of single nucleotide substitutions. Clin Biochem Rev 27:63–75
7. Nazarenko IA, Bhatnagar SK, Hohman RJ (1997) A closed tube format for amplification and detection of DNA based on energy transfer. Nucleic Acids Res 25:2516–2521. https://doi.org/10.1093/nar/25.12.2516
8. Khripin Y (2006) High-throughput genotyping with energy transfer-labeled primers. In: Didenko VV (ed) Fluorescent energy transfer nucleic acid probes: designs and protocols, methods in molecular biology, vol 335. Humana, Totowa, pp 215–240
9. Kalendar R, Baidyussen A, Serikbay D, Zotova L, Khassanova G, Kuzbakova M et al (2022) Modified "Allele-specific qPCR" method for SNP genotyping based on FRET. Front Plant Sci 12:747886. https://doi.org/10.3389/fpls.2021.747886
10. Kalendar R, Shustov AV, Akhmetollayev I, Kairov U (2022) Designing Allele-specific competitive-extension PCR-based assays for high-throughput genotyping and gene characterization. Front Mol Biosci 9:773956. https://doi.org/10.3389/fmolb.2022.773956

11. Yerzhebayeva R, Abekova A, Konysbekov K, Bastaubayeva S, Kabdrakhmanova A, Absattarova A et al (2018) Two sugar beet chitinase genes, *BvSP2* and *BvSE2*, analysed with SNP Amplifluor-like markers, are highly expressed after Fusarium root rot inoculations and field susceptibility trial. PeerJ 6:e5127. https://doi.org/10.7717/peerj.5127

12. Baidyussen A, Jatayev S, Khassanova G, Amantayev B, Sereda G, Sereda S et al (2021) Expression of specific alleles of zinc-finger transcription factors, *HvSAP8* and *HvSAP16*, and corresponding SNP markers, are associated with drought tolerance in barley populations. Int J Mol Sci 22:12156. https://doi.org/10.3390/ijms222212156

Chapter 17

Allele-Specific Mutation Genotyping with Mismatches in Primer Design

Yutaro Saito, Fumito Tada, Tadashi Takashina, and Hidetoshi Ikegami

Abstract

Genotyping technologies for single nucleotide polymorphisms (SNPs) and other mutation types have evolved to become essential tools in various fields. Although high-throughput genotyping technologies occupy a key position in handling large amounts of SNP data, simple, low-cost, and conventional genotyping technologies remain in demand. Allele-specific (AS) polymerase chain reaction (PCR) and its related improved methods can effectively identify target SNPs and allele types using AS primers that introduce instability through mismatched bases at and around the SNP site. In this chapter, we present what is known from the literature on primer design with mismatches for AS-PCR and describe three cases of mutation detection (SNPs and insertions/deletions) associated with functional genes of crop species, which could be useful to guide future AS-PCR experiments.

Key words SNP, Allele specific, PCR, Genotyping, Mismatch, Design

1 Introduction

Analyzing mutations and their genomic locations is one of the most basic and indispensable approaches to research in various industries and a wide range of academic fields. For example, in agriculture, it is used to test plants for agronomically important traits, such as high yield, quality, and disease resistance. Single nucleotide polymorphisms (SNPs) are particularly valuable mutation types because they are numerous and scattered throughout the genome; therefore, SNP genotyping techniques are in high demand in a variety of settings [1, 2].

The recent rapid progress of high-throughput sequencing applications, including next-generation sequencing and third-generation sequencing, has greatly reduced the costs for single-base determination and SNP genotyping. These approaches are very powerful, efficient, and highly compatible with information technology for further analytical scheme development. However,

Yuri Shavrukov (ed.), *Plant Genotyping: Methods and Protocols*, Methods in Molecular Biology, vol. 2638,
https://doi.org/10.1007/978-1-0716-3024-2_17,

they require the use of expensive fluorescent probes and microchips or specialized and expensive equipment and reagents, which is impractical in many laboratories.

Sometimes, the analysis of large numbers of targets is not required, but rather, multiple samples of a single target site need to be processed. In these situations, low-throughput simplex PCR assays applying conventional agarose electrophoresis without the need for specialist equipment, facilities, or reagents can be valuable for allele identification and are still in use. These low-throughput applications have also been linked to the development of new high-throughput genotyping technologies, such as kompetitive allele-specific PCR (KASP) and PCR allelic competitive extension.

Allele-specific PCR (AS-PCR) is the simplest low-throughput technique and the basis for SNP mutation detection [3, 4]. Comparable or similar techniques include PCR amplification of specific alleles (PASA) [5, 6], the amplification refractory mutation system (ARMS) [7, 8], and allele-specific amplification (ASA). Subsequently, a series of methods have been developed to design primer mismatches near or at an SNP site, while primer specificity for alleles and allele detection efficiency have been greatly improved [9].

In this chapter, we describe current basic knowledge about primer design in the AS-PCR method and present practical cases in which specific mutation targets, SNPs and insertions/deletions (InDels), were successfully detected.

Primer design is one of the most critical factors affecting AS-PCR outcomes. Mismatches between primers and the target deoxyribonucleic acid (DNA) can affect duplex stability, which impacts template DNA amplification [10]. To overcome the problem of low SNP detection efficiency in AS-PCR, it may be beneficial to add an artificial mismatch within three bases of the 3′-end of the primers [11, 12]. However, designing highly specific primers for SNP detection is typically time- and labor-consuming because of the repetitive nature of exploring appropriate mismatches and base positions. To address this point, the study on the base mismatch impact of primers on allele detection sensitivity was conducted using the *Brassica oleracea* genome [9]. In this report, it was concluded that (1) the highest specificity was obtained by introducing a mismatch at the second site from the 3′-end for C/G type SNPs (with the SNP site as the first site) and at the third site for non-C/G types; (2) in the case of introducing a mismatch at the second site, the polymorphism detection efficiency for all types of mismatches is almost equal; and (3) in the case of introducing a mismatch at the third site, the polymorphism detection efficiency is the highest for C/A type SNPs. As an example, if the target SNP type is A/T, introducing a mismatch at the third site from the 3′-end showed a maximum polymorphic rate of 81.9% (average polymorphic rate: 29.1%).

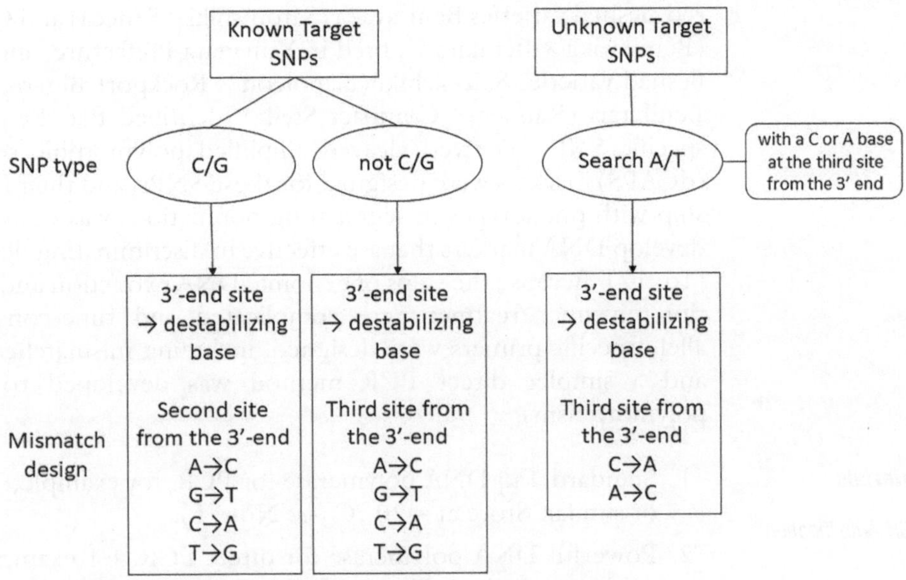

Fig. 1 An optimized procedure in the design of an AS-PCR primer sequence

The procedure for primer sequence design was statistically most validated based on results published earlier [9], and it can be described in four major points as follows. (1) For fixed target SNPs of the C/G type, identify the base at the second site from the 3′-end, and for other fixed target SNP types, identify the base at the third site. (2) If the base in point 1 is "A," select "C" as a mismatched base; if "G," select "T"; if "C," select "A"; and, finally, if "T," select "G." (3) Design primers incorporating the mismatched bases at the third site of point 2 and the mismatched base with strong destabilization at the 3′-end. (4) If the target SNP site is not fixed, select an A/T SNP with "C" or "A" base two bases inside from the SNP site (the third site from the SNP) and design primers incorporating a destabilizing mismatched base at the SNP site and a mismatched base at the third site (*see* Fig. 1). It should be noted that this procedure does not guarantee good results and that the PCR reaction should be tested to ensure that it works.

2 Example A. Detection of SNPs Specific for the Red Flesh Trait in Sweet Cherry Cultivars

Flesh color in sweet cherry is regulated by a single gene that shows red flesh dominance [13, 14]. Sooriyapathirana et al. [15] described *PavMYB10.1*, located in linkage group 3, as a candidate gene controlling fruit skin and flesh color [15]. A comparison of nucleotide sequences in the promoter of *PavMYB10* between red-fleshed varieties Benisayaka (Satonishiki × Seneca) and C-225-4 (Benisayaka × Benikirari), bred in Yamagata Prefecture, and white-fleshed varieties Satonishiki (Napoleon × Rockport Bigarreau) and Benikirari (Rainier × Compact Stella) identified two Benisayaka-specific SNPs. Derived cleaved amplified polymorphic sequence (dCAPS) markers were designed for these SNPs, and their relationship with phenotypes in segregating populations was examined to develop DNA markers that are effective in discriminating flesh traits [16, 17]. Because the steps of genomic DNA extraction and restriction enzyme treatment are complicated and time-consuming, allele-specific primers were designed, including mismatched bases, and a simpler direct PCR method was developed to detect polymorphisms.

2.1 Materials

2.1.1 PCR Amplification

1. Standard Taq DNA polymerase for PCR, for example, TaKaRa or similar. Store at −20 °C (*see* **Note 1**).
2. Powerful DNA polymerase for direct PCR, for example, Shimadzu Ampdirect Plus or similar. Store at −20 °C.
3. Primers diluted to 10 μM in water and stored at −20 °C.
4. Thermocycler.

2.1.2 Gel Electrophoresis

1. Agarose, molecular grade. Store at room temperature.
2. Buffer for gel preparation and electrophoresis, Tris–acetate–EDTA (TAE): 40 mM Tris–acetate, pH 8.0, and 1 mM EDTA in distilled water. Store at room temperature.
3. 10× loading buffer. Store at room temperature.
4. 100 bp DNA ladder. Store at −20 °C.
5. Ethidium bromide (EtBr) solution at 0.5 μg/mL in distilled water. Store at room temperature in the dark.
6. Standard horizontal gel electrophoresis apparatus.
7. Ultraviolet (UV) transilluminator.

2.2 Methods

2.2.1 Primer Design

To detect the two SNPs specific to the red-fleshed variety Benisayaka, a forward primer was designed, including the SNP sequence at the 3′-end, and an A-C, T-G mismatch was placed 3 bp upstream of the SNP sequence, referring to ref. 18. The reverse primer was designed approximately 430 bp downstream (*see* Fig. 2). The design of forward primers 1 and 2 and reverse primer sequences is shown in Table 1. The primer set *PavPAL1-1* was used as a control (*see* **Notes 2–5**).

2.2.2 Standard PCR Amplification

1. PCRs are performed in a total volume of 10 μL, including genomic DNA (obtained using any commercially available kit) diluted in water at 10 ng/μL (*see* **Note 6**). The reaction mixtures are prepared on ice.

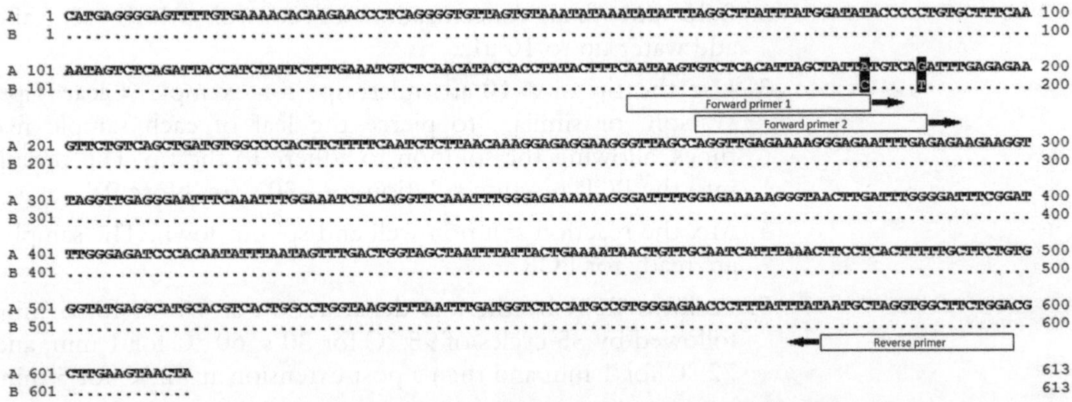

Fig. 2 Design of ASP primers in the *PavMYB10.1* promoter region (example A). (**a**) Sequences specific to white flesh varieties (Benikirari, Beniyutaka, and Satonishiki). (**b**) Sequences specific to red flesh varieties (Benisayaka and C-225-4). Genotype *AA*: Benikirari, Beniyutaka, and Satonishiki; genotype *AB*: Benisayaka and C-225-4

Table 1
Primer sequences in example A. Bold font indicates SNP sites; underlining indicates mismatched sites

Primer set	Forward primer	Reverse primer
Set 1	5′-AATAAGTGTCTCACATTAGCTCTT**C**-3′	5′-TCCAGAAGCCACCTAGCATT-3′
Set 2	5′-GTCTCACATTAGCTATT**C**TGGCAT-3′	
PavPAL1-1	5′-GATATGCTCCGAACTTCACC-3′	5′-CTCTCCATTAACCCCAGTTGTC-3′

2. Prepare a PCR mix with 1 μL 10× PCR buffer (Mg^{2+} plus) (1× in the PCR), 0.8 μL 2.5 mM dNTP mixture (final concentration: 0.2 mM of each), 0.5 μL 10 μM AS primer set (final concentration: 0.5 μM of each), 0.25 μL 10 μM *PavPAL1-1* primer set (final concentration: 0.25 μM), 0.05 μL 5 U/μL Taq polymerase (0.025 U/μL), and 5.9 μL water.
3. Add 9 μL PCR mix to 1 μL DNA (10 ng). Mix well and spin down. The samples are ready for PCR.
4. Perform PCR amplification under standard conditions using the following steps: 5 min of denaturation at 95 °C; 35 cycles of 95 °C for 15 s, 60 °C (annealing) for 15 s, and 72 °C for 30 s; and a final extension at 72 °C for 5 min (*see* **Notes 7** and **8**).

2.2.3 Direct PCR

1. PCRs are performed in a reaction volume of 10 μL per sample.
2. Prepare a PCR mix with 5 μL buffer, 0.05 μL powerful DNA polymerase, 0.5 μM final concentration of AS primer set,

0.25 μM final concentration of control primer set,. and add water up to 10 μL.

3. Use the tip of a 10 μL microtip (for example, Clear Tips, Watson, or similar) to pierce the leaf of each sample five times, allowing the solution to adhere to the tip. Dip the tip into the PCR reaction solution for ~30 s (*see* **Note 9**).
4. Mix the reaction solution well and spin it down. The samples are ready for PCR.
5. Perform PCR by thermal denaturation at 95 °C for 5 min, followed by 35 cycles of 95 °C for 30 s, 60 °C for 1 min, and 72 °C for 1 min and then a post extension at 72 °C for 5 min.

2.2.4 Gel Electrophoresis

1. Separate amplification products on a 1.5% agarose gel prepared (*see* **Note 10**) and electrophoresed in 1× TAE buffer.
2. Add the 50 ml electrophoresis buffer (1× TAE buffer) and 0.75 g agarose to a beaker two to four times the volume of the agarose solution.
3. Heat it in a microwave oven at 800 W for 60–70 s until boiling point and stir occasionally to dissolve all particles.
4. Cool the beaker at room temperature for 5–10 min while stirring until the solution reaches about 50 °C.
5. Pour the solution into the gel maker, add the comb, and allow it to set at room temperature.
6. Add 1 μL 10× loading buffer to 10 μL of each PCR product.
7. Load 5 μL of each sample per well as well as 3 μL of 100 bp DNA ladder to check the size of PCR fragments.
8. Run the gel at 100 V constant voltage for about 30 min.
9. Immerse the gel in an EtBr solution for 15 min at room temperature.
10. Visualize the products by illumination with 312 nm UV light (*see* Figs. 3 and 4 and **Note 11**).

3 Example B. Detection of Sex Trait-Specific SNP in Figs

The common fig (*Ficus carica*) has male and female plants, of which only the female is edible and desirable for selection. The fig's sex trait is controlled by a single gene and can be accurately identified by the *FcRAN1* genotype, as reported in ref. 19. The *FcRAN1* genotype can be identified by a CAPS marker, but this is time and labor intensive and requires the use of a rare restriction enzyme. An example of an AS-PCR marker developing is presented below based on SNP in the *FcRAN1* gene.

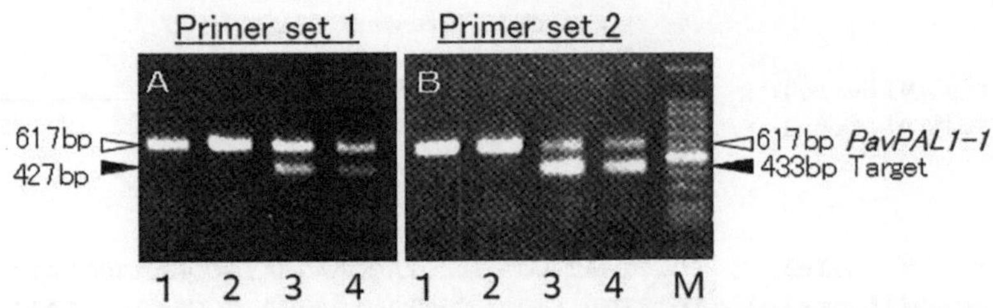

Fig. 3 Electrophoresis of PCR products (example A). (**a**) Primer set 1. (**b**) Primer set 2. Lane 1: Benikirari; Lane 2: Beniyutaka (white-fleshed varieties); Lane 3: Benisayaka; Lane 4: C-225-4 (red-fleshed varieties); M: 100 bp DNA ladder

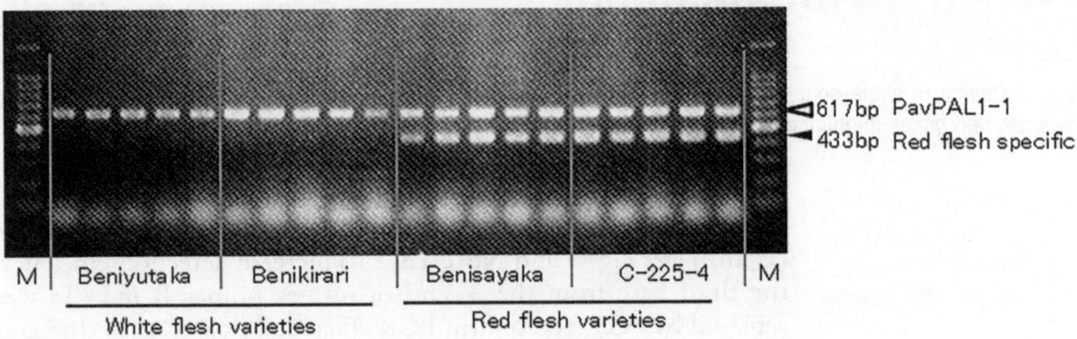

Fig. 4 Direct PCR using five independent samples (example B). M: 100 bp DNA ladder

3.1 Materials

3.1.1 PCR Amplification

1. Taq DNA polymerase for PCR, like Quick Taq polymerase with HS DyeMix produced by Toyobo Co. or similar. Store at −20 °C. DNA polymerase for colony direct PCR to reduce costs in multisample analysis (*see* **Note 1**).
2. Primers diluted to 10 μM in water. Store at −20 °C.
3. Thermocycler.

3.1.2 Gel Electrophoresis

1. Standard agarose, molecular grade. Store at room temperature.
2. Buffer for gel preparation and electrophoresis: 10× Tris–borate–EDTA (TBE): 90 mM Tris–borate, pH 8.0, and 2 mM EDTA in distilled water. Store at room temperature.
3. 100 bp DNA ladder. Store at −20 °C.
4. EtBr solution at 0.5 μg/mL in distilled water. Store at room temperature in the dark.
5. Standard horizontal gel electrophoresis apparatus.
6. UV transilluminator.

3.2 Methods

3.2.1 Primer Design

Sequence information was obtained from ref. 19. The target SNP site (seq000259_9876) was located in the seventh exon of *FcRAN1*. Two primer sets were designed to amplify an SNP detection product and a control product (*see* **Notes 2–5** and **12**). For

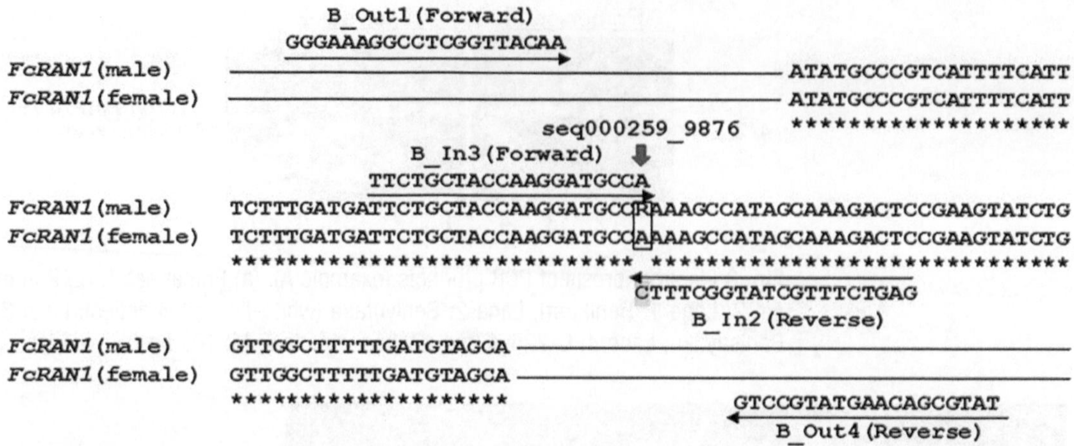

Fig. 5 Design of allele-specific primers for *FcRAN1* SNP detection (example B). The B_In2 primer has two mismatched bases (highlighted) at the SNP site and the third site from the SNP

SNP site mismatch of the first product (primer set B_Out1 and B_In2), nucleotide "C" was used at the 3′-end of the reverse primer (B_In2) (*see* Fig. 5 and **Note 13**). To increase detection efficiency, the third base from the 3′-end of reverse primer B_in2 (T) was replaced by "C." To confirm PCR efficiency, primer set B_In3 and B_Out4 amplifying the downstream region of the SNP site was used as a control.

3.2.2 PCR Amplification

1. PCRs are performed in a total volume of 10 μL using genomic DNA obtained with the use of any commercially available kit diluted in water at 10 ng/μL (*see* **Note 6**). The reaction mixtures are prepared on ice.
2. Prepare a PCR mix with 5 μL 2× PCR reaction buffer, like Toyobo 2× Quick Taq DyeMix; 1.0 μL 2.5 μM normal forward primer B_Out1 and allele-specific forward primer B_In2 (final concentration: 0.25 μM); 1.0 μL 2.5 μM normal reverse primer B_Out4 and allele-specific reverse primer B_In3 (final concentration: 0.25 μM); and 1.0 μL distilled water.
3. Add 9 μL of the PCR mix to 1 μL of DNA (10 ng). Mix well and spin it down. The samples are ready for PCR.
4. Perform PCR amplification using standard conditions as follows: 3 min of denaturation at 94 °C; 40 cycles of 94 °C for 30 s, 55 °C (annealing) for 20 s, and 68 °C for 1 s; and a final extension at 68 °C for 1 min (total ~ 1 h) (*see* **Notes** 7 and **8**).

3.2.3 Gel Electrophoresis

1. Separate amplification products on a 1% agarose gel prepared (*see* **Note 10**) and run in 1× TBE buffer.

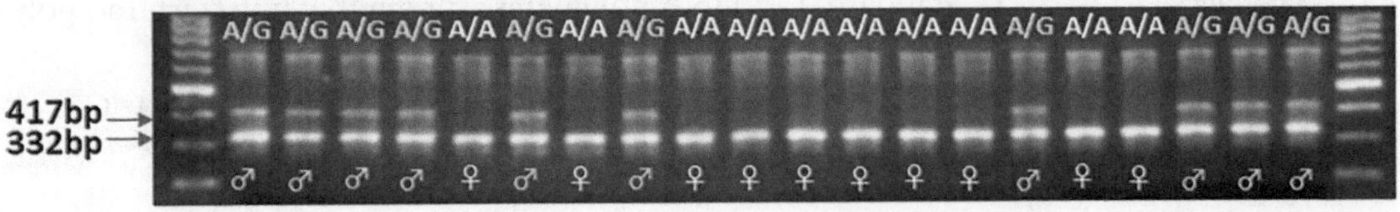

Fig. 6 Agarose gel electrophoresis of AS-PCR amplification products from *F. carica* (example B). The bigger PCR product for 417 bp was amplified by primers B_Out1 and B_In2. The smaller (332 bp) product was amplified by primers B_In3 and B_Out4. Male plants have the A/G genotype, whereas female plants have the A/A genotype

2. Add the 100 ml electrophoresis buffer (1× TBE buffer) and 1.0 g agarose to a beaker two to four times the volume of the agarose solution.
3. Heat in a microwave oven, stirring occasionally to dissolve all particles.
4. Cool the beaker under tap water while stirring until the gel solution is approximately at body temperature.
5. Pour the solution into the gel maker, add the comb, and allow setting at room temperature.
6. Load 9 μL of each sample per well and 4 μL of 100 bp DNA ladder to check the size of PCR fragments.
7. Run the gel at 100 V constant voltage for about 30–40 min.
8. Immerse the gel in an EtBr solution for 15 min at room temperature.
9. Visualize the products by illumination with 312 nm UV light (*see* Fig. 6 and **Note 11**).

4 Example C. Detection of a Self-Incompatible-Specific InDel in Sweet Cherry

Sweet cherry is a generally self-incompatible plant species, but the mutant $SFB^{4'}$, which has four bases deleted from pollen side factor SFB^{4}, resulted in self-fertility or self-compatibility. $SFB^{4'}$ individuals are cross-compatible with all sweet cherries and can be widely used as pollinators. However, the detection of $SFB^{4'}$ requires fragment length determination by DNA sequencers, which are expensive to analyze [20]. Although dCAPS markers can be detected at low cost through agarose gel electrophoresis [21], they require restriction enzyme digestion, and the small difference in fragment length is difficult to resolve. Therefore, a low-cost method to detect $SFB^{4'}$ through agarose gel electrophoresis was developed using allele-specific primers based on the 4 bp InDel between SFB^{4} and $SFB^{4'}$.

4.1 Materials

4.1.1 PCR Amplification

1. Standard Taq DNA polymerase, preferably hot-start Taq polymerase. Store at −20 °C (*see* **Note 1**).
2. Primers diluted to 10 μM in water and stored at −20 °C.
3. Thermocycler.

4.1.2 Gel Electrophoresis

1. Standard agarose, molecular grade. Store at room temperature.
2. TAE buffer for gel preparation and electrophoresis: 40 mM Tris–acetate, pH 8.0, and 1 mM EDTA in distilled water. Store at room temperature.
3. 10× loading buffer. Store at room temperature.
4. 100 bp DNA ladder. Store at −20 °C.
5. EtBr solution at 0.5 μg/mL in distilled water. Store at room temperature in the dark.
6. Standard horizontal gel electrophoresis apparatus.
7. UV transilluminator.

4.2 Methods

4.2.1 Primer Design

Sequence information was obtained from GenBank, and the base showing the difference between *SFB*[4] and *SFB*[4′] sequences was placed at the 3′-end of the primer. To increase primer specificity, a mismatch was introduced where the third base from the 3′-end of the primer is "A" for "C" ("C" for "A") or "G" for "T" ("T" for "G"). Two primer sets were designed, one with mismatches introduced into the forward primer and the other with mismatches introduced into the reverse primer (*see* Fig. 7 and **Notes 2–5**). Primers for the phenylalanine ammonia-lyase gene *PavPAL1-1* were used as an internal control.

4.2.2 PCR Amplification

1. PCRs are performed in a total volume of 10 μL, including genomic DNA extracted and purified using any commercially available kit diluted in water at 10 ng/μL (*see* **Note 6**). The reaction mixtures are prepared on ice.
2. Prepare a PCR mix with 1 μL 10× Ex Taq buffer (1× in the PCR); 0.8 μL 2.5 mM dNTP mixture (final concentration: 0.2 mM each); 0.5 μL 10 μM *SFB* primer (final concentration: 0.5 μM); 0.05 μL 10 μM *PaPAL1-1* primer (final concentration: 0.05 μM); 0.05 μL 5 U/μL Taq polymerase, like TaKaRa Ex HS or similar (0.025 U/μL); and 6.05 μL water.
3. Add 9 μL PCR mix to 1 μL DNA (10 ng). Mix well and spin it down. The samples are ready for PCR.
4. Perform the PCR amplification using standard conditions. The suggested protocol is as follows: 2 min of denaturation at 94 °C; 40 cycles at 94 °C for 30 s, 60 °C (annealing) for 30 s, and 72 °C for 40 s; and a final extension at 72 °C for 5 min (*see* **Notes 7** and **8**).

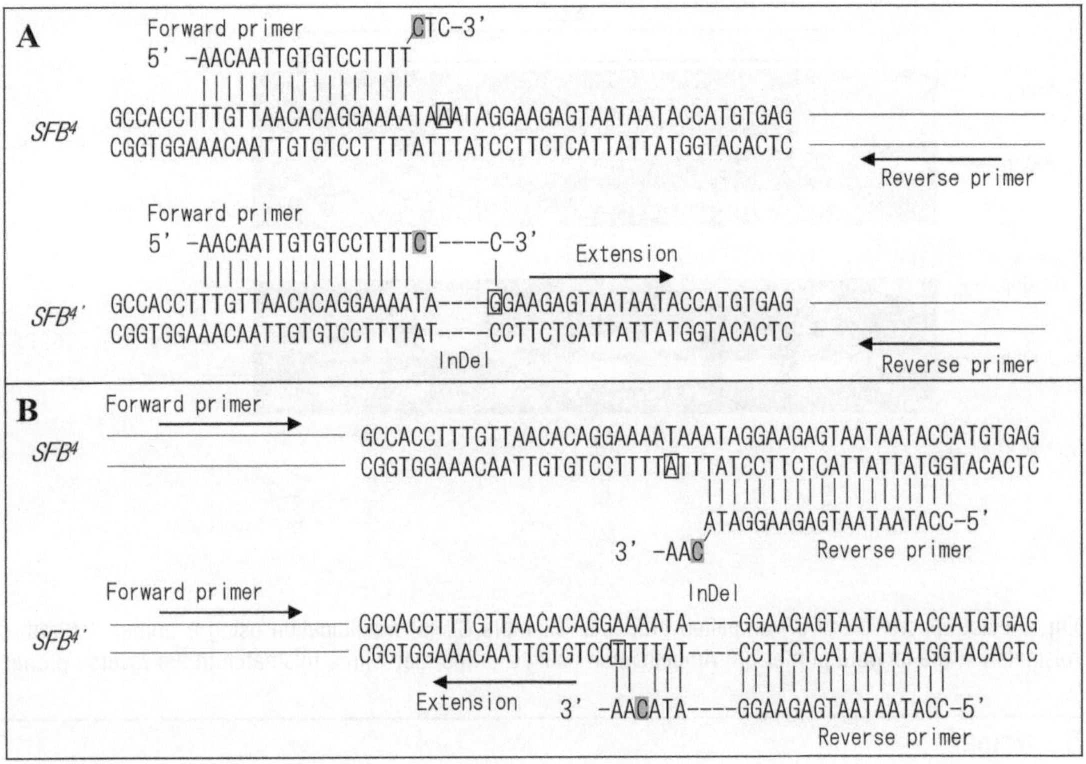

Fig. 7 Design of AS primers for *SFB*$^{4'}$ InDel detection (example C). (**a**) Using a primer set with a mismatched base in the forward primer. (**b**) Using a primer set with a mismatched base in the reverse primer. Box shows nucleotides where sequence differences appear for the first time. Gray shading represents mismatched sites

4.2.3 Gel Electrophoresis

1. Separate amplification products on a 1% agarose gel prepared (*see* **Note 10**) and run in 1× TAE buffer.
2. Add the 50 ml electrophoresis buffer (1× TAE buffer) and 0.75 g agarose to a beaker two to four times the volume of the agarose solution.
3. Heat in a microwave oven at 800 W for 60–70 s until boiling point and stir occasionally to dissolve all particles.
4. Cool the beaker under tap water while stirring until the gel solution is approximately at body temperature.
5. Pour the solution into a gel maker, add the comb, and allow setting at room temperature.
6. Add 1 μL 10× loading buffer to 10 μL of each PCR product.
7. Load 5 μL of each sample and 3 μL 100 bp DNA ladder to check the size of the PCR fragments.
8. Run the gel at 100 V constant voltage for about 30 min.
9. Immerse the gel in an EtBr solution for 15 min at room temperature.
10. Visualize the products by illumination with 312 nm UV light (*see* Fig. 8 and **Note 11**).

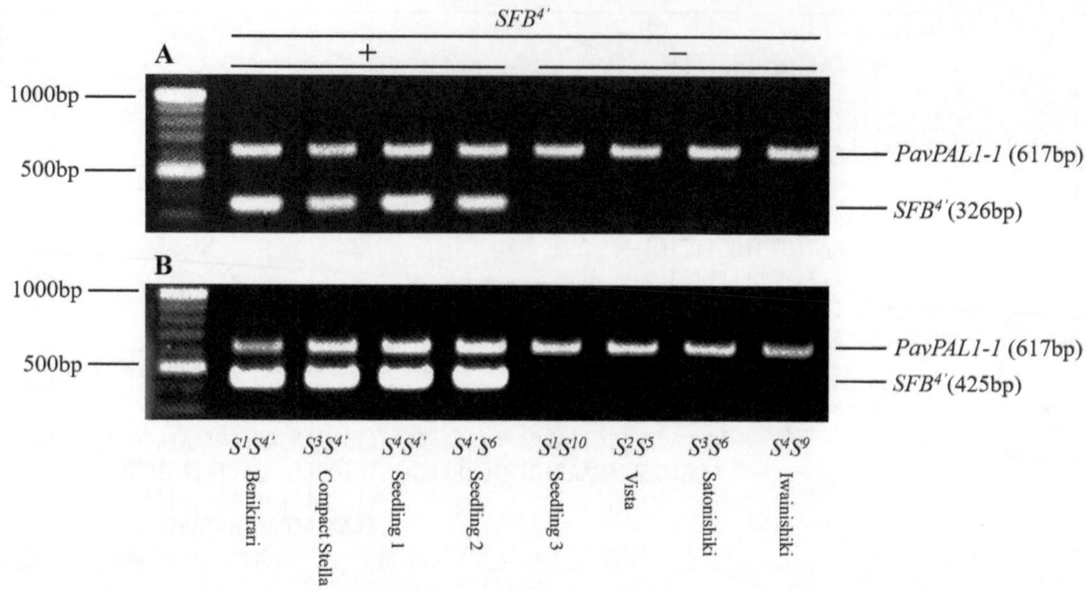

Fig. 8 Electrophoresis of PCR-amplified products (example C). (**a**) Amplification using a primer set with a mismatch in the forward primer. (**b**) Amplification using a primer set with a mismatch in the reverse primer

5 Notes

1. Reagents for PCR should always be used with appropriate temperature control.
2. The Tm values of several primers used in the AS-PCR should be as close as possible because of the uneven signal intensity of PCR products. However, if the difference is in the range of 5 °C, it is usually possible to explore appropriate reaction temperatures.
3. When designing your reaction tube or plate setup, be sure to include positive control samples as well as negative ones for easier troubleshooting.
4. It is recommended to design and try several primer sets with different sequences as often designing only one primer set does not guarantee that it will work.
5. Primer length is usually designed around 20–25 bp. The level of purification for primer synthesis can be either desalted grade or column purification grade.
6. Genomic DNA can be extracted using standard DNA extraction protocols that yield high-quality DNA. Ethanol contamination during DNA extraction is one possible DNA quality loss factor. The application of high-quality and optimal concentrations of template DNA will provide more reliable results.
7. If the activity of the polymerase is strong, better results may be expected by shortening or omitting the time of each PCR step.

8. If the AS-PCR is not successful, the first step is to estimate and understand the reasons for the failure and then modify those factors.
9. In general, performing direct PCR using a crudely prepared sample, such as a collection from a leaf, as a template reduces the success rate of the PCR reaction. However, if this is necessary, try to reduce the amount of sample to be tested and the number of times a leaf is poked. When poking a leaf during direct PCR, tear off the leaf and place a piece of filter paper or plastic wrap under the leaf for easy poking. A 10 μl tip is best for poking a leaf; a 200 μl or 1000 μl tip may contain too much leaf juice, which may interfere with the PCR reaction. When the tips are immersed in the reaction solution, the reaction solution may flow back into the tips. In this case, the reaction solution in the tip can be pipetted out.
10. All used primers can be designed in the exon region to potentially create more universal SNP markers. In this case, it is easier to design primers if the SNP targeted for detection is located in relatively long exon regions.
11. The agarose concentration of the gel for electrophoresis depends on the size of the PCR products to be detected and is usually within 1–2%.
12. First, use primers that introduce the first mismatch at the SNP site, but if that does not work well, introduce the second mismatch into the primers. If a mismatch is used inside of the SNP site, it is better to use primer combinations that have been confirmed to work to some extent for mismatches at SNP sites.
13. Researchers should determine AS-PCR reaction conditions such as polymerase type, cocktail-mix composition, the number of reaction cycles, extension time, reaction temperature, etc. according to the rules assumed in PCR reagent manuals. Among these factors, priority should be given to determining the upper-level factors (polymerase type, for example) that have a greater impact on the PCR results.

References

1. Schramm C, Shavrukov Y, Anderson P, Kurishbaev A, Jatayev S (2019) Development of single nucleotide polymorphism (SNP) markers for cereal breeding and crop research: current methods and future prospects. In: Ordon F, Friedt W (eds) Advances in breeding techniques for cereal crops. BD Publishing, Cambridge, pp 327–362. https://doi.org/10.19103/AS.2019.0051.16
2. Morgil H, Gercek YC, Tulum I (2020) Single nucleotide polymorphisms (SNPs) in plant genetics and breeding. In: Çalişkan M (ed) The recent topics in genetic polymorphisms. InTech Open, London, pp 825–400. https://doi.org/10.5772/intechopen.91886
3. Ugozzoli L, Yam P, Petz LD, Ferrara GB, Champlin RE, Forman SJ et al (1991) Amplification by the polymerase chain reaction of hypervariable regions of the human genome for evaluation of chimerism after bone marrow transplantation. Blood 77:1607–1615.

https://doi.org/10.1182/blood.v77.7.1607.1607

4. Bottema CD, Sommer SS (1993) PCR amplification of specific alleles: rapid detection of known mutations and polymorphisms. Mutat Res 288:93–102. https://doi.org/10.1016/0027-5107(93)90211-W
5. Sommer SS, Groszbach AR, Bottema CD (1992) PCR amplification of specific alleles (PASA) is a general method for rapidly detecting known single-base changes. BioTechniques 12:82–87
6. Sarkar G, Cassady J, Bottema CDK, Sommer SS (1990) Characterization of polymerase chain reaction amplification of specific alleles. Anal Biochem 186:64–68. https://doi.org/10.1016/0003-2697(90)90573-R
7. Wu DY, Ugozzoli L, Pal BK, Wallace RB (1989) Allele-specific enzymatic amplification of β-globin genomic DNA for diagnosis of sickle cell anemia. Proc Natl Acad Sci U S A 86:2757–2760. https://doi.org/10.1073/pnas.86.8.2757
8. Newton CR, Heptinstall LE, Summers C, Super M, Schwarz M, Anwar R et al (1989) Amplification refractory mutation system for prenatal diagnosis and carrier assessment in cystic fibrosis. Lancet 334:1481–1483. https://doi.org/10.1016/S0140-6736(89)92931-0
9. Liu J, Huang S, Sun M, Liu S, Liu Y, Wang W et al (2012) An improved allele-specific PCR primer design method for SNP marker analysis and its application. Plant Methods 8:34. https://doi.org/10.1186/1746-4811-8-34
10. Bru D, Martin-Laurent F, Philippot L (2008) Quantification of the detrimental effect of a single primer-template mismatch by real-time PCR using the 16S rRNA gene as an example. Appl Environ Microbiol 74:1660–1663. https://doi.org/10.1128/AEM.02403-07
11. Cha RS, Zarbl H, Keohavong P, Thilly WG (1992) Mismatch amplification mutation assay (MAMA): application to the c-H-*ras* gene. Genome Res 2:14–20. https://doi.org/10.1101/gr.2.1.14
12. Kwok S, Chang SY, Sninsky JJ, Wang A (1994) A guide to the design and use of mismatched and degenerate primers. Genome Res 3:539–547. https://doi.org/10.1101/gr.3.4.S39
13. Fogle HW (1958) Inheritance of fruit color in sweet cherries (*Prunus avium*). J Hered 49: 294–298. https://doi.org/10.1093/oxfordjournals.jhered.a106830
14. Schmidt H (1998) On the genetics of fruit colour in sweet cherries. Acta Hortic 468:77–81. https://doi.org/10.17660/actahortic.1998.468.6
15. Sooriyapathirana SS, Khan A, Sebolt AM, Wang D, Bushakra JM, Lin-Wang K et al (2010) QTL analysis and candidate gene mapping for skin and flesh color in sweet cherry fruit (*Prunus avium* L.). Tree Genet Genomes 6:821–832. https://doi.org/10.1007/s11295-010-0294-x
16. Saito Y, Adachi E, Isuzugawa K, Kurosaka S, Takashina T (2017) Development of DNA marker for flesh color derived from sweet cherry (*Prunus avium* L.) cultivar 'Benisayaka'. Bull Yamagata Pref Agric Res 9:33–41
17. Saito Y, Adachi E, Isuzugawa K, Shirasawa K, Ikenaga M, Sato H (2017) Development of DNA marker for distinguishing flesh colour in sweet cherry (*Prunus avium* L.). Acta Hortic 1235:15–25. https://doi.org/10.17660/ActaHortic.2019.1235.2
18. Hayashi K, Hashimoto N, Daigen M, Ashikawa I (2004) Development of PCR-based SNP markers for rice blast resistance genes at the *Piz* locus. Theor Appl Genet 108:1212–1220. https://doi.org/10.1007/s00122-003-1553-0
19. Mori K, Shirasawa K, Nogata H, Hirata C, Tashiro K, Habu T et al (2017) Identification of *RAN1* orthologue associated with sex determination through whole genome sequencing analysis in fig (*Ficus carica* L.). Sci Rep 7:41124. https://doi.org/10.1038/srep41124
20. Takahashi Y, Yamamoto T, Adachi E, Ikeda K, Sato Y, Noguchi K (2018) Development of self-incompatibility (*S*) alleles identification methods and *S* genotyping in sweet cherry (*Prunus avium* L.) cultivars. Bull Agric Res Yamagata Prefect 10:15–27
21. Ikeda K, Watari A, Ushijima K, Yamane H, Hauck NR, Iezzoni AF et al (2004) Molecular markers for the self-compatible $S^{4'}$-haplotype, a pollen-part mutant in sweet cherry (*Prunus avium* L.). J Am Soc Hortic Sci 129:724–728. https://doi.org/10.21273/jashs.129.5.0724

Chapter 18

PCR Allele Competitive Extension (PACE)

Daniel von Maydell

Abstract

PCR allele competitive extension (PACE™) is a marker system to analyze single nucleotide polymorphisms (SNPs). As strongly related technologies like kompetitive allele-specific PCR (KASP™) or Amplifluor®, PACE is based on a polymerase chain reaction (PCR) with two competing allele-specific primers and an endpoint fluorescent measurement. The system requires sequence information surrounding the SNP of interest and a qPCR machine or a fluorescent plate reader.

Key words PCR allele competitive extension (PACE), Single nucleotide polymorphisms (SNPs), Codominant marker system

1 Introduction

With the rise of next-generation sequencing (NGS) techniques, single nucleotide polymorphisms (SNPs) became frequently used markers in the molecular-genetic research of plants. SNPs are codominant markers and are highly abundant across the genome. Therefore, they have been used for various genetic analyses, including studies on genetic diversity and population structure, linkage mapping, quantitative trait loci (QTL) mapping, genome-wide association studies (GWAS), marker-assisted selection (MAS), and genomic selection (GS) [1].

NGS or chip-based methods can be used to discover or recover a vast number of SNPs in one run. In contrast, marker systems like PCR allele competitive extension (PACE, 3CR Bioscience) have been developed to reanalyze single biallelic SNPs. Consequently, PACE should be most efficient if a small number of SNPs but a high number of genotypes is to be analyzed. For instance, PACE has been used to genetically differentiate between sexes in plants [2, 3], to identify crossings [4], or for genetic mapping [5]. Verified markers can be used in marker-assisted selection. Interestingly, a PACE product has been designed to detect variants of the SARS-Cov-

Yuri Shavrukov (ed.), *Plant Genotyping: Methods and Protocols*, Methods in Molecular Biology, vol. 2638,
https://doi.org/10.1007/978-1-0716-3024-2_18,

2 virus [6]. PACE requires at least short sequence information surrounding the SNP and a qPCR machine or a fluorescent plate reader. A reference genome is not mandatory, so this method can be of interest for the genetic analysis of plants with little prior research. For small projects, currently, reagent prices range from approximately \$0.1 to \$0.24 per sample (before tax, depending on the chosen reaction volume and other factors). Upscaled projects might achieve lower prices.

The chemistry of PACE is nearly identical to the chemistry of KASP, the procedure for which is described in detail in the literature [7]. Briefly, PACE consists of a polymerase chain reaction (PCR) with two competing allele-specific primers and one common reverse primer and an endpoint fluorescent measurement. The allele-specific primers differ in their 3′ end nucleotides (the SNP) and in a tail sequence at the 5′ end. If allele-specific primers bind to the target sequence, these tails are incorporated into the amplified sequences. The PACE genotyping master mix contains complement sequences of those tails, which quench either the fluorescein amidite (FAM) or hexachloro-fluorescein (HEX) fluorescent dyes. During further PCR, the complement sequences are incorporated as well, and the fluorescent dyes are released. Depending on the allelic constitution of the locus, either the FAM or the HEX (respective homozygous genotypes) or both dyes (heterozygous genotypes) are released. There are no studies available that compared the efficiency of PACE with related methods like KASP. According to the manufacturer, the PACE genotyping master mix is compatible with the KASP and Amplifluor systems.

The following chapter provides a description of the PACE genotyping protocol, including the development of new markers. Several additional remarks might help new users to increase the probability of a successful implementation of this marker system, especially in small projects on minor crops.

2 Materials

2.1 Primer Design

1. Deoxyribonucleic acid (DNA) sequence containing the SNP of interest (*see* **Note 1**).
2. Software for primer design or alternatively using the assay design of the manufacturer (3CR Bioscience) (*see* **Note 2**).

2.2 PACE Assay

1. 1–10 ng genomic DNA per sample (*see* **Note 3**).
2. 2 × PACE genotyping master mix, which contains fluorescent reporting cassette, Taq polymerase, deoxynucleotide triphosphates (dNTPs), buffer, $MgCl_2$ at 4.4 mM, performance enhancers, and the reference dye ROX (5-carboxyl-X-rhodamine, succinimidyl ester) (*see* **Note 4**).

3. Two allele-specific primers and one common primer (100 μM).
4. Nuclease-free water.
5. qPCR machine (e.g., CFX96 real-time system, Bio-Rad) or fluorescent plate reader, optionally with specialized clustering software for data interpretation (e.g., CFX Maestro software, Bio-Rad).
6. PCR plate and optical clear seal (*see* **Note 5**).

3 Methods

3.1 Primer Design

If not yet available, primer design is the first step before starting a PACE assay. For this purpose, a list of sequences containing SNPs for marker design can be sent to the manufacturer (*see* **Note 6**). Primers can be ordered from any oligo-nucleotide manufacturer (e.g., Invitrogen, Thermo Fisher Scientific, or any local companies) (*see* **Note 7**).

New markers should be tested for functionality through the analysis of the samples of both homozygous and heterozygous genotypes. If necessary, a mixture of both genotypes can be used. A negative control (buffer in which DNA is dissolved) has to be used instead of template DNA. Rather frequently, a test of functionality might fail. This can be caused by multiple reasons, which sometimes remain obscure, yet some considerations are given in **Note 8**.

3.2 PACE Assay

General comments: All assays should include negative controls, which contain the buffer, in which DNA is dissolved and reagents. If available, it is recommended to add positive controls for both homozygous genotypes and optionally for the heterozygous genotypes as well.

1. Prepare the assay mix containing both allele-specific primers, the common (reverse) primer, and the nuclease-free water according to Table 1 (*see* **Note 9**).
2. Choose a plate scale suitable for the genotyping project and conduct the following steps in accordance with the chosen scale (*see* Table 2 and **Notes 10** and **11**).
3. Pipette DNA samples in wells and optionally dry the DNA (e.g., with a speed vac).
4. Prepare a mixture of the genotyping master mix and the assay mix (and nuclease-free water if the DNA was dried).
5. Pipette this mixture to the DNA samples (the volume of the assay mix can be ignored).
6. Cover the plate with an optical seal.
7. Briefly centrifuge the plate.

Table 1
Preparation of 100 μl of the PACE assay mix

	Starting concentration (μM)	Final concentration (μM)	Volume (μl)
Allele-specific primer 1	100	12	12
Allele-specific primer 2	100	12	12
Common primer	100	30	30
Nuclease-free water			46

Table 2
Reagent volumes for the total PACE reaction mix for different plate scales and for hydrated and dried DNA according to the manufacturer's protocol. GMM, genotyping master mix; AM, assay mix. The low proportion of the AM is ignored for the calculation of the total reaction volume

	Hydrated DNA			Dried DNA			
	96-well plate	384-well plate	384-well Array Tape	96-well plate	384-well plate	1536-well plate	384-well Array Tape
2 × PACE GMM	5	2.5	0.8	5	2.5	0.5	0.4
PACE GAM	0.138	0.069	0.022	0.138	0.069	0.011	0.011
Water	0	0	0	5	2.5	0.5	0.4
DNA	5	2.5	0.8	Dried	Dried	Dried	Dried
Total	10	5	1.6	10	5	1	0.8

Table 3
Thermal cycling conditions for PACE genotyping assays

Step	Temperature	Time	Cycles
Enzyme activation	94 °C	15 min	1
Denaturation Annealing and elongation	94 °C 65–57 °C (−0.8 °C each cycle)	20 s 60 s	10
Denaturation Annealing and elongation	94 °C 57 °C	20 s 60 s	30
Cooling down and measurement	35 °C	60 s	1

8. Start the PCR amplification with thermal cycling conditions as present in Table 3. Briefly, the PCR starts with Taq-polymerase enzyme activation, followed by a "touchdown" temperature schedule from 65 to 57 °C in ten steps. Then amplification is continued at 57 °C for 30 cycles. Fluorescence is measured after cooling down to 35 °C (*see* **Note 12**).

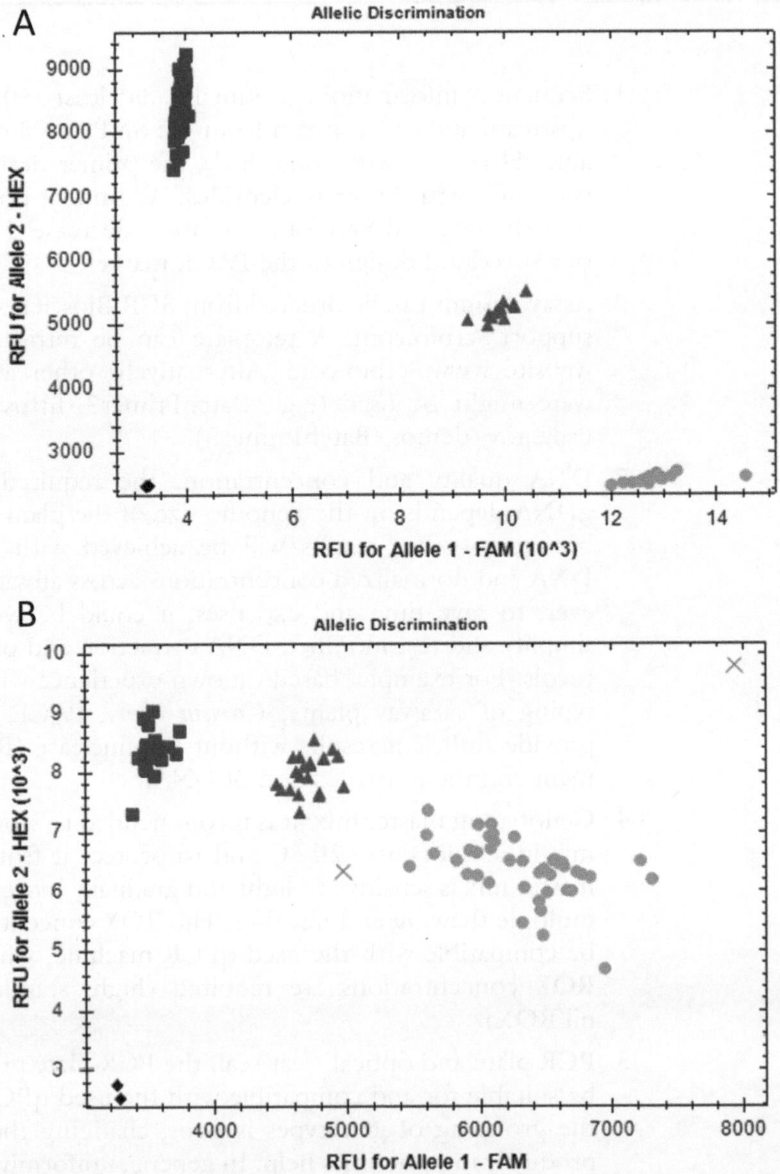

Fig. 1 Examples of allele discrimination using PACE assay in a 96-well plate format. Clustering was conducted using CFX Maestro software (Bio-Rad). Orange dots, homozygous allele 1; blue squares, homozygous allele 2; green triangles, heterozygous genotypes; black diamonds, negative controls; red crosses, undetermined samples. (**a**) Optimal result with a perfect grouping for allele discrimination. (**b**) Result with some issues treated with partial manual annotation. Stronger HEX signal accompanied by a weak grouping of homozygous allele 1. One sample (red cross in the middle) has a rather insufficient association with a cluster of genotypes, and another sample (red cross in the top-right corner) has an unusually high signal

9. Interpret the data by using clustering software (*see* **Note 13** and Fig. 1).
10. If necessary, repeat PCR with ten additional cycles at 57 °C (*see* **Note 14**).

4 Notes

1. Sequence information: optimally, at least 50 nucleotides upstream and downstream from the SNP site should be available. However, with some luck, the primer design might be successful with fewer nucleotides. Additional polymorphisms near the targeted SNP of interest can decrease the probability of a successful design of the PACE marker.
2. Assay designs can be ordered from 3CR Bioscience via email to support.3crbio.com. A template can be retrieved from the website www.3crbio.com. Alternatively, other available software might be used (e.g., BatchPrimer3, https://wheat.pw.usda.gov/demos/BatchPrimer3).
3. DNA quality and concentration: the required amount of gDNA depends on the genome size of the plant species. The best genotyping results will be achieved with well-purified DNA and normalized concentrations across all samples. However, to save time and expenses, it could be worthwhile to simplify and test modified DNA extraction and purifying protocols. For example, based on own experience with the genotyping of caraway plants, *Carum carvi*, PACE assays could provide sufficient results without ribonuclease (RNase) treatment and the normalization of DNA.
4. Genotyping master mix: it is recommended to store the master mix in aliquots at −20 °C and to protect it from light. The master mix is sensitive to light and gradually loose activity after multiple thawing and freezing. The ROX concentration has to be compatible with the used qPCR machine, where different ROX concentrations are required (high, standard, low, or nil ROX).
5. PCR plate and optical clear seal: the PCR plate product has to be suitable for and compatible with the used qPCR system. If the grouping of genotypes is poor, changing the PCR plate product can sometimes help. In general, uniformly white plates increase fluorescent detection. Choose an optical seal that sticks very tightly.
6. Potentially, any software for primer design can be used. Alternatively, the currently free assay design of the 3CR Bioscience manufacturer is very suitable. A list of sequences with several thousand SNPs can be sent to 3CR Bioscience for primer design. A researcher can decide later whether PACE markers have to be designed for the selected SNPs. If a sequence is too long, a researcher may cut a fragment of about 50 nucleotides upstream and downstream of the targeted SNP site. If an SNP-containing sequence is too short and insufficient for PACE marker development, the primer cannot be developed

using the manufacturer or other software. In the submitted sequences, the position of the targeted SNPs of interest has to be written in brackets (e.g., "[T/C]"), whereas other polymorphisms have to be indicated with a nucleotide ambiguity IUPAC code (standard code for degenerative primers). A list of primer sequences with both allele-specific primers and one common primer per each SNP will be returned to the customer. The sequences of the allele-specific primers will also contain the oligo-tails associated with the fluorescent HEX (GAAGGTCGGAGTCAACGGATT) or FAM (GAAGGTGAC CAAGTTCATGCT). These tails have to be added manually if the primers are designed "in-house" by a researcher.

7. Primer manufacturing: as an inexpensive solution for small projects, ordering approximately 25 nmol desalted primer oligos can be sufficient. Yet allele-specific primers, including the tail, are very long (around 50 nucleotides). Thus, for allele-specific primers, sometimes higher amounts of oligos (about 50 nmol) are necessary to be ordered.
8. Success of the marker design: firstly, good PACE marker development depends on the quality of sequence information. In particular, when working with short-sequence data of plant species without a reference genome, the risk is higher that erroneous sequence information might cause an unsuccessful marker design. For example, based on own experience in experiments with the plant species caraway, *Carum carvi*, PACE markers with good amplification were observed, but all samples showed a heterozygous genotype. This indicated the potential presence of homologous loci instead of an SNP at one locus. In other experiments, PACE markers with amplification for one homozygous genotype only were observed. In some cases, a rare allele of an SNP might be due to a simple sequencing error. Additionally, other polymorphisms near the targeted SNP of interest can impair the functionality of a PACE marker. Other polymorphisms or homologous regions might be reasons for preferable signals of HEX or FAM so that allele discrimination can show weak clustering.
9. Assay mix: the assay mix of the three primers can be prepared for several plates and can be frozen and defrosted several times.
10. Multiple PACE markers on one plate: if the functionality of several new markers has to be tested or if only a few samples have to be tested with several markers, it is possible to analyze several markers on the same plate (but in separate wells). However, markers will run with different efficiency. Thus, the interpretation of results has to be carried out for each marker separately.

11. Modifying the concentrations or amounts of reagents and plate scales: to reduce costs, a reduction of the reaction volume might be reasonable. For example, in our experiments, in a 96-well plate format, a reduction from 10 μl to 5 μl was successful (2.5 μl DNA, 2.5 μl GMM, genotyping master mix, and 0.069 μl assay mix (AM)) [4]. Increasing primer concentration might sometimes help if allelic discrimination in the genotype separation is weak. Scales of 384- and 1536-well plates require much fewer reagents and DNA so that expenses can be considerably reduced. Yet robotic pipetting systems are recommendable here. According to the manufacturer, 1536-well plates can only be used with dried DNA. In general, it has to be considered that liquids can evaporate differently across the plate if the plate remains uncovered for too long a time.
12. Modifying cycling conditions: in general, a higher temperature should increase the specificity of primers and, hence, allelic discrimination. The manufacturer will provide melting temperatures for each primer. Therefore, if the separation of genotypes is poor, a researcher may check whether increasing the annealing temperature for amplification is possible. However, based on own experience, better results were rarely received with higher annealing temperatures.
13. Data interpretation: qPCR systems with supplementary built-in clustering software are helpful to gain results shortly after PCR. Yet, frequently, manual corrections are necessary (compare Fig. 1b): some samples can show considerably higher signals. Even if these signals seem reliable, it might be better to exclude such samples and to get better clustering for the rest of the samples. Sometimes, one fluorescent dye signal is stronger, and therefore, heterozygous genotypes can cluster near a group of homozygous genotypes. In this case, automatic clustering might not recognize two groups. As long as different groups visually exist, manual correction should be acceptable.
14. Additional cycles: depending on several factors (DNA quantity and quality, primer efficiency, used PCR plates, qPCR machine, etc.) 30 cycles might not be sufficient to discriminate genotypes. Often, merely a few samples need additional PCR cycles, for instance, because of low DNA concentration or pipetting errors.

References

1. Semagn K, Babu R, Hearne S, Olsen M (2014) Single nucleotide polymorphism genotyping using Kompetitive Allele Specific PCR (KASP): overview of the technology and its application in crop improvement. Mol Breed 33:1–14. https://doi.org/10.1007/s11032-013-9917-x
2. Toth JA, Stack GM, Cala AR, Carlson CH, Wilk RL, Crawford JL et al (2020) Development and validation of genetic markers for sex and cannabinoid chemotype in *Cannabis sativa* L. GCB Bioenergy 12:213–222. https://doi.org/10.1111/gcbb.12667

3. Varela ES, Bekaert M, Ganeco-Kirschnik LN, Torati LS, Shiotsuki L, de Almeida FL et al (2021) A high-density linkage map and sex-linked markers for the Amazon Tambaqui *Colossoma macropomum*. BMC Genomics 22: 709. https://doi.org/10.1186/s12864-021-08037-8
4. von Maydell D, Brandes J, Lehnert H, Junghanns W, Marthe F (2020) Breeding synthetic varieties in annual caraway: observations on the outcrossing rate in a polycross using a high-throughput genotyping system. Euphytica 217:1. https://doi.org/10.1007/s10681-020-02732-5
5. Pidon H, Wendler N, Habekuβ A, Maasberg A, Ruge-Wehling B, Perovic D et al (2021) High-resolution mapping of *Rym14Hb*, a wild relative resistance gene to barley yellow mosaic disease. Theor Appl Genet 134:823–833. https://doi.org/10.1007/s00122-020-03733-7
6. Harper H, Burridge A, Winfield M, Finn A, Davidson A, Matthews D et al (2021) Detecting SARS-CoV-2 variants with SNP genotyping. PLoS One 16:e0243185. https://doi.org/10.1371/journal.pone.0243185
7. Alvarez-Fernandez A, Bernal MJ, Fradejas I, Martin Ramírez A, Md Yusuf NA, Lanza M et al (2021) KASP: a genotyping method to rapid identification of resistance in *Plasmodium falciparum*. Malar J 20:16. https://doi.org/10.1186/s12936-020-03544-7

Chapter 19

Molecular Beacon Probe (MBP)-Based Real-Time PCR

Gopal Kumar Prajapati, Ashutosh Kumar, Aakanksha Wany, and Dev Mani Pandey

Abstract

In the advancement of molecular biology techniques, several probe-based techniques, like molecular beacon probe (MBP) assay, TaqMan probe, and minor groove binder (MGB) probe assay, have been reported to identify specific sequences through real-time polymerase chain reaction (PCR). All probe-based methods are more sensitive than the conventional PCR for the detection and quantification of target genes. MBP is a hydrolysis probe that emits fluorescence when getting the specific sequences on the gene. Here, we describe the application of MBP for the identification of the motif sequences present in the promoters of differentially expressed genes.

Key words Molecular beacon probe, Real-time PCR, Identification of sequence, Motif

1 Introduction

The advancement of molecular techniques has made a significant development in the detection of genes and their regulatory mechanisms. Furthermore, many more studies have been done for the identification of differentially expressed genes during their growth, development, and responses to the environment. However, the identification of each individual component of a regulatory mechanism in a metabolic pathway is scanty. Advanced molecular biology techniques, like microarray and next-generation sequencing (NGS), generate enormous throughput data concerning sequences/genes, but to understand the regulatory mechanism of differentially coexpressed genes, probe-based molecular techniques have significantly proven an important role. In recent times, probe-based real-time PCR techniques have gained momentum and have become the favored method for the detection and quantification of regulatory elements (such as *cis*-elements, enhancers, and repressors). Molecular beacon probe (MBP), TaqMan probe, and minor groove binder (MGB) probe-based

Yuri Shavrukov (ed.), *Plant Genotyping: Methods and Protocols*, Methods in Molecular Biology, vol. 2638, https://doi.org/10.1007/978-1-0716-3024-2_19,

real-time PCR are used for the detection of specific sequences in nucleic acid with high accuracy [1, 2]. These probes are also used for the detection of continuous mutation during amplification in a closed system [3].

Real-time PCR assays are using fluorescence resonance energy transfer (FRET) probes. Resonance energy transfer is a nonradioactive process, occurs without molecular collision, and does not produce heat [4]. FRET is a distance-dependent process where energy is transferred from an excited molecular fluorophore (the donor) to another fluorophore (the acceptor) through intermolecular long-range dipole–dipole coupling [5]. When energy transfer occurs, the acceptor molecule quenches the donor molecule fluorescence, which is captured and analyzed through light microscopy imaging [6, 7] (Fig. 1). The MBP method was first developed by Tyagi and Kramer in 1996, which is based on FRET, and was used to identify gene mutation, pathogens, and in vivo and in vitro genetic studies [8–10]. MBPs were further used to detect specific sequences of target genes with more specificity and precision [11].

Molecular beacons are 15- to 25-nucleotide-long hairpin-loop-shaped probes that hybridize to target a sequence (having the complementarity of a target sequence) and become fluorescent. MBPs consist of four parts: the complementary stem (with a length of 5–7 nucleotides), pinhead, fluorescent dye at 5′-end, and quencher at 3′-end. The complementary stems bring the two ends of the molecule, having a fluorophore and a quencher (Table 1), in close vicinity. The pinhead of an MBP contains probe sequences that are complementary to the target sequence. Though stems have a short length, they are able to keep the fluorophore and quencher in close proximity so that the fluorescence of the fluorophore is to be masked by the quencher. However, when an MBP hybridizes with a template/target sequence, the hairpin-loop structure gets open (Fig. 2). The separation of the 5′-end fluorophore and the 3′-end quencher causes the fluorophore to restore fluorescence, and the signal is perceived by the detector. The strength of the received fluorescence signal implies its proportionality to the amount of target sequence [8–10, 12–14]. Moreover, MBPs with different-colored fluorophores are used together to concurrently identify multiple targets in the same reaction [9, 15]. Furthermore, MBPs are commercially available and can be procured, with different combinations of the 5′-end fluorophore and the 3′-end quencher (Table 1), from many companies like PREMIER Biosoft, USA; IDT, USA; GeneLink, USA; Merck, Germany; and others. Also, to design MBPs and related primers, many software programs are available. Though most prominently software Beacon Designer (PREMIER Biosoft, USA) is used to design MBPs, some others are also available, such as Visual OMP (DNA Software, Michigan Life Science and Innovation Center, Plymouth, USA; https://www.dnasoftware.com/

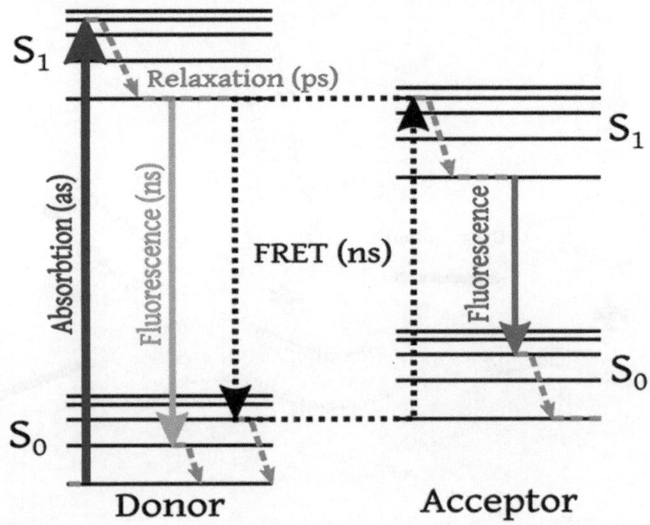

Fig. 1 Jablonski diagram showing the coupled transitions involved between donor emission and acceptor absorbance during FRET. Absorption is represented by a blue arrow, and emission transitions are represented by green and red arrows, whereas vibrational relaxation is indicated by dashed orange arrows. The black dashed line indicates a virtual photon. In the presence of a suitable acceptor, the donor fluorophore can transfer excited state energy directly to the acceptor without emitting a photon (black arrow). (The figure was adapted from https:/www.olympus-lifescience.com/en/microscope-resource/primer/techniques/fluorescence/fret/fretintro/)

Table 1
List of commercially available dyes used as fluorophores and their compatible quenchers

Dyes	
Fluorophore	**Compatible quencher**
6-FAM™	BHQ™-1[a], DABCYL
JOE™	BHQ™-1[a], DABCYL
TET™	BHQ™-1[a], DABCYL
HEX™	BHQ™-1[a], DABCYL
Cyanine 3	BHQ™-2[a], DABCYL
ROX™	BHQ™-2[a], DABCYL
TxRd (Sulforhodamine 101-X)	BHQ™-2[a], DABCYL
Cyanine 5	BHQ™-3[a], DABCYL
Cyanine 5.5	BHQ™-3[a], DABCYL

[a]Black hole quencher

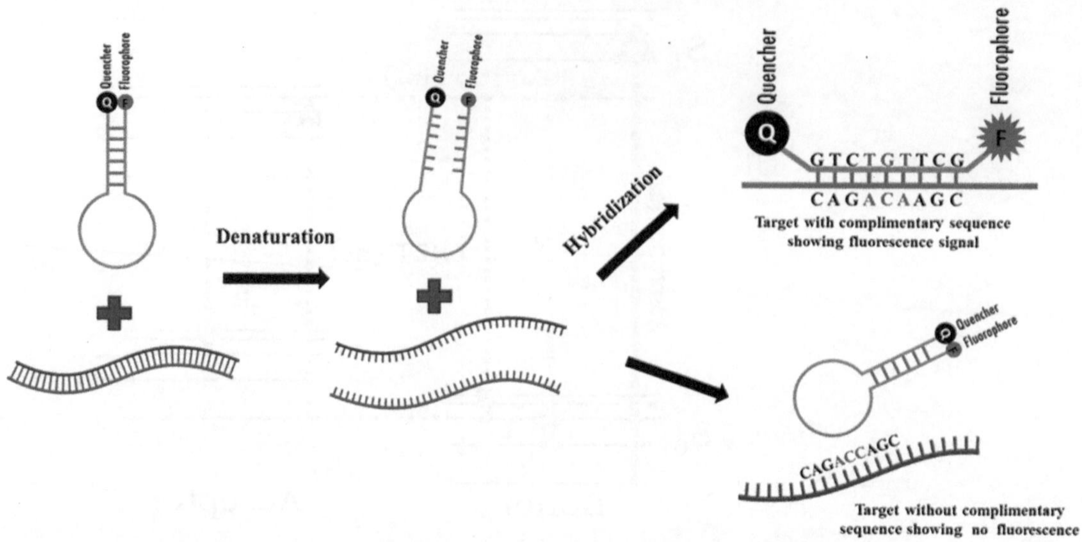

Fig. 2 Schematic procedural representation of MBP. As a free probe, no fluorescence is observed due to the proximity of the fluorophore (attached at one end of the MBP) and quencher (attached at another end of the MBP). The hybridization of probes with (specific) target sequence releases fluorophores and quenchers and causes fluorescence [8]

products/visual-omp), a Python-based program PinMol, which is developed to design molecular beacons for live-cell imaging [16], and custom-designed fluorophore-quencher probes, which are made available by Biolegio (Nijmegen, Netherlands; https://www.biolegio.com/shop/categories/probes-for-qpcr).

The additional advantages of MBP-based qPCRs over the conventional PCR are increased sensitivity and accuracy. Thus, designed MBPs are preferably a more powerful tool to detect and quantify target sequences [17]. MBPs are highly specific and can even detect a single base-pair mismatch, thereby decreasing the overall hybridization affinity of MBPs with a target sequence, making them a perfect candidate for single-nucleotide polymorphism (SNP) detection [9, 10, 18, 19]. Furthermore, earlier MBPs were used to detect point mutation in the *rpoB* gene in *Mycobacterium tuberculosis* [20], point mutation in the *methylene tetra hydro folate reductase* gene responsible for cardiovascular diseases and neural tube defects [21], and the allele-specific analysis of *β-chemokine receptor* 5, deducing the vulnerability of HIV-I [10]. MBPs also have been developed as DNA biosensors due to their advantageous real-time fluorescence-based hybridization [22–25]. These biosensors have been used for the in vivo detection of nonlabeled nucleic acid targets in real time [26]. Not only biosensors but MBPs have also been used for the in vivo detection of mutated messenger ribonucleic acids (mRNAs) transcribed from a defective gene, which helped in the early detection of cancer in real time. A 15-bp-long MBP, an antisense sequence for human basic fibroblast growth factor, was developed to detect the mutated mRNA for a faulty protein in a living cell [27]. Further, an MBP microinjected into K562 human leukemia cells visualized their real-time interaction with mRNA [21, 28]. These numerous applications of MBPs made them a dynamic probe for target sequence detections with ease of handling.

The current chapter describes the procedure for the application of molecular beacons as a detection probe. Here, we are elaborating with an example of the first-ever report about the use of MBPs in plant-based assays. We used MBPs for the validation of specific *cis* elements/motif sequences present in the promoter of differentially expressed genes (DEGs) during anoxia in rice plants [29]. Earlier, our in silico studies have identified the presence of a consensus promoter motif GCC- box (GCCGCC) in the upregulated DEGs while a TCC-box (TCCTCC) in the downregulated DEGs in anoxic rice coleoptile [30]. Specific MBPs were designed for GCC- and TCC-boxes with the help of the online available software Beacon Designer (PREMIER Biosoft, USA) and procured. Consequently, we validated the presence of GCC- and TCC-boxes in the promoter of upregulated and downregulated DEGs, respectively, using specifically designed GCC- and TCC-box MBPs through a FRET-based real-time PCR assay protocol, described here. Although the described protocol is used for the validation of the presence of specific target sequence/s in a single plant genotype, its application is not restricted to any single system. The FRET-based real-time PCR assay may prove its utility as a strong and convenient tool for the identification of target sequences in any biological system.

2 Materials

2.1 Plant Material

1. Plant tissue of the studied genotype, for example, rice seedling (*Oryza sativa japonica* cv. Azucena) is used for this chapter presentation (*see* Subheading 3.1 and **Note 1**).
2. Petri dishes.
3. Sterile filter paper.
4. Suitable plant tissue harvesting protocol.

2.2 Genomic DNA Isolation (See Note 2)

1. Sterile scissors.
2. Sterile spatula.
3. Microcentrifuge tubes (2 ml).
4. 50 ml centrifuge tubes.
5. 2% (w/v) cetyl trimethyl ammonium bromide (CTAB).

6. 1.5 M NaCl.
7. 25 mM ethylenediaminetetraacetic acid (EDTA) (pH 8.0).
8. 100 mM Tris–HCl (pH 7.5).
9. 3% polyvinylpyrrolidone (PVP).
10. 0.3% (v/v) β-mercaptoethanol.
11. Milli-Q water (volume makeup).
12. Chloroform:isoamyl alcohol, 24:1.
13. Isopropanol (chilled).
14. 95% (v/v) ethanol.
15. 70% (v/v) ethanol.
16. Tris–EDTA (TE) buffer: 10 mM Tris–HCl (pH 7.5) and 0.1 mM EDTA.

2.3 Purification of Genomic DNA

1. Ribonuclease (RNAse) A stock solution (10 mg/ml).
2. 3 M sodium acetate.
3. Absolute ethanol (100%).
4. 70% ethanol.

2.4 Molecular Beacon Probe Based Real-Time PCR

1. PCR tube stand.
2. PCR tubes (0.2 ml).
3. 1× Taq buffer (provided with Taq polymerase).
4. 0.1–0.5 mM $MgCl_2$.
5. 200 μM deoxynucleotide triphosphates (dNTPs).
6. 0.1–0.5 μM target-sequence-specific forward primer.
7. 0.1–0.5 μM target-sequence-specific reverse primer.
8. 0.05 U/μl Taq polymerase (any suitable one).
9. 25–50 ng purified template DNA.
10. 0.3 μM molecular beacon Probe (procured from any specialized MBP synthesizing company).

2.5 Equipment

1. Mortar and pestle.
2. Water baths or incubator (65 °C and 37 °C).
3. Bench centrifuge.
4. Refrigerated centrifuge (for 2 ml tubes).
5. −20 °C freezer.
6. Gel electrophoresis system.
7. Ultraviolet-visible (UV/Vis) spectrophotometer.
8. 96-well PCR plate.

9. Transparent 96-well PCR film.
10. Real-time PCR system (from any suitable company).

3 Methods

3.1 Plant Material and Tissue Collection

1. The protocol described here is based on an experimental study using a genotype of plant species. However, the authors' work is primarily specialized on *Oryza sativa japonica* cv. Azucena, grown under lab conditions.
2. Seeds of rice were surface sterilized using 0.02% sodium hypochlorite for 10 min under laminar airflow.
3. The sterilized seeds were washed with autoclaved Milli-Q water three to four times until the smell of hypochlorite disappears.
4. The Petri dishes were lined with three to four layers of sterile filter paper and were moisturized with sterile Milli-Q water.
5. With the help of sterile forceps, the rice seeds were placed in the Petri dishes (ten seeds/Petri dish).
6. The Petri dishes were then kept under dark conditions for 48 h to promote seed germination; after that, they were transferred to normal lab conditions.
7. Rice seeds were observed until two to three leaves had grown, and then the leaf tissue was harvested using sterile scissors and immediately stored in liquid nitrogen.

3.2 Genomic DNA Isolation Using the CTAB Method

1. Approximately, 100 mg of specific tissue was used for genomic DNA isolation.
2. Prechill the mortar and pestle (to minimize freeze thaw) and the 95% ethanol solution at −20 °C.
3. Also, preheat the water baths or incubator (65 °C and 37 °C) before beginning the extraction.
4. Prepare the CTAB extraction buffer (10 ml/g of leaf tissue) by adding the components given in Subheading 2.2 (**steps 5–10**) and adding 0.3% (v/v) β-mercaptoethanol in a 50 ml centrifuge tube, and then preheat it in the 65 °C water bath or incubator (*see* **Note 2**).
5. Grind the leaf tissues using liquid nitrogen and transfer the fine powder into a prechilled 2 ml microcentrifuge tube.
6. To this, add 800 μl of 2× prewarmed CTAB buffer and incubate in a 65 °C water bath or incubator and mix by inversion every 5 min for 30 min–1 h.
7. After incubation, cool the tubes at room temperature (RT) and add 750 μl chloroform and isoamyl alcohol (24:1).
8. Mix gently by inversion and incubate for 10 min at RT.

9. Centrifuge the samples for 10,000 rpm (15,000 × *g*) for 15 min at 4 °C.
10. Carefully decant the supernatant into a fresh tube.
11. Again, add an equal volume of 24:1 chloroform:isoamyl alcohol and mix by gentle inversion for 5 min.
12. Centrifuge the sample at 10,000 rpm (15,000 × *g*) for 5 min and pipette out the upper aqueous phase into a new microcentrifuge tube. Avoid the organic interphase layer.
13. Add 0.6 volumes of chilled isopropanol, mix, and incubate at –20 °C for 1 h.
14. Centrifuge at 12,000 rpm (18,000 × *g*) for 5 min at 4 °C.
15. Decant the isopropanol and wash the pellet with 1 ml of 70% ethanol and gently mix by inversion.
16. Centrifuge at 10,000 rpm (15,000 × *g*) for 5 min at 4 °C.
17. Drain out the excess ethanol and air-dry the pellet at RT under laminar air flow only.
18. Dissolve the pellet in minimum quantity (50–100 μl of TE buffer at pH 8.0).
19. Keep the tubes overnight at 4 °C to maximize solubilization and then store them at −20 °C until further use.

3.3 DNA Quality and Quantity Assessment

1. Assess the quality of the extracted DNA using a UV/Vis spectrophotometer.
2. The quality and quantity are calculated, taking the absorbance peak at A_{260} nm, and then assess through the A_{260}/A_{280} absorbance ratio. The quantity is usually expressed in micrograms by the spectrophotometer.
3. However, if the A_{260}/A_{280} ratio is <1.8 or >2.2, it might have protein or RNA contaminations. If so, proceed to RNAse treatment (*see* **Note 3**).

3.4 Purification of DNA by RNAse Treatment

1. Aliquot 25 μl of isolated genomic DNA into a new microcentrifuge tube and add 1.0 μl of RNase (10 mg/ml). Incubate at 37 °C for 30 min.
2. After incubation, add 0.1 volume of 3 M sodium acetate (chilled) and 2.5 volume of 100% ethanol. Mix gently by inversion and incubate at −20 °C for 1 h or overnight.
3. Centrifuge at 14,000 rpm (21,000 × *g*) for 15 min at 4 °C.
4. Decant the supernatant and rinse the pellet with 70% ethanol at 10,000 rpm (15,000 × *g*) for 5 min at 4 °C.
5. Remove ethanol and its traces by inverting the tube over a clean tissue paper and let the pellet air-dry.

6. Dissolve the pellet in 20 μl TE buffer (pH 8.0) and keep it at 4 °C to solubilize the pellet.
7. Check the quantity and quality of purified DNA through the spectrophotometer (*see* **Note 3**).
8. Store the purified DNA at −20 °C.

3.5 Design of Molecular Beacon Probes (MBPs) and Primers

1. The molecular beacon probe design is very specific and has selective target detection. For this, we may use any one of the abovementioned software to design MBPs with their prescribed parameters.
2. Keep primer length between 18 and 24 nucleotides long, amplicon size varying from 100 to 300 nucleotides in length, while T_m of the target probe sequence with 8–10 °C higher temperature than the annealing temperature.
3. Keep the GC content in the stem sequence ~65–70%, while the stem sequences should be in such a manner that it cannot be complimentary to the probe sequence, but forms a hairpin-like structure.
4. Attach FAM (6-fluorescein amidite (6-FAM)) at the 5′-end of the designed MBP and BHQ (black hole quencher) at the 3′-end of the designed MBP.
5. Procure designed MBPs and primers from any suitable molecular companies specialized in regular and modified oligos (for example, here, we designed two different MBPs and target-sequence-specific primers for their detection in two different target sequences, as listed in Tables 2 and 3, respectively) (*see* **Note 4**).
6. Use the designed MBPs to identify any targeted genetic fragments, for example, the presence of GCC- and TCC-boxes in the promoter regions of the upregulated or downregulated DEGs of the studied genotypes, respectively. Similarly, other fragments of the coding or noncoding regions of the targeted genes can also be used for MBP design and their validation.

3.6 Validation of Molecular Beacon Probe Based on Real-Time PCR Assay

1. Prepare the master mix with a total reaction volume of 15 μl (Table 4).
2. Set real-time PCR conditions (Table 5) and perform amplification using a real-time PCR system.
3. Use "No Template Control" (NTC) for PCR amplification as the control.
4. The PCR amplification assay will confirm the presence of target fragments. Here, we are showing an example of the PCR

Table 2
Design of molecular beacon probe sequence for two DEGs, *MT* (*methyltransferase domain-containing protein*) and *RG* (*rho*-GAP *domain-containing protein*), up- and downregulated in rice seedlings in anoxia conditions

DEGs	Length of MBP (nt)	MBP with stem sequence (bold and underlined)
MT (Os06g05910)	24	5′-**CGCGATC**GCCGCCGCCG**GATCGCG**-3′
RG (Os12g05900)	29	5′-**CGCGATC**CTCCTCCTCCTCCTC**GATCGCG**-3′

Table 3
Design of primer sequence for two DEGs, *MT* (*methyltransferase domain-containing protein*) and *RG* (*rho*-GAP *domain-containing protein*), up- and downregulated in rice seedlings in anoxia conditions

DEGs	Left primer (5′–3′)	Right primer (5′–3′)	Amplicon size
MT (Os06g05910)	CCTCCTAGTTCGTCCG TCAA	TCGAGCCTGGAC TTCACC	107
RG (Os12g05900)	CATCATTAGCGGAGGATT	CGGAGGTGGCTAAATAAC	162

Table 4
Master mix for molecular beacon probe assay genotyping and validation

Component	Final concentration
Taq buffer	1×
Taq polymerase (1 unit)	5 u/μl
dNTPs	0.2 mM
$MgCl_2$	3 mM
Primers	0.45 μM
gDNA	3 ng
MBP	0.3 μM

Table 5
Real-time PCR conditions used for amplification

Cycle numbers	Temperature	Time
1	95 °C	4–5 min
35–40	95 °C 55–65 °C 72 °C	15–20 s 35–40 s 40–45 s

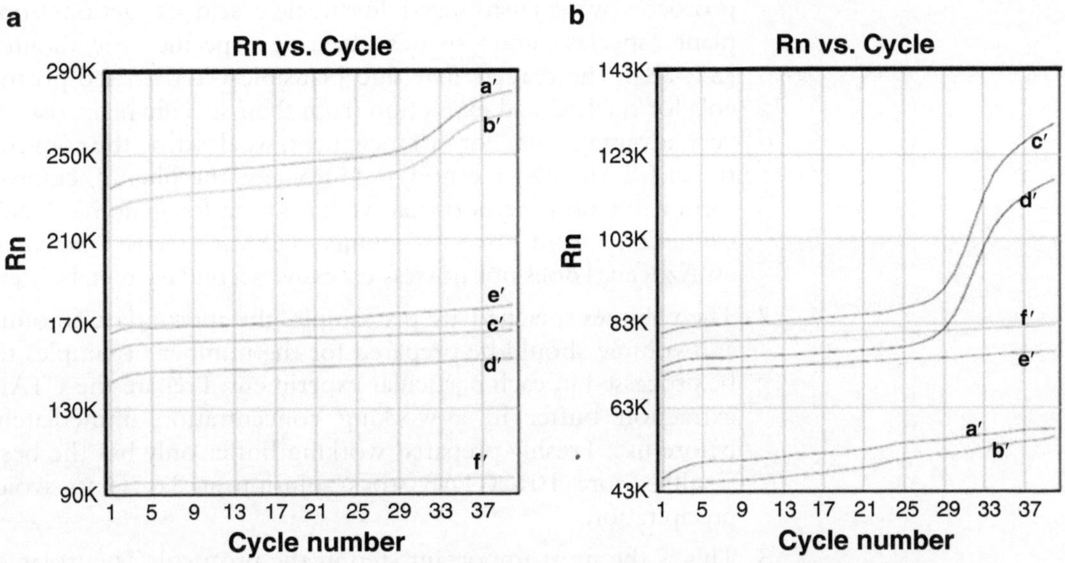

Fig. 3 (**a**) MBP amplification curve for the detection of a consensus GCC promoter motif in the *MT* gene. The graph indicates the Rn vs cycle number. Curves *a′* and *b′* depict the amplification of the promoter region with a GCC probe. Curves *c′* and d′ are the amplification of the TCC probe, and *e′* and *f′* are amplifications of the NTC. (**b**) MBP amplification curve for the detection of a consensus TCC promoter motif in the *RG* gene. The graph indicates the Rn vs cycle number. Curves *a′* and *b′* depict the amplification of the promoter region with a GCC probe. Curves *c′* and *d′* are the amplification of the TCC probe, and *e′* and *f′* are amplifications of the NTC

amplification assay in a rice genotype, which confirmed the presence of two promoter motifs, GCC-box and TCC-box, in up- and downregulated DEGs, respectively [29, 31] (Fig. 3) (*see* **Note 5**).

4 Notes

1. In molecular studies, nucleic acid extraction from different plant tissue is a very crucial step. Various protocols were developed for extracting good-quality nucleic acids from plant tissues [32–34]. The quality of nucleic acid extraction is extremely affected by the presence of polysaccharides, polyphenols, and other secondary metabolites, which are specific to the plant tissues. These metabolites hamper the quality and quantity of extracted nucleic acids [35, 36]. Therefore, it is very important to know the presence of specific metabolites in plant tissues and to consider different parameters during their harvesting. The harvested tissue may be flash frozen by using liquid nitrogen (below −180 °C), a temperature at which all the metabolic activities of cells are ceased. Consequently, the well-known cetyltrimethylammonium bromide (CTAB)-based

protocols were customized for nucleic acid extraction from plant species and tissues bearing specific metabolites [37, 38]. The readers may find plausible CTAB-based protocols for nucleic acid extraction from their specific plant tissues very advantageous. For instance, here, we describe the conventional CTAB (2%) method. Additionally, the phenol/chloroform extraction method that we have used for genomic DNA extraction from rice seedlings (*Oryza sativa japonica* cv. Azucena) does not possess excessive secondary metabolites.

2. The volumes specified are per sample; the appropriate amount and volume should be prepared for the number of samples to be processed in each particular experiment. Prepare the CTAB extraction buffer in a working concentration immediately before use. Freshly prepared working buffer only has the best results. Store 10% CTAB stock solution at 37 °C to avoid precipitation.
3. This is the most important step in the protocol. The isolated genomic DNA should be free from RNA, for high quality. The quality and quantity of the RNA-free genomic DNA are estimated through spectrophotometry (A_{260} nm, A_{280} nm, and A_{230} nm). A purity of DNA sample is considered optimal if no DNA degradation is observed below 10 K bp, and the value of A_{260}/A_{280} and A_{260}/A_{230} should be ≥1.8 and ≥2.0, respectively, which means there is no evidence of RNA or polysaccharide contamination. The isolated DNA from each sample is analyzed by 0.7% agarose gel electrophoresis to check DNA shearing.
4. The described protocol is to validate specific MBPs on the example of consensus motifs present in the promoter of DEGs. For example, in our experiments, rice coleoptiles were collected from plants grown in anoxic conditions. One of the selected upregulated genes was *methyltransferase domain-containing protein* (*MT*; Os06g05910), having a GCC-box. Another downregulated gene, *rho*-GAP *domain-containing protein* (*RG*; Os12g05900), having a TCC-box in their promoter regions, was selected. Gene-specific MBPs (Table 3) and primers (Table 4) were designed as described earlier for this method [9, 15] and procured [31].
5. The protocol described here is a FRET-based real-time PCR assay. The result is observed in the form of a graph generated by the inbuilt software of the real-time PCR machine. Here, we have performed the procedure for detecting the consensus promoter motif by using GCC-box-containing MBP, while TCC-box-containing MBP is used as a reference and vice versa. The presence of a GCC-box in the promoter of the *MT* gene is observed with average Ct values of 30.62 (Fig. 3a).

However, amplification of the NTC and negative control as a TCC-box-containing MBP is undetected, confirming the presence of GCC-box in the *MT* gene (Fig. 3a). Similarly, the presence of TCC-box in the promoter of the *RG* gene is observed with average Ct values of 30.13 (Fig. 3b). Amplification of the NTC and negative control as a GCC-box-containing MBP is undetected, confirming the presence of TCC-box in the *RG* gene (Fig. 3b) [31].

Acknowledgments

Permission to use the Beacon Designer 7 software to design probes was generously granted by PREMIER Biosoft, USA. Central Rainfed Upland Rice Research Station (CRURRS), Hazaribagh-825301, Jharkhand, India, is gratefully acknowledged for providing the Azucena rice seeds used in this experiment. This work was financially supported by the University Grants Commission (UGC), GOI, New Delhi, India, as a major research project [F. No. 37-113/2009 (SR)]. The financial assistant provided to Mr. Gopal Kumar Prajapati in the form of a project fellow from the abovementioned UGC-sponsored project is gratefully acknowledged.

References

1. Dummit B, Chang YH (2006) Molecular beacons for DNA binding proteins: an emerging technology for detection of DNA binding proteins and their ligands. Assay Drug Dev Technol 4:343–349. https://doi.org/10.1089/adt.2006.4.343
2. Yao Y, Nellåker C, Karlsson H (2006) Evaluation of minor groove binding probe and Taqman probe PCR assays: influence of mismatches and template complexity on quantification. Mol Cell Probes 20:311–316. https://doi.org/10.1016/j.mcp.2006.03.003
3. Hur D, Kim MS, Song M, Jung J, Park H (2015) Detection of genetic variation using dual-labeled peptide nucleic acid (PNA) probe-based melting point analysis. Biol Proced Online 17:14. https://doi.org/10.1186/s12575-015-0027-5
4. Stryer L (1978) Fluorescence energy transfer as a spectroscopic ruler. Annu Rev Biochem 47:819–846. https://doi.org/10.1146/annurev.bi.47.070178.004131
5. Sekar RB, Periasamy A (2003) Fluorescence resonance energy transfer (FRET) microscopy imaging of live cell protein localizations. J Cell Biol 160:629–633. https://doi.org/10.1083/jcb.200210140
6. Förster T (1965) Delocalized excitation and excitation transfer. In: Sinanoglu O (ed) Modern quantum chemistry, vol 3. Academic Press Inc., New York, pp 93–137
7. Elangovan M, Day RN, Periasamy A (2002) Nanosecond fluorescence resonance energy transfer-fluorescence lifetime imaging microscopy to localize the protein interactions in a single living cell. J Microsc 205:3–14. https://doi.org/10.1046/j.0022-2720.2001.00984.x
8. Tyagi S, Kramer FR (1996) Molecular beacons: probes that fluoresce upon hybridization. Nat Biotechnol 14:303–308. https://doi.org/10.1038/nbt0396-303
9. Tyagi S, Bratu DP, Kramer FR (1998) Multicolor molecular beacons for allele discrimination. Nat Biotechnol 16:49–53. https://doi.org/10.1038/nbt0198-49
10. Kostrikis LG, Tyagi S, Mhlanga MM, Ho DD, Kramer FR (1998) Spectral genotyping of human alleles. Science 279(5354):1228–1229. https://doi.org/10.1126/science.279.5354.1228

11. Göhring J, Jacak J, Barta A (2014) Imaging of endogenous messenger RNA splice variants in living cells reveals nuclear retention of transcripts inaccessible to nonsense-mediated decay in Arabidopsis. Plant Cell 26:754–764. https://doi.org/10.1105/tpc.113.118075
12. Ortiz E, Estrada G, Lizardi PM (1998) PNA molecular beacons for rapid detection of PCR amplicons. Mol Cell Probes 12:219–226. https://doi.org/10.1006/mcpr.1998.0175
13. Marras SAE, Kramer FR, Tyagi S (2002) Efficiencies of fluorescence resonance energy transfer and contact-mediated quenching in oligonucleotide probes. Nucleic Acids Res 30: e122. https://doi.org/10.1093/nar/gnf121
14. Wu CS, Peng L, You M, Han D, Chen T, Williams KR et al (2012) Engineering molecular beacons for intracellular imaging. Int J Mol Imaging 2012:501579. https://doi.org/10.1155/2012/501579
15. Tyagi S, Kramer FR (2012) Molecular beacons in diagnostics. F1000 Med Rep 4:10. https://doi.org/10.3410/M4-10
16. Bayer LV, Omar OS, Bratu DP, Catrina IE (2019) PinMol: Python application for designing molecular beacons for live cell imaging of endogenous mRNAs. RNA 25:305–318. https://doi.org/10.1261/rna.069542.118
17. Goel G, Kumar A, Puniya AK, Chen W, Singh K (2005) Molecular beacon: a multitask probe. J Appl Microbiol 99:435–442. https://doi.org/10.1111/j.1365-2672.2005.02663.x
18. Marras SA, Kramer FR, Tyagi S (1999) Multiplex detection of single-nucleotide variations using molecular beacons. Genet Anal 14:151–156. https://doi.org/10.1016/s1050-3862(98)00018-7
19. Tan W, Fang X, Li J, Liu X (2000) Molecular beacons: a novel DNA probe for nucleic acid and protein studies. Chemistry 6:1107–1111. https://doi.org/10.1002/(sici)1521-3765(20000403)6:7<1107::aid-chem1107>3.3.co;2-0
20. Piatek AS, Tyagi S, Pol AC, Telenti A, Miller LP, Kramer FR et al (1998) Molecular beacon sequence analysis for detecting drug resistance in *Mycobacterium tuberculosis*. Nat Biotechnol 16:359–363. https://doi.org/10.1038/nbt0498-359
21. Giesendorf BA, Vet JA, Tyagi S, Mensink EJ, Trijbels FJ, Blom HJ (1998) Molecular beacons: a new approach for semiautomated mutation analysis. Clin Chem 44:482–486. https://doi.org/10.1093/clinchem/44.3.482
22. Liu X, Tan W (1999) A fiber-optic evanescent wave DNA biosensor based on novel molecular beacons. Anal Chem 71:5054–5059. https://doi.org/10.1021/ac990561c
23. Liu X, Farmerie W, Schuster S, Tan W (2000) Molecular beacons for DNA biosensors with micrometer to submicrometer dimensions. Anal Biochem 283:56–63. https://doi.org/10.1006/abio.2000.4656
24. Steemers FJ, Ferguson JA, Walt DR (2000) Screening unlabeled DNA targets with randomly ordered fiber-optic gene arrays. Nat Biotechnol 18:91–94. https://doi.org/10.1038/72006
25. Brown LJ, Cummins J, Hamilton A, Brown T (2000) Molecular beacons attached to glass beads fluoresce upon hybridization to target DNA. Chem Commun 7:621–622. https://doi.org/10.1039/B000389L
26. Li J, Tan W, Wang K, Xiao D, Yang X, He X et al (2001) Ultrasensitive optical DNA biosensor based on surface immobilization of molecular beacon by a bridge structure. Anal Sci 17: 1149–1153. https://doi.org/10.2116/analsci.17.1149
27. Matsuo T (1998) *In situ* visualization of mRNA for basic fibroblast growth factor in living cells. Biochim Biophys Acta 1379:178–184. https://doi.org/10.1016/s0304-4165(97)00090-1
28. Bratu DP, Cha BJ, Mhlanga MM, Kramer FR, Tyagi S (2003) Visualizing the distribution and transport of mRNAs in living cells. Proc Natl Acad Sci U S A 100:13308–13313. https://doi.org/10.1073/pnas.2233244100
29. Prajapati GK, Kashayp N, Kumar A, Pandey DM (2013) Identification of GCC box in the promoter region of ubiquinol cytochrome c chaperone gene using molecular beacon probe and its *in silico* protein-DNA interaction study in rice (*Oryza sativa* L.). Int J Comp Bioinfo In Silico Mod 2:213–222
30. Kumar A, Smita S, Sahu N, Sharma V, Shankaracharya, Vidyarthi AS et al (2009) *In silico* analysis of motifs in promoters of differentially expressed genes in rice (*Oryza sativa* L.) under anoxia. Int J Bioinforma Res Appl 5:525–547. https://doi.org/10.1504/IJBRA.2009.028681
31. Prajapati GK, Pandey DM (2015) Molecular beacon probe based promoter motifs validation in anoxia responsive differentially expressed genes and their in silico interaction studies with AP2/EREBP TF in rice (*Oryza sativa* L.). Int J Pharm Pharm Sci 7:123–130
32. Porebski S, Bailey LG, Baum BR (1997) Modification of a CTAB DNA extraction protocol for plants containing high polysaccharide and polyphenol components. Plant Mol Biol Rep

15:8–15. https://doi.org/10.1007/BF02772108

33. Wany A, Kumar A, Nigam VK, Pandey DM (2013) Isolation and *in silico* characterization of geraniol dehydrogenase encoding gene using modified DNA extraction protocol from American Basil (*Ocimum americanum* L.). Int J Pharm Bio Sci 4:365–379
34. Aboul-Maaty NAF, Oraby HAS (2019) Extraction of high-quality genomic DNA from different plant orders applying a modified CTAB-based method. Bull Natl Res Cent 43:25. https://doi.org/10.1186/s42269-019-0066-1
35. Scott KD, Playford J (1996) DNA extraction technique for PCR in rain forest plant species. Biotechniques 20:974. https://doi.org/10.2144/96206bm07
36. Khanuja SPS, Shasany AK, Darokar MP, Kumar S (1999) Rapid isolation of DNA from dry and fresh samples of plants producing large amounts of secondary metabolites and essential oils. Plant Mol Biol Rep 17:1. https://doi.org/10.1023/A:1007528101452
37. Moreira PA, Oliveira DA (2011) Leaf age affects the quality of DNA extracted from *Dimorphandra mollis* (Fabaceae), atropical tree species from the Cerrado region of Brazil. Genet Mol Res 10:353. https://doi.org/10.4238/vol10-1gmr1030
38. Wang X, Xiao H, Zhao X, Li C, Ren J, Wang F et al (2012) Isolation of high-quality DNA from a desert plant *Reaumuria soongorica*, genetic diversity in plants. In: Çalişkan M (ed) Genetic diversity in plants. InTech, London. https://doi.org/10.5772/38367

Chapter 20

Molecular Beacons – Loop-Mediated Amplification (MB-LAMP)

Patrick Hardinge

Abstract

High specificity has been demonstrated in polymerase chain reaction (PCR) with the use of molecular beacons (MBs) to detect amplified sequences containing mutations or single-nucleotide polymorphisms (SNPs). MBs have been adapted for use with the isothermal nucleic acid amplification technology loop-mediated amplification (LAMP) by targeting single-stranded loop sequences under optimized conditions to enable applications such as plant genotyping. LAMP has several benefits over PCR, such as rapid amplification, single-temperature reaction conditions enabling low-cost equipment, and robustness to certain PCR inhibitors. However, and despite the increased number of primers required, the specificity of LAMP is limited, and false positive results can be problematic. In this chapter, design considerations for molecular beacons in LAMP assays are described, as well as a method for MB-LAMP amplification and detection, with an example of gene sequences in genetically modified (GM) maize samples.

Key words Molecular beacons (MBs), Loop-mediated amplification (LAMP), Genotyping, DNA, Isothermal nucleic acid amplification technology (iNAAT), Single-nucleotide polymorphisms (SNPs), Locking nucleic acids (LNAs)

1 Introduction

In this chapter, the use of molecular beacons is introduced as described earlier [1], which were initially designed for real-time polymerase chain reaction (PCR) experiments, coupled with the isothermal nucleic acid amplification technology and LAMP (loop-mediated amplification) [2], the aim of which is to provide rapid nucleic acid amplification at a single temperature with highly specific fluorescence detection of genes, mutations, and single-nucleotide polymorphism (SNP) for plant genotyping.

The first described molecular beacon for real-time PCR had a loop structure of 15 nucleotides specific to the target sequence and two complementary arms of five nucleotides each. To the 5′-end of one of these arms is attached a fluorophore, which is quenched by fluorescence resonance energy transfer (FRET) due to the

Yuri Shavrukov (ed.), *Plant Genotyping: Methods and Protocols*, Methods in Molecular Biology, vol. 2638,
https://doi.org/10.1007/978-1-0716-3024-2_20,

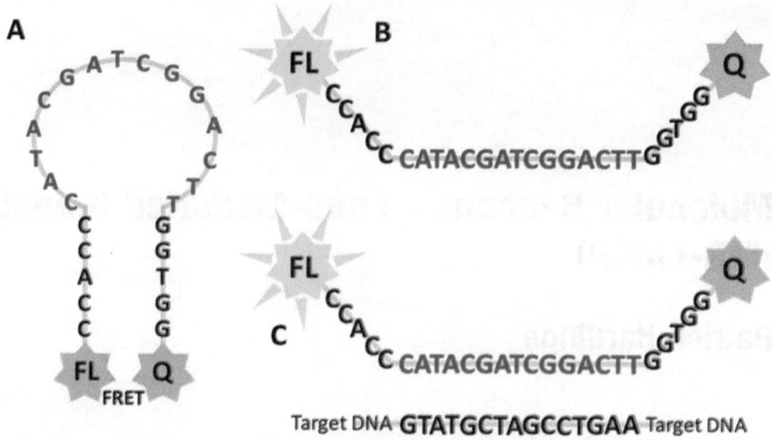

Fig. 1 Molecular beacon structural formations. (**a**, **b**) represent single-stranded DNA, "closed" and "open" molecular beacons consisting of five-nucleotide complementary arms, and a 15-nucleotide probe specific for the target DNA sequence. In the "closed" configuration, fluorescence is quenched as the complementary arms bring the fluorophore and quencher into proximity. In the presence of the target DNA sequence (**c**), the probe sequence anneals to the template in a more thermostable arrangement, resulting in unquenching as the arms are separated and an increase in fluorescence. During isothermal amplification, (**a**, **b**) formations are in equilibrium

proximity of a quenching molecule on the 3′-end of the other arm. During the annealing stage of PCR, the molecular beacon, under favorable reaction conditions and suitable beacon design, would bind to the complementary target sequence, and the fluorophore would be unquenched. The increase in fluorescence provides evidence of specific amplification (*see* Fig. 1).

The application of molecular beacons with LAMP is more complex due to the isothermal reaction and the use of strand displacement polymerase. A single-stranded sequence of nucleotides is required to specifically anneal the molecular beacon's loop structure and suitable regions as it includes the "stem" region of LAMP amplicons, but this is a highly dynamic area rapidly switching between single and double stranded during amplification [3]. Perhaps more favorable are the "loop" regions of loop-mediated isothermal amplification, which have received the most attention in the scientific literature [4–6]. The specificity of MB-LAMP has been improved using locking nucleic acids (LNAs) in the probe sequence [7–9] and through the optimization of reaction conditions [10, 11].

Loop-mediated isothermal amplification (LAMP) requires displacement primers (F3 and B3), which strand invade the double-stranded deoxyribonucleic acid (DNA) to initiate amplification and the LAMP primers (forward inner primer (FIP) and backward inner primer (BIP)), which are required to form hairpin loops, which are

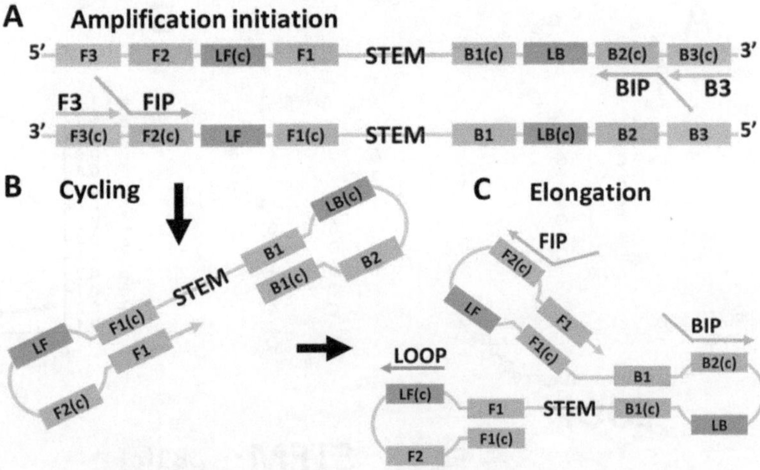

Fig. 2 Key stages in the LAMP reaction. During the initiation of the amplification, displacement and LAMP primers are required to strand invade the double-stranded DNA target sequence (**a**). A polymerase with displacement activity creates the nascent strands with loops in "dumbbell" formations, which are required for the cycling stage (**b**). Further priming with LAMP primers from "dumbbells" creates highly looped structures in the elongation stage (**c**). Displacement primer sequences are shown in orange (F3 and B3). Loop primer sequences are shown in grey (LF and LB). LAMP primers (FIP and BIP) are composed of two sequences (F1, F2 and B1, B2). Complementary sequences are marked with c

essential for the LAMP stages of cycling and elongation (*see* Fig. 2). The LAMP reaction is rapid and specific due to the four primers involved. Other primers can be added that target the "stem" region of LAMP amplicons [12] or the "loop" regions (LF and LB) [13] to further accelerate the reaction. LAMP is an isothermal reaction, typically at a constant temperature between 60 and 65 °C; therefore, expensive thermocycling is not required.

A basic requirement for LAMP is a heat block to maintain a constant temperature typically between 60 and 65 °C, depending on the assay. The assay temperature will impact the equilibrium between the "open" and "closed" states of the molecular beacon, with a higher temperature pushing the balance toward the "open" state and an increase in background fluorescence. The detection requires the excitation of the selected fluorophore, for example, FAM or JOE, and appropriate emission filters for that fluorophore. Quantitative PCR devices can be used for this (*see* Fig. 3). The duration of the MB-LAMP assay is dependent on template concentration and the reaction kinetics of the assay; lower concentrations of DNA templates provide longer cycle threshold (Ct) times.

MB-LAMP molecular diagnostics have been used to detect genetically modified (GM) maize [3]. LAMP molecular beacons for the cauliflower mosaic virus (CaMV) 35S promoter (35Sp) and

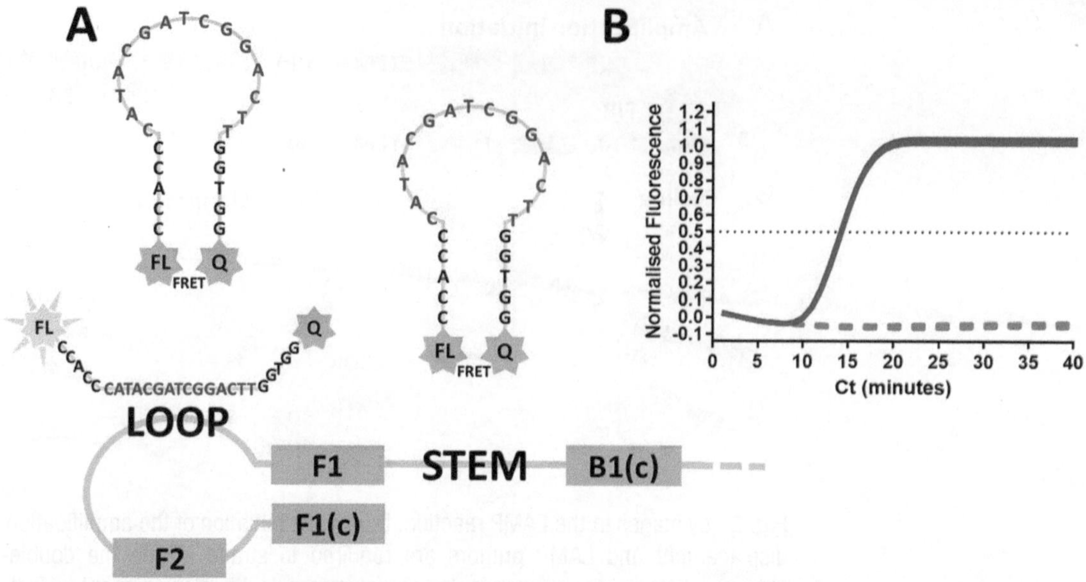

Fig. 3 MB-LAMP fluorescent detection. In (**a**), the "closed" and "open" molecular beacons in the reaction mix anneal specifically to the target sequence contained within the single-stranded LOOP (in this example) in a thermodynamically stable configuration. The unquenching of the fluorophore increases fluorescent emission in response to appropriate excitation. An example trace is shown in (**b**) with positive results in red and no-template controls (NTCs) in blue

the nopaline synthase terminator (NOSt) from *Agrobacterium tumefaciens* successfully detected the genes, but the specificity was insufficient for a multiplex assay. More recently, the specificity of LAMP molecular beacons has been enhanced using locking nucleic acids (LNAs) in the probe sequence of the MB to enable the multiplexing of SARS-CoV-2 mutations [9]. However, this method is applicable to any other plant material with clear differences in the genome based on genetic transformation, mutations, or natural genetic polymorphism in various plant species without a special restriction for genotypes.

2 Materials

LAMP amplification rapidly produces high concentrations of amplified template DNA. LAMP is also highly sensitive to DNA template contamination (sample and amplified); therefore, stringent anticontamination procedures are required (*see* **Note 1**). Unless otherwise stated, all reagents are prepared at room temperature (RT).

2.1 Requirements for Molecular Beacons

1. Design of the molecular beacon: as a rule, 15–30 nucleotides should form the loop of the MB, and this sequence must be complementary to the target sequence for specific detection. The target sequence can be the stem or loop regions of LAMP amplicons, but the loop regions are favored. When designing the LAMP primers, the region between F1 and F2 and B1 and B2 should be of a sufficient sequence length to enable the MB to hybridize without interfering with LAMP primer binding. Each arm of the molecular beacon should be complementary to each other and have a guanine–cytosine (GC) content below 80%. Arm lengths between five and seven nucleotides are most appropriate. The fluorophore FAM is most used in MB-LAMP assays and is attached to the 5′-arm with a suitable quencher attached to the 3′-end of the other arm. The terminal nucleotide of the 5′-arm should not be guanine due to possible quenching of the fluorophore. The melt temperature of the arms should favor the "closed" formation of the MB at the assay temperature. This is usually between 57 and 65 °C; otherwise, background fluorescence will be high.
2. Modifications to the LAMP molecular beacons: to improve the specificity of the LAMP molecular beacons, LAMP loop sequences with high GC content can be selected in preference to high adenine–thymine (AT) content. Also, locking nucleic acids (LNAs) can be added to the loop of the molecular beacon for stronger hybridization to the target sequence.

2.2 Reagents for MB-LAMP

1. Salmon sperm carrier DNA: supplied at 10 mg/mL and required at 100 ng per reaction. Dilute from stock to 1 mg/mL and 100 μg/mL with molecular-grade water (*see* **Note 2**) and store the labeled dilutions at −2 °C.
2. Isothermal buffer (*see* **Note 3**): supplied at 10× concentration for direct usage. Store at −20 °C.
3. 1,4-dithiothreitol (DTT): weigh out 1.54 g of DTT in a falcon tube and add 10 mL of molecular-grade water. Thoroughly mix and prepare 100-μL-labeled single-use aliquots of 1 M DTT. Store at −20 °C.
4. Polyvinylpyrrolidone (PVP): weigh out 0.4 g of PVP in a falcon tube and add 10 mL of molecular-grade water. Thoroughly mix and prepare 100-μL-labeled aliquots of 40 mg/mL PVP. Store at −2 °C.
5. Potassium chloride (KCl): weigh out 0.895 g of KCl in a falcon tube and add 10 mL of molecular-grade water. Thoroughly mix and prepare 100-μL-labeled aliquots of 1.2 M KCl. Store at −20 °C.
6. Deoxyribonucleotide triphosphates (dNTPs) (*see* **Note 4**): add 10 μL deoxyadenosine triphosphate (dATP), deoxycytidine

triphosphate (dCTP), deoxyguanosine triphosphate (dGTP), and deoxythymidine triphosphate (dTTP) to 160 μL of molecular-grade water. Mix and prepare 65-μL-labeled aliquots of 5 mM dNTPs. Store at −20 °C.

7. Betaine: supplied at 5 M concentration and used directly to give a final concentration of 0.8 M (*see* **Note 5**). Store in fridge at 4 °C.
8. Bst polymerase (*see* **Note 6**): supplied at 8000 U/μL and used directly to give a final concentration of 0.32 U/μL. Store at −20 °C.
9. LAMP primers (*see* **Note 7**): prepared fresh at the time of the MB-LAMP assay from the defrosted stocks of each primer. To make 100 μL 10× primer mix, add 2 μL of each displacement primer (F3 and B3), 4 μL of each loop primer (LF and LB), and 8 μL of each LAMP primer (FIP and BIP) to 72 μL of molecular-grade water.
10. Molecular beacon (*see* **Note 8**): defrosted from the frozen stock to give a final concentration between 125 nM and 2.5 μM.

2.3 Consumables and Specialist Equipment

1. Nonstick 1.5 ml tubes (*see* **Note 9**).
2. Maximum recovery pipette tips (*see* **Note 9**).
3. qPCR tubes or 96-well microtiter plate (*see* **Note 10**).
4. Vortex.
5. Microcentrifuge.
6. Temperature-controlled device with a fluorescence excitation source and optical filters, such as a Rotor-Gene qPCR thermocycler (*see* **Note 11**).

2.4 Consumables for Genome Extraction from Plant Material

1. Genomic DNA purification kit containing nucleic lysis solution, ribonuclease (RNase), protein precipitation solution, and DNA rehydration solution.
2. Freshly prepared 70% ethanol solution.
3. Reference material: GM maize event BT11 containing a 35S promoter and NOS terminator DNA sequences at a defined w/w ratio with wild-type maize. Other plant materials that have been ground to a powder can be used.

3 Methods

The preparation of the MB-LAMP assay must be in an area where contamination from the DNA template is controlled (*see* **Note 1**). All procedures can take place at RT if warm-start polymerase is used, unless otherwise specified. If another type of polymerase is used, the preparation of the MB-LAMP mix should be on ice.

3.1 Preparation of LAMP Primers and Molecular Beacon

1. To prepare 150 μL of the primer mix for 75 MB-LAMP reactions of 20 μL, thaw the 100 μM LAMP primers (FIP and BIP), loop (LF and LB), and displacement primers (F3 and B3) and briefly vortex and pulse spin.
2. Add 3 μL of each displacement primer, 6 μL of each loop primer, and 12 μL of each LAMP primer to a nonstick 1.5 mL tube using fresh pipette tips. Return the primers to storage at −20 °C.
3. Add 108 μL of molecular-grade water, briefly vortex, and pulse spin. Label the primer mix.
4. Keep the primer mix on ice until use.
5. Thaw the molecular beacon, briefly vortex, and pulse spin. Keep it on ice until use.

3.2 Preparation of Plant Samples

1. Extract genomic DNA from the maize GM reference material using, for example, the Promega genomic DNA purification kit or another appropriate commercial kit with high-quality extracted and purified DNA.
2. Weigh out 40 mg of the reference GM maize powder or ground plant material and add 600 μL of the nucleic lysis solution, then vortex. Incubate for 30 min at 65 °C. Briefly vortex every 2 min during incubation.
3. Add 3 μL of RNase. Mix by inverting the tube repeatedly and incubate for 15 min at 37 °C.
4. Add 200 μL of the protein precipitation solution. Fully mix with a vortex for 20 s, then centrifuge for 3 min at 13,000 rpm (approx. 19,000 × *g* using a microcentrifuge with a rotor radius of 10 cm).
5. Transfer the supernatant and discard the pellet. To the supernatant, add 600 μL of RT isopropanol and mix by inversion several times. Centrifuge for 1 min at 13,000 rpm.
6. Remove the supernatant and wash the pellet by inversion with 600 μL RT 70% ethanol. Centrifuge for 1 min at 13,000 rpm; remove the supernatant; and air-dry the DNA pellet for 15 min.
7. Add 100 μL of the DNA rehydration solution to the dried DNA pellet. Store the DNA extract at 4 °C.

3.3 Preparation of the LAMP-MB Mix

1. For 72 reactions of 20 μL reaction volumes (if using a Rotor-Gene qPCR device), prepare 1125 μL of the MB-LAMP mix for 15 μL loads with 5 μL samples according to the following protocol. Ensure that the required MB-LAMP assay components are thawed, briefly vortexed, and pulse spun.
2. To a nonstick 1.5 mL tube, add 7.5 μL of the 1/10 dilution of salmon sperm DNA for a final concentration of 100 ng per

reaction. Add 272.5 μL of molecular-grade water (*see* **Note 12**) and 150 μL of the 10× isothermal buffer slowly.

3. Add 15 μL DTT to give a final concentration of 10 mM.
4. Slowly draw up and add 15 μL PVP for a final concentration of 400 μg/mL.
5. Add 75 μL KCl to give a final concentration of 60 mM.
6. Add 90 μL of the combined dNTPs for a final concentration of 300 μM with respect to each dNTP.
7. Add 240 μL betaine solution to give a final concentration of 0.8 M.
8. Add 60 μL of warm-start Bst polymerase to give a final concentration of 0.32 U/μL. Slowly draw up the polymerase due to viscosity.
9. Add 150 μL of the LAMP primer mix and gently pipette up and down to ensure the homogeneity of the reactants (*see* **Note 13**).
10. Add 50 μL of the LAMP molecular beacon to give a final concentration of 2.5 μM (*see* **Note 8**).

3.4 LAMP-MB Assay

1. Add 15 μL of the LAMP-MB mix to each of the reaction tubes.
2. Add 5 μL of the sample or no-template control (NTC) as required.
3. Transfer the tubes to the heat block with fluorescence detection and start the reaction and data collection.
4. Set the assay parameters depending on assay optimization, temperature (between 60 and 65 °C), acquisition interval, and the length of the assay (between 30 and 60 min).

4 Notes

1. A single copy of DNA template can amplify using the MB-LAMP method; therefore, a high risk of contamination needs to be mitigated. To reduce the risk of contamination, aliquots of the key reagents for MB-LAMP can be prepared in advance in single-use batches. This also reduces the MB-LAMP assay preparation time. A dedicated area free from amplified DNA and other nucleic acid templates is an essential requirement for reagent preparation and assay setup. Negative template controls (NTCs) should be included to monitor environmental contamination. All equipment, reagents, and plasticware should be specifically dedicated to these procedures. Surfaces should be wiped with a 70% ethanol solution from spray bottles. Ultraviolet (UV) cross-linking should be

used for pipettes and plasticware. A class II biological safety cabinet provides an ideal place to prepare reagents, primers, and MB-LAMP assays with a reduced risk of contamination.

2. Although the use of salmon sperm DNA with MB-LAMP has not previously been described, there is evidence that the carrier DNA enhances LAMP sensitivity and reaction kinetics [14]. Fully defrost the salmon sperm DNA, vortex, and pulse spin to draw the liquid down from the lid and homogenize the solution. It is highly advisable to slowly pipette the required volume and slowly release the salmon sperm DNA into the molecular-grade water diluent. Improved LAMP assay sensitivity and reaction kinetics have been demonstrated with salmon sperm DNA from Invitrogen. However, other sources of salmon sperm DNA and other carriers of DNA solutions are available.

3. Fully defrost the 10× isothermal buffer and thoroughly mix using a vortex to ensure that any precipitation returns to the solution. The 10× isothermal buffer can be purchased from New England Biolabs or prepared in-house with 20 mM Tris–HCl, 10 mM $(NH_4)_2SO_4$, 50 mM KCl, 2 mM $MgSO_4$ and 0.1% Tween 20 at pH 8.8 (at 25 °C).

4. The deoxyribonucleotide triphosphates (dNTPs) are diluted from 100 mM stock (available from Invitrogen), 10 μL of each dNTP, in 160 μL molecular-grade water. For large total volume assays, each dNTP can be used directly for a final concentration of 300 mM with respect to each dNTP.

5. Betaine can be purchased as a 5 M solution. There is some debate as to the benefit of 0.8 M betaine in LAMP and MB-LAMP assays [10]. It is primarily used to decrease the stability of GC base pairs to aid amplification but may not be essential in an MB-LAMP assay.

6. Bst polymerase can be procured from New England Biolabs (NEB) as a warm-start enzyme. An aptamer inhibits polymerase activity below 45 °C, enabling LAMP assays to be set up at RT. This polymerase can be used for MB-LAMP, but alternative versions, enzymes, and suppliers are available.

7. Synthesized LAMP primers are stored at −20 °C at a stock concentration of 100 μM. The 10×LAMP primer mix contains 8 μM of both the forward inner primer (FIP) and backward inner primer (BIP), 4 μM of the forward loop (LF) and backward loop (LB), and 2 μM of the forward displacement primer (F3) and backward displacement primer (B3). Twice the concentration of the LAMP primers (FIP and BIP) or twice the concentration of all the primers can be used. Other ratios of primers are likely to be suboptimal.

8. The synthesized molecular beacons are stored at −20 °C at a stock concentration of 100 μM. The final concentration of the MB will depend on assay optimization with reported concentrations ranging from 0.125 μM [3] to 2.5 μM [8].
9. The binding of the DNA template to the plasticware during MB-LAMP preparation could reduce the availability of the template for amplification. To reduce this risk, nonstick 1.5 mL tubes and maximum recovery pipette tips are recommended throughout the MB-LAMP preparation.
10. If using a Rotor-Gene thermocycler, the appropriate 0.1 mL tubes and caps should be used. When loading the MB-LAMP reaction mix, ensure that the tip of the pipette is angled and is deep into the tube to prevent loss. If using 96-well microtiter plates, ensure that they are appropriate for the wells of the heat block.
11. If using a Rotor-Gene thermocycler, set the assay parameters to acquire the channel corresponding to the fluorophore of the molecular beacon (for example, the green channel for the fluorophore FAM). Define the assay temperature (between 60 and 65 °C), the length of the assay (between 30 and 60 min, depending on the LAMP assay and the concentration of the DNA template), and the acquisition time (between 30 and 60 s). Alternative devices can be used that allow these parameters to be defined.
12. The volume of molecular-grade water should be increased if a lower concentration of the molecular beacon is used. In this example, 272.5 μL of molecular-grade water was used because 50 μL of the MB was added (to give a final concentration of 2.5 μM). An MB concentration of 125 nM would only require 2.5 μL of the MB and an increase in the molecular-grade water added to 320 μL.
13. To reduce the risk of bubble and micro-bubble formations in the MB-LAMP mix, avoid using the vortex. As an alternative, use the pipette to slowly draw up and down repeatedly to ensure thorough mixing. Bubbles and micro-bubbles could interfere with the reactions.

Acknowledgments

The research into molecular beacons in LAMP assays was funded by BBSRC Stand Alone LINK grant BB/L022346/1, a BBSRC Industrial CASE award BB/G017336/1, and the European Regional Development Fund with Cardiff University 80762-CU-085 (East).

References

1. Tyagi S, Kramer FR (1996) Molecular beacons: probes that fluoresce upon hybridisation. Nat Biotechnol 14:303–308. https://doi.org/10.1038/nbt0396-303
2. Notomi T, Okayama H, Masubuchi H, Yonekawa T, Watanabe K, Amino N et al (2000) Loop-mediated isothermal amplification of DNA. Nucleic Acids Res 28:E63. https://doi.org/10.1093/nar/28.12.e63
3. Hardinge P, Murray JAH (2019) Lack of specificity associated with using molecular beacons in loop mediated amplification assays. BMC Biotechnol 19:55. https://doi.org/10.1186/s12896-019-0549-z
4. Liu W, Huang S, Liu N, Dong D, Yang Z, Tang Y et al (2017) Establishment of an accurate and fast detection method using molecular beacons in loop-mediated isothermal amplification assay. Sci Rep 7:40125. https://doi.org/10.1038/srep40125
5. Su Y, Huang S, Hong L, Zou D, Tang Y, Chao S et al (2019) Establishment of the molecular beacon-loop-mediated isothermal amplification method for the rapid detection of *Porphyromonas gingivalis*. J Microbiol Methods 160:68–72. https://doi.org/10.1016/j.mimet.2019.01.013
6. Wan L, Chen T, Gao J, Dong C, Wong AH, Jia Y et al (2017) A digital microfluidic system for loop-mediated isothermal amplification and sequence specific pathogen detection. Sci Rep 7:14586. https://doi.org/10.1038/s41598-017-14698-x
7. Bakthavathsalam PC, Longatte G, Jensen SO, Manefield M, Gooding JJ (2018) Locked nucleic acid molecular beacon for multiplex detection of loop mediated isothermal amplification. Sens Actuators B Chem 268:255–263. https://doi.org/10.1016/j.snb.2018.04.081
8. Sherrill-Mix S, Van Duyne GD, Bushman FD (2021) Molecular beacons allow specific RT-LAMP detection of B.1.1.7 variant SARS-CoV-2. J Biomol Tech 32:98–101. https://doi.org/10.7171/jbt.21-3203-004
9. Sherrill-Mix S, Hwang Y, Roche AM, Glascock A, Weiss SR, Li Y et al (2021) Detection of SARS-CoV-2 RNA using RT-LAMP and molecular beacons. Genome Biol 22:169. https://doi.org/10.1186/s13059-021-02387-y
10. Varona M, Eitzmann DR, Pagariya D, Anand RK, Anderson JL (2020) Solid-phase microextraction enables isolation of BRAF V600E circulating tumour DNA from human plasma for detection with a molecular beacon loop-mediated isothermal amplification assay. Anal Chem 92:3346–3353. https://doi.org/10.1021/acs.analchem.9b05323
11. Carvalho J, Lopes-Nunes J, Figueiredo J, Santos T, Miranda A, Riscado M et al (2021) Molecular beacon assay development for severe acute respiratory syndrome coronavirus 2 detection. Sensors 21:7015. https://doi.org/10.3390/s21217015
12. Gandelman O, Jackson R, Kiddle G, Tisi L (2011) Loop-mediated amplification accelerated by stem primers. Int J Mol Sci 12:9108–9124. https://doi.org/10.3390/ijms12129108
13. Nagamine K, Hase T, Notomi T (2002) Accelerated reaction by loop-mediated isothermal amplification using loop primers. Mol Cell Probes 16:223–229. https://doi.org/10.1006/mcpr.2002.0415
14. Kiddle G, Hardinge P, Buttigieg N, Gandelman O, Pereira C, McElgunn CJ et al (2012) GMO detection using a bioluminescent real time reporter (BART) of loop mediated isothermal amplification (LAMP) suitable for field use. BMC Biotechnol 12:15. https://doi.org/10.1186/1472-6750-12-15

Chapter 21

TaqMan Probes for Plant Species Identification and Quantification in Food and Feed Traceability

Maria Doroteia Campos, Catarina Campos, and Hélia Cardoso

Abstract

In the last few years, the traceability and labeling of processed food and feeds have gained increasing importance due to the impact that mislabeling and product fraud may have on human/animal health or on the quality of final products, such as milk, cheese, and meat, as a consequence of animal dietary. The presence of contaminants or possible frauds due to the use of alternative plant materials in food and feeds can greatly impact the economy; therefore, they are becoming important targets for product certification by competent institutional services. This is especially relevant when complex matrixes are considered, in which the visual identification of the different components is quite difficult or even impossible. Despite the existence of mandatory traceability requirements for the analysis of feed/food composition addressed by European Community regulations, the labels do not always provide a sufficient guarantee about the ingredients and additive composition of those products. In this sense, the development of new methodologies that aim to assess the traceability of feed and food complex matrixes is crucial. In this chapter, a general protocol is presented for the establishment of quantitative real-time PCR-based techniques based on TaqMan assays applied to feed/food traceability, with a special focus on applications in the areas of food and feed security (e.g., for the detection of plant species involved in allergenic reactions), fraud detection (e.g., genetically modified organisms), and certification (e.g., protected denomination of origin).

Key words Traceability, Authenticity, Contaminant detection, Certification, PCR-based method

1 Introduction

Food and feed authentication techniques are gaining popularity, mostly due to increasing public awareness concerning their quality and safety. The authentication of food and feed is defined as the analytical process for verifying that the products follow their label description [1]. Food and feed products that are of high value and undergo several processing steps are often targets of fraudulent labeling. Authenticity has been a major concern for producers, regulators, and consumers around the world. For this reason, there is a growing need for reliable analytical techniques to provide a decisive answer to the authenticity of food and feedstuffs.

Yuri Shavrukov (ed.), *Plant Genotyping: Methods and Protocols*, Methods in Molecular Biology, vol. 2638,
https://doi.org/10.1007/978-1-0716-3024-2_21,

Enzyme-linked immune sorbent assays (ELISAs) have been used for a long time for allergen detection associated with the presence of specific plant species [2]. More recently, different spectroscopy-based methods have been developed that focus on food/feed traceability, such as hyperspectral imaging, visible and near-infrared (Vis/NIR) spectroscopy, Fourier transform infrared (FTIR) spectroscopy, near-infrared (NIR) spectroscopy, and Fourier transform near-infrared spectroscopy (FT-NIRS) [3–7]. Additionally, elemental analysis/isotopic ratio mass spectrometry (EA-IRMS) [8–10] has been developed, aiming at the identification of the geographical origin of several products to detect fraud in plant products with the denomination of protected origin. Nevertheless, at the EU level, optical microscope analysis performed after mesh fractionation steps remains one of the official methods for ascertaining the botanical identity of raw materials and the composition of processed material or complex feed mixtures, though it has been recently flanked by deoxyribonucleic acid (DNA)-based methods [5]. The high stability of DNA made DNA-based methods the most reliable method for the analysis of processed and nonprocessed food and feed products for the screening of their ingredients as well as the tracing of the existence of putative contaminants. DNA-based methods depend on the use of efficient molecular tools, which include methodologies for species authentication based on highly specific amplification of DNA fragments through polymerase chain reaction (PCR), detected through either conventional end-point PCR, followed by agarose gel electrophoresis, or the so-called quantitative real-time PCR (qPCR) [11]. Several protocols relying on both methodologies were already used to detect plant species in food and feed samples. The target species might prove to be an undesirable contaminant (quite relevant in the case of allergenic species, in which low amounts are often required to provoke a reaction) or a confirmation of the labeling [2, 12–16]. Also, DNA-based analytical methods have been widely used for the detection and quantification of genetically modified organism (GMO) contaminations in non-GMO feeds and for the identification of animal DNA [13] and the presence and/or quantity of GMOs in raw agricultural materials and in processed and refined ingredients [11, 17].

When compared to qPCR, conventional end-point PCR is more accessible, requiring less expensive equipment, but its lower cost cannot compensate for the extreme sensitivity and quantitative properties of qPCR, which allow to achieve lower limits of detection. This fact has led in recent years to the increased use of qPCR methods for the detection of ingredients in complex matrixes. qPCR relies on probes that hybridize to a site within the target sequence, and during amplification, the number of cycles required to reach a threshold level of fluorescence is used to estimate the starting number of target sequences [18–20]. qPCR instrumentation requires considerably short hands-on time, and the detection of amplified products is automated, simple, and reproducible [21]. Two reagents are widely used for amplicon quantification: (1) double-stranded (ds) DNA-binding agents, such as SYBR® Green I, and (2) fluorescent dyes, such as those used in TaqMan probes. SYBR Green can be used to quantify the amount of amplicon during the course of the PCR process by tracking the overall fluorescence emission. The dye binds into the minor groove of dsDNA and provides sensitive detection. During PCR, as the target sequence is amplified, SYBR Green binds to each new copy of dsDNA during the extension step. The recognition of products requires melting curve analysis, but a major disadvantage of this method is that its binding is nonspecific [18, 21, 22]. This fact might be extremely problematic in the case of similar sequences across species and also if the target gene belongs to an extended gene family, within species. Often, probes are designed in polymorphic regions across species but conserved within the target species. Protein-coding sequences are the most selected regions mostly due to the high availability of transcript sequences in public databases. Nevertheless, it is of extreme importance to know well the selected target gene—whether it is a member of a gene family or a single gene. Usually, protein-coding sequences are very conserved across gene family members [23, 24], which can represent an additional difficulty in the specificity of probes. The use of fluorescent probe technology allows increased sensitivity and specificity compared to SYBR Green. TaqMan probes, often designed as TaqMan assays, are fluorogenic oligomers that are intended to hybridize to an internal site within the target sequence delimited by forward and reverse primers, as defined by conventional PCR. The 5′-end of the probe is labeled with a fluorescent reporter dye, and the 3′-end carries a fluorescent quencher dye. During qPCR, the 5′ exonuclease activity of *Taq* polymerase digests the hybridized probe, leading to a separation of the two dyes and, consequently, to increased fluorescence proportional to the amount of the specific PCR product [18, 21]. TaqMan minor groove binder (MGB) chemistry incorporates two further technologies: the addition to the probe of an MGB moiety that highly increases the stability and specificity of probe hybridization, and a nonfluorescently quenched (NFQ) that enhances spectral performance. NFQ offers the advantage of a lower background signal, which results in better precision in quantitation and allows shorter probes to be used on hybridization-based assays [25]. MGB stabilizes the hybridized probe and effectively raises the melting temperature, ensuring higher efficiency and specificity with shorter sequences [25]. TaqMan MGB probes are effective over a broad wavelength range (FAM, VIC, TET, and NED reporter dyes). qPCR combined with the chemistry of TaqMan MGB probes represents the most specific and sensitive detection system, even when few target DNAs are present, and is here applied to feed/food traceability, with a special focus on application in the areas of food and feed security, fraud detection, and certification.

2 Materials

2.1 Equipment and General Laboratory Supplies

1. Chamber flow cabinet with an ultraviolet (UV) lamp.
2. Micropipettes with different volume ranges (at least three are required: 1–10 μL, 10–100 μL, and 100–1000 μL).
3. 15 mL falcon tubes.
4. Filtered pipette tips covering the different micropipette volume ranges.
5. Spectrophotometer (NanoDrop-2000C, Thermo Fisher Scientific, or similar).
6. Optical paper for spectrophotometer cleaning.
7. Vortex.
8. Nuclease-free standard microfuge tubes (0.2 and 1.5 mL).
9. Centrifuge for standard microfuge tubes.
10. Support for standard microfuge tubes.
11. Centrifuge for 96-well plates.
12. Real-time qPCR equipment (for example, 7500 Real-Time PCR System, Applied Biosystems, or similar).
13. Standard 96-well polypropylene PCR detection plates compatible with the qPCR equipment available.
14. Plate seals compatible with 96-well plates and with the qPCR equipment.
15. Microplate sealer.
16. Support for 96-well plates.
17. Thermal cycler (Bio-Rad, Thermo Fisher Scientific, or similar).
18. Electrophoresis system.
19. Gel imaging system.
20. Microwave (for gel electrophoresis preparation).
21. Dry bath (required for DNA gel extraction and competent cell transformation).
22. Petri dishes (9 cm diameter).
23. Balance (scale).
24. Gloves.

2.2 Reagents

1. Nuclease-free water.
2. 2× Luminaris Color Probe Low ROX qPCR Master Mix.
3. TaqMan gene expression assays (each assay includes two species-specific primers and a specific probe).
4. λDNA.
5. Agarose for gel electrophoresis.
6. Electrophoresis buffer (0.5% TAE buffer).
7. Greensafe solution.
8. Taq DNA polymerase.
9. Deoxynucleotide triphosphates (dNTPs) set.
10. Primers for the amplification of the target region.
11. DNA ladder.
12. pGEM-T Easy Vector System I (Promega) or similar.
13. *Escherichia coli* competent cells.
14. LB Broth high salt and LB Agar high salt.
15. Ampicillin (or others, considering the specificity of the cloning vector).
16. DNA plasmid extraction kit.
17. Gel DNA recovery kit.
18. Restriction enzyme(s) located at multiple cloning sites of the cloning vector (for example, EcoRI if the pGEM-T Easy Vector System I is used).

2.3 Software

1. Software aligner (e.g., MEGA, publicly available at https://www.megasoftware.net/; BioEdit, publicly available at https://thalljiscience.github.io/; MUSCLE, public available at https://www.ebi.ac.uk/Tools/msa/muscle/).
2. Primer designer software (e.g., Primer3Plus, publicly available at https://www.bioinformatics.nl/cgi-bin/primer3plus/primer3plus.cgi, Primer Select (Lasergene), Primer Express 2.0 (Invitrogen)).

3 Methods

3.1 Design of Specific Primers and TaqMan Probes

1. Search at a public database the target gene sequence (at the genomic level) for the species of interest (e.g., NCBI, Ensembl Plants, Phytozome, Plaza, and species-specific genome databases).
2. Use the target gene sequence previously identified to perform a BLAST search in the available databases (e.g., NCBI, https://blast.ncbi.nlm.nih.gov/Blast.cgi; Ensembl Plants, https://plants.ensembl.org/Multi/Tools/Blast; Phytozome, https://

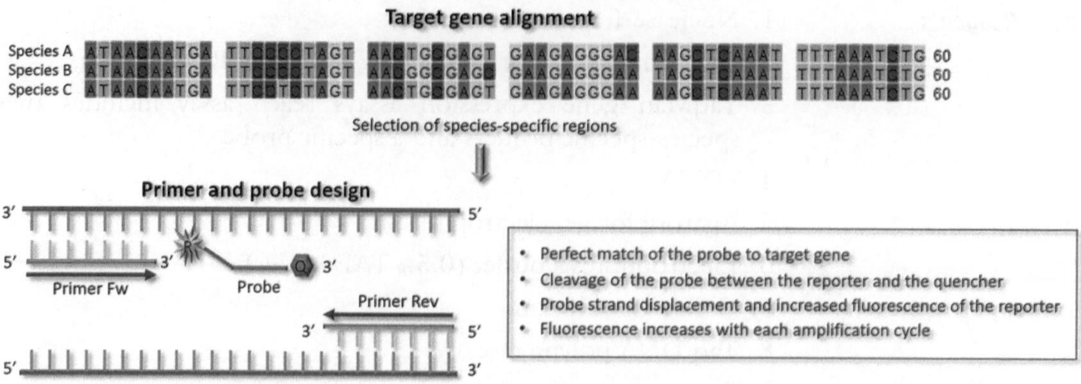

Fig. 1 Schematic illustration of the workflow of qPCR with TaqMan probes

phytozome-next.jgi.doe.gov/blast-search; Plaza, https://bioinformatics.psb.ugent.be/plaza/versions/plaza_v1/blast/index, and species-specific genome databases) in order to retrieve all target-homologous sequences. Special attention must be given to closely related species and to nondesirable and/or contaminant species.

3. Perform a sequence alignment using available sequence aligner software. A bootstrap analysis based on 1000 replicates using the neighbor-joining method should be followed.

4. Identify a polymorphic region across sequences (*see* Fig. 1 and **Note 1**) and design target-specific primers and probes using available software.

5. Perform in silico confirmation of the specificity of each TaqMan probe assay by doing searches against public databases (same as in **step 1** above). The total specificity of the probe is a crucial step of the procedure.

6. Order the primers and probes from a commercial company. For probes, indicate the TaqMan MGB technology with the use of an NFQ, labeled with the respective 5′-end fluorescent reporter dye (e.g., FAM, VIC, TET, NED).

3.2 Evaluation of Assay Specificity

1. Select the nondesirable plant species and/or contaminants (according to the goal of the study; *see* **Note 2**) to be detected/quantified.

2. Extract genomic DNA (gDNA) from each plant species previously selected and also from each mixture, feed, or food sample. One of the available methods can be used to obtain high-quality and high-yield gDNA (*see* Fig. 2 and **Note 3**).

3. Quantify and assess gDNA purity by measuring absorbance (Ab) at 230, 260, and 280 nm in a NanoDrop™ 2000c spectrophotometer or similar. High-quality gDNA must show a sigmoid curve with a value of $Ab_{260/280}$ between 1.8 and 2.0 and a higher value of $Ab_{260/230}$.

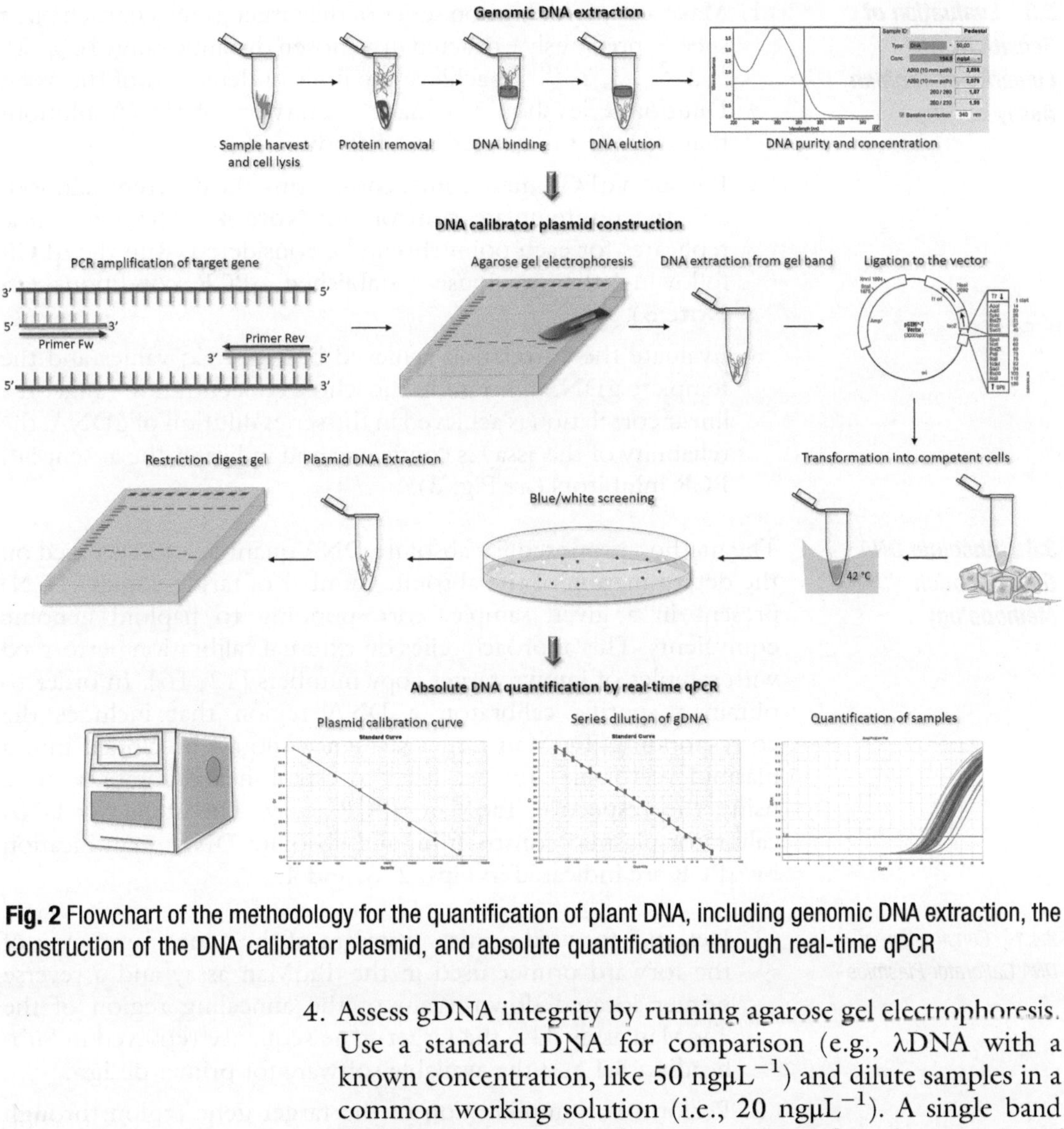

Fig. 2 Flowchart of the methodology for the quantification of plant DNA, including genomic DNA extraction, the construction of the DNA calibrator plasmid, and absolute quantification through real-time qPCR

4. Assess gDNA integrity by running agarose gel electrophoresis. Use a standard DNA for comparison (e.g., λDNA with a known concentration, like 50 ngμL^{-1}) and dilute samples in a common working solution (i.e., 20 ngμL^{-1}). A single band without the presence of a smear will confirm the integrity of the gDNA.
5. Run a qPCR by using 80–100 ng of gDNA as a template in a total volume of 20 μL (*see* **Note 4**), and usually, three technical replicates are considered. Positive target controls (gDNA of each target plant species) to assess the specificity of the conditions and no-template controls to assess the presence of contaminations should be included in each qPCR plate. Amplification must be specific to the specific target species (*see* **Note 5**).

3.3 Evaluation of Sensitivity and Linearity of TaqMan Assay

1. Make a twofold dilution series of the target gDNA of each plant species previously extracted in a chosen dynamic range (e.g., 2^{-1}, 2^{-2}, . . ., 2^{-16}) to achieve the limit of detection of the assay. Dilution series should be made in a mixture of gDNA solutions that exclude the species under study.
2. Prepare a qPCR master mix, considering the different dilutions and the nontemplate control (*see* **Note 4**). Three technical replicates for each point should be considered. Run the qPCR following the previously established qPCR conditions (*see* **Note 5**).
3. Evaluate the correlation achieved between Cq values and the template gDNA amount in the whole concentration range. If a linear correlation is achieved in the series dilution of gDNA, the reliability of the assay is confirmed and indicates the absence of PCR inhibitors (*see* Fig. 2).

3.4 Absolute DNA Quantification Methodology

The method proposed for absolute DNA quantification is based on the determination of the absolute number of target copies (TCN) present in a given sample, corresponding to haploid genome equivalents. This approach relies on external calibration performed with samples of known target copy numbers [12, 16]. In order to obtain a specific calibrator, a DNA region that includes the corresponding TaqMan target sequence should be cloned into a plasmid vector and further used to establish a calibration curve using the respective TaqMan qPCR assay. The steps for DNA calibrator plasmid construction and absolute DNA quantification by qPCR are indicated in Figs. 2, 3, and 4.

3.4.1 Construction of DNA Calibrator Plasmids

1. Design a forward primer upstream of the annealing region of the forward primer used in the TaqMan assay and a reverse primer located downstream of the annealing region of the TaqMan assay. Use the target-gene sequence retrieved in Subheading 3.1 and the available software for primer design.
2. Perform the amplification of the target gene region through end-point PCR using Taq DNA polymerase (*see* **Note 6**).
3. Separate the generated amplicons by agarose gel electrophoresis (1.2% agarose (w/v) can be used in 0.5% TAE (v/v)). A DNA ladder must be loaded into the gel to further allow the extrapolation of the band size.
4. Extract the expected amplicon from the agarose gel using a scalpel blade and further perform DNA gel purification. Check the quantification of purified DNA by spectroscopy (Ab 260 nm).
5. Insert the purified DNA into a cloning vector and use the product of the ligation reaction to transform *Escherichia coli* competent cells (several commercial strains are available for this

1) Calculation of mass of plasmid + insert

Vector (pGEM®-T Easy) + insert	3015 bp + 500 bp
Mass of one bp	1,10E-21
Mass of one plasmid molecule (g)	= (3015+500)*1,10E-21 = 3,85E-18

2) Relation of plasmid mass to copy number

# Copies plasmid DNA	Mass of plasmid + insert (g)	Mass of plasmid DNA (g)
1,0E+09		3,85E-09
1,0E+08		3,85E-10
1,0E+07		3,85E-11
1,0E+06		3,85E-12
100000	3,85E-18	3,85E-13
10000		3,85E-14
1000		3,85E-15
100		3,85E-16
10		3,85E-17
1		3,85E-18

3) Concentrations of plasmid DNA needed to achieve the copy # of interest

Copy # plasmid DNA	Mass of plasmid needed (g)	Volume pippeted*	Final [ ] of plasmid DNA (g/ul)
1,0E+09	3,85E-09	5	=3,85E-9/5 = 7,70E-10
1,0E+08	3,85E-10	5	7,70E-11
1,0E+07	3,85E-11	5	7,70E-12
1,0E+06	3,85E-12	5	7,70E-13
100000	3,85E-13	5	7,70E-14
10000	3,85E-14	5	7,70E-15
1000	3,85E-15	5	7,70E-16
100	3,85E-16	5	7,70E-17
10	3,85E-17	5	7,70E-18
1	3,85E-18	5	7,70E-19

*In this example, 5µL of plasmid DNA solution is pipetted into each PCR reaction.

Fig. 3 Methodology for (1) the calculation of the mass of plasmid + insert (calibrator), (2) the relation of plasmid + insert mass to the number of copies of the target gene in a tenfold series of dilutions, (3) the final concentration of plasmid DNA needed to achieve the copy numbers desired of the plasmid DNA calibrator

purpose) by following standard procedures (e.g., heat shock treatment, electroporation). After the transformation procedure, bacteria must be plated in a solid LB culture medium, added with the respective antibiotic (the specificity of the antibiotic is given by the cloning vector).

6. Select putative recombinant clones (usually white colonies if the multiple cloning site of the cloning vector is located within the *LacZ* gene) and inoculate them in the liquid LB culture medium, added with the same antibiotic used in **step 5** (overnight under 37 °C and 225 rpm).

7. Perform DNA plasmid extraction with a commercial kit or following the alkaline lysis protocol [26].

8. Perform the screening of the selected putative recombinant clones by using restriction enzymes or through PCR (using

Copy #	Log copies	Cq value
1,0E+08	8,00	13,21
1,0E+07	7,00	16,08
1,0E+06	6,00	19,37
1,0E+05	5,00	23,07
10000	4,00	26,65
1000	3,00	29,58
100	2,00	33,74
10	1,00	35,41

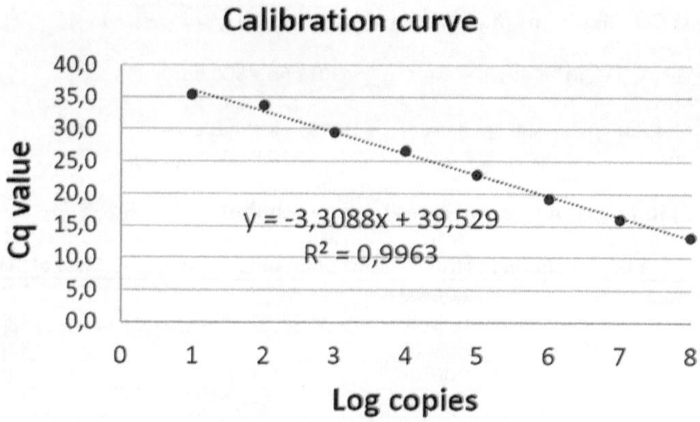

Food/feed sample	Cq value	Log copies target species	TCN80*
A	29,34	=(29,34-39,529)/-3,3088 = 3,08	=10^(3,08) = 1202,26
B	30,52	2,72	526,94
C	37,54	0,60	3,99
D	36,73	0,84	6,99
E	ND	ND	ND
F	30,94	2,60	393,90
G	30,75	2,65	450,18

*In this example, TCN80 is the target copy number in food/feed sample using 80 ng of gDNA.
ND - not detected

Fig. 4 Example of a calibration curve from a TaqMan assay generated from dilution series of a recombinant plasmid DNA showing the Cq values retrieved from qPCR. The calibration curve equation is used to infer the gDNA amount of the target species in terms of the target copy number (TCN) to assess the traceability of feed and food samples

primers that flank the insertion region or restriction enzymes located at 5′ and 3′ of the insertion site). The result of this screening is achieved by agarose gel electrophoresis following the procedure described above (*see* **step 3**).

9. Select the recombinant clones based on the band pattern achieved at the electrophoresis gel. Clones that exhibit a band with a similar size to the previous amplified fragment, besides the band that corresponds to the digested vector, should be selected (confirm it in **step 3**).

10. Confirm the sequence of the inserts of selected recombinant clones by Sanger sequencing using the T7 and SP6 universal primers located at the cloning vector (flanking the insert at 5′ and 3′). One bacterial clone harboring each of the recombinant plasmids is chosen.

3.4.2 Quantification of the Target gDNA Amount

1. Establish a calibration curve of the tenfold dilution series of the selected recombinant plasmid DNA by using the TaqMan qPCR assay previously established. An example of the establishment of a calibration curve using the pGEM®-T Easy

Vector and an insert of 500 bp in the dynamic range of 1E9 to 1E1 is demonstrated in Fig. 3. The calibration curve allows the validation of the probe in the dynamic range chosen for the assay (*see* Fig. 2). Run the qPCR as previously described (*see* also **Notes 4** and **5**).

2. Calculate the amplification efficiency for each probe using the equation $E = \left(10^{\left(-\frac{1}{\text{slope}}\right)} - 1\right) * 100$. The values of efficiency and linearity (the coefficient of determination R^2) must fall into the acceptance criteria [27].
3. The calibration curve from each TaqMan assay can now be used to infer the gDNA amount of the target species in terms of TCN, in order to assess the traceability of feed and food samples, as shown in Fig. 4.

4 Notes

1. The specificity of primers and probes is a crucial step for the success of the procedure (*see* Fig. 1). The region considered for primer and probe design needs to be specific to the target gene (considering the cases of complex gene families) and the species to guarantee the total accuracy of the method. A crucial point that should be considered for the selection of the gene/gene region/member of a gene family to be targeted through qPCR is its availability in the databases of homologous gene sequences across a high diversity of plant species. The consideration of a high number of sequences for the construction of a sequence alignment allows for better identification of specific regions to avoid the existence of apparent specificity due to the lack of available sequences. Furthermore, intronic and untranslated (UTR) regions might be considered for primer and probe design due to the high variability detected across species [16, 24, 28, 29]. If the experiment is to be performed using a previously defined target gene, a confirmation of the existence of available sequences in databases is required; otherwise isolation of the target sequence from the desired species as well as from other species, in particular from closely related species, possible contaminants, or labeled ingredients, must be performed.
2. The target species might be used for species traceability, which aims to identify a possible undesirable contaminant that could have a negative impact on human health (allergens) or animal product quality (e.g., meat and milk quality), or for fraud detection, which aims to confirm the feed/food composition indicated in the label. If the intention of the study is to identify and quantify the plants in a complex mixture, an assay needs to

be developed for each target species, which should be tested against all the species of the mixture.

3. Most of the gDNA extraction methods are based on cetyltrimethylammonium bromide (CTAB) extraction, with some crop-to-crop specification. Several commercial kits are also available and can be used for this purpose.
4. Concentration of primers and probes might require optimization, but as a starting point, it is suggested 400 nM of each primer and 100 nM of probe, in a reaction mixture including the probe master mix, performing a total volume of 20 μL. The volume of the reaction can be adjusted depending on the master mix used; in some cases, it is possible to reduce it to up to 16 μL.
5. The conditions for qPCR may require optimization; however, it is suggested to use the default conditions of the software as a starting point (e.g., 10 min at 95 °C for initial denaturation, an amplification program of 40 cycles at 95 °C for 15 s and 60 °C for 1 min). Adjusting manually the fluorescence threshold set above the background level is required. Cq values between 34 and 35 are taken as indicative of trace amounts, while a value of Cq = 35 is considered the cutoff limit, defining no detection [16, 19].
6. End-point PCRs are carried out in a thermocycler and with GoTaq polymerase following the manufacturer's instructions. PCR conditions may require optimization, but as a starting point, it is proposed to be 35 cycles, each cycle consisting of a denaturation step for 30 s at 94 °C, followed by the annealing step for 30 s, which occurs at variable temperatures depending on the species-specific primers, and the extension step at 72 °C. An initial step at 94 °C for 5 min and a final step at 72 °C for 10 min is considered.

Acknowledgments

This work was supported by EU funds through the FP7-SME program [THEME (SME-2012-2)] under the Project Feed-Code (grant agreement 315464) and national funds through FCT under the Project UIDB/05183/2020.

References

1. Danezis GP, Tsagkaris AS, Camin F, Brusic V, Georgiou CA (2016) Food authentication: techniques, trends and emerging approaches. Trends Anal Chem 85:123–132. https://doi.org/10.1016/j.trac.2016.02.026
2. Lopez-Calleja IM, La Cruz SD, Pegels N, Gonzalez I, Garcia T, Martin R (2013) High resolution TaqMan real-time PCR approach to detect hazelnut DNA encoding for ITS rDNA in foods. Food Chem 141:1872–1880.

https://doi.org/10.1016/j.foodchem.2013.05.076

3. Ren G, Wang Y, Ning J, Zhang Z (2020) Using near-infrared hyperspectral imaging with multiple decision tree methods to delineate black tea quality. Spectrochim Acta A Mol Biomol Spectrosc 237:118407. https://doi.org/10.1016/j.saa.2020.118407
4. Zhang B, Jiang X, Shen F, He X, Fang Y, Hu Q (2021) Rapid screening of DON contamination in whole wheat meals by Vis/NIR spectroscopy and computer vision coupling technology. Int J Food Sci Technol 56:2588–2595. https://doi.org/10.1111/ijfs.14775
5. Gagneten M, del Pilar BM, Rodríguez SD (2021) Evaluation of SIMCA and PLS algorithms to detect adulterants in canola oil by FT-IR. Int J Food Sci Technol 56:2596–2603. https://doi.org/10.1111/ijfs.14866
6. Nogales-Bueno J, Feliz L, Baca-Bocanegra B, Hernández-Hierro JM, Heredia FJ, Barroso JM et al (2020) Comparative study on the use of three different near infrared spectroscopy recording methodologies for varietal discrimination of walnuts. Talanta 206:120189. https://doi.org/10.1016/j.talanta.2019.120189
7. Milinovic J, Garcia R, Rato AE, Cabrita MJ (2019) Rapid assessment of monovarietal portuguese extra virgin olive oil's (EVOO's) fatty acids by fourier-transform near-infrared spectroscopy (FT-NIRS). Eur J Lipid Sci Technol 121:1800392. https://doi.org/10.1002/ejlt.201800392
8. Liu Z, Zhang W, Zhang Y, Chen T, Shao S, Zhou L et al (2019) Assuring food safety and traceability of polished rice from different production regions in China and Southeast Asia using chemometric models. Food Control 99: 1–10. https://doi.org/10.1016/j.foodcont.2018.12.011
9. Jiménez-Morillo NT, Palma V, Garcia R, Dias CB, Cabrita MJ (2020) Combination of stable isotope analysis and chemometrics to discriminate geoclimatically and temporally the virgin olive oils from three mediterranean countries. Foods 9:9121855. https://doi.org/10.3390/foods9121855
10. Liu H, Qin Y, Ma Q, Zhao Q, Guo X, Ma L et al (2021) Discrimination the geographical origin of Yanchi Tan Lamb with different muscle sections by stable isotopic ratios and elemental profiles. Int J Food Sci Technol 56: 2604–2611. https://doi.org/10.1111/ijfs.14900
11. Anklam E, Gadani F, Heinze P, Pijnenburg H, Van Den Eede G (2002) Analytical methods for detection and determination of genetically modified organisms in agricultural crops and plant-derived food products. Eur Food Res Technol 214:3–26. https://doi.org/10.1007/s002170100415
12. Braglia L, Gianì S, Breviario D, Gavazzi F, Mastromauro F, Morello L (2016) Development and validation of the modular Feedcode method for qualitative and quantitative determination of feed botanical composition. Anal Bioanal Chem 408:8299–8316. https://doi.org/10.1007/s00216-016-9943-8
13. Braglia L, Morello L, Gavazzi F, Gianì S, Mastromauro F, Breviario D et al (2018) Interlaboratory comparison of methods determining the botanical composition of animal feed. J AOAC Int 101:227–234. https://doi.org/10.5740/jaoacint.17-0150
14. Holzhauser T, Wangorsch A, Vieths S (2000) Polymerase chain reaction (PCR) for detection of potentially allergenic hazelnut residues in complex food matrixes. Eur Food Res Technol 211:360–365. https://doi.org/10.1007/s002170000152
15. Herman L, de Block J, Viane R (2003) Detection of hazelnut DNA traces in chocolate by PCR. Int J Food Sci Technol 38:633–640. https://doi.org/10.1046/j.1365-2621.2003.00722.x
16. Campos MD, Valadas V, Campos C, Morello L, Braglia L, Breviario D et al (2018) A TaqMan real-time PCR method based on alternative oxidase genes for detection of plant species in animal feed samples. PLoS One 13: 0190668. https://doi.org/10.1371/journal.pone.0190668
17. Cottenet G, Blancpain C, Sonnard V, Chuah PF (2013) Development and validation of a multiplex real-time PCR method to simultaneously detect 47 targets for the identification of genetically modified organisms. Anal Bioanal Chem 405:6831–6844. https://doi.org/10.1007/s00216-013-7125-5
18. Marmiroli N, Maestri E (2007) Polymerase chain reaction (PCR). In: Picó Y (ed) Food toxicants analysis: techniques, strategies and developments. Elsevier BV, Amstardam et al, pp 147–187. https://doi.org/10.1016/B978-044452843-8/50007-9
19. Campos MD, Patanita M, Campos C, Materatski P, Varanda CMR, Brito I et al (2019) Detection and quantification of *Fusarium* spp. (*F. oxysporum*, *F. verticillioides*, *F. graminearum*) and *Magnaporthiopsis maydis* in maize using real-time PCR targeting the ITS region. Agronomy 9:9020045. https://doi.org/10.3390/agronomy9020045
20. Ribeiro JA, Albuquerque A, Materatski P, Patanita M, Varanda CMR, Félix MDR et al

(2022) Tomato response to *Fusarium* spp. infection under field conditions: study of potential genes involved. Horticulturae 8: 8050433. https://doi.org/10.3390/horticulturae8050433

21. Espy MJ, Uhl JR, Sloan LM, Buckwalter SP, Jones MF, Vetter EA et al (2006) Real-time PCR in clinical microbiology: applications for routine laboratory testing. Clin Microbiol Rev 19:165–256. https://doi.org/10.1128/CMR.00022-06
22. Johnson G, Nolan T, Bustin SA (2013) Real-time quantitative PCR, pathogen detection and MIQE. In: Wilks M (ed) PCR detection of microbial pathogens, Methods in molecular biology, vol 943. Humana, Totowa, pp 1–16. https://doi.org/10.1007/978-1-60327-353-4_1
23. Cardoso HG, Nogales A, Frederico AM, Svensson JT, Macedo ES, Valadas V et al (2015) Natural *AOX* gene diversity. In: Gupta KJ, Mur LAJ, Neelwarne B (eds) Alternative respiratory pathways in higher plants. Wiley Blackwell, West Sussex, pp 241–254
24. Cardoso H, Campos MD, Nothnagel T, Arnholdt-Schmitt B (2011) Polymorphisms in intron 1 of carrot *AOX2b-a* useful tool to develop a functional marker? Plant Genet Resour 9:177–180. https://doi.org/10.1017/S1479262111000591
25. Kutyavin IV, Afonina IA, Mills A, Gorn VV, Lukhtanov EA, Belousov ES et al (2000) 3′-minor groove binder-DNA probes increase sequence specificity at PCR extension temperatures. Nucleic Acids Res 28:655–661. https://doi.org/10.1093/nar/28.2.655
26. Birnboim HC, Doly J (1979) A rapid alkaline extraction procedure for screening recombinant plasmid DNA. Nucleic Acids Res 7: 1513–1523. https://doi.org/10.1093/nar/7.6.1513
27. Broeders S, Huber I, Grohmann L, Berben G, Taverniers I, Mazzara M et al (2014) Guidelines for validation of qualitative real-time PCR methods. Trends Food Sci Technol 37:115–126. https://doi.org/10.1016/j.tifs.2014.03.008
28. Cardoso H, Campos MD, Costa AR, Campos MC, Nothnagel T, Arnholdt-Schmitt B (2009) Carrot alternative oxidase gene *AOX2a* demonstrates allelic and genotypic polymorphisms in intron 3. Physiol Plant. https://doi.org/10.1111/j.1399-3054.2009.01299.x
29. Nogales A, Nobre T, Cardoso HG, Muñoz-Sanhueza L, Valadas V, Campos MD et al (2016) Allelic variation on *DcAOX1* gene in carrot (*Daucus carota* L.): an interesting simple sequence repeat in a highly variable intron. Plant Gene 5:49–55. https://doi.org/10.1016/j.plgene.2015.11.001

Chapter 22

Tetra-Primer Amplification Refractory Mutation System (T-ARMS)

Arnab Mukherjee and Tirthartha Chattopadhyay

Abstract

Single-nucleotide polymorphisms (SNPs), the most abundant genetic variation in the population, have become the molecular marker of choice. Generally, the efficient detection of SNPs requires specialized costly equipment. Although there are a few strategies for detecting SNPs through polymerase chain reaction, followed by restriction enzyme digestion and agarose gel electrophoresis, these methods are time-consuming and might be less diagnostic. Interestingly, the tetra primer amplification refractory mutation system (T-ARMS) strategy utilizes a pair of allele-specific primers in a single PCR for the diagnostic detection of SNPs in a codominant manner through standard agarose gel electrophoresis. The simplicity and robustness of the strategy have inspired the researchers to adopt this low-cost method of SNP detection in different crop plants. Here, we have described the principle, methods, and conditions for the T-ARMS strategy. The described methodology starts from the isolation of genomic DNA and ends with the post-PCR analysis of refractory amplicons in standard agarose gel electrophoresis. The limitations and future perspectives are also discussed. Taken together, T-ARMS evolves as a method of choice for low-cost SNP detection in plants.

Key words Allele-specific amplicon, Codominant marker, Genotyping, Point mutation, Polymerase chain reaction, Refractory amplification

1 Introduction

Single-nucleotide polymorphisms (SNPs) have been considered as the most abundant and stable molecular markers in genomes wherein the frequency of their variation is quite high in the population [1]. The abundance of SNPs in genomes was noticed with the progression of the Human Genome Project and the International HapMap Project [2, 3]. With the advancement in next-generation sequencing (NGS) technology, a lot of crop genomes are being sequenced and resequenced, which has led to the identification of numerous important SNPs. More interestingly, functional SNPs directly affecting phenotypes are being rapidly identified through different methods, including genotyping by

Yuri Shavrukov (ed.), *Plant Genotyping: Methods and Protocols*, Methods in Molecular Biology, vol. 2638, https://doi.org/10.1007/978-1-0716-3024-2_22,

sequencing (GBS) and genome-wide association studies (GWAS). Naturally, the identification of allelic variants arising from SNPs is of pivotal importance.

Although there are high-throughput methods based on fluorescence detection for SNPs, the cost and requirement of specific equipment have made them out of the reach of common laboratories working with limited funds. As a change in a single nucleotide does not change the size of the amplicon generated by flanking markers, the identification of SNP alleles through agarose gel electrophoresis has always remained challenging. If the site of the SNP is within a restriction enzyme recognition site, then a fragment containing the SNP can be amplified by flanking primers through polymerase chain reaction (PCR), and the resulting amplicon can be digested by a specific restriction enzyme to identify the allelic variation. This method, known as cleaved amplified polymorphic sequence (CAPS), generates a codominant marker for the SNP, which can be analyzed through standard agarose gel electrophoresis after the restriction digestion of the amplified fragment. However, the possibility of incomplete restriction digestion (particularly in the PCR buffer) is of major concern, which can adversely affect the analysis of the results. A variant of the CAPS technology is the derived CAPS (dCAPS) strategy, where through modification in the designed primers, a partial restriction site is incorporated. Depending on the nature of the SNP, only one of the two alleles develops a perfect restriction site and gets cleaved through restriction digestion. However, just like CAPS, the dCAPS strategy also faces the same limitation originating from the possibility of incomplete restriction digestion. Furthermore, post-PCR restriction digestion increases the cost and time requirements for SNP detection.

Keeping the limitations of the CAPS and dCAPS strategies in mind, emphasis has been given to the detection of SNPs through direct PCR and agarose gel electrophoresis. These allele-specific amplification technologies include the amplification refractory mutation system (ARMS) [4], tetra-primer PCR [5], bidirectional PCR amplification of specific alleles (Bi-PASA) [6], and the tetra-primer amplification refractory mutation system (T-ARMS) [7]. In the present chapter, we will mainly focus on the description of the T-ARMS method.

The T-ARMS strategy [7] utilizes four primers for a single locus in one reaction (*see* Fig. 1). The outer forward and outer reverse primer pair amplifies the common genomic region containing the SNP. The inner primers are allele specific as only one of the inner primers, depending upon the SNP, will coamplify an inner fragment in combination with either the outer forward or the reverse primer. Thus, the inner primers become specific for one allele and refractory for the other allele during PCR amplification. As the orientation of the inner primers is opposite to each other and the

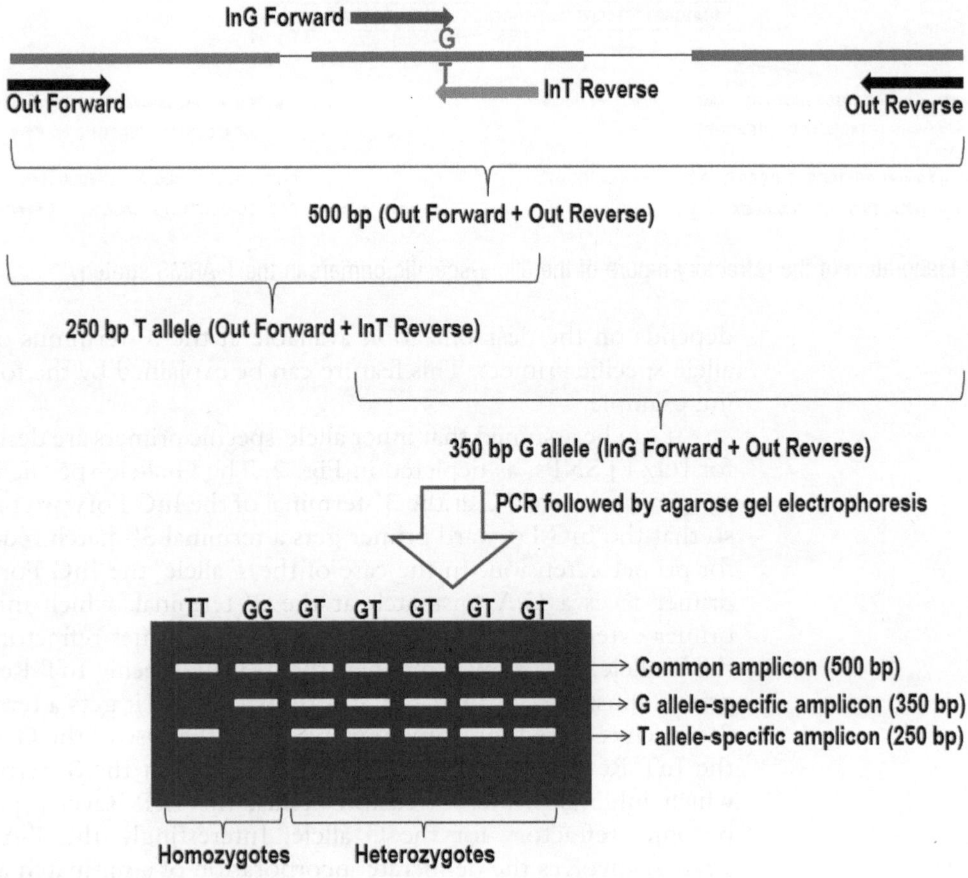

Fig. 1 Schematic diagram of the T-ARMS strategy

distance of the SNP from the two termini of the target common region is different, the allele-specific amplicons become easily separable through agarose gel electrophoresis (*see* Fig. 1). In the case of homozygotes, carrying either the wild-type or the mutant allele, an allele-specific amplicon is generated along with the common amplicon. In the case of the heterozygotes, carrying the wild-type and mutant alleles, the common amplicon is accompanied by both allele-specific amplicons. Thus, the T-ARMS strategy is perfect for agarose-gel-based SNP detection in a codominant manner (*see* Fig. 1).

The success of the T-ARMS strategy depends on the allelic discrimination revealed by the inner primers. In this case, the requirement of perfect matching at the terminal 3′-base for primer extension by the *Taq* DNA polymerase in PCR is exploited to achieve allele-specific amplification. Furthermore, a deliberate mismatch at the −2 position (in reference to the 3′-terminal) of the inner primers is incorporated to increase primer specificity [8]. The nature of the mismatch at the −2 position that is to be incorporated

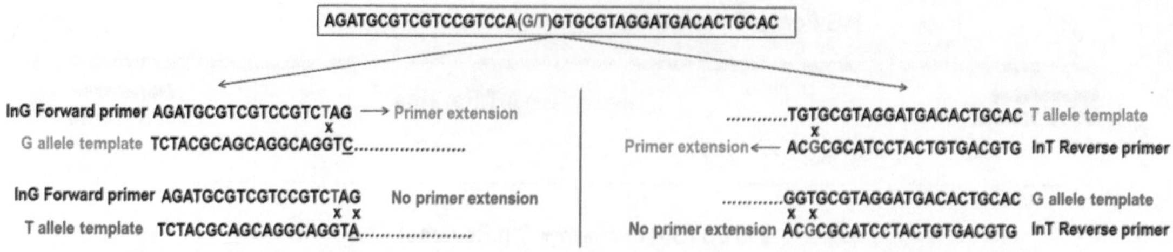

Fig. 2 Elaboration of the refractory nature of the allele-specific primers in the T-ARMS strategy

depends on the destabilization available at the 3′-terminus of the allele-specific primers. This feature can be explained by the following example.

It can be assumed that inner allele-specific primers are designed for [G/T] SNPs, as depicted in Fig. 2. The G-allele-specific inner primer must have a G at the 3′-terminal of the InG Forward primer so that the InG Forward primer gets a terminal 3′-match required for primer extension. In the case of the T allele, the InG Forward primer faces a G:A mismatch at the 3′-terminal, which inhibits primer extension. Thus, the InG Forward becomes refractory for the T allele. In a similar manner, the T-allele-specific InT Reverse primer must have an A at the 3′-terminal so that it gets a terminal 3′-match required for primer extension. In the case of the G allele, the InT Reverse primer faces a G:A mismatch at the 3′-terminal, which inhibits primer extension. Thus, the InT Reverse primer becomes refractory for the G allele. Interestingly, the T-ARMS strategy involves the deliberate incorporation of a mismatch at the −2 position of the allele-specific inner primers. The mismatch at the 3′-terminal and the incorporated deliberate mismatch at the −2 position should ensure allele specificity but must not introduce excessive disability, which can inhibit the amplification of the target allele. A deliberate mismatch at the −2 position should be carefully decided on the basis of the strength of the mismatch at the 3′-terminal of the refractory allele. The strengths of the primer nucleotide mismatches are summarized in Table 1. For example, the InG Forward primer is refractory for the T allele due to a strong G:A mismatch (*see* Fig. 2). Therefore, a weak mismatch (G:T in this case) at the −2 position should be incorporated into the InG Forward primer. Similarly, the InT Reverse primer is refractory for the G allele due to a strong G:A mismatch (*see* Fig. 2). Hence, the weak G:T mismatch is incorporated at the −2 position into the InT Reverse primer. For each and every T-ARMS assay, the overall destabilization of the inner allele-specific primers should be almost equal. For example, a maximum or strong mismatch at the 3′-terminal should accompany a weak mismatch at the −2 position and vice versa. On the other hand, medium-strong mismatches at the 3′-terminal should accompany a similar medium-strong mismatch at the −2 position. The success of a T-ARMS assay largely depends on this critical principle of primer designing.

Table 1
PCR components of a standard T-ARMS assay

Component	Volume (μl)
Genomic DNA	2.00
Out-forward primer (10 μM)	0.25
Out-reverse primer (10 μM)	0.25
In-forward primer (10 μM)	0.25
In-reverse primer (10 μM)	0.25
2× *Taq* master mix	2.50
Molecular-biology-grade water	4.50
Total	*10.00*

2 Materials

The materials (equipment, reagents, chemicals, and plasticware) required for T-ARMS analysis are as follows.

2.1 Equipment

1. Table-top centrifuge.
2. Vortex machine.
3. Dry bath.
4. Automated thermal cycler.
5. Gel electrophoresis unit with a power pack.
6. Ultraviolet (UV) transilluminator/gel documentation system.

2.2 Chemicals and Reagents

1. Customized oligonucleotide primers.
2. *Taq* DNA polymerase-based PCR master mix.
3. Agarose.
4. Sodium borate.
5. Sodium dodecyl sulphate (SDS).
6. Potassium acetate.
7. Sodium chloride.
8. Ethylenediaminetetraacetic acid (EDTA).
9. β-mercaptoethanol (β-ME).
10. Isopropanol.
11. Ethanol.
12. Molecular biology grade water.
13. Ethidium bromide.

2.3 Plasticware

1. Variable volume micropipettes.
2. Tips for micropipettes.
3. Microcentrifuge tubes (1.5 ml).
4. Micropestles.
5. 0.2 ml PCR tubes.

2.4 Buffers and Stock Solutions

1. DNA extraction buffer [100 mM Tris–HCl (pH 8.0), 50 mM EDTA, 500 mM NaCl, 1% (w/v) sodium dodecyl sulfate (SDS), and 0.1% (v/v) β-ME (added just before use)].
2. 5 M potassium acetate solution.
3. 70% (v/v) ethanol.
4. 10 mM sodium borate buffer (20×).
5. Ethidium bromide solution (10 mg/ml).

3 Methods

3.1 Designing of Primers

The designing of primers is the main crucial part of a successful T-ARMS assay (*see* **Note 1**). For this purpose, the Primer1 web tool (http://primer1.soton.ac.uk/primer1.html) is the best option. The user has to input the sequence (up to 1000 bp) that contains the target SNP site. The position and alternate alleles are to be submitted. The user can set different parameters, like outer product length, inner product length, the relative size difference between the inner and outer products, prime melting temperature (Tm), and the maximum allowed complementarity among the primers. Depending upon the input parameters, multiple primer sets can be obtained, from which users can select the best one as per their specific requirements. The outer and inner primer molar ratio is also particularly important (*see* **Note 2**).

3.2 Genomic DNA Isolation

The following protocol is based on previous studies [9], which has been found to yield a sufficient amount of genomic DNA with quality suitable for the T-ARMS assay:

1. Take a punch of leaf disk using the lid of the 1.5 ml microcentrifuge tube. This allows users to collect multiple samples without any chance of contamination. Any other suitable plant tissue (like seedlings, cotyledonary leaves, etc.) can be taken; however, the weight of the sample must not exceed ~50 mg.
2. Crush the leaf disk in the microcentrifuge tube in 400 μl of DNA extraction buffer using a micropestle, to complete tissue homogenization.
3. Add 128 μl of 5 M potassium acetate solution and mix thoroughly by brief vortexing (1–2 s).

4. Spin the tubes in the table-top centrifuge at 10,000 rpm (8000 × *g*) for 3 min.
5. Collect the supernatant (~400 μl) in a fresh microcentrifuge tube. For some tissues rich in polysaccharides, proteins, and other metabolites that can interfere with PCR, add an equal volume of chloroform:isoamyl alcohol (24:1), mix gently, and spin at 10,000 rpm (8000 × *g*) for 1 min, and then collect the upper layer in a fresh microcentrifuge tube.
6. Add an equal volume (400 μl) of cold isopropanol and mix gently. Let the tubes stand at room temperature for ~2 min.
7. Spin the tubes in the centrifuge at 10,000 rpm (8000 × *g*) for 3 min.
8. Discard the supernatant and add 100 μl of 70% ethanol to wash the pellet.
9. Spin the tubes at 10,000 rpm (8000 × *g*) for 10 s, discard the supernatant, and dry the pellets at 65 °C for 7 min in a dry bath.
10. Dissolve the dried pellet in 200 μl of molecular-biology-grade water. From this genomic DNA preparation, use 2 μl in PCR.

3.3 PCR and Gel Electrophoresis

1. Successful T-ARMS assays need properly optimized PCR conditions. A standard 10 μl reaction volume contains 2 μl of genomic DNA as a template; optimum concentrations of out-forward, out-reverse, in-forward, and in-reverse primers; an appropriate volume of *Taq* DNA polymerase-based PCR master mix; and molecular-biology-grade water up to 10 μl. If a T-ARMS assay is optimized with equal molar ratio of all four primers, then the PCR components will be as shown in Table 1.
2. The thermal profile of the PCR also needs optimization on a case-by-case basis, with a particular focus on annealing temperature (*see* **Note 3**). A standard thermal profile for the T-ARMS assay consists of initial denaturation at 94 °C for 4 min, followed by 35 cycles of denaturation at 94 °C for 30 s, annealing at 64 °C (or other optimized annealing temperature) for 30 s, and extension at 72 °C for 30 s (if the size of the largest target amplicon is <500 bp). The thermal profile should end with a final extension at 72 °C for 7 min, followed by holding at 4 °C for 2 min.
3. For *Taq* master mixes, which contain premixed loading dyes, the amplified products can be directly loaded in agarose gel. The suitable percentage of agarose gel depends on the relative size differences of the amplicons. Generally, 1.5–2.5% gels are found most suitable for most of the cases. Forty milliliters of 1.5% agarose gel contains 600 mg agarose, 2 ml of 20× sodium

borate buffer, and 38 ml of water. After melting and cooling, the gel is poured, with the addition of 2 μl of ethidium bromide solution (10 mg/ml). As standard practice, gel electrophoresis is carried out in 1× sodium borate buffer at 120 V for 20 min. Following electrophoresis, the results can be visualized and/or captured using a UV transilluminator/gel documentation system.

4. Finally, sequence features and DNA quality remain critical points, which are important to avoid troubles as described in **Note 4**.

4 Notes

1. The primer design is very important for the T-ARMS strategy, and it has been successfully applied in many crops, including model crops of rice [10–13] and tomato [14–17]. The protocol is based on the successful T-ARMS assay implementation based on correct primer design.
2. Outer and inner primer molar ratio: in the original article describing the T-ARMS strategy [7], a 1:10 molar ratio for outer and inner primers was recommended to ensure the amplification of allele-specific bands. However, this ratio might be variable, as reported by several other studies. For example, a 1:1 molar ratio of outer and inner primers successfully genotypes allelic variants at the *OVATE* locus in tomatoes [15]. On the other hand, a 1:2 molar ratio was found most suitable in the case of allelic discrimination of the receptor-like protein gene in tomatoes [17]. Interestingly, sometimes one of the two allele-specific amplicons has been found to fetch preferential amplification in the T-ARMS assay. In that case, the amount of the particular allele-specific primer has to be increased. This strategy has been applied while genotyping tomato lines for the *DEETIOLATED 1* locus, where the authors used 7.5 pmol of outer primers, 5 pmol of wild-type allele-specific primer, and 10 pmol of mutant allele-specific primer [16]. These observations advocate that the appropriate molar ratio for each of the four primers in the T-ARMS assay should be empirically optimized to obtain satisfactory results.
3. Optimization of annealing temperature in PCR: just like the primer molar ratio, the annealing temperature for the PCR has a profound influence on the success of the T-ARMS strategy. All four designed primers should have similar guanine–cytosine (GC) content and melting temperatures (Tm). Furthermore, there should be less chance of the formation of primer self and cross dimers. The allele-specific amplicons are generally amplified using the amplicon generated by the outer primers, as a

template. Naturally, the proper annealing temperature of the outer primers that allows for the specific and sufficient amplification of the common fragment (*see* Fig. 1) must be empirically decided. Next, the annealing temperature using one outer primer and its corresponding inner primer (which generates one allele-specific amplicon) should be optimized. Similarly, the annealing temperature using the other outer primer and its corresponding inner primer (which generates the other allele-specific amplicon) should be separately standardized. The optimum annealing temperature for the final T-ARMS assay should be decided on the basis of all the previous observations.

4. Sequence features and DNA quality: though simple in its principle, the development of a successful T-ARMS assay may face different problems. For example, the T-ARMS assay has not been recommended for genomic regions rich in GC content [18]. Moreover, the quality of the template DNA has been advocated to play a crucial role, where the presence of contaminating proteins and ribonucleic acid (RNA) can limit the success of the amplification. However, in contrast to this, genomic DNA isolated from tomato leaves using a rapid method [9] has been found to be quite sufficient for the T-ARMS assay, as has been reported during the detection of three different SNPs in tomatoes [15–17]. Another problem is associated with a typical situation when the primer and target template face a weak G/T or A/C mismatch; in these cases, often a failure in achieving allele-specific refractory amplicons is found. Moreover, lengthy optimization steps for the annealing temperature and primer molar ratio (the two major factors behind a successful T-ARMS strategy) are challenging.

5 Conclusion and Future Perspective

Like all assays, T-ARMS has its advantages and disadvantages. However, when proper care is taken during the designing of primers and the optimization of amplification conditions, T-ARMS offers the most straightforward low-cost approach for the diagnostic detection of SNP alleles. Although the detection of some specific mismatches (like G/T and A/C) are challenging, advanced strategies for designing more specific primers [19], when coupled with the T-ARMS strategy, should overcome this challenge. Moreover, the use of strand displacement polymerase in place of the standard *Taq* DNA polymerase in T-ARMS has evidenced the improvement of the strategy [20]. Taken together, T-ARMS is a novel strategy with ample scope for improvement, which can be adopted even by those laboratories lacking specialized costly equipment for SNP detection.

Acknowledgments

The authors thank Dr. Shirin Akhtar and Dr. Deepak Maurya for their help and support. We acknowledge funding from Bihar Agricultural University (Project code: SNP/CI/Rabi/2018-5).

References

1. Brookes AJ (1999) The essence of SNPs. Gene 234:177–186. https://doi.org/10.1016/s0378-1119(99)00219-x
2. The International HapMap Consortium (2005) A haplotype map of the human genome. Nature 437(7063):1299–1320. https://doi.org/10.1038/nature04226
3. Collins FS, Morgan M, Patrinos A (2003) The human genome project: lessons from large-scale biology. Science 300(5617):286–290. https://doi.org/10.1126/science.1084564
4. Newton CR, Graham A, Heptinstall LE, Powell SJ, Summers C, Kalsheker N et al (1989) Analysis of any point mutation in DNA. The amplification refractory mutation system (ARMS). Nucleic Acids Res 17:2503–2516. https://doi.org/10.1093/nar/17.7.2503
5. Ye S, Humphries S, Green F (1992) Allele specific amplification by tetra-primer PCR. Nucleic Acids Res 20:1152. https://doi.org/10.1093/nar/20.5.1152
6. Liu Q, Thorland EC, Heit JA, Sommer SS (1997) Overlapping PCR for bidirectional PCR amplification of specific alleles: a rapid one-tube method for simultaneously differentiating homozygotes and heterozygotes. Genome Res 7:389–398. https://doi.org/10.1101/gr.7.4.389
7. Ye S, Dhillon S, Ke X, Collins AR, Day IN (2001) An efficient procedure for genotyping single nucleotide polymorphisms. Nucleic Acids Res 29:e88. https://doi.org/10.1093/nar/29.17.e88
8. Chiu RWK, Murphy MF, Fidler C, Zee BCY, Wainscoat JS, Lo YMD (2001) Determination of RhD zygosity: comparison of a double amplification refractory mutation system approach and a multiplex real-time quantitative PCR approach. Clin Chem 47:667–672. https://doi.org/10.1093/clinchem/47.4.667
9. Kumar V, Kumar P, Chattopadhyay T (2017) A rapid and reproducible method for isolating genomic DNA from a few crop plants suitable for polymerase chain reaction-based genotyping. J Appl Nat Sci 9:1119–1122. https://doi.org/10.31018/jans.v9i2.1332
10. Ponjaroenkit S, Janpen K, Sakulsingharoj C, Sangtong V (2017) Development of allele-specific SNP markers for *PPR10* gene at *Rf4* locus of fertility restorer gene for identification of maintainer and restorer lines. Genom Genet 10:38–45. https://doi.org/10.14456/gag.2017.6
11. Liu Y, Zhang A, Wang F, Wang J, Bi J, Kong D et al (2019) Development and validation of a PCR-based functional marker system for identifying the low amylose content-associated gene Wx^{hp} in rice. Breed Sci 69:702–706. https://doi.org/10.1270/jsbbs.19043
12. Pereira A, Tcacenco FA, Klabunde GHF, de Andrade A (2019) Detecting acetyl-coenzyme a carboxylase resistance gene in rice (*Oryza sativa* L.). Mol Biol Rep 46:6271–6276. https://doi.org/10.1007/s11033-019-05068-z
13. Das D, Sen P, Purkayastha S, Saha AK, Roy A, Rai P et al (2021) A perfect PCR based co-dominant marker for low grain-arsenic accumulation genotyping in rice. Ecotoxicol Environ Saf 212:111960. https://doi.org/10.1016/j.ecoenv.2021.111960
14. Arens P, Mansilla C, Deinum D, Cavellini L, Moretti A, Rollan S et al (2010) Development and evaluation of robust molecular markers linked to disease resistance in tomato for distinctness, uniformity and stability testing. Theor Appl Genet 120:655–664. https://doi.org/10.1007/s00122-009-1183-2
15. Maurya D, Mukherjee A, Akhtar S, Chattopadhyay T (2021) Development and validation of the *OVATE* gene-based functional marker to assist fruit shape selection in tomato. 3 Biotech 11:474. https://doi.org/10.1007/s13205-021-03029-7
16. Mukherjee A, Akhtar S, Chattopadhyay T (2021) Development of tetra-primer amplicon refractory mutation system (T-ARMS) strategy for identification of the *dark green* mutant allele in tomato. 3 Biotech 11:182. https://doi.org/10.1007/s13205-021-02732-9
17. Chattopadhyay T, Maurya D, Shree B, Akhtar S, Roy C (2022) Development of polymerase chain reaction-based marker for identification of a nonsense mutation in a

Pseudomonas-responsive receptor like protein gene of tomato. J Hortic Sci Biotechnol 97: 113–121. https://doi.org/10.1080/14620316.2021.1947905

18. Medrano RFV, de Oliveira CA (2014) Guidelines for the tetra-primer ARMS–PCR technique development. Mol Biotechnol 56:599–608. https://doi.org/10.1007/s12033-014-9734-4
19. Touroutine D, Tanis JE (2020) A rapid, superselective method for detection of single nucleotide variants in *Caenorhabditis elegans*. Genetics 216:343–352. https://doi.org/10.1534/genetics.120.303553
20. Alyethodi RR, Singh U, Kumar S, Alex R, Deb R, Sengar GS et al (2018) T-ARMS PCR genotyping of SNP rs445709131 using thermostable strand displacement polymerase. BMC Res Notes 11:132. https://doi.org/10.1186/s13104-018-3236-6

Chapter 23

Penta-Primer Amplification Refractory Mutation System (PARMS) with Direct PCR-Based SNP Marker-Assisted Selection (D-MAS)

Chao Tan and Yanyu Yang

Abstract

The penta-primer amplification refractory mutation system (PARMS) is a high-throughput, low-cost, and automated genotyping assay system that utilizes competitive allele-specific polymerase chain reaction (AS-PCR) combined with a homogeneous fluorescence-based reporting system to detect genetic variation occurring at single-nucleotide polymorphism (SNP). It is flexible in terms of the number of SNPs and samples to be analyzed, and the whole process only needs standard liquid handling, thermal cycling instruments, and plate reading instruments, which are present in many labs. Its compatibility with DNA samples prepared from a variety of sources and extraction technologies, such as alkaline lysis, makes it suitable for a direct PCR-based SNP marker-assisted selection system (D-MAS), a simple, cost- and labor-saving, and robust SNP genotyping system. It combines rapid and high-throughput DNA extraction through modified alkaline lysis with PARMS to dramatically reduce the time of manual operation and result analysis in the molecular breeding of major crops.

Key words PARMS, D-MAS, High-throughput, SNP, Cost-saving and labor-saving

1 Introduction

Single-nucleotide polymorphism (SNP) refers to a single base change at a certain position on the genome that is stable and occurring in more than 1% of the population. Although four possible nucleotide bases can be present, SNPs are usually biallelic [1, 2]. Because SNPs are stable, widespread on genomes, associated with many complex phenotypes, and amenable to a wide range of genotyping platforms available and it is easy to manage the data, SNP markers are considered the preferred marker system in many fields, such as drug development, genetic disease diagnosis, genetic analysis, and molecular breeding [3–7].

PARMS (primer amplification refractory mutation system) was developed by Gentides Co., China, and utilizes competitive

Yuri Shavrukov (ed.), *Plant Genotyping: Methods and Protocols*, Methods in Molecular Biology, vol. 2638,
https://doi.org/10.1007/978-1-0716-3024-2_23,

allele-specific polymerase chain reaction (AS-PCR) [8, 9] combined with a homogeneous fluorescence-based reporting system to detect SNPs and has been used in molecular breeding and genetic analysis [10–16]. Primers used in PARMS include two allele-specific primers (ASPs) corresponding to two SNP alleles at the 3′ terminal, respectively; two common fluorescent primers (CFPs) labeled with fluorophore FAM and HEX, respectively; and one common locus-specific primer (LSP) paired with each ASP or CFP to amplify the corresponding DNA template. It is compatible with DNA samples prepared from a variety of sources and extraction technologies.

SNP markers have been extensively developed for further application in rice, wheat, and maize breeding as well as genetic analysis [17–21] with the development of crop-breeding methods, from phenotype selection to marker-assisted selection and genomic breeding, which are more and more depending on genotyping. The more accurate the breeding is, the more genotyping is needed. The two major automated and high-throughput SNP genotyping methods, TaqMan (Thermo Fisher, USA) and KASP (Kompetitive Allele-Specific PCR, LGC Genomics, UK), need high-quality DNA. The traditional cetyltrimethylammonium bromide (CTAB) [22] and sodium dodecyl sulfate (SDS) [23] plant DNA extraction methods are slow and exhausting because of the tedious steps. On the other hand, magnetic bead kits (LGC Genomics, UK) are high-throughput but expensive.

D-MAS (direct PCR-based SNP marker-assisted selection system) combined rapid and high-throughput DNA extraction by modified alkaline lysis with a high-throughput SNP genotyping assay by PARMS or KASP [24]. Figure 1 shows the whole procedure of D-MARS for crops. Firstly, seeds have to be sowed in 96 plugs and grown in the greenhouse. Secondly, young leaves are sampled into deep 96-well plates. The DNA of the samples is extracted by alkaline lysis and then diluted. Finally, the genotyping assay is carried out by taking the diluted DNA samples in an alkaline lysis solution as the template through PARMS. In such a system, the genotyping of 1536 samples can be completed in about 8 h, including 3 h for leaf sample collection, 2 h for DNA extraction, and 3 h for genotyping assay and data analysis [12, 25].

2 Materials

2.1 Reagents and Solutions

All the solutions should be prepared with ultrapure water and analytical-grade reagents and stored as indicated below.

1. Lysis buffer: 0.3 M NaOH solution. It can be stored at room temperature for no more than 2 months.
2. Neutralization buffer: 200 mM Tris–HCl (pH 7.0). It can be stored at room temperature for no more than 2 months.

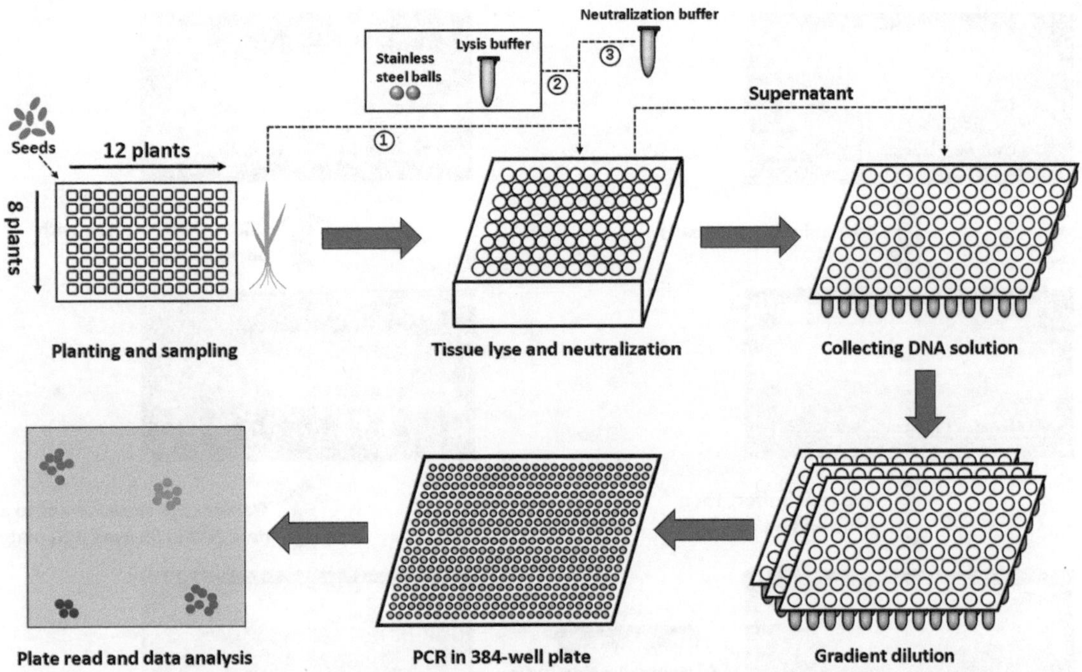

Fig. 1 The scheme of PARMS application for D-MAS. The red and blue dots represent different homozygous genotypes, and the green dots represent the heterozygous genotypes. The black dots are NTCs

3. PARMS master mix is provided in a ready-to-use 2× format containing chlorpyrifos (CPF) as well as rhodamine X (ROX) dyes as the passive reference. It can be stored at 4 °C for 2 weeks or at −20 °C for long-term storage without repeated freezing and thawing more than three times.
4. The primer can be stored at 4 °C for 1 year or at −20 °C for long-term storage (*see* **Notes 1–3**).

2.2 Consumables and Equipment

1. Deep 96-well plates for sampling and DNA extraction (*see* Fig. 2a–c).
2. 96-well PCR plate (for 200 μL) for DNA collection and dilution (*see* Fig. 2d–f).
3. Stainless steel ball bearing, 4 mm in diameter, used for tissue grinding.
4. TissueLyser for grinding.
5. PCR microtiter plates with 96-, 384-, or 1536-wells are suitable for PARMS (*see* **Note 4**).
6. Reagent dispensing: reagent dispensing can be carried out robotically with dispensing instruments or manually with a suitable pipette, depending on the plate type, sample quantities, and lab facilities.

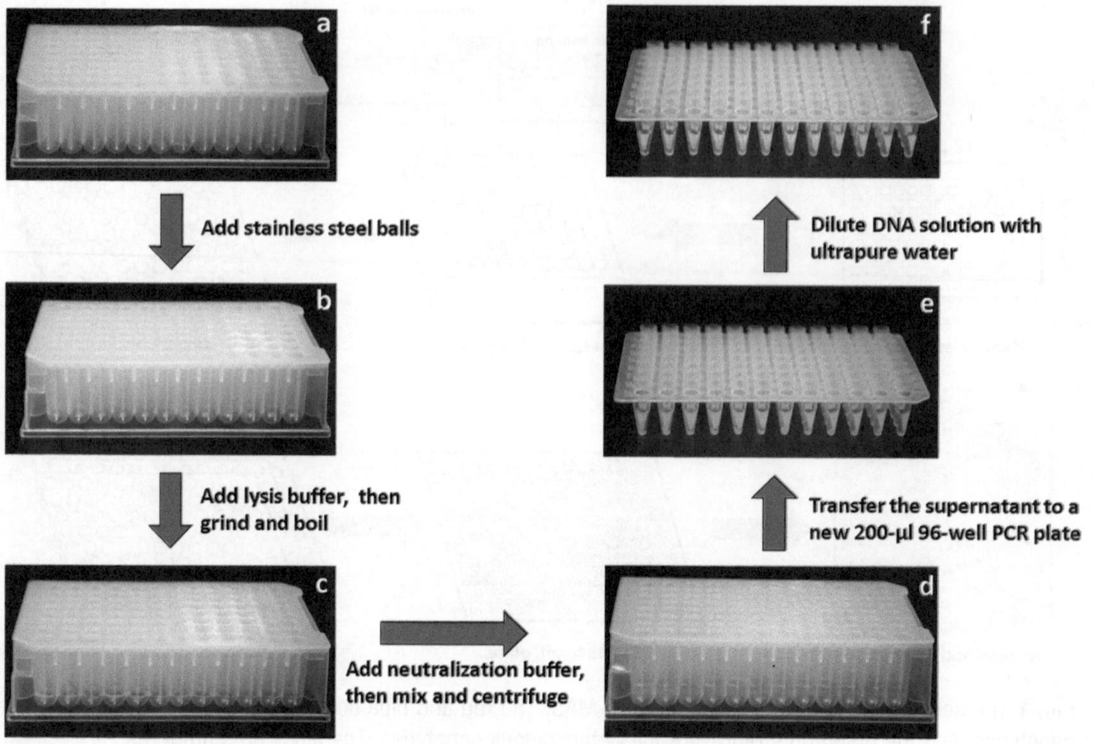

Fig. 2 The procedure for DNA extraction using the alkaline lysis approach

7. Plate sealing: plates can be sealed with a plastic membrane by hand or a sealing instrument. For non-water-bath-based thermal cyclers, protease-free and deoxyribonuclease (DNase)-free mineral oil can also be used to reduce cost (*see* **Note 4**).
8. Thermal cycler: all standard thermal cyclers, including Peltier-block-based and water-bath-based thermal cyclers, are suitable for PARMS.
9. Plate reader: any multichannel real-time PCR instrument or fluorescent plate reader that can detect FAM, HEX, and ROX fluorescent signals is suitable.

3 Methods

3.1 Plants and Sampling

Seeds have to be sown in 96 plugs, 12 × 8 small cubes arranged together in one tray. Plants are grown in the greenhouse until the three-leaf stage for rice, wheat, and maize, while rapeseed plants are grown until the four-leaf stage and ready for sampling (*see* **Note 5**).

3.2 DNA Extraction

All DNA samples are extracted in deep 96-well plates using the alkaline lysis approach with the following procedure:

1. Young leaves of about 1 cm^2 are sampled into deep 96-well plates, separately, one sample per well (*see* Fig. 2a).
2. Add two 4 mm stainless steel balls per well in deep 96-well plates with leaf samples (*see* Fig. 2b).
3. Add 100 μL of lysis buffer per sample in each well. After grinding at a high speed (50 Hz) for 1 min in TissueLyser, the plate is incubated at 96 °C for 1 min in a water bath (*see* **Note 6** and Fig. 2c).
4. Add 200 μL of neutralization buffer per sample in each well and mix thoroughly by inverting the plates several times (*see* **Note 7** and Fig. 2d).
5. After centrifugation at 3000 × *g* for 1 min, transfer the supernatant to a new 200 μL 96-well PCR plate and make a series of dilutions (*see* Fig. 2e, f).
6. Gradient dilution of DNA in the alkaline lysis solution has to be made with ultrapure water from lowest to highest dilution, from 1:5 to 1:30, respectively, with a total range of dilutions for five-, ten-, 20-, and 30-folds, prior to using the diluted DNA for PCR templates (*see* **Notes 8** and **9**). The DNA samples in the alkaline lysis solution could be stored for at least 2 months at −20 °C (*see* **Note 10**).

3.3 SNP Genotyping Assay

The SNP genotyping assay is carried out through PARMS, and this is similar to those in the KASP system (*see* **Note 11**). Components of the PARMS genotyping assay reaction mixture are shown in Table 1. The whole procedure for PARMS is indicated below and should be carried out at room temperature unless indicated.

The primer mix is prepared according to Table 2. The primer can be designed using the online software SNP Primer (http://www.snpway.com/) based on the SNP sequence data submitted by users, and two ASPs and one LSP will be generated, or the software RealTimeDesign (https://www.biosearchtech.com/support/tools/design-software/realtimedesign-software) or the online tool on WheatOmics 1.0 (http://202.194.139.32/snprimer).

1. Choose the appropriate microtiter plates according to the number of SNPs and samples. Set up the reaction volume as indicated in Table 3.
2. Array DNA samples and at least two NTCs (no-template controls) into microtiter plates, centrifuge, and dry them in an incubator at 65 °C for 30 min until the DNA samples are completely dried.
3. Prepare the PARMS assay. Dispense the PARMS assay reagents and the corresponding primer mix into wells with dried DNA samples, then seal the microtiter plate with a plastic membrane or mineral oil.

Table 1
Assembly of the PARMS genotyping assay reaction mixture

Components	Microtiter plates (capacity per well)			
	96-well (100/200 μL)	384-well (20/40 μL)	384-well (2/2.5 μL)	1536-well (2 μL)
2× PARMS Master mix	10 μL	5 μL	1 μL	1 μL
10× primer mix	2 μL	1 μL	0.2 μL	0.2 μL
DNA (alkaline lysis extract)	1–2 μL	1–2 μL	1–2 μL	1–2 μL
Ultra-pure water	To 20 μL	To 10 μL	To 2 μL	To 2 μL
Total volume	20 μL	10 μL	2 μL	2 μL

Table 2
Assembly of the 10× primer mix

Components	Concentration (μM)	Volume of 10× primer mix	Note
ASP1	10	150 μL	Corresponding to FAM-labeled CFP
ASP2	10	150 μL	Corresponding to HEX-labeled CFP
LSP	10	400 μL	Common reverse primer for ASPs and CFPs
H_2O	–	300 μL	Ultra-pure water
Total volume	–	1 mL	–

Table 3
Reaction volume for different kinds of microtiter plates

Plate well density	Capacity per well (μL)	Recommended reaction volume (μL)	Minimum reaction volume (μL)
96-well	100/200	20	10
384-well	20/40	10	5
384-well	2/2.5	2	0.8
1536-well	2	2	1

4. Run PCR on a thermal cycling instrument according to the program described in Table 4 (*see* **Note 12**).
5. Read the plates to gain fluoresnet intensity values of FAM, HEX and ROX using a fluorescent plate reader according to the parameters indicated in Table 5. The fluorophore ROX is

Table 4
Thermal cycling program for the PARMS genotyping system

Stage	Step	Temperature (°C)	Time	Cycles per stage
1	1	94	15 min	1
2	2 3	94 65 to 57	20 s 1 min (drop 0.8 °C per cycle)	10
3	4 5	94 57	20 s 1 min	28

Table 5
Functional parameters of fluorophores and fluorescent dyes

Fluorophores/fluorescent dyes	Application	Excitation wavelength (nm)	Emission wavelength (nm)
FAM	Detecting SNP allele 1	492	518
HEX	Detecting SNP allele 2	533	559
ROX	Passive reference	578	604

Table 6
Additional thermal cycling program for the PARMS genotyping system

Stage	Step	Temperature (°C)	Time	Cycles per stage
1	1 2	94 57	20 s 1 min	3

used as internal passive dye in the same samples for normalizing the fluoresnet intensity of allelic reporter FAM and HEX.

6. The data output from the fluorescent plate reader should be analyzed with any online software, like SNP Decoder (http://www.snpway.com/) and SNPviewer (https://www.biosearchtech.com/products/pcr-reagents-kits-and-instruments/pcr-instruments-and-software/genotyping-and-lims-software/snpviewer). Each sample individual will be represented as a single dot and should be crowed together into clusters which could be classfied as diffevent genotypes (see Fig.1). If the dots are not clustered clearly, additional thermal cycling should be taken, according to program mentioned in Table 6 and read the plate again until clear and tight genotyping clusters are obtained (*see* **Note 13**).

4 Notes

1. Desalted purified primers are suitable, while polyacrylamide gel electrophoresis (PAGE)-purified and high-performance liquid chromatography (HPLC)-purified primers may improve the results. Some primers may become disabled after being used for a time, so it is preferred to dispense the primer solution in small volumes in microtubes after dissolving.
2. A newly designed primer set needs to be verified. Generally, 12–24 DNA samples are sufficient for primer verification, including three to seven genotypes with identified SNP alleles in each of the three groups (both homozygotes and heterozygotes) with additional two to three samples of NTCs. If F_2 or another segregating population is used for primer verification, it is advised to increase the sample capacity and to use at least one genotype for each homo- and heterozygote and two to three NTCs as controls.
3. Because the ethylenediaminetetraacetic acid (EDTA) in Tris–EDTA (TE) buffer can chelate Mg^{2+} and affect PCR, it is highly recommended to dissolve or dilute the primers and DNA samples in ultra-pure water. If primers and DNA samples are dissolved or diluted in TE buffer, it is suggested to add extra Mg^{2+} in the reaction for compensation. The amount of Mg^{2+} added should be equal to the EDTA in the reaction.
4. To reduce the cost of genotyping, it is recommended to use cheaper disposable 384-well PCR plates and mineral oil to seal the plate instead of a plastic membrane, which is more expensive.
5. According to our tests, old leaves of rice at the flower stage as well as cotton are also amenable to D-MAS.
6. To ensure fully grinding, add 50 μL of 0.3 M NaOH solution for leaf samples smaller than 1 cm^2 or 200 μL of 0.3 M NaOH solution for large leaf samples.
7. The volume of neutralization buffer added can vary relative to the amount of the lysis buffer added. Generally, the volume of the neutralization buffer added has to be twice bigger than volume of the added lysis buffer.
8. Using DNA samples in the alkaline lysis solution without dilution may cause the failure of PARMS. All gradient dilutions from five- to 30-fold of DNA in the alkaline lysis solution achieves a successful genotyping result, although over 20-fold dilution may cause a slight decline in genotyping fluorescence intensity.

9. The series of multiple dilutions can vary depending on leaf age and leaf size of sampling. For old and big leaves, 20-fold dilution is recommended. For younger leaves, tenfold dilution works better. For cucumber, tomato, sunflower, and rapeseed, fivefold dilution, or tenfold dilution for big leaf, is recommended.
10. For wheat, DNA samples in the alkaline lysis solution can be stored at −20 °C for no longer than 2 weeks.
11. KASP can also be used for SNP genotyping assay in D-MAS, but it is recommended to use PARMS to carry out the genotyping assay. According to our test results, PARMS is less sensitive to PCR inhibitors in DNA samples in the alkaline lysis solution and shows tighter genotyping clusters than those used through KASP [24]. For the KASP system, please refer to the manufacturers' protocol.
12. The best results can be obtained when reactions for most samples enter the plateau stage with no amplification occurring in NTC. For most primer sets, the reactions enter the plateau stage after 28–30 cycles (*see* Table 4, stage 3). The best cycle numbers can be confirmed through real-time PCR. If such instruments are not available, the fluorescent signal intensities can be analyzed after PCR, and the best cycle numbers is confirmed with analysis after each additional thermal cycling reaction.
13. Some NTCs may generate signals after 40 cycles (*see* Table 4, stage 3). If the genotyping result is not good after 40 cycles, the primers are recommended to be changed.

References

1. Vignal A, Milan D, SanCristobal M, Eggen A (2002) A review on SNP and other types of molecular markers and their use in animal genetics. Genet Sel Evol 34:275–305. https://doi.org/10.1186/1297-9686-34-3-275
2. Perkel J (2008) SNP genotyping: six technologies that keyed a revolution. Nat Methods 5: 447–453. https://doi.org/10.1038/nmeth0508-447
3. Jenkins S, Gibson N (2002) High-throughput SNP genotyping. Comp Funct Genomics 3: 57–66. https://doi.org/10.1002/cfg.130
4. Kim S, Misra A (2007) SNP genotyping: technologies and biomedical applications. Annu Rev Biomed Eng 9:289–320. https://doi.org/10.1146/annurev.bioeng.9.060906.152037
5. Nijman IJ, Kuipers S, Verheul M, Guryev V, Cuppen E (2008) A genome-wide SNP panel for mapping and association studies in the rat. BMC Genomics 9:95. https://doi.org/10.1186/1471-2164-9-95
6. Delannay X, McLaren G, Ribaut JM (2012) Fostering molecular breeding in developing countries. Mol Breed 29:857–873. https://doi.org/10.1007/s11032-011-9611-9
7. Thomson MJ (2014) High-throughput SNP genotyping to accelerate crop improvement. Plant Breed Biotech 2:195–212. https://doi.org/10.9787/PBB.2014.2.3.195
8. Newton CR, Graham A, Heptinstall LE, Powell SJ, Summers C, Kalsheker N et al (1989) Analysis of any point mutation in DNA. The amplification refractory mutation system (ARMS). Nucleic Acids Res 17:2503–2516. https://doi.org/10.1093/nar/17.7.2503
9. Heim M, Meyer UA (1990) Genotyping of poor metabolisers of debrisoquine by allele-

specific PCR amplification. Lancet 336(8714): 529–532. https://doi.org/10.1016/0140-6736(90)92086-w

10. Zhang B, Zhao N, Liu YY, Jia L, Fu Y, He XX et al (2019) Novel molecular markers for high-throughput sex characterization of *Cynoglossus semilaevis*. Aquaculture 513:734331. https://doi.org/10.1016/j.aquaculture.2019.734331
11. Wang XD, Zheng M, Liu HF, Zhang L, Chen F, Zhang W et al (2020) Fine-mapping and transcriptome analysis of a candidate gene controlling plant height in *Brassica napus* L. Biotechnol Biofuels 13:42. https://doi.org/10.1186/s13068-020-01687-y
12. Xu Q, Xu FC, Qin DD, Li MF, Fedak G, Cao WG et al (2020) Molecular mapping of QTLs conferring Fusarium head blight resistance in Chinese wheat cultivar Jingzhou 66. Plan Theory 9:1021. https://doi.org/10.3390/plants9081021
13. Huang J, Gao LJ, Luo SM, Liu KQ, Qin DJ, Pan YH et al (2022) The genetic editing of *GS3* via CRISPR/Cas9 accelerates the breeding of three-line hybrid rice with superior yield and grain quality. Mol Breed 42:22. https://doi.org/10.1007/s11032-022-01290-z
14. Gao J, Liang HF, Huang J, Qing DJ, Wu H, Zhou WY et al (2021) Development of the PARMS marker of the *TAC1* gene and its utilization in rice plant architecture breeding. Euphytica 217:49. https://doi.org/10.1007/s10681-020-02747-y
15. Chai L, Feng B, Liu X, Jiang LC, Yuan S, Zhang ZW et al (2021) Fine mapping of a locus underlying the ectopic blade-like outgrowths on leaf and screening its candidate genes in rapeseed (*Brassica napus* L.). Front Plant Sci 11:616844. https://doi.org/10.3389/fpls.2020.616844
16. Li Z, Yuan R, Wang M, Hong MY, Zhu L, Li XF et al (2022) Development of the PARMS marker of the dominant genic male sterility (DGMS) line and its utilization in rapeseed (*Brassica napus* L.) breeding. Plan Theory 11:421. https://doi.org/10.3390/plants11030421
17. Rasheed A, Wen WE, Gao FM, Zhai SN, Jin H, Liu JD et al (2016) Development and validation of KASP assays for genes underpinning key economic traits in bread wheat. Theor Appl Genet 129:1843–1860. https://doi.org/10.1007/s00122-016-2743-x
18. Ma CY, Ma XN, Yao LS, Liu YJ, Du FL, Yang XH et al (2017) *qRfg3*, a novel quantitative resistance locus against *Gibberella* stalk rot in maize. Theor Appl Genet 130:1723–1734. https://doi.org/10.1007/s00122-017-2921-5
19. Pan QC, Xu YC, Li K, Peng Y, Zhan W, Li WQ et al (2017) The genetic basis of plant architecture in 10 maize recombinant inbred line populations. Plant Physiol 175:858–873. https://doi.org/10.1104/pp.17.00709
20. Yang GL, Chen SP, Chen LK, Sun K, Huang CH, Zhou DH et al (2019) Development of a core SNP arrays based on the KASP method for molecular breeding of rice. Rice 12:21. https://doi.org/10.1186/s12284-019-0272-3
21. Zhao JJ, Wang ZW, Liu HX, Zhao J, Li T, Hou J et al (2019) Global status of 47 major wheat loci controlling yield, quality, adaptation and stress resistance selected over the last century. BMC Plant Biol 19:5. https://doi.org/10.1186/s12870-018-1612-y
22. Murray MG, Thompson WF (1980) Rapid isolation of high molecular weight plant DNA. Nucleic Acids Res 8(19):4321–4325. https://doi.org/10.1093/nar/8.19.4321
23. Dellaporta SL, Wood J, Hicks JB (1983) A plant DNA minipreparation: version II. Plant Mol Biol Rep 1:19–21. https://doi.org/10.1007/BF02712670
24. Lu J, Hou J, Ouyang YD, Luo H, Zhao JH, Mao C et al (2020) A direct PCR-based SNP marker-assisted selection system (D-MAS) for different crops. Mol Breed 40:9. https://doi.org/10.1007/s11032-019-1091-3
25. Zhang JC, Zhang DJ, Fan YW, Li CC, Xu PK, Li W et al (2021) The identification of grain size genes by RapMap reveals directional selection during rice domestication. Nat Commun 12:5673. https://doi.org/10.1038/s41467-021-25961-1

Chapter 24

High-Resolution Melting (HRM) Genotyping

Nayoung Kim, Ji-Su Kwon, Won-Hee Kang, and Seon-In Yeom

Abstract

High-resolution melting (HRM) analysis is a simple, fast, and inexpensive real-time polymerase chain reaction (PCR)-based method used to identify genetic variation between populations and detect single-nucleotide polymorphisms (SNPs) in nucleic acid sequences. HRM is a powerful technique that detects the differences between SNP allele melting temperatures by using a fluorescent dye inserted into the duplex deoxyribonucleic acid (DNA) structure. Prior to performing HRM analysis, optimizing the primer design, PCR mixture, and software settings is essential to obtain accurate and reliable results. In this chapter, we describe a detailed SNP genotyping method that includes primer design and the analysis of the shapes and positions of the melt curve of the luminescence intensity of the fluorescent dye attached to amplified DNA using software of qPCR instruments. This protocol is applicable for genotyping germplasm, genetic mapping, and marker-assisted breeding in plants.

Key words High-resolution melting analysis, HRM, Single-nucleotide polymorphism, SNP genotyping, Genetic variation, Real-time PCR

1 Introduction

High-resolution melting (HRM) analysis is a simple, rapid, and cost-effective method for identifying genetic variation [1]. This technique was first introduced in 2003 by the University of Utah, Medical School [2]. Polymerase chain reaction (PCR) is used prior to performing HRM to obtain amplified double-stranded deoxyribonucleic acid (DNA), also known as amplicon, with the fluorescent intercalating dye on the DNA region of interest, including single-nucleotide polymorphisms (SNPs). HRM is a method based on the dissociation of the PCR product from the double-stranded DNA (dsDNA) to the single-stranded DNA (ssDNA), where the fluorescent signal decreases as the real-time PCR instrumentation increases the temperature. During HRM, the dissociation of duplex DNA is measured in real time by the HRM machine, and DNA polymorphisms are easily monitored based on the shape and position of the melt curve. Therefore, genotyping can be performed by

Yuri Shavrukov (ed.), *Plant Genotyping: Methods and Protocols*, Methods in Molecular Biology, vol. 2638,
https://doi.org/10.1007/978-1-0716-3024-2_24,

comparing melting curves between genotypes according to the differences in melting temperature (T_m), which depends on sequence variation. For example, the different T_m between two homozygous genotypes can be distinguished by the temperature shift on the melting curves. However, in contrast to homozygous genotypes, heterozygous genotypes can be easily differentiated by the shape change of the melt curve (*see* Fig. 1).

Although HRM is already widely used for plant genotyping, the assay lacks a standardized or structured protocol [3]. This chapter was written for researchers or plant genotyping beginners for ease and convenience in HRM analysis. Obtaining optimal results for SNP detection necessitates that HRM experiments be designed with many conditions in consideration, including primer design (guanine–cytosine (GC) content, primer size, and T_m), template DNA, PCR cycling conditions, PCR amplicon parameters (amplicon size and the number of SNPs in the amplicon sequences), and melting parameters (melting temperature range). The results are then analyzed using a software tool [4]. Thus, the HRM assay procedure was described using an SYBR Green fluorescence binding dye kit and examples of a Rotor-Gene Q 2plex HRM platform machine and a CFX Opus 96 Real-Time PCR System. This assay includes four steps: (1) primer design to amplify the DNA region, including SNP alleles for genotyping; (2) HRM reaction sample preparation; (3) real-time PCR program parameters and instructions; and (4) HRM data analysis using a software program. All of the HRM steps are summarized in Fig. 1. In this chapter, the materials, methods, and general considerations for the four main steps in HRM are described "in-depth," together with troubleshooting examples. This chapter could be helpful in facilitating genotyping, high-throughput screening of germplasm, genetic mapping, SNP discovery, and marker-assisted breeding in plants [5].

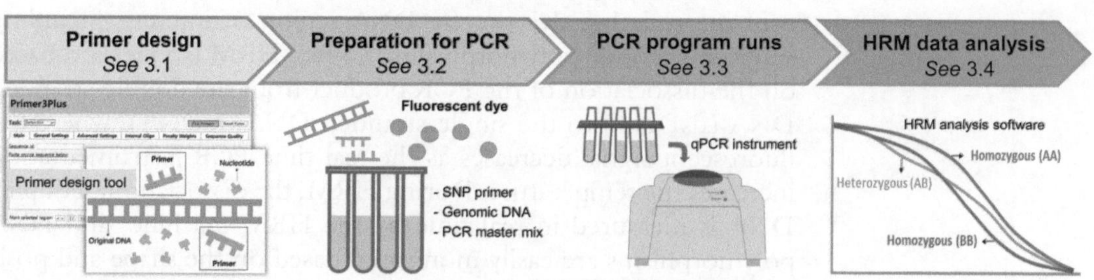

Fig. 1 Summary of the HRM genotyping procedure. The schematic diagram of the HRM genotyping procedure in this chapter consists of four main steps: (1) primer design using primer design software, (2) preparation for PCR running using fluorescence binding dye kit, (3) PCR program runs using qPCR, (4) HRM data analysis using qPCR software

2 Materials

2.1 Consumables and Components of the Real Time-PCR Thermocycler

1. Sterile 0.1 mL strip tubes and caps.
2. Loading block.
3. 72-well rotor with a locking ring (QUIAZEN, Germany) or Hard-Shell® 96-Well PCR Plates (Bio-Rad, USA) and other plates suitable for the researcher's qPCR instruments.
4. Sterile 1.5 mL microcentrifuge tubes.
5. Rotor-Gene Q 2plex HRM platform (QIAGEN, Germany) or CFX Opus 96 Real-Time PCR System (Bio-Rad, USA) and other instruments for qPCR.

2.2 HRM PCR Reagent

1. Sterile water.
2. 2× QuantiNova SYBR Green PCR master mix (QIAGEN, QuantiNova SYBR Green PCR Kit, Germany) or iTaq™ Universal SYBR Green Supermix (Bio-Rad, USA) and other similar kit products with dyes suitable for qPCR.
3. 50 ng/μL genomic DNA.
4. Two SNP primers (10 pm/μL each).

2.3 Software

1. Primer3Plus software, Primer3, Primer3web, and NCBI Primer-BLAST software (*see* **Note 1**).
2. Rotor-Gene Q series software (QUAZEN, version 2.3.1.49), Precision Melt Analysis software (Bio-Rad, version 1.3), and other software suitable for the researcher's qPCR instruments.

3 Methods

3.1 Primer Design

1. Search for polymorphisms in the target gene sequences of two homozygous populations. Align the target gene's FASTA format sequences to identify the target gene's location on the genome's chromosomes using BLASTN at NCBI (https://blast.ncbi.nlm.nih.gov) or NCBI BLAST+ at EMBL-EBI (https://www.ebi.ac.uk/Tools/sss/ncbiblast/nucleotide.html).
2. Identify base variation in a single DNA nucleotide between two genotypes to develop SNP markers using the target sequences (*see* **Note 2**). To ensure accurate result analysis, identify the target sequence with only one SNP (target sequence variation) in the PCR amplicon region.
3. Design primers for the target sequences using primer design tools such as Primer3Plus, Primer3, Primer3web, and NCBI Primer-BLAST software (*see* **Note 1**) [6]. To obtain the best

genotyping results, a PCR product size range of 100–300 base pairs (bp) is recommended, which includes primer sequences for HRM analysis (*see* **Note 3**). Primers should be designed in such a way that the SNP position is located in the center of the expected PCR product, with no bias at either end. The ideal primer size should be 18–30 bp. The optimal primer T_m is between 58 and 60 °C, and the recommended primer GC content is between 40% and 60%.

4. The primer design tool generates primer sequences based on specified criteria. Select an appropriate primer set based on length, melting temperature, GC content, and amplicon size.

3.2 Preparation of Reaction Samples for HRM

1. After PCR using the designed primers with experimental samples and controls, and before beginning HRM analysis, the single band and size of each PCR product can be confirmed with electrophoresis as an optional step. Two different homozygous samples are required as controls to determine the genotype of all samples (*see* **Note 4**).
2. Prepare on ice the genomic DNA, primers, and PCR master mix (for example, 2× QuantiNova SYBR Green or iTaq™ Universal SYBR Green Supermix). The composition of the PCR mixture is detailed in Table 1 if using the PCR master mix or a separate fluorescent dye (*see* Table 1 and **Note 5**). The tube with a PCR mixture should be placed on ice and protected from light due to the fluorescent dye.
3. Transfer the loading block onto the ice. Place the 100 μL strip tubes into the chilled loading block (*see* **Note 6**).
4. Briefly vortex the genomic DNA (10 ng/μL) and aliquot (*see* **Note 7**).
5. Load 5 μL of DNA into each strip tube (*see* **Note 8**).
6. Add 2 μL of primers (1 μL each of forward and reverse primers) directly to the bottom wall of the 100 μL strip tubes. While pipetting, make the tip stand vertically and aliquot without bubbles at the bottom of the strip tubes.
7. In a 1.5 mL microcentrifuge tube, combine the PCR master mix (for example, 2× QuantiNova SYBR Green or iTaq™ Universal SYBR Green Supermix) with sterile water. Briefly vortex and add 13 μL of the mixture into the strip tubes while avoiding bubbles. Keep the loading block chilled while adding the reagents to preserve their activity (*see* **Note 9**).
8. Remove the loading block from the ice and cap the 100 μL strip tubes. Place the samples into qPCR instruments and fasten it (*see* **Note 10**).

Table 1
Composition of PCR mixture

PCR master mix used		The separate fluorescent dye used	
Component	**Volume per sample (μL)**	**Component**	**Volume per sample (μL)**
Genomic DNA (10 ng/μL)	5	Genomic DNA (10 ng/μL)	5
Forward primer (10 pm/μL)	1	Forward primer (10 pm/μL)	1
Reverse primer (10 pm/μL)	1	Reverse primer (10 pm/μL)	1
PCR master mix	10	DNA Taq polymerase (0.5 unit) 10× Taq polymerase buffer dNTPs (2.5 mM) Fluorescent dye (5 mM)	0.1 2 1 1
Sterilized water	3	Sterilized water	8.9
Total	20	Total	20

3.3 Real-Time PCR Running Program Parameters and Instructions

1. Before performing HRM analysis, make sure to determine and use the appropriate software for the real-time PCR machine in your laboratory. Perform the melt calibration setup prior to HRM analysis according to real-time PCR or HRM instrument manual or protocol (*see* **Note 11**). A detailed description of how to run both instruments (2plex HRM platform and CFX Opus 96 Real-time PCR System) is wroted in the Notes (*see* **Notes 16** and **17**).
2. Power on the computer and turn on the switch of the qPCR instrument. Open the machine and insert a 96-well plate or 72-well rotor containing the prepared samples into the qPCR instrument (*see* **Note 12**).
3. Set the PCR cycle on the software (*see* Table 2). Select the temperature and time of initial denaturation/enzyme activation according to the type of master mix to be used (*see* **Notes 13** and **14**). The melt calibration condition (HRM cycle step) can be modified according to the amplicon, but specific modification is unnecessary (*see* **Note 15**).
4. After setting the PCR running condition, check the list of fluorescent dyes on the software and select (click) the fluorescent used for analysis.

Table 2
PCR cycle settings for HRM analysis

Cycle step	Temperature (°C)	Time	Cycles
Initial denaturation (hold)	95	2 min	1
Denaturation Annealing Extension	95 50–60 72	20 s 20–30 s 10–30 s	30 ~ 40
Hold 2 (optional step)	95	1 min	1
Hold 3 (optional step)	40	1 min	1
HRM	70–95 0.1 °C interval	2 s per step	1

5. After checking all the set values and conditions, click the "Start Run" button. The data file is automatically saved in the computer for HRM analysis (*see* **Notes 16** and **17**).

3.4 HRM Data Analysis

1. HRM data are analyzed with the instrument's software (*see* **Note 18**). When the running of the instrument is complete, the data file is automatically saved in the software. Check the amplification rate and melt curve graph file to ensure that all samples display a uniform amplification and a single melt peak (*see* **Notes 19** and **20**).
2. The typical progression of HRM analysis needs normalizing by setting the melting area where differences between samples appear as they are melted (*see* Fig. 2a, b) (*see* **Note 21**). When the value of the fluorescence graph lines decreases to 0 (*see* Fig. 2b), the reaction is complete as no fluorescence and all samples become single-stranded.
3. Before confirming the genotypes, it should be verified that each positive and negative replicate samples (indicated by two red and blue lines) is similar in the melting profile graph (*see* Fig. 2b, c). If you have checked the normalized melt plot, open the difference plot window (*see* **Note 22**). After that, clearly designate differences in the melting plot between the two homozygous samples that are to be used as controls (*see* Fig. 2c, d). Normalized melt curve plots and difference plots determine the genotypes of each sample (*see* **Note 23**). In the case of the Rotor-Gene Q software, the samples that do not match the set threshold value are marked as "Variation," to be reconfirmed (*see* Fig. 2d).
4. Based on normalized melt plots and different graph plots (*see* **Notes 24** and **25**), genetic variants can be determined in all samples, which are represented in the results of the various

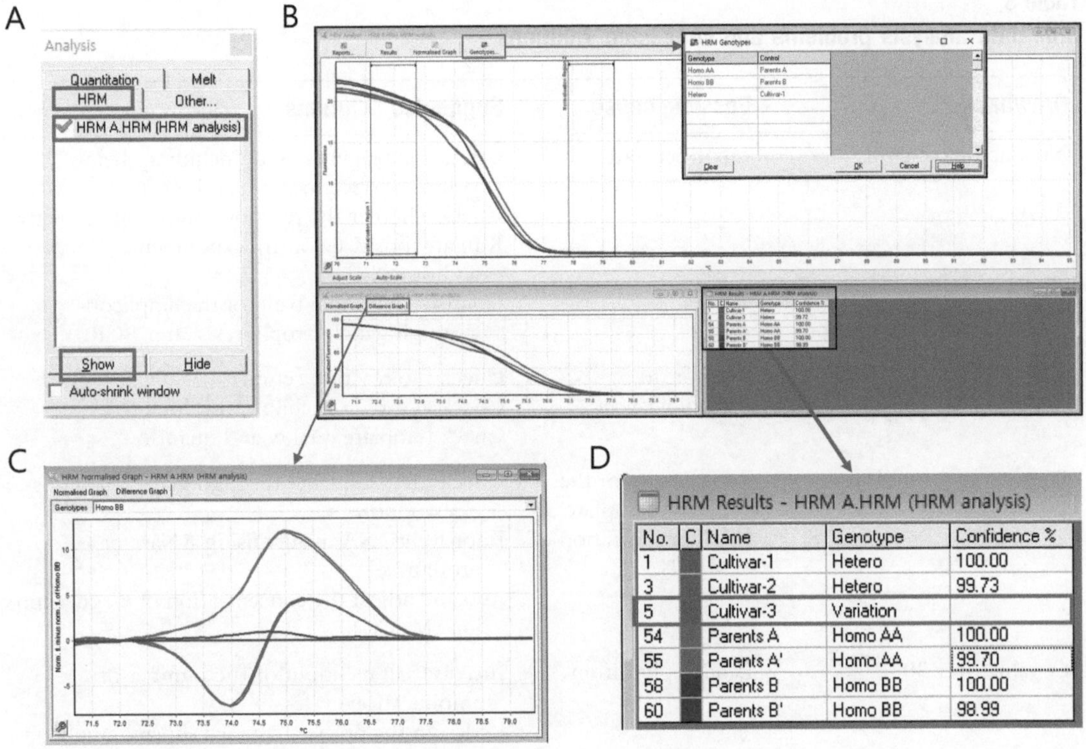

Fig. 2 Four main windows from running HRM result analysis with Rotor-Gene Q software. (**a**) A small window appears after clicking on the 'Analysis' palette. (**b**) The 'HRM Analysis', 'HRM Normalized Graph', and 'HRM Results' windows represent the raw data melt curve. (**c**) The 'Difference Graph' result is represented by the 'HRM Normalized Graph' window. (**d**) The expanded 'HRM Results' window is shown. Red boxes in the figure highlight the selected button and their indicated details

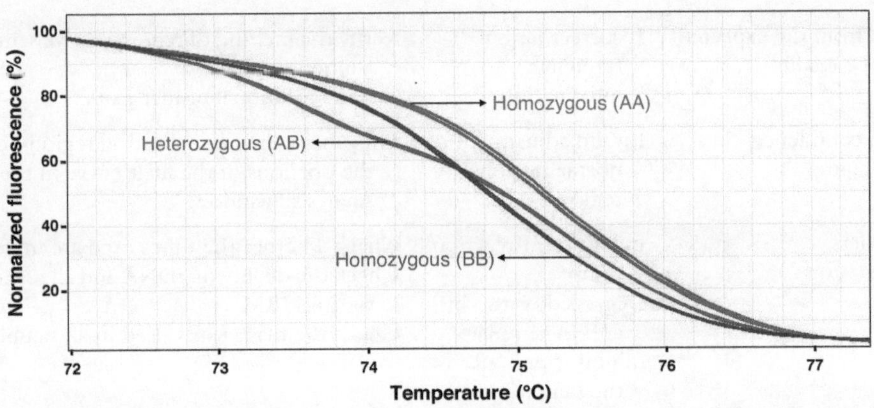

Fig. 3 High-resolution melt curve with a single-point mutation of the SNP marker. This example result represents an aligned melt curve. The plots demonstrate the sharp decrease in fluorescence when the double-stranded DNA melts into its single-stranded form as the temperature increases. The red line is a homozygous *AA* genotype, the blue line is a homozygous *BB* genotype, and the green line is a heterozygous *AB* genotype

Table 3
HRM data analysis problems and suggested solutions

Drawback	Possible cause	Suggested solutions
No amplified curve	Incorrect setup	Check the reagents used, including their volume, concentration, and purity. Check whether the reaction was set up correctly. Run any positive control experiment additionally. Confirm the single band of the amplicon through gel electrophoresis after PCR.
Jagged plot	Poor or no amplification	Check the reaction setup. Check and reset cycling conditions. Check template quality and quantity.
Poor curve resolution	Poor quality of the gDNA template Low concentration of $MgCl_2$	Check gDNA quality through gel electrophoresis. Improve gDNA quality using RNase or proteinase. Add and adjust the concentration of magnesium ion ($MgCl_2$).
Ct value over 30	Late amplification	Increase sample input or the number of amplification cycles. Redesign the primer to gain a shorter amplicon.
Cluster overlap of melt curves	Long amplicon length	Redesign the primer to gain a shorter amplicon.
Amplification or melt plots are saturated early or late	Gain setting too high or too low	Select Auto-Gain Optimization when setting the program's running conditions.
Melting results are not matched with the corresponding samples	Inconsistent concentration of DNA and primers	Adjust reaction components to be the same volume and concentration in each sample.
Variation from the expected genotype results	Incorrect target sequence information	Verify the melting of representative samples with Sanger sequencing. Design additional primer pairs.
Modified confidence percentages	Invalid adjustment of normalization regions	Only move the normalization regions, excluding the portions of the melt curve in the "HRM Analysis" window.
Data outliers	Inconsistent reaction setup Degraded or too little gDNA template Inhibitor present in the sample	Check whether the tubes used are uniform. Check the quality of gDNA and all reagents used to run HRM. Check the master mix used in all samples.

samples by curve shifting (for homozygotes) and a curve shape change (for heterozygotes) (*see* Fig. 3).

5. The potential problems in HRM data analysis and their suggested solutions are presented in Table 3.

4 Notes

1. In this chapter, the Primer3Plus software (https://www.bioinformatics.nl/cgi-bin/primer3plus/primer3plus.cgi) for primer design was used. Primers could also be designed using alternative free web-based tools such as Primer3 (https://bioinfo.ut.ee/primer3/), Primer3web (https://primer3.ut.ee/), and NCBI Primer-BLAST (https://www.ncbi.nlm.nih.gov/tools/primer-blast/).
2. A base change of C/T, G/A, C/A, and G/T between two populations can precisely detect T_m (melting temperature) curves during HRM analysis. It is recommended to make a base change to guanine (G) for high SNP stability.
3. This genotyping method can detect SNPs when the PCR product size is up to 1000 bp and is highly accurate at <400 bp. However, amplicons smaller than 250 bp are the most successfully detected SNPs, with 100–300 bp being the optimal product size.
4. By adding the same amount of genomic DNA from two homozygous control samples, an artificial heterozygous sample can be used as a control. An artificial control sample can be analyzed as a representative control for other samples.
5. If using a separate fluorescent dye such as SYTO9 or SYBR Green, prepare DNA Taq polymerase, Taq polymerase buffer, and dNTPs.
6. If the loading block has cooled sufficiently, you may remove it from the ice and run the PCR on the bench instead.
7. Depending on the genome size of the species, the amount of DNA template needs to be adjusted.
8. Check whether the same concentration and amount of gDNA template are used in each sample. Large differences in the gDNA final concentration will affect the shift of melting curves between samples.
9. Only one reaction mixture should be used for all samples to reduce the variation in melting curve results between samples.
10. Be careful not to allow the sample liquid to stick to the tube wall as this will have an impact when closing the tube cab and could ultimately cause problems during the PCR run.
11. The Rotor-Gene Q series software (version 2.3.1.49) [7] and the Precision Melt Analysis™ software (version 1.3) [8] are provided by the Rotor-Gene Q 2plex HRM platform and Bio-Rad Laboratories, respectively. Before performing HRM analysis, download and set up the appropriate software for the real-time PCR machine in your laboratory.

12. In the case of the rotor and plate, both devices assign a unique number and place to the sample, which can memorize the type of sample as well as the number.
13. PCR running condition should be modified depending on the enzyme, reagent (master mix), or primer used. The 2× QuantiNova SYBR Green PCR master mix and iTaq™ Universal SYBR Green Supermix require a 2 min initial denaturation.
14. After HRM running, a single and specific size band can be confirmed through agarose gel electrophoresis using PCR reaction products.
15. The default ramp for Rotor-Gene Q is 0.1 °C with a hold of 2 s, and Bio-Rad is 0.2 °C with a hold of 10 s. It is sufficient to perform the melting cycling step with the default value, which can be adjusted.
16. A brief procedure for the Rotor-Gene Q series software (version 2.3.1.49) is as follows. For more details, refer to the software manuals and QIAGEN instrument protocols. Double-click on the icon to run the software. Select "New" and follow "High-Resolution Melt Run" to make a new file. Select the "72-Well Rotor" option in the "Rotor Type" box, check the "Locking Ring Attached" box below, and click "Next >>." Fill out the name in the "Operator" box and write any notes related to the experiment in the "Notes" box below. Select or write the 20 μL value in the "Reaction Volume" box, and for "Sample Layout," select the preferred default value (serial 1, 2, 3...). Click "Next >>" to set the PCR running conditions and click the "Edit Profile..." option in the middle of the window. Click "Hold" to set an appropriate initial hold time, then set the "Hold Temperature" and "Hold Time." Click "Cycling," drag the red line to adjust the temperature, and set the denaturing and annealing times according to product size. Then click on the "Acquiring to Cycling A" block at the bottom and select the channel corresponding to the dye being used by referring to the "Dye Channel Selection Chart," which is accessed by clicking the "Dye Chart >>." Enter the selected dye color in the "Acquiring Channels" box and click "OK." The run condition can be modified according to the amplicon by clicking "HRM"; specific modification is unnecessary. When the profile setup is complete, check the "Gain Optimization..." white box and click the "Optimize Acquiring" button to set the gain value automatically. After that, click the "OK" box. After checking all the set values and conditions, click the "Start Run" button. You can save this run file to your computer using "Save Template."

17. A brief procedure for the Precision Melt Analysis™ software (version 1.3) is as follows. For more details, refer to the software manuals and Bio-Rad instrument protocols. Double-click on the icon (Bio-Rad Precision Melt Analysis desktop) to start the software. To open the Run setup window, click "User-defined" on the "Run setup" tab of the Startup Wizard. Click "Create New" or "Edit Selected" to set the PCR running conditions and set the cycling steps and temperature on the Protocol Editor window. When the profile setup is complete, click "OK" and save the protocol. Then return to the Experiment Setup window and click the "Next >>" button. Click "Edit Selected" for the setting plate. Click "Select Fluorophores" and verify the list of fluorophores used and check the Selected box. After selecting the number of plates to be analyzed by dragging, check the fluorophores in the "load" option. Fluorophores and target names can be specified for each plate site. When all setting is complete, click "OK." Click the "Start Run" tab and then the checkbox on Start Run on Selected Blocks. After clicking the "Start Run" button, running starts, and data can be saved as a file.
18. Each instrument has different software packages, respectively. Details and specific analytical methods can be found in the corporation's instrument manuals or software protocols.
19. The uniformity of the sample is indicated by a value called the cycle threshold or the crossing point, which value depends on the instrument and software's computational methods.
20. The melt curve represents the characteristics of dsDNA during the heating process. Melting temperature is the temperature at which 50% of DNA is denatured and the confirmation of single-nucleotide polymorphisms (SNP) can be inferred for each sample based on this information.
21. When setting the premelt and postmelt areas to check the Normalized Melt Curve, in the case of the Rotor-Gene Q, they are displayed as Normalized 1 and 2, and in the case of Bio-Rad, they are represented by the green and red regions. Premelt and postmelt regions can be adjusted by dragging the line and moving the sliders by holding down the mouse cursor. All data outside the defined regions are excluded, and expanding the area alters the baseline slope in the Normalized Graph (or Normalized Melt Curve). Samples can be analyzed by comparing graph plots with the control samples.
22. In the case of the Rotor-Gene Q software, the normalized melt plots and difference plots are shown in the Normalized Graph and Difference Graph windows, and in the case of the Bio-Rad software, they are shown in the Normalized Melt Curve and Difference Curve windows, respectively.

23. The curve difference between the melting plots among other samples can be clearly distinguished for genotyping based on the melting curves of the two homozygous samples.
24. A brief procedure for the Rotor-Gene Q series software (version 2.3.1.49) is as follows. For more details, refer to the software manuals and QIAGEN instrument protocols. Double-click the Rotor-Gene Q program that has finished running, then click the "Run in Virtual Mode" middle block to analyze the HRM data. Click the "Analysis" button at the top left of the window, then select the HRM option (*see* Fig. 2a). Next, click the "Show" button below. "HRM Analysis" (top graph), "HRM Normalized Graph" (bottom-left graph), and "HRM Result" (bottom-right table) should appear (*see* Fig. 2b). In the "HRM Analysis" window, click the upper right "Genotypes..." button. A small new window will appear. Fill out each genotype name onto each field and indicate the representative samples in the "Control" category (*see* Fig. 2b). Click the "Difference Graph" palette in the "HRM Normalized Graph" window to indicate the graph plot differences of all samples. Next, select the desired genotype to compare against all other samples through the "Genotypes" drop-down menu. The sample selected as a representative in the 'Genoyptes' panel becomes the control and is comparable to the plots of all other samples (*see* Fig. 2c). The genotype results for all samples are displayed in tabular format in the "HRM Results" window at the bottom right. Confidence values are automatically measured and represented on the table, and the threshold value can be modified.
25. A brief procedure for the Precision Melt Analysis™ software (version 1.3) is as follows. For more details, refer to the software manuals and Bio-Rad instrument protocols. Double-click the Precision Melt Analysis™ software icon that has finished running, click "File" at the top left of the window, select "Open," and then click "Melt File." Click the "Precision Melt" tab on the top left of the window, and "Normalized Melt Curve" (top left), "Difference Curve" (top right), "Well selector" (bottom left), and "Spreadsheet" (bottom right) windows will appear. In the "Difference Curve" window (top right), select the reference cluster in the "Reference cluster" menu in the drop-down. Clustering samples is also possible manually, so two different homo control samples can be assigned. Changes made to the cluster are applied to all samples in the cluster. Following the clustering results, the genotype for all samples can be designated in tabular format at the bottom right of the spreadsheet window. The Percent Confidence values are automatically measured and represented on the table.

Acknowledgments

This work was supported by the National Research Foundation of Korea (NRF), funded by the Korean Government [NRF-2015R1A6A1A03031413, 2017R1E1A1A01072843, and 2019R1C1C1007472].

References

1. Gundry CN, Vandersteen JG, Reed GH, Pryor RJ, Chen J, Wittwer CT (2003) Amplicon melting analysis with labeled primers: a closed-tube method for differentiating homozygotes and heterozygotes. Clin Chem 49:396–406. https://doi.org/10.1373/49.3.396
2. Wittwer CT, Reed GH, Gundry CN, Vandersteen JG, Pryor RJ (2003) High-resolution genotyping by amplicon melting analysis using LCGreen. Clin Chem 49:853–860. https://doi.org/10.1373/49.6.853
3. Tucker EJ, Huynh BL (2014) Genotyping by high-resolution melting analysis. In: Fleury D, Whitford R (eds) Crop breeding: methods and protocols, vol 1145. Humana, New York, pp 59–66. https://doi.org/10.1007/978-1-4939-0446-4_5
4. Park SW, An SJ, Yang HB, Kwon JK, Kang BC (2009) Optimization of high resolution melting analysis and discovery of single nucleotide polymorphism in *Capsicum*. Hortic Environ Biotechnol 50:31–39
5. Kim N, Kang WH, Lee J, Yeom SI (2019) Development of clustered resistance gene analogs-based markers of resistance to *Phytophthora capsici* in chili pepper. Biomed Res Int 2019:1093186. https://doi.org/10.1155/2019/1093186
6. Hung JH, Weng Z (2016) Designing polymerase chain reaction primers using Primer3Plus. Cold Spring Harb Protoc 2016(9):821–826. https://doi.org/10.1101/pdb.prot093096
7. QUIAZEN (2012) QUIAZEN Rotor-Gene® Q User Manuals. https://www.qiagen.com/us/products/discovery-and-translational-research/epigenetics/dna-methylation/methylation-specific-pcr/rotor-gene-q/. Accessed 10 May 2022
8. Bio-Rad (2017) Precision Melt Analysis™ Software User Guide. https://www.bio-rad.com/ko-kr/product/precision-melt-analysis-software?ID=df190aee-f184-497e-bfb6-b6dd632c99b5. Accessed 2 June 2022

Chapter 25

Modified High-Resolution Melting (HRM) Marker Systems Increasing Discriminability Between Homozygous Alleles

Satoshi Watanabe, Yoshiyuki Yamagata, and Nobuhiro Kotoda

Abstract

Targeted single-nucleotide polymorphism (SNP) genotyping, especially for functional nucleotide polymorphism, is widely used for current breeding programs in crops. One of the cost- and time-effective approaches for genotyping is high-resolution melting (HRM) analysis for polymerase chain reaction (PCR) amplicons, including target SNP. The reliability of a genotype obtained from an HRM marker depends on the difference in T_m values between two amplicons. Increasing the reliability of HRM marker genotypes could be archived with the selection of the best nearest neighboring nucleotide substitution (NNNs) in primer sequences surrounding SNPs. This chapter provides an easy-way protocol to design primer sequences for NNNs-HRM markers with table and web service, as well as several tips to develop HRM markers that distinguish between homozygous alleles (e.g., between A/A and C/C).

Key words Molecular marker, SNPs, High-resolution melting (HRM) analysis, PCR, Marker-assisted selection

1 Introduction

1.1 Plant Genotyping and the Role of the HRM Method

The genotype information obtained from a deoxyribonucleic acid (DNA) marker linking to a target gene (locus) is essential for marker-assisted selection in recent plant breeding programs [1]. The large number of sequence reads obtained from next-generation sequencing (NGS) technologies can decrease the cost for genotyping of individual and single NGS analysis and can obtain genotype data for massive number of loci at one time [2]. Especially, single-nucleotide polymorphism (SNP) is more prevalent compared to insertion- or deletion-type mutation. There are four categories of SNP types: class I (noncomplementary transitions, cytosine (C) to thymine (T), and guanine (G) to adenine (A)), class II (noncomplementary transversions, C to A and G to T), class III (C to G), and class IV (A to T). The class I and II types of SNPs are quite frequent in general plant species compared to class

Yuri Shavrukov (ed.), *Plant Genotyping: Methods and Protocols*, Methods in Molecular Biology, vol. 2638,
https://doi.org/10.1007/978-1-0716-3024-2_25,

III and class IV. In addition, NGS analysis targeting partial genomic regions with restriction enzymes (RAD-seq analysis) [3] and random primers (GRAS-Di) [4] decreases the cost of sequencing per sample. These techniques can reduce the total cost of time and labor for the genotyping of individual plants. Developing a linkage map covering whole-target genomes and finding quantitative trait loci (QTL) become easier with target sequencing. Moreover, the graphical genotype of each individual could be used for genomic selection and marker-assisted recurrent selection for the breeding of quantitative traits. Electrophoresis-based techniques, such as SSR (simple sequence repeat), CAPS (cleaved amplified polymorphic sequence or PCR-RFLP), and dCAPS (derived CAPS), are widely used in many laboratories for the detection of polymorphism between two alleles. However, genotyping a massive number of plants with few loci is time-consuming, and it should be improved when it comes to the aspects of time and cost. High-resolution melting (HRM) analysis based on the difference in melting temperature between two polymerase chain reaction (PCR) products having single-nucleotide polymorphism can save time during post-PCR experiments, including during electrophoresis, gel staining, and the recording of genotypes. We also proposed new techniques to increase the robustness of HRM markers with the nearest neighboring nucleotide substitution (NNNs) in primer sequences to increase the difference in melting temperatures between two PCR products, including SNP [5]. This chapter provides an easy guideline protocol for developing NNNs-HRM markers for readers, with simple charts and web services. Several tips are also provided for designing good primers for NNNs-HRM markers to increase discriminability between homozygous alleles.

1.2 The Principal Behind of NNNs-HRM Markers

A HRM marker can distinguish the SNP in PCR amplicons with the difference of melting temperature reflected in thermal dissociation from double strand DNA (dsDNA) to single strand DNA, with the decrease of the fluorescence dye binding dsDNA. The main parameters determining the thermal dynamics of dsDNA are the number of hydrogen bonds between complementary bases (A to T or G to C) and the stacking effects between neighboring nucleotides (NNs). The primers of an HRM marker are generally designed to amplify small PCR fragments (<100 bp), including SNPs (*see* Fig. 1a). The T_m values (melting temperature of dsDNA) are calculated from their sequence with NN thermal parameters [6], and a large difference in T_m values between two amplicons, including SNPs, is essential for precise genotyping. A previous study [5] showed that the sequence of three nucleotides surrounding SNP (two bases just before and after SNP) is one of the key factors for determining the difference in T_m values between two amplicons (*see* Fig. 1b). There is high correlation ($r = 0.65$) observed between the difference of T_m values and the difference of predicted Gibbs free

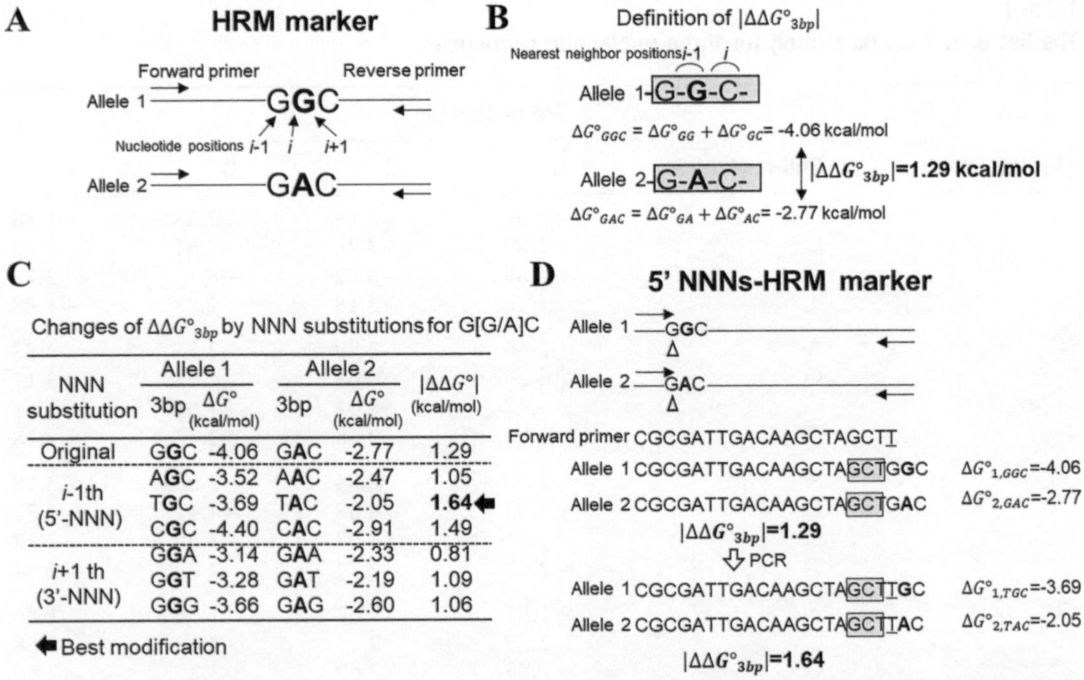

Fig. 1 Example calculation of $|\Delta\Delta G^{\circ}_{3bp}|$ and concept of NNNs-HRM marker. (**a**) Primer location for conventional HRM analysis. (**b**) ΔG°_{3bp} calculation with thermos dynamic parameters of nearest neighboring nucleotides and the definition of $|\Delta\Delta G^{\circ}_{3bp}|$. (**c**) A List of NNNs for sample SNP (B: G to A) indicates highest $|\Delta\Delta G^{\circ}_{3bp}|$ in case of T nucleotide substitution at 5′NNN. (**d**) Site directed mutagenesis in primer sequence maximize $|\Delta\Delta G^{\circ}_{3bp}|$ after PCR. This figure is reprint of previous study with the publisher permission [5]

energy of three nucleotides ($|\Delta\Delta G^{\circ}_{3bp}|$). Maximizing the difference of Tm values $|\Delta\Delta G^{\circ}_{3bp}|$ between two amplicons could be achieved with the suitable nucleotide substitution in the primer sequence neighboring target SNP (*see* Fig. 1c, d). We designated this type of HRM marker as NNNs (nearest neighboring nucleotide substitution)-HRM marker and showed that this technique can increase the robustness for discriminability of genotyping between homozygous alleles.

2 Methods

2.1 Suitable NNNs Selection

The point for designing an NNNs-HRM marker is to select a nucleotide sequence that shows the highest difference in T_m values between two PCR products harboring SNP (*see* Fig. 1c). Table 1 shows the ΔG°_{3bp} (kcal/mol) for all three nucleotide sequences. The value of $|\Delta\Delta G^{\circ}_{3bp}|$ can be calculated as described below.

Two nucleotides play very important roles in NNNs-HRM preparation. If "SNP−1 nucleotide (the first nucleotide prior the SNP)" and "SNP+1 nucleotide (the first nucleotide following after the SNP)" are the same "A," and the A/G site for SNP in the

Table 1
The list of ΔG°_{3bp} (kcal/mol) for three nucleotide sequence

		3rd nucleotide			
1st nucleotide	2nd nucleotide	A	C	G	T
A	A	−2.00[a]	−2.44	−2.28	−1.88
	C	−2.89	−3.28	−3.61	−2.72
	G	−2.58	−3.52	−3.12	−2.72
	T	−1.46	−2.18	−2.33	−1.88
C	A	−2.45	−2.89	−2.73	−2.33
	C	−3.29	−3.68	−4.01	−3.12
	G	−3.47	−4.41	−4.01	−3.61
	T	−1.86	−2.58	−2.73	−2.28
G	A	−2.30	−2.74	−2.58	−2.18
	C	−3.69	−4.08	−4.41	−3.52
	G	−3.14	−4.08	−3.68	−3.28
	T	−2.02	−2.74	−2.89	−2.44
T	A	−1.58	−2.02	−1.86	−1.46
	C	−2.75	−3.14	−3.47	−2.58
	G	−2.75	−3.69	−3.29	−2.89
	T	−1.58	−2.30	−2.45	−2.00

[a]Each score was calculated from the data of "Unified oligonucleotide DH° and DS° NN parameters [6]"

middle will result in either "AAA" or "AGA" sequences surrounding the SNP. The absolute difference in ΔG°_{3bp} for these sequences, calculated from Table 1, is 0.58 ($|(-2.00) - (-2.58)|$). Considering all the combinations for neighboring nucleotide substitution, six patterns of sequences could become candidates ("TAA" and "TGA," "AAT" and "AGT," etc.). In this case, the combination of "TAA" and "TGA," having nucleotide substitution from "A" to "T" at the position of the first nucleotide prior to SNP, showed the highest absolute difference in ΔG°_{3bp} (1.17). In the case of designed NNNs-HRM primers for SNPs categorized into class III and class IV with the condition of 50–100 bp PCR product size, the difference in the predicted T_m values is less than 0.5°. Moreover, based on our experimental observation, a difference of T_m value over 0.5° of HRM marker guaranteed accurate genotyping, for example using Roche LightCycler 96 system or other similar PCR amplificators. With this restriction, most SNPs grouped into class III and class IV are not suitable for the target of HRM marker [5].

The procedure to choose suitable NNNs, can made automatically with any computer programs, such as "Perl" or "R environment." Perl script sample is also provided to pick up and design primer pairs in previous study [5] (*see* **Note 1**). However, a simple way that anyone can use it, would be helpful. In the presented protocol, easy way is provided to design primer pairs for SNPs

Table 2
Table for the selection of best suitable nucleotide substitution surrounding class I and II SNP

one SNP nucleotide	Another SNP nucleotide	SNP−1 nucleotide	SNP+1 nucleotide A	C	G	T
A	C	A	GAA_5′ NNNs	GAC_5′ NNNs	GAG_5′ NNNs	GAT_5′ NNNs
		C	GAA_5′ NNNs	GAC_5′ NNNs	GAG_5′ NNNs	GAT_5′ NNNs
		G	GAG_3′ NNNs	GAG_3′ NNNs	GAG_NS	GAG_3′ NNNs
		T	TAG_3′ NNNs	TAG_3′ NNNs	GAG_5′ NNNs	TAG_3′ NNNs
	G	A	TAA_5′ NNNs	TAC_5′ NNNs	TAG_5′ NNNs	TAT_5′ NNNs
		C	CAC_3′ NNNs	TAC_5′ NNNs	CAC_3′ NNNs	CAC_3′ NNNs
		G	GAC_3′ NNNs	TAC_5′ NNNs	TAG_5′ NNNs	TAT_5′ NNNs
		T	TAC_3′ NNNs	TAC_NS	TAC_3′ NNNs	TAC_3′ NNNs
C	A	A	GCA_5′ NNNs	GCC_5′ NNNs	GCG_5′ NNNs	GCT_5′ NNNs
		C	GCA_5′ NNNs	GCC_5′ NNNs	GCG_5′ NNNs	GCT_5′ NNNs
		G	GCG_3′ NNNs	GCG_3′ NNNs	GCG_NS	GCG_3′ NNNs
		T	TCG_3′ NNNs	TCG_3′ NNNs	GCG_5′ NNNs	TCG_3′ NNNs
	T	A	GCA_5′ NNNs	ACA_3′ NNNs	GCG_5′ NNNs	ACA_3′ NNNs
		C	GCA_5′ NNNs	CCA_3′ NNNs	GCG_5′ NNNs	CCA_3′ NNNs
		G	GCA_NS	GCA_3′ NNNs	GCA_3′ NNNs	GCA_3′ NNNs
		T	GCA_5′ NNNs	GCC_5′ NNNs	GCG_5′ NNNs	TCA_3′ NNNs
G	A	A	TGA_5′ NNNs	TGC_5′ NNNs	TGG_5′ NNNs	TGT_5′ NNNs
		C	CGC_3′ NNNs	TGC_5′ NNNs	CGC_3′ NNNs	CGC_3′ NNNs
		G	GGC_3′ NNNs	TGC_5′ NNNs	TGG_5′ NNNs	TGT_5′ NNNs
		T	TGC_3′ NNNs	TGC_NS	TGC_3′ NNNs	TGC_3′ NNNs
	T	A	CGA_5′ NNNs	CGC_5′ NNNs	AGC_3′ NNNs	AGC_3′ NNNs
		C	CGC_3′ NNNs	CGC_NS	CGC_3′ NNNs	CGC_3′ NNNs
		G	CGA_5′ NNNs	CGC_5′ NNNs	GGC_3′ NNNs	GGC_3′ NNNs
		T	CGA_5′ NNNs	CGC_5′ NNNs	TGC_3′ NNNs	TGC_3′ NNNs
T	C	A				

(continued)

Table 2
(continued)

one SNP nucleotide	Another SNP nucleotide	SNP−1 nucleotide	SNP+1 nucleotide A	C	G	T
			GTA_5′ NNNs	ATA_3′ NNNs	GTG_5′ NNNs	ATA_3′ NNNs
		C	GTA_5′ NNNs	CTA_3′ NNNs	GTG_5′ NNNs	CTA_3′ NNNs
		G	GTA_NS	GTA_3′ NNNs	GTA_3′ NNNs	GTA_3′ NNNs
		T	GTA_5′ NNNs	GTC_5′ NNNs	GTG_5′ NNNs	TTA_3′ NNNs
	G	A	CTA_5′ NNNs	CTC_5′ NNNs	ATC_3′ NNNs	ATC_3′ NNNs
		C	CTC_3′ NNNs	CTC_NS	CTC_3′ NNNs	CTC_3′ NNNs
		G	CTA_5′ NNNs	CTC_5′ NNNs	GTC_3′ NNNs	GTC_3′ NNNs
		T	CTA_5′ NNNs	CTC_5′ NNNs	TTC_3′ NNNs	TTC_3′ NNNs

Note: For example, one allele is represented by "TAA," and another allele has a "TCA" sequence. "one SNP nucleotide" is "A" and "another SNP nucleotide" is "C," then "SNP−1 nucleotide" is "T" and "SNP+1 nucleotide" is "A" that shows TAG_3′NNNs. The 3′NNN fragment indicates the substitution of the first nucleotide following SNP. In this case, a nucleotide "A" of SNP+1 position is recommended to be changed to the nucleotide "G". The 5′NNN in other cells indicates the first nucleotide substitution prior to SNP. NS (no substitution) indicates that substitution of the original sequence is unnecessary

grouped into class I and II. Table 2 display a simple chart to select suitable nucleotide sequence for the highest $|\Delta\Delta G^{\circ}_{3bp}|$.

In another example, two individuals have "C/A" SNP. "SNP−1 nucleotide" and "SNP+1 nucleotide" are "T" and "A," respectively. In this case, one allele is represented by "TCA," and another allele has a "TAA" sequence. In Table 2 "one SNP nucleotide" is "C" and "Another SNP nucleotide" is "A," then "SNP−1 nucleotide" is "T" and "SNP+1 nucleotide" is "A" that shows TCG_3′ NNNs in table. The 3′NNN fragment indicates the substitution of the first nucleotide following SNP. The word 5′NNN indicates the first nucleotide substitution prior to SNP. In this case, the sequence of "TCA" is better to be changed to "TCG" with substitution from "A" to "G" at "SNP+1 nucleotide" site. NS (no substitution) indicates that substitution of the original sequence is unnecessary. The range of $|\Delta\Delta G^{\circ}_{3bp}|$ scores calculated from the figures in Table 1 are from 1.17 to 1.83, and the average score is 1.57. All nucleotide sequences shown in Table 2 guarantee large differences in T_m values for PCR amplicons. The procedure shown here is a very easy way to select suitable substitution sequences to maximize the difference in T_m values between two PCR products, including SNPs. The copy and paste of recommended 3 nucleotide sequence on to actual sequence is ready for next primer design step.

2.2 The Design of Primers for NNNs-HRM Markers

The "Primer3" is generally used to design primer pairs or probes for target sequences [7]. The stand-alone Primer3 program could also be available if wanting to design primers on your own PC. Some tips for using stand-alone Primer3 are discussed in **Note 1.** To design the NNNs-HRM marker, one primer needs to contain nucleotide substitution compared to the original sequence surrounding SNP. Therefore, the positions of primer end have to be fixed at the nucleotide just before or after a target SNP. If Primer3 site (https://bioinfo.ut.ee/primer3-0.4.0/) is used with retrieved sequence surrounding SNP (approximately 200 bp sequence that SNP are located at the position of center), paste this 200 bp sequence harboring replaced 3 nucleotide (suitable NNNs) in the "input box." Then paste the 20–25 bp of primer sequence that is just before SNP including NNNs in the "Pick left primer, or use left primer below" box. The option of "Product Size Ranges" box is set with "50–100" range (*see* **Note 2**). Then click the "Pick Primer" bottom to design primer at opposite side. When 3′NNNs substitution is used to design primer, complementary sequence is pasted in the "Pick right primer, or use right primer below (5′–3′ on opposite strand)" box. We also provide web service, "NNNs-Finder (https://nnns-hrm.ag.saga-u.ac.jp)" to pick up best NNNs and prepare the sequence for designing primer.

The designed primer sequences using "NNNS-Finder" and "Primer3" can be used for the synthesis of primers. It is highly recommended to check the redundancy of primer sequences for the target genome. Basic local alignment search tool (BLAST) search is recommended in case with reference genome sequence for target species is available. Standalone program of BLAST + [8] also can conduct local BLAST search. The sequences that show unique in their genome are recommended for the ordering of primers.

2.3 PCR Conditions

The PCR conditions for NNNs-HRM are almost the same as for the general PCR experiment except for adding a fluorescent dye, such as EvaGreen Dye (Biotium Inc.). The concentration of DNA suitable for PCR depends on the genome size of each species. Soybean, rice, citrus, and olive DNA can be used at a concentration of 10–20 ng per reaction (*see* **Note 3**). The amount of forward and reverse primers is 1 pmol each, and the reaction volume of PCR is 10 μL. Hot-start-activated *Taq* DNA polymerase is recommended to avoid unspecific amplification (*see* **Note 4**). Forty cycles, including denaturation (95 °C for 15 s), annealing (58–60 °C for 15 s), and extension (72 °C for 20 s), are enough for amplification. The conditions for melting after amplification are as follows: 95 °C for 10 s, 65 °C for 1 min, and then increasing temperatures to 97 °C with a slope of 0.1 °C/s with continuous measurement of the fluorescence of the PCR product. Software is provided by the manufacturer for the postrun analysis of HRM data (*see* **Note 5**).

2.4 Examples of HRM Pattern

Several studies have been conducted using NNNs-HRM markers to detect polymorphism between Japanese soybean cultivars. Figure 2 displays an ideal pattern with normal amplification with an easily distinguishable melting peak. Sigmoid amplification pattern appeared around 20–22 cycles, and the fluorescent intensity reached a plateau level within 40 cycles. The normalized melting peak of homozygotes (T/T) and (C/C) showed a single peak isolated of each other. Meanwhile, the heterozygotes (T/C) showed a bimodal pattern in the graph of "normalized melting

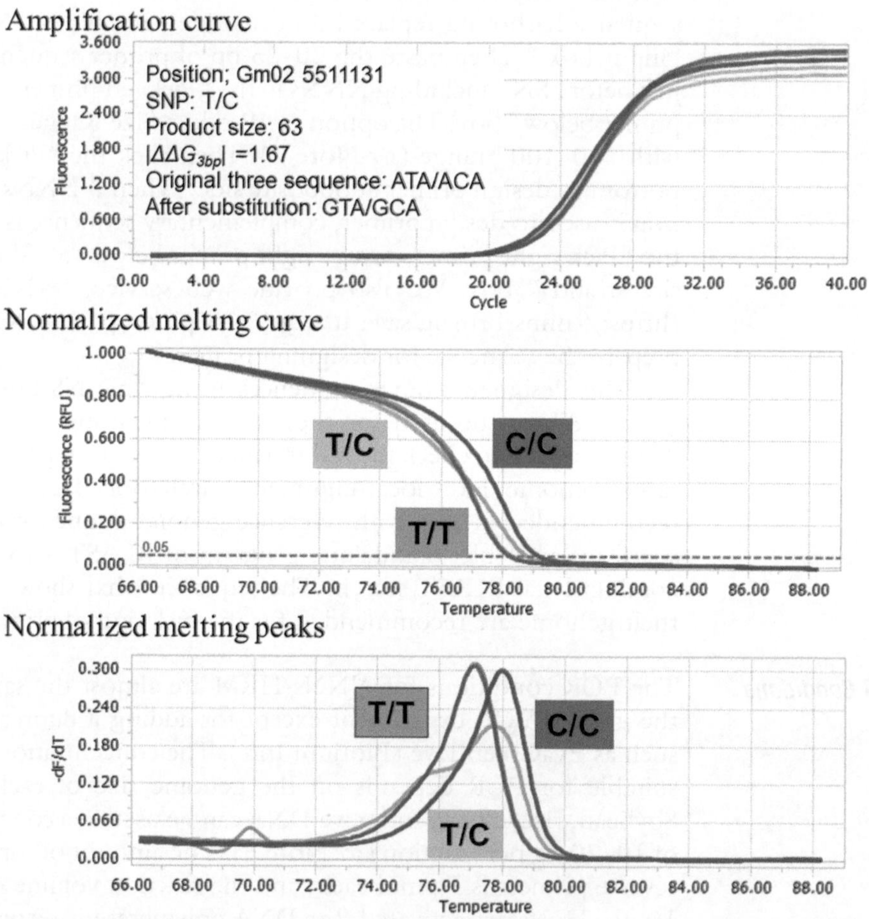

Fig. 2 The ideal results of NNNs-HRM marker genotyping. Three panels represent analytical results as follows. "Amplification curve," "Normalized melting curve," and "Normalized melting peaks" were produced and displayed with the supplementary software to LightCycler 96 system (Roche). The blue, red, and orange lines indicate the three types of genotypes, homozygotes for alleles 1 and 2, and heterozygotes, respectively. The detailed information about NNNs-HRM marker, position (chromosome name and physical position), nucleotide of SNP, product size of amplicon, $|\Delta\Delta G^{\circ}_{3bp}|$ after substitution, original and suitable NNNs sequence, is displayed in "Amplification curve" panel

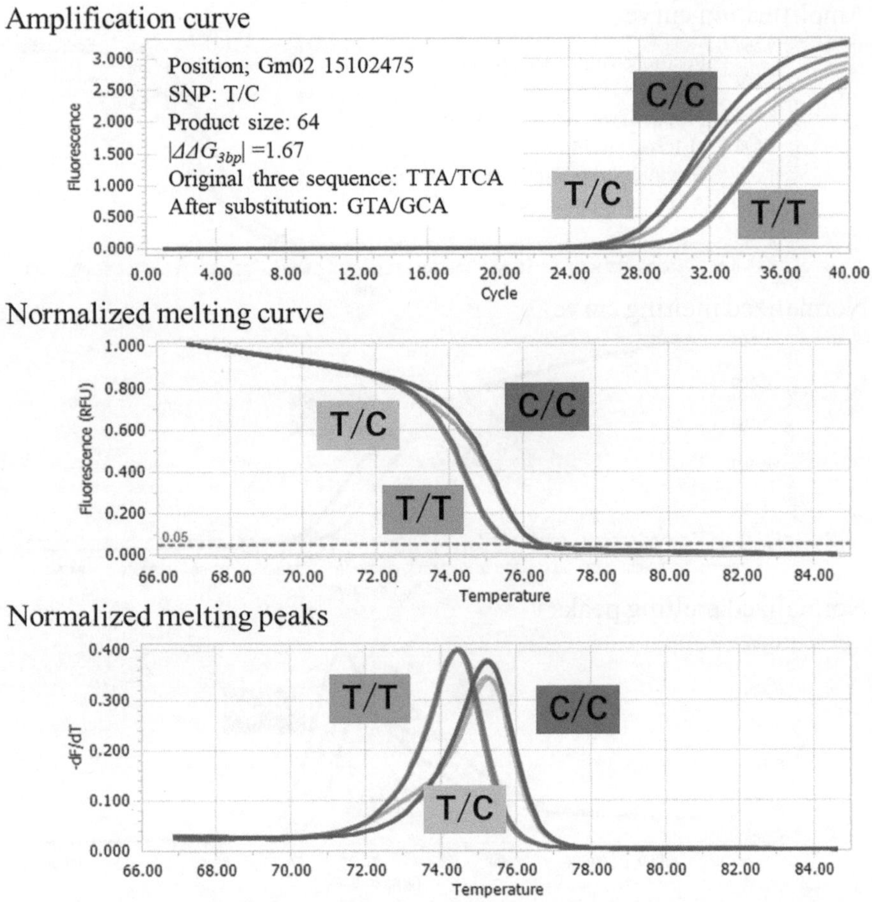

Fig. 3 The result of NNNs-HRM analysis showing the difference in amplification efficiency among genotypes with different alleles. The other explanations are the same as in Fig. 2

peaks" with earlier decreasing of fluorescent in the "normalized melting curves." These features indicate normal amplification of a single product in each PCR.

Figure 3 showed the difference in amplification curve among alleles, and the T/T homozygote showed lower amplification efficiency than the C/C and T/C genotypes. The high amplification efficiency is observed in the order of C/C > T/C > T/T, and the melting peak of the heterozygous allele (T/C) is similar to that of the C/C homozygous allele. This pattern indicates the lack of uniformity between two amplicons; one amplicon has C and another has T, and one cultivar probably has an extra polymorphism in primer annealing sequences. In such cases, the heterozygotes genotype has the possibility of genotyping error.

Figure 4 showed the normal amplification, but the temperature range of normalized melting peaks was expanding widely compared to the ideal one (*see* Fig. 2). This expanding peak indicates the

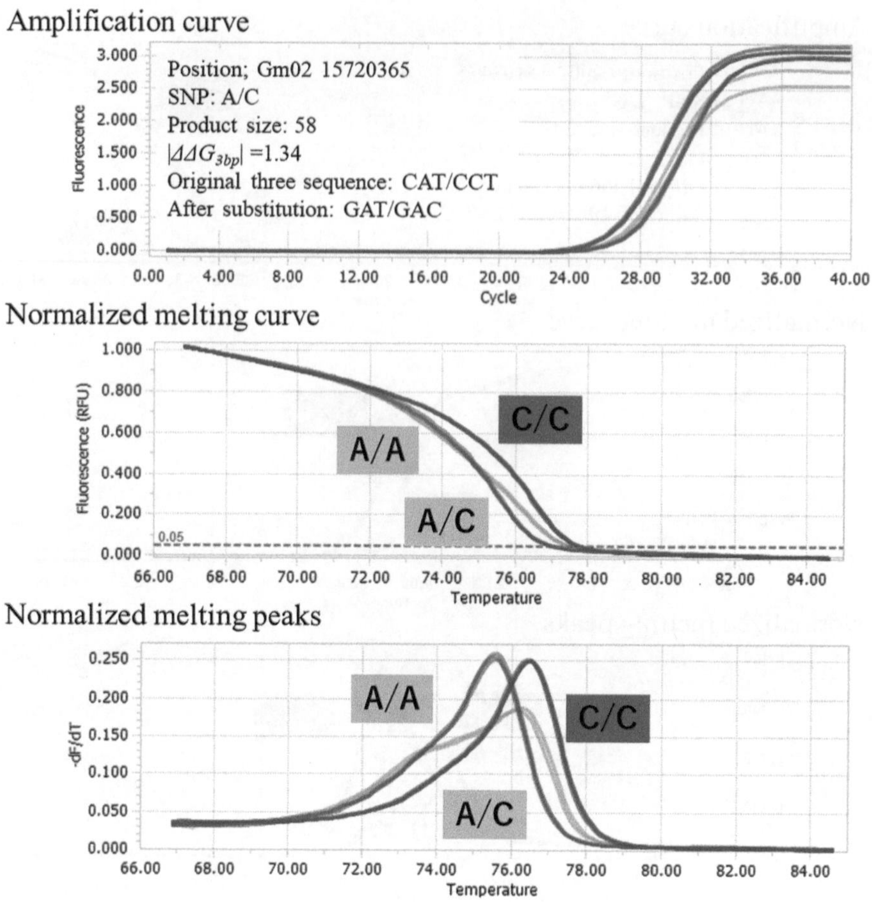

Fig. 4 The result of NNNs-HRM marker showing unspecific amplification in their PCR products. The other explanations are the same as in Fig. 2

unspecific amplification in PCR products. When clear difference is observed in the melting peaks among genotypes, there is no matter for possibility of genotype error in this case.

2.5 Conclusion

HRM analysis without post-PCR experiments has more advantages compared to other gel-based polymorphic DNA band-detecting techniques. As shown in this chapter, NNNs-HRM marker method can increase the robustness for the discrimination of SNP genotypes. The helpful application of such an HRM marker system for mapping experiments and the analysis of genetic diversity among germplasms [5, 9, 10] has been already confirmed in rice (*Oryza sativa*), soybean (*Glycine max*), citrus (*Citrus*), and olive (*Olea europaea*). The combination of NNNs-HRM markers and bulk-segregant analysis [11] or the selective genotyping approach [12] facilitates the identification of locus-related agronomic traits in a segregated population at lesser costs and time [13]. The reducing the cost for genotyping with HRM markers would be also valuable for general breeding programs to screen massive number of plants with marker-assisted selection.

3 Notes

1. The example of Perl scripts to design NNNs-HRM primers are displayed in previous study [5]. Some Perl scripts, however, do not work because of update of programs used in this script. Therefore, the key steps are briefly pointed out below. The original Variant Call Format (VCF) files generally contain the information for SNP with the locations (chromosome and physical position), target nucleotide of reference genome and alternative nucleotide detected in analyzed samples, coverages, and additional statistical parameters for sample genotypes. The sequence surrounding SNP sites have to be obtained to design NNNs-HRM primer. There are many techniques to retrieve the target sequences from a database made from FASTA format file containing genome sequences. BLAST+ provides command-line-based programs for the retrieval of target sequences, including SNPs [8]. Before using BLAST+, a reader needs to install this program according to the instruction of NCBI website (https://blast.ncbi.nlm.nih.gov/Blast.cgi). First step is setting up the genome database with "make-blastdb.exe." After making database, the user can freely retrieve sequence like with "blastdbcmd." The 100 bp upward and downward sequence for target SNP is sufficient to design primer for HRM marker. After obtaining the sequence including SNP, "VCF_to_HRMprimer_NNNsHRM.pl" with standalone Primer3 can select suitable NNNs and then design primers sequence as shown in previous study [5]. The input file for Primer3 can set the option of "SEQUENCE_FORCE_LEFT_END" or "SEQUENCE_FORCE_RIGHT_END" that can fix the position of the end of primer at the SNP neighboring site.
2. The effect of single nucleotide change in prime sequences for T_m value is inversely proportional to the size of PCR amplicon. The size of PCR amplicon <80 bp is recommended to guarantee the over 0.5-degree difference in T_m values. Meanwhile, short PCR product size, ex. 50 bp product size, does not affect amplification efficiency and melting pattern from the empirical observation and experience.
3. The high-quality DNA can provide clear and reproducible results for genotyping. In the case of soybean, DNA extracted from leaf using the general cetyltrimethylammonium bromide (CTAB) method is sufficient to obtain a clear melting pattern for HRM analysis. The DNA extracted from seed powder,

sometime, show inconstant shape for melting pattern among same genotype group. Impurities (protein and other substrates) in the DNA solution could affect the amount of PCR products and the shape of the melting pattern.

4. It is highly recommended to use the *Taq* DNA polymerase without proofreading activity (3′–5′ exonuclease activity) for NNNs-HRM analysis. There is a possibility that mismatch at primer end site would be removed by proofreading activity. Antibody or other modification for "hot-start" to avoid unspecific reaction during preparation of mixture is also recommended.
5. There are several ways to evaluate the fluorescent reduction of PCR product for genotyping with melting curves. The comparison of "normalized melting peak" is better than "relative difference plot in fluorescence" or "normalized melting curves" to judge the genotype of individual.

Acknowledgments

We are grateful to Tokiko Kitajima (Saga University) for technical support and for showing the HRM data examples.

References

1. Collard BCY, Mackill DJ (2008) Marker-assisted selection: an approach for precision plant breeding in the twenty-first century. Philos Trans R Soc Lond Ser B Biol Sci 363: 557–572. https://doi.org/10.1098/rstb.2007.2170
2. Jaganathan D, Bohra A, Thudi M, Varshney RK (2020) Fine mapping and gene cloning in the post-NGS era: advances and prospects. Theor Appl Genet 133:1791–1810. https://doi.org/10.1007/s00122-020-03560-w
3. Baird NA, Etter PD, Atwood TS, Currey MC, Shiver AL, Lewis ZA et al (2008) Rapid SNP discovery and genetic mapping using sequenced RAD markers. PLoS One 3:1–7. https://doi.org/10.1371/journal.pone.0003376
4. Miki Y, Yoshida K, Enoki H, Komura S, Suzuki K, Inamori M et al (2020) GRAS-Di system facilitates high-density genetic map construction and QTL identification in recombinant inbred lines of the wheat progenitor *Aegilops tauschii*. Sci Rep 10:21455. https://doi.org/10.1038/s41598-020-78589-4
5. Yamagata Y, Yoshimura A, Anai T, Watanabe S (2018) Selection criteria for SNP loci to maximize robustness of high-resolution melting analysis for plant breeding. Breed Sci 68:488–498. https://doi.org/10.1270/jsbbs.18048
6. SantaLucia J (1998) A unified view of polymer, dumbbell, and oligonucleotide DNA nearest-neighbor thermodynamics. Proc Natl Acad Sci U S A 95:1460–1465. https://doi.org/10.1073/pnas.95.4.146
7. Untergasser A, Cutcutache I, Koressaar T, Ye J, Faircloth BC, Remm M et al (2012) Primer3—new capabilities and interfaces. Nucleic Acids Res 40:e115. https://doi.org/10.1093/nar/gks596
8. Camacho C, Coulouris G, Avagyan V, Ma N, Papadopoulos J, Bealer K et al (2009) BLAST+: architecture and applications. BMC Bioinformatics 10:421. https://doi.org/10.1186/1471-2105-10-421
9. Watanabe S, Yamada R, Kanetake H, Kaga A, Anai T (2019) Identification and characterization of a major QTL underlying soybean isoflavone malonylglycitin content. Breed Sci 69:

564–572. https://doi.org/10.1270/jsbbs.19027

10. Matsumoto Y, Kubo T, Itami Y, Islam MZ, Watanabe S, Kotoda N (2022) QTL mapping of polymethoxyflavone (PMF) accumulation in citrus. Tree Genet Genomes 18:1–13. https://doi.org/10.1007/s11295-022-01539-w
11. Quarrie SA, Lazic-Jancic V, Kovacevic D, Steed A, Pekic S (1999) Bulk segregant analysis with molecular markers and its use for improving drought resistance in maize. J Exp Bot 50: 1299–1306. https://doi.org/10.1093/jxb/50.337.1299
12. Darvasi A (1997) The effect of selective genotyping on QTL mapping accuracy. Mamm Genome 8:67–68
13. Sarkar MAR, Otsu W, Suzuki A, Hashimoto F, Anai T, Watanabe S (2020) Single-base deletion in *GmCHR5* increases the genistein-to-daidzein ratio in soybean seed. Breed Sci 70: 265–276. https://doi.org/10.1270/jsbbs.19134

Chapter 26

A New SNP Genotyping Technology by Target SNP-Seq

Jian Zhang, Jingjing Yang, and Changlong Wen

Abstract

In order to promote the widespread application of single-nucleotide polymorphism (SNP)-based genotyping, a new method was developed and called target SNP-seq which combined the advantages of multiplex polymerase chain reaction (PCR) amplification and high throughput sequencing on Illumina X Ten platform. Compared with kompetitive allele-specific PCR (KASP), microchips, and genotyping by sequencing (GBS), target SNP-seq uses perfect SNPs based on the analysis of variome (whole-genome sequence data of different accessions) and is flexible, cost-effective, and highly accurate for genotyping middle-scale SNPs. It could genotype hundreds of SNPs in massive DNA samples within 3 days at the cost of $7 for each DNA sample. The high efficiency and low cost of target SNP-seq make it more competitive than current SNP genotyping methods, and it has excellent potential for application in genetic research, as well as in promoting plant-breeding processes in the near future.

Key words Perfect SNP, Multiplex PCR, Target SNP-seq, High-throughput sequencing

1 Introduction

Single-nucleotide polymorphisms (SNPs) are widely used for genetic mapping, in evolutionary studies and genome-wide association studies, for detecting genetic diversity, and as genetic markers in marker-assisted selection. They are the most abundant genetic variations in plant genomes (several SNPs can occur per 1000 nucleotides). SNPs are highly related to phenotypic variation and disease resistance as they may alter protein structure and function, influence the binding affinity of transcription factors, modify alternative splicing, and regulate noncoding ribonucleic acid (RNA).

Several traditional SNP genotyping methods based on electrophoresis systems, such as CAPS (cleaved amplified polymorphic sequence), dCAPs (derived CAPS), and AS-PCR (allele-specific PCR), are commonly used in genetic research. It is difficult to meet the rapidly increasing need for SNP genotyping. Recently, the continuous progress in high-throughput sequencing has facilitated the development of SNP genotyping methods, such as the

Yuri Shavrukov (ed.), *Plant Genotyping: Methods and Protocols*, Methods in Molecular Biology, vol. 2638,
https://doi.org/10.1007/978-1-0716-3024-2_26,

gene chip microarray or the KASP (kompetitive allele-specific PCR) platform. Chip-based genotyping platforms are suitable for genotyping hundreds of samples with thousands of SNPs, while the KASP platform is appropriate for thousands of samples with a few SNPs [1]. Moreover, the huge cost of one-time investment in instruments or expensive microarrays also limits the use of SNPs in most laboratories. Therefore, it is urgent to discover a new technology that is cost-effective and highly accurate for genotyping middle-scale genome-wide SNPs.

At present, whole-genome sequences of hundreds of species have been assembled, which has promoted large-scale resequencing and produced millions of SNPs for genetic research [2]. Next-generation sequencing (NGS) technologies have enabled genome resequencing for exploring genome-wide polymorphisms among individuals, as well as targeted resequencing for the rapid and simultaneous detection of polymorphisms in genes associated with various biological functions [3]. However, a certain number of SNPs that are not detected by PCR or validated experimentally because of false SNPs due to sequencing errors and misalignments, high variation in the SNP flanking sequence, the location in the repeat regions of the genome, and high or low GC content, which may cause false positive or false negative results in SNP genotyping. To the best of our knowledge, a few studies have focused on developing perfect SNP loci, which harbor conserved flanking regions and have unique PCR amplification in the genome. Herein, we developed a new method for SNP genotyping which called Target SNP-seq. It combines the advantages of high-throughput sequencing and multiplex PCR amplification by using genome-wide perfect SNPs and meets the requirements of harboring stable flanking sequences and being captured uniquely in PCR amplification [4]. Furthermore, this technology could gain hundreds of SNP genotypes by sequencing the SNP region with an average depth of 1000 × in a short time and at low cost. Target SNP-seq has excellent potential for application in genetic research and is already used for the identification of cucumber, eggplant, pepper, and watermelon varieties [4–7].

2 Materials

2.1 Plasticware and Consumables

1. 96 deep well plate (1.5 mL).
2. 1.5 mL screw-capped sterile conical plastic tube.
3. PCR plates, 96-well.
4. Sterile deionized water.
5. 1× Tris–EDTA (TE): 10 mM Tris–HCl, pH 7.5; 1 mM ethylenediaminetetraacetic acid (EDTA), sterile.
6. 100% absolute ethanol.

7. 70% ethanol (freshly made).
8. 1 mM EDTA.
9. Magnetic bead suspension liquid (AMPure).

2.2 Enzymes and Kits

1. DNA magnetic bead kit (LGC, England).
2. GenoPlexs 3 × M enzyme (MolBreeding Biotechnology).
3. Qubit®2.0 dsDNA HS assay kit (Thermo Fisher Scientific).
4. Primer panel mix (MolBreeding Biotechnology).

2.3 Equipment

1. Top-bench centrifuge.
2. Veriti 96 PCR amplifier (Applied Biosystems).
3. Microspectrophotometer, like NanoDrop 2000.
4. DNA sequencer (Illumina X Ten platform).

3 Methods

3.1 Discovery of Genome-Wide Perfect SNPs for Target SNP-seq

The genome sequencing data or genotype VCF (Variant Call Format) files of various accessions could be downloaded from the NCBI SRA (Sequence Read Archive). Adaptors and low-quality sequences in the raw data were filtered using Trimmomatic. Clean reads were mapped to the reference genome using Burrows–Wheeler Alignment (v0.7.15), and raw variant calling was performed using the Haplotype Caller module in the Genome Analysis Toolkit (v3.8). SNPs and short insertions and deletions (InDels) were separated from the raw variant data and then filtered into high-quality (HQ) SNPs and InDels using the variant filtration module of the Genome Analysis Toolkit (GAT)K. The genome-wide perfect SNPs were obtained as follows. (1) HQSNPs were filtered into candidate SNPs based on the following parameters: MAF > 0.05, missing rate <0.1, and heterozygosity <0.1. (2) HQ SNPs and InDels were used as constraints during filtration to isolate SNPs with no variation in the 30 bp flanking sequence of the candidate SNPs, which formed the conserved SNP data set. (3) The 30 bp SNP flanking sequence was extracted and was aligned to the reference genome using the Basic Local Alignment Search Tool (BLAST) (*E* value = 1e-15). SNPs with a unique flanking sequence alignment to the genome were retained, forming a unique SNP data set. Finally, genome wide perfect SNP loci were collected and detected by multiple polymerase chain reaction (multiplex PCR) technique (*see* Fig. 1).

3.2 DNA Extraction and Quality Testing

Total genomic DNA was isolated from freeze-dried tissue stored in 96-well plates (*see* **Note 1**) using a plant DNA magnetic bead kit (LGC Genomics) according to the manufacturer's protocol. The

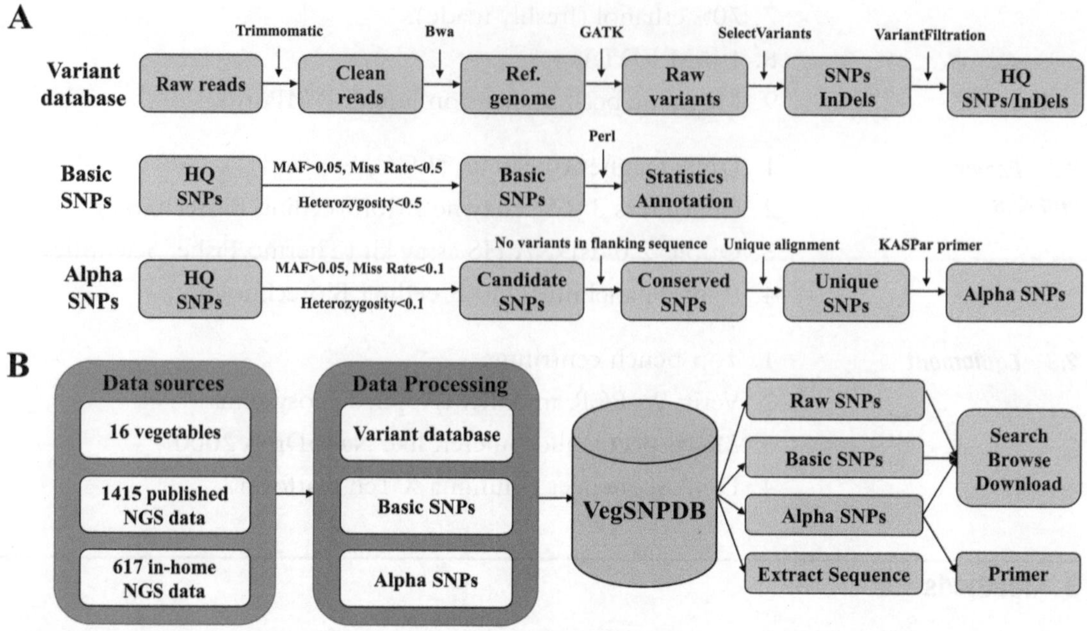

Fig. 1 Methodology for data processing to obtain the variant database (**a**) and perfect SNPs (**b**)

concentration of all extracted DNA was quantified using a Qubit2.0 fluorometer with a Quant-iT dsDNA HS assay kit (Thermo Fisher Scientific) before the targeted SNP sequencing. DNA should meet the following requirements: A260/A280 > 1.8, A260/A230 > 1.8, DNA concentration >50 ng/μL.

3.3 Construction of Target SNP-seq Library

The library construction of target SNP-seq consisted of two rounds of PCR (*see* Fig. 2). The first round of PCR was to capture the target SNP's locus and 30 bp flanking regions in DNA samples through multiplex PCR. The second round of PCR aimed to distinguish each DNA sample by adding a unique barcode adaptor.

The first round of PCR is recommended to be performed in a total volume of 30 μL mixture containing 50 ng DNA, 8 μL SNP primer mix (0.2 μmol/L), and 10 μL 3M enzymes (*see* **Note 2**) from MolBreeding Biotechnology Company, China, or another similar supplier. The thermal cycling regime should be as follows: 95 °C for 5 min, followed by 17 cycles at 95 °C for 30 s and annealing at 60 °C for 4 min, and final extension at 72 °C for 4 min (*see* **Note 3**). Then the PCR products are collected through magnetic bead suspension and purified by 80% (*v*/*v*) ethanol.

For the second round of PCR, the 30 μL PCR mixture consisted of 11 μL purified PCR product from the first round of PCR, 10 μL Taq enzyme, 18 μL pure water, and 1 μL barcode adaptors. The adaptors are composed of two components: (1) Forward

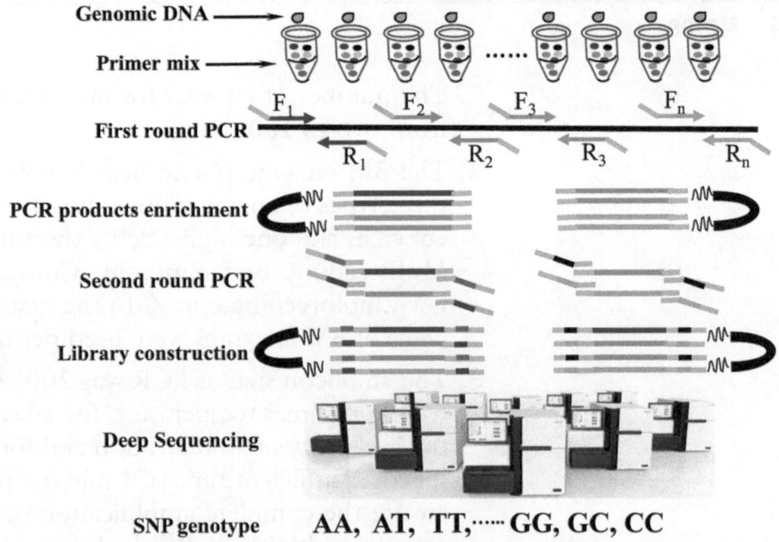

Fig. 2 The procedure for target SNP-seq genotyping

5′-AATGATACGGCGACCA-CCGAGATCTACACTCTTTCCC-TACACGACGCTCTTCCG-3′ and (2) Reverse 5′-CAAGCAGA AGACGGCATACGAGATXXXXXXXXGTGACTGGAGTTCCTT GGCACCCGAGA-3′ (the underlined eight-base sequences indicate barcodes). PCR was run at 95 °C for 3 min, followed by seven cycles of 95 °C for 15 s, 58 °C for 15 s, 72 °C for 30 s, and the final extension at 72 °C for 4 min (*see* **Note 4**). The products from the second round of PCR have to be enriched by magnetic beads, along with washing three times with 100 μL 80% (*v/v*) ethanol and 23 μL of Tris–HCl buffer (10 mM, pH 8.0–8.5). Hereafter, the DNA library should be sequenced on an Illumina X Ten platform in any suitable sequencing provider, for example, MolBreeding Biotechnology Company (Shijiazhuang, China).

3.4 Target SNP Genotype Calling

Sequencing reads from all DNA samples were aligned to the reference genome by using BWA (Burrows–Wheeler Alignment Tool) to determine the physical location of each target SNP amplicon. SNP genotypes were called by GATK (the Genome Analysis Toolkit). In our experiments, SNP base with higher frequency of sequence reads at a SNP locus were taken as the major allele. To ensure the accuracy of SNP genotypes, sequencing data were filtered with major alleles being covered by less than 20 sequence depth in a DNA sample. For heterozygous DNA sample, a ratio of major and minor alleles under 0.7 was treated as a heterozygous genotype (*see* **Note 5**). Finally, specific SNP variant bases for each variety were identified by analyzing the in-house barcodes assigned raw sequence reads.

4 Notes

1. The number of samples for target SNP-seq is flexible, ranging from two to 1000 samples.
2. The 3M enzyme (GenoPlexs 3 × M enzyme) is a mixture of three types of enzymes, including two mutant Taq-polymerase enzymes and one high-fidelity thermostable DNA polymerase. MolBreeding Biotechnology Company's website is http://www.molbreeding.com/. In the first round of PCR, two units (2U) of 3M enzymes were used per reaction.
3. The amplicon size of PCR was 200–280 bp. Therefore, 60 °C was an optimal temperature for enzymes, both for increasing the specificity of amplification and for stabilizing the extension speed. A sufficient time of 4 min for annealing was necessary to ensure the complete amplification of different DNAs and primers in multiplex PCR, which could warrant the uniformity of amplicons.
4. The thermal cycling regime was as follows: 95 °C for 5 min, followed by 17 cycles (95 °C for 30 s, 60 °C for 4 min) and the final extension at 72 °C for 4 min. So the duration of one single cycle takes 4.5 min.
5. The ratio of two major alleles was calculated to determine the heterozygous genotype of one SNP.

Acknowledgments

Authors would like to thank Jianan Zhang (MolBreeding Biotechnology Co., Ltd., Shijiazhuang, China) for their valuable comments and corrections to this protocol.

References

1. Semagn K, Babu R, Hearne S, Olsen M (2014) Single nucleotide polymorphism genotyping using Kompetitive Allele Specific PCR (KASP): overview of the technology and its application in crop improvement. Mol Breed 33:1–14. https://doi.org/10.1007/s11032-013-9917-x
2. Yang J, Zhang J, Du H, Zhao H, Li H, Xu Y et al (2022) The vegetable SNP database: an integrated resource for plant breeders and scientists. Genomics 114:110348. https://doi.org/10.1016/j.ygeno.2022.110348
3. Onda Y, Takahagi K, Shimizu M, Inoue K, Mochida K (2018) Multiplex PCR targeted amplicon sequencing (MTA-Seq): simple, flexible, and versatile SNP genotyping by highly multiplexed PCR amplicon sequencing. Front Plant Sci 9:201. https://doi.org/10.3389/fpls.2018.00201
4. Zhang J, Yang J, Zhang L, Luo J, Zhao H, Zhang J et al (2020) A new SNP genotyping technology target SNP-seq and its application in genetic analysis of cucumber varieties. Sci Rep 10:5623. https://doi.org/10.1038/s41598-020-62518-6
5. Liu W, Qian Z, Zhang J, Yang J, Wu M, Barchi L et al (2019) Impact of fruit shape selection on genetic structure and diversity uncovered from genome-wide perfect SNPs genotyping in eggplant. Mol Breed 39:140. https://doi.org/10.1007/s11032-019-1051-y

6. Du H, Yang J, Chen B, Zhang X, Zhang J, Yang K et al (2019) Target sequencing reveals genetic diversity, population structure, core-SNP markers, and fruit shape-associated loci in pepper varieties. BMC Plant Biol 19:578. https://doi.org/10.1186/s12870-019-2122-2
7. Yang J, Zhang J, Du H, Zhao H, Mao A, Zhang X et al (2021) Genetic relationship and pedigree of Chinese watermelon varieties based on diversity of perfect SNPs. Hortic Plant J 8:489–498. https://doi.org/10.1016/j.hpj.2021.09.004

Chapter 27

Derived Polymorphic Amplified Cleaved Sequence (dPACS) Assay

Shiv Shankhar Kaundun, Sarah-Jane Hutchings, Joe Downes, and Ken Baker

Abstract

The derived polymorphic amplified cleaved sequence (dPACS) assay is a simple polymerase chain reaction/restriction fragment length polymorphism (PCR-RFLP)-based procedure for detecting known single-nucleotide polymorphisms (SNPs) and deletion–insertion polymorphisms (DIPs). It is relatively straightforward to carry out using basic and commonly available molecular biology kits. The method differs from other PCR-RFLP assays in that it employs 35–55 bp primer pairs that encompass the entire targeted DNA region except for a few diagnostic nucleotides being examined. In so doing, it allows for the introduction of nucleotide mismatches in one or both primers for differentiating wild from mutant sequences following polymerase chain reaction, restriction digestion and MetaPhor gel electrophoresis. Primer design and the selection of discriminating enzymes are achieved with the help of the dPACS 1.0 program. The method is exemplified here with the positive detection of serine 264-*psb*A, a key determinant for the effective binding of some photosystem II inhibitors to their target. A serine-to-glycine mutation at codon 264 of *psb*A causes resistance to serine-binding photosystem II herbicides in several grasses and broad-leaf weeds, including *Amaranthus retroflexus*, which is employed in this study.

Key words Derived polymorphic amplified cleaved sequence, dPACS, SNPs, INDELs, PCR-RFLP, Genotyping, *Amaranthus retroflexus*

1 Introduction

Single-nucleotide polymorphisms (SNPs) and deletion–insertion polymorphisms (DIPs) are abundantly present in the genome [1, 2]. When they occur in exons and alter protein structure and function, they can have a profound effect on the organism [3]. Base changes that are located in noncoding regions may still be consequential when they affect gene splicing and transcription factor binding [4]. Mutations in intergenic regions have no direct impact on the phenotype but can nevertheless be used as markers in phylogenetics, genome-wide association, and population genetics studies, among others [5]. Over the years, several increasingly

Yuri Shavrukov (ed.), *Plant Genotyping: Methods and Protocols*, Methods in Molecular Biology, vol. 2638, https://doi.org/10.1007/978-1-0716-3024-2_27,

sophisticated procedures have been developed for the investigation of sequence variation in deoxyribonucleic acid (DNA). It is now possible to analyze several thousands of SNPs in a few individuals with CHIPS-based procedures [6]. When a large number of samples and only a few SNPs/DIPs are involved, the single-step and close-tube TaqMan or kompetitive allele-specific (KASP) technology is favored [7, 8]. Amplicon-seq and pyrosequencing are also useful tools for detecting any new polymorphisms at and around the targeted nucleotide bases [9, 10]. However, the major limitations of all these advanced methods are consumable and instrumentation costs as well as bioinformatics requirements [11, 12]. Other cheaper alternatives include the allele-specific assay (ASA), the cleaved amplified polymorphic sequence (CAPS), and derived cleaved amplified polymorphic sequence (dCAPS) assays [13–15]. ASA is simple and proceeds in two steps, namely, polymerase chain reaction (PCR), followed by horizontal gel electrophoresis. However, ASA is often ambiguous as the identification of wild and mutant alleles often relies on the differential priming of a single base at the 3′-end of the PCR primer [15]. The CAPS and dCAPS assays are more reliable but can be limited in their applications due to the lack of discriminating restriction enzymes and ambiguity resulting from sequence variation around the SNP/DIP being investigated [13, 14]. In this chapter, we describe a variant of the polymerase chain reaction/restriction fragment length polymorphism (PCR-RFLP) approach, denoted as the derived polymorphic amplified cleaved sequence (dPACS) method [16, 17]. The dPACS approach proceeds in three main steps, namely, PCR using primers that completely encompass the whole DNA region of interest except for the SNPs/DIPs being interrogated, restriction digestion with discriminating enzymes, and horizontal gel electrophoresis. As with other PCR-RFLP approaches, allele detection is permitted by the gain of a restriction site for the wild or mutant DNA. The dPACS assay compares favorably with other PCR-based RFLP procedures because it is less sensitive to other nucleotide sequence variations around the SNP/DIP being analyzed. As the primers are in close proximity to each other and are often adjacent to the SNP/DIP in question, it permits the introduction of nucleotide mismatches in one or both PCR primers and the selection of a wider range of restriction enzymes for allele identification. The primers, with forced mutations as required, and discriminant restriction enzymes are chosen with a custom-developed dPACS 1.0 freeware (http://opendata.syngenta.agroknow.com/models/dpacs).

2 Materials

2.1 Template DNA in Tris–EDTA (TE) Buffer or Sterile Water

DNA from wild (S264-*psb*A) and mutant (G264-*psb*A) *Amaranthus retroflexus* plants was used here at around 10 ng/μL.

2.2 Polymerase Chain Reaction

1. PuReTaq Ready-To-Go PCR beads or a similar type.
2. PCR primers in TE or sterile water.
3. Deoxyribonuclease (DNase)-free sterile water.
4. PCR thermal cycler.
5. Pipettors with filter tips or autoclaved tips.

2.3 Restriction Digestion

1. Selected restriction enzyme for the identification of the SNP or INDEL (insertion/deletion) of interest.
2. 10 × reaction buffer for the chosen restriction enzyme.
3. 6 × gel loading dye: 0.4% orange G, 0.03% bromophenol blue, 0.03% xylene cyanol FF, 15% Ficoll 400, 10 mM Tris–HCl (pH 7.5), and 50 mM EDTA (pH 8.0).
4. Incubation chamber or water bath.

2.4 MetaPhor Gel Electrophoresis

1. MetaPhor™ agarose (Lonza, Walkersville, MD, USA) (*see* **Note 1**).
2. 1 × Tris–borate–EDTA (TBE) buffer.
3. DNA ladder (50–1000 bp).
4. Ethidium bromide: 10 mg/mL stock solution.
5. Horizontal gel-electrophoresis apparatus.
6. Gel documentation system with ultraviolet (UV) transilluminator.

3 Methods

3.1 Enzyme Selection and Primer Design with the dPACS 1.0 Program

Enzyme selection and primer design with the dPACS 1.0 program (http://opendata.syngenta.agroknow.com/models/dpacs) is illustrated here with the detection of a key S264G mutation in the chloroplastic *psb*A gene (maternally inherited in most species, including *A. retroflexus*), which encodes the D1 protein. The S264G mutation results from a single adenine-to-guanine change in the first base of the 264-codon triplet (AGT to GGT) of *psb*A.

3.1.1 Inputs into the dPACS Program

Wild and mutant sequences (A, T, G, C only) differing at a single or few nucleotides are entered in the dPACS 1.0 program, as well as the number of desired mismatches on the primers. Each missing base of INDELs is entered as hyphens in place of nucleotides.

```
Wild type: tggtcgattgatcttccaatatgctagtttcaacaactctcgttctttaca
Mutant:    tggtcgattgatcttccaatatgctggtttcaacaactctcgttctttaca
```

Fig. 1 Wild (agt) and mutant (ggt) *psbA* sequences around critical codon 264 (in bold)

Ideally, the submitted sequence should consist of around 25 base pairs on each side of the SNP/INDEL as the restriction sites of some enzymes can be several bases away from their nucleotide recognition sequences. The wild and mutant *A. retroflexus psbA* sequences (GenBank reference: DQ887375.1) entered in the dPACS programs were as shown in Fig. 1.

The focus here was to find restriction enzymes that would selectively cleave the wild-type sequence only. The number of desired mismatches ranged from zero to two, although more forced mutations on the primers could be envisaged.

3.1.2 Outputs from the dPACS Program

The dPACS program generates a number of restriction enzymes and corresponding mismatches to be included in the primers that will allow differentiation between wild and mutant sequences. The outputs can be filtered by the restriction of the wild-type or mutant sequence and the location of forced mismatches on the forward, the reverse, or both primers. The program will also highlight, in red, any nonspecific additional restriction sites present on both wild and mutant sequences for a particular enzyme so that these can be eliminated manually with further mismatches on the primers. The results generated by the dPACS program for the S264G *psbA* mutation are summarized in Table 1.

With zero mismatches, the dPACS program identified *Mae*I as the only enzyme that discriminates the wild from the mutant sequence (*see* Table 1, Fig. 2a). The wild-type sequence would be digested with *Mae*I as it contains the enzyme restriction site CTAG. In contrast, the mutant sequence (CTGG) would not be restricted due to the loss of the *Mae*I recognition site. *Spe*I (ACTAGT) is an alternative enzyme for restricting the wild-type DNA but requires a single forced guanine-to-adenine change on the forward primer at the N-3 position with respect to the SNP being examined (*see* Table 1, Fig. 2b). The dPACS program generated another 27 enzyme options for differentially restricting the wild-type *psbA* DNA. These include four different enzymes with one forced mutation on the reverse primer and six and three options with two mismatches on the forward and reverse primers, respectively. One mismatch each on the forward and reverse primers yielded as many as 14 different enzyme options with different nucleotide mismatches on the primers (*see* Table 1).

Table 1
List of enzymes and corresponding mismatches (underlined) to introduce on the forward, the reverse, or both primers for selectively restricting the wild *psb*A sequence

Number of mismatches	Location of mismatches	Restriction enzyme	Enzyme recognition sequence
0	–	*Mae*I	CTAG
1	Forward primer	*Spe*I	ACTAGT
1	Reverse primer	*Alu*I	AGCT
1	Reverse primer	*Nhe*I	GCTAGC
1	Reverse primer	*Set*I	ASST
1	Reverse primer	*Tsp*EI	AATT
2	Forward primer	*Ags*I	TTSAA
2	Forward primer	*Asp*BHI	YSCNS
2	Forward primer	*Dde*I	CTNAG
2	Forward primer	*Hae*I	WGGCCW
2	Forward primer	*Hyp188*I	TCNGA
2	Forward primer	*Taq*I	TCGA
2	Reverse primer	*Fai*I	YATR
2	Reverse primer	*Mae*II	ACGT
2	Reverse primer	*Tsp*4CI	ACNGT
2	Both primers	*Apo*I	RAATTY
2	Both primers	*Avr*II	CCTAGG
2	Both primers	*Bts*IMutI	CAGTG
2	Both primers	*Eco*RII	CCWGG
2	Both primers	*Hind*III	AAGCTT
2	Both primers	*Hinf*I	GANTC
2	Both primers	*Mae*III	GTNAC
2	Both primers	*Mse*I	TTAA
2	Both primers	*Ple*I	GAGTC
2	Both primers	*Rsa*I	GTAC
2	Both primers	*Sty*I	CCWWGG
2	Both primers	*Tsp*EI	AATT
2	Both primers	*Tsp*RI	CASTGNN
2	Both primers	*Xba*I	TCTAGA

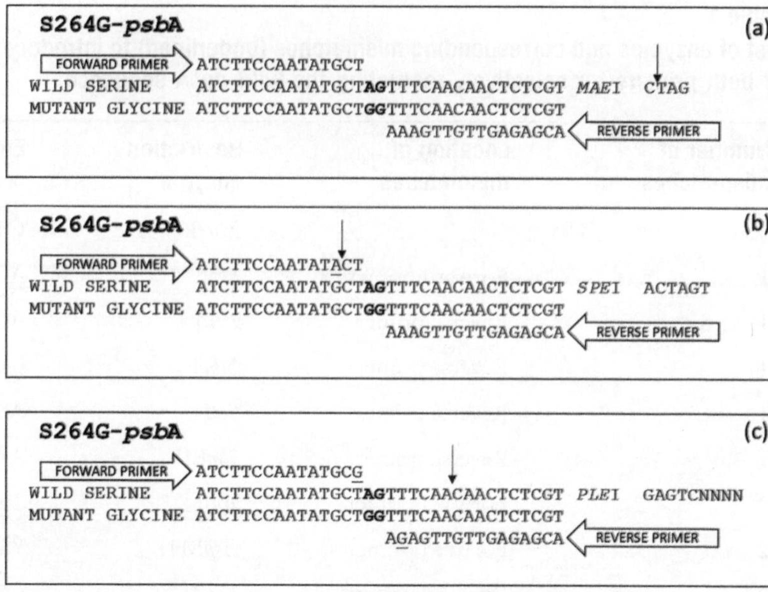

Fig. 2 Location of PCR primers on the targeted *psb*A gene for (**a**) *Mae*I, (**b**) *Spe*I, and (**c**) *Ple*I. The diagnostic dinucleotides being examined are in bold. Forced mismatches on the forward and reverse primers are underlined. Vertical arrows indicate the enzyme restriction site. It is noteworthy that the complement bases are used on the reverse primers

3.1.3 Choice of Restriction Enzyme

Various criteria should be taken into consideration for choosing the right restriction enzyme for use in the dPACS assay. The restriction site of the selected enzyme should be contained in the primer (s) and the diagnostic base(s) being investigated to avoid false negatives, which may result from additional SNPs among the many individual samples being examined. Similarly, enzymes that contain degenerate restriction sites, such as *Set*I and *Ags*I, should be excluded to avoid false positives for the targeted SNP or INDEL. Restriction enzymes that require fewer forced mutations and are as far away as possible from the 3′-end of the primers should be preferred to ensure better levels of PCR amplification. The chosen enzyme should be commercially available, inexpensive, and highly effective to prevent uncertainty, which could result from the partial digestion of PCR products. *Ple*I (*see* Table 1, Fig. 2c) was selected for positively identifying the wild-type *psb*A as it met all the important criteria mentioned above (*see* **Note 2**).

3.1.4 Primer Design

Contrary to other PCR methods, the primers are fixed around the SNP in question and are designed manually with the inclusion of the forced mutation(s) as suggested by the dPACS 1.0 program. The PCR primers encompass the whole DNA region being targeted except for the single or few diagnostic bases being analyzed. The primers should be between 35 and 55 bp to generate a PCR product of around 80–100 base pairs that can be visualized on MetaPhor gel (*see* **Note 3**). The length of the primers should be adjusted to provide at least a-five base-pair difference between the two digested DNA bands following restriction with the selected enzyme. The band difference should be 20 bp or more between the undigested PCR product and the larger restricted fragment to allow adequate resolution on MetaPhor gel electrophoresis. The *Ple*I forward (5′ TGCTTCATGGTTACTTTGGTCGATT GATCTTCCAATATGC G3′) and reverse (5′ AAGCAGCTAA GAAAAAGTGTAAAGAACGAGAGTTGTTGA GA3′) primers for S264G mutation detection were 40 bp and 41 bp respectively. The 3′-end of the forward primer was located on the last base of the 263 *psb*A codon triplet, while the 3′-end of the reverse codon primed the third base of the 264 codon (*see* Fig. 2c). With this strategy, a PCR product of 83 bp (40 + 41 + 2 base pairs) is expected. The cleavage of wild-type S264 *psb*A with *Ple*I (GAGTCN$_4$|) would generate two smaller fragments of 48 and 35 bp as the restriction site of the enzyme is six bases inside the 41 base pair reverse primer. On the other hand, the 264 *psb*A mutant sequence would be undigested and would appear as the original 83 bp PCR fragment on MetaPhor gel.

3.2 DNA Template Preparation

1. DNA can be manually isolated from fresh, dried, or frozen plant tissues with the commonly employed phenol/chloroform or cetyltrimethylammonium bromide (CTAB) method [18]. DNA extraction can also be automated using a number of commercially available equipment, such as the KingFisher Flex Purification system (Thermo Fisher Scientific, UK) and the Wizard Magnetic DNA Plant System kit (Promega, USA) [17, 19].
2. Assess the purity of the extracted DNA dissolved in TE buffer or sterile water by measuring the optical density at 260 and 280 nm. An A_{260}/A_{280} ratio of ~1.8 is indicative of good-quality DNA suitable for downstream polymerase chain reaction (*see* **Note 4**).

3.3 Polymerase Chain Reaction (PCR)

1. Taq polymerase and other PCR components can be sourced from different manufacturers. For convenience, we generally use PuReTaq Ready-To-Go PCR beads (Amersham Biosciences, UK) that contain all necessary stabilizers, buffers, Taq polymerase, and dNTPs for PCR amplification (*see* **Note 5**).
2. Set up a PCR mixture that contains the forward and reverse primers in sterile water for use with PuReTaq Ready-To-Go PCR beads (*see* Table 2). Produce 10% more reaction mix than is required for ease of pipetting, especially for the last samples.

Table 2
PCR reaction mix for 32 samples

Component	Volume (μL) 1 sample	Volume (μL) 32 samples	Final concentration (μM)
Forward primer (0.1 nmol/μL)	0.2	7.2	0.8
Reverse primer (0.1 nmol/μL)	0.2	7.2	0.8
Sterile water	19.6	705.6	
Total	20	720	

3. Remove caps and add 20 μL PCR reaction mix to individual Ready-To-Go PCR tubes. Add 5 μL of DNA template (~50 ng) per sample. Snap caps back onto tubes, mix well by vortexing gently, and centrifuge briefly. Place tubes on ice or in a cold block. Include a negative control (sterile water instead of DNA template) to check for any contamination or interfering primers/dimers during PCR (*see* **Note 6**).
4. An automated thermal cycler is used to perform PCR. The amplification parameters are often based on the manufacturer's recommendation for Taq polymerase and the primers employed in PCR. Similarly, the temperatures and cycling times depend on the template and primers used. Typical cycling conditions for Master Cycle Gradient Thermocycler Model 96 (Eppendorf, UK), PuReTaq Ready-To-Go PCR, and *psb*A target sequence were as follows: a denaturation step at 95 °C for 4 min, followed by 30 cycles of 30 s at 95 °C, 30 s at 60 °C, and 1 min at 72 °C. A final extension step for 10 min at 72 °C was also included.

3.4 Restriction Digestion of PCR Products

Restriction digestion takes advantage of naturally occurring enzymes that cleave DNA at specific sequences. This is achieved by simply adding the amplified PCR product to a reaction mix that contains the discriminant restriction enzyme and corresponding buffer as specified by the manufacturer. The restriction digestion is carried out for a set amount of time at the optimal temperature for the selected enzyme in capped tubes or any appropriate well plates in a total of 20 μL (*see* **Note 7**).

1. Set up a reaction mix for the restriction digestion as indicated in Table 3.
2. Pipette 14 μL restriction reaction mix in each tube or well.
3. Add 6 μL of PCR product to the reaction mix and mix gently by pipetting.

Table 3
Reaction mix for restriction digestion of 32 PCR samples

Component	Volume (μL) 1 sample	Volume (μL) 32 samples
10× enzyme buffer	1	36
Restriction enzyme (5 U/μL)	2	72
Sterile water	11	396
Total	14	504

As for PCR, the restriction digestion mix is produced in excess of around 10%. The rCutSmart Buffer from New England Biolabs was employed for the restriction of the S264G-*Ple*I assay developed here

4. Incubate the tubes or well plate at optimal temperature (usually 37 °C or otherwise as suggested by the manufacturer) in the incubation chamber or water bath for 1 h.
5. After 1 h, add 3 μL loading dye to stop the reaction prior to gel electrophoresis (*see* **Note 8**).

3.5 Horizontal MetaPhor Gel Electrophoresis

MetaPhor gel electrophoresis is highly effective for separating DNA fragments that differ by a few bases only. The protocol can be divided into three main stages: (1) preparation of MetaPhor gel, (2) loading of restriction digests into individual wells and the running of the gel at an appropriate voltage and period of time, (3) staining and visualization of the gel under UV light and documentation of the results.

3.5.1 Preparing the MetaPhor Gel

1. Add MetaPhor to 1 × TBE. For 20 × 24 cm gel size, add 1.6 g MetaPhor to 400 mL TBE to make 4% gel in a 1 L flask.
2. Melt MetaPhor in a microwave oven, mixing several times during heating.
3. Let the MetaPhor cool down to around 55 °C.
4. Pour the MetaPhor into a tray, remove the bubbles with a pipette tip as necessary, insert combs, and allow to set.
5. After solidification, remove the combs and put the gel in the gel box. Pour enough 1 × TBE buffer into the gel box to cover the gel by at least 0.5 cm.

3.5.2 Loading Samples and Running Electrophoresis

1. Load 1 μg DNA ladder in the first well, followed by 10 μL restriction digest samples in subsequent wells. The negative control should also be loaded in one of the wells to check for DNA contamination during PCR setup and cycling. Take care not to mix samples between wells.
2. Run electrophoresis at 20 V for 1 h (*see* **Note 9**).
3. Remove the gel from the electrophoresis tank for staining.

3.5.3 Staining and Visualization of the Gel

1. Stain the gel in 1 µg/mL ethidium bromide for 20 min (*see* **Note 10**).
2. Rinse the gel for 20 min in 1000 mL ddH2O.
3. Put the gel onto the UV transilluminator and take a photo.

3.6 dPACS Results for S264G Mutation Analysis

PCR generated an expected 83 bp fragment for all plants. Upon restriction with *Ple*I and electrophoresis on 4% MetaPhor gel, plants bearing the wild-type serine allele generated two smaller fragments of 45 and 38 bp, which could be easily resolved from the mutant 83 bp undigested PCR fragment (*see* Fig. 3). The clear and unambiguous results generated here demonstrate the reliability and robustness of the dPACS approach. The dPACS method proved to be useful for identifying several other difficult-to-genotype SNPs and INDELs (*see* **Note 11**).

4 Notes

1. It is imperative to employ MetaPhor instead of agarose to resolve the relatively small size difference between restricted and nonrestricted PCR-RFLP fragments.
2. In spite of necessitating a thymine-to-guanine mismatch on the 3′-end base of the forward primer and an adenine-to-guanine change on the penultimate base at the 3′-end of the reverse primer, *Ple*I primers were chosen over *Mae*I and *Spe*I, which required zero and one mismatch on the primers, respectively (*see* Fig. 2c). This is because the *Ple*I was half the price of the latter two restriction enzymes, and the corresponding *Ple*I-primers generated a strong band upon PCR. The dPACS assay appears to tolerate more mismatches on the primers compared to other PCR-based methods because it uses relatively long and stable primers to target a short DNA fragment.
3. Since oligonucleotides are synthesized in the 3′- to 5′--direction, the longer the primers are, the more prone they are to nucleotide errors at the critical 3′-position. The dPACS primers should be at least 35 bp long to allow a clear differentiation between restricted and nonrestricted PCR fragments but not above 55 bp to avoid unnecessary costs due to an extra purification step following chemical synthesis by primer providers.
4. As PCR targets fragment sizes of less than 100 bp, the dPACS procedure is amenable to relatively low amounts/quality of DNA. For instance, the method was shown to be applicable to DNA extracted from single Amaranth seeds, which weigh 0.8 mg on average [16].

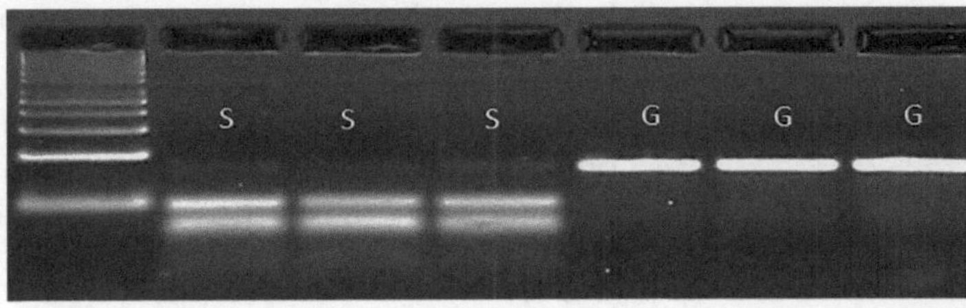

Fig. 3 Typical dPACS profiles for wild S264 and mutant G264 *psbA* in *Amaranthus. retroflexus* individuals. Lane 1: DNA ladder; lanes 2, 3, 4 *PleI* restricted S264 wild haplotype; lanes 5, 6, and 7 *PleI* unrestricted G264 mutant haplotype

5. When a Ready-To-Go PCR bead is reconstituted to a 25 μL final volume, the concentration of each dNTP is 200 μM in 10 mM Tris–HCl (pH 9.0 at room temperature), 50 mM KCl, and 1.5 mM $MgCl_2$.
6. In the case of the formation of primer-dimers (which could appear as long as the undigested PCR products, thus rendering agarose gel electrophoresis difficult to interpret) in significant amounts, a "hot-start" approach should be envisaged for the PCR [20].
7. The activity of restriction enzymes can decrease rapidly if not maintained in optimal buffer and temperature conditions. Restriction enzymes should be kept on ice during handling and placed immediately at −20 °C then after. A common issue with PCR-RFLP methods is partial DNA restriction. This is manifested by smaller restricted DNA fragments appearing more intense than larger nonrestricted fragments on MetaPhor gel. To ensure good restriction digestion by the selected enzymes, the DNA concentration should be between 20 and 100 ng/μL in the final reaction mixture. At least five to ten units of enzymes should be employed per μg of DNA. The optimal buffer and reaction temperature as suggested by the manufacturers should be employed. Excess evaporation should be avoided during incubation as an increased salt concentration in the buffer can inhibit enzyme performance. Nonspecific digestion and star activity are other problems that may arise with restriction enzymes. These can occur in high enzyme: DNA ratio conditions and in prolonged (e.g., overnight) incubations. Incubation of PCR products with the selected enzymes should not be allowed to proceed for more than 1 h to avoid nonspecific restriction digestion. Other contributing factors include high glycerol concentrations, high pH or low ionic strength, and the presence of organic solvents. Suppliers that have optimized enzymes and buffers to avoid star activity are recommended. The recommendations provided with each enzyme for optimal activity, including the use of the correct

buffer, enzyme amount, and reaction time for the enzyme, should be followed.

8. The loading dye also allows DNA tracking during gel electrophoresis. DNA bands (~80–100 bp) in the dPACS procedure will migrate between the blue (bromophenol blue) and orange (orange G) dyes.

9. Gel electrophoresis should be allowed to run for the shortest time that permits clear fragment resolution between digested and undigested bands but not any further to prevent band diffusion following long exposure in the gel tank.

10. Staining can also be achieved by directly adding ethidium bromide to the gel (1 μL of a 10 mg/mL stock solution for every 100 mL TBE) and to the buffer in the gel tank.

11. For simplicity, the method was exemplified here by genotyping a mutation in the uniparentally inherited *psb*A gene. Therefore, individual plants contained either the wild or mutant allele. The dPACS assay is a codominant marker method. As such, it can clearly distinguish between homozygous and heterozygous individuals for biparentally inherited genes [16, 17]. As the primers are in close proximity to the SNP being analyzed, it is sometimes possible to positively identify both wild and mutant alleles in a single assay that uses a unique PCR product and a cocktail of two different discriminating restriction enzymes [21].

Acknowledgments

The authors are grateful to colleagues in the plant production team at the Syngenta, Jealott's Hill International Research Centre, for growing the *A. retroflexus* plants used in this study.

References

1. Brookes AJ (1999) The essence of SNPs. Gene 234:177–186. https://doi.org/10.1016/s0378-1119(99)00219-x
2. Montgomery S, Goode D, Kvikstad E, Albers C, Zhang Z, Mu X et al (2013) The origin, evolution, and functional impact of short insertion-deletion variants identified in 179 human genomes. Genome Res 23:749–761. https://doi.org/10.1101/gr.148718.112
3. Wang L, Guo W, Fang C, Feng W, Huang Y, Zhang X et al (2021) Functional characterization of a loss-of-function mutant I324M of arginine vasopressin receptor 2 in X-linked nephrogenic diabetes insipidus. Sci Rep 11:11057. https://doi.org/10.1038/s41598-021-90736-z
4. Safi A, Medici A, Szponarski W, Martin F, Clément-Vidal A, Marshall-Colon A et al (2021) GARP transcription factors repress Arabidopsis nitrogen starvation response via ROS-dependent and -independent pathways. J Exp Bot 72:3881–3901. https://doi.org/10.1093/jxb/erab114
5. Veitia RA (2022) Who ever thought genetic mutations were random? Trends Plant Sci 27:733–735. https://doi.org/10.1016/j.tplants.2022.03.003
6. Teumer A, Ernst FD, Wiechert A, Uhr K, Nauck M, Petersmann A et al (2013)

Comparison of genotyping using pooled DNA samples (allelotyping) and individual genotyping using the affymetrix genome-wide human SNP array 6.0. BMC Genomics 14:506. https://doi.org/10.1186/1471-2164-14-506

7. Hidaka A, Sasazuki S, Matsuo K, Ito H, Charvat H, Sawada N et al (2016) *CYP1A1*, *GSTM1* and *GSTT1* genetic polymorphisms and gastric cancer risk among Japanese: a nested case–control study within a large-scale population-based prospective study. Int J Cancer 139:759–768. https://doi.org/10.1002/ijc.30130
8. Semagn K, Babu R, Hearne S, Olsen M (2014) Single nucleotide polymorphism genotyping using Kompetitive Allele Specific PCR (KASP): overview of the technology and its application in crop improvement. Mol Breed 33:1–14. https://doi.org/10.1007/s11032-013-9917-x
9. Arita H, Narita Y, Matsushita Y, Fukushima S, Yoshida A, Takami H et al (2015) Development of a robust and sensitive pyrosequencing assay for the detection of IDH1/2 mutations in gliomas. Brain Tumor Pathol 32:22–30. https://doi.org/10.1007/s10014-014-0186-0
10. Li F, Henderson G, Sun X, Cox F, Janssen PH, Guan LL (2016) Taxonomic assessment of rumen microbiota using total RNA and targeted amplicon sequencing approaches. Front Microbiol 7:987. https://doi.org/10.3389/fmicb.2016.00987
11. Griffin TJ, Smith LM (2000) Single-nucleotide polymorphism analysis by MALDI–TOF mass spectrometry. Trends Biotechnol 18:77–84. https://doi.org/10.1016/s0167-7799(99)01401-8
12. Muir P, Li S, Lou S, Wang D, Spakowicz DJ, Salichos L et al (2016) The real cost of sequencing: scaling computation to keep pace with data generation. Genome Biol 17:53. https://doi.org/10.1186/s13059-016-0917-0
13. Neff MM, Neff JD, Chory J, Pepper AE (1998) dCAPS, a simple technique for the genetic analysis of single nucleotide polymorphisms: experimental applications in *Arabidposis thaliana* genetics. Plant J 14:387–392. https://doi.org/10.1046/j.1365-313x.1998.00124.x
14. Ota M, Asamura H, Oki T, Sada M (2009) Restriction enzyme analysis of PCR products. In: Komar AA (ed) Single nucleotide polymorphisms: methods and protocols. Humana, Totowa, pp 405–414. https://doi.org/10.1007/978-1-60327-411-1_25
15. Bottema C, Sommer S (1993) PCR amplification of specific alleles: rapid detection of known mutations and polymorphisms. Mutat Res 288:93–102. https://doi.org/10.1016/0027-5107(93)90211-w
16. Kaundun SS, Hutchings SJ, Marchegiani E, Rauser R, Jackson LV (2020) A derived Polymorphic Amplified Cleaved Sequence assay for detecting the Δ210 *PPX2L* codon deletion conferring target-site resistance to protoporphyrinogen oxidase-inhibiting herbicides. Pest Manag Sci 76:789–796. https://doi.org/10.1002/ps.5581
17. Kaundun SS, Marchegiani E, Hutchings SJ, Baker K (2019) Derived polymorphic amplified cleaved sequence (dPACS): a novel PCR-RFLP procedure for detecting known single nucleotide and deletion-insertion polymorphisms. Int J Mol Sci 20:3193. https://doi.org/10.3390/ijms20133193
18. Rogers SO, Bendich AJ (1989) Extraction of DNA from plant tissues. In: Gelvin SB, Schilperoort RA, Verma DPS (eds) Plant molecular biology manual. Springer, Dordrecht, pp 73–83. https://doi.org/10.1007/978-94-009-0951-9_6
19. Otto P (2002) MagneSil™ paramagnetic particles: magnetics for DNA purification. JALA: J Assoc Lab Autom 7:34–37. https://doi.org/10.1016/S1535-5535-04-00191-1
20. Green MR, Sambrook J (2018) Hot start polymerase chain reaction (PCR). Cold Spring Harb Protoc 5:pdb.prot095125. https://doi.org/10.1101/pdb.prot095125
21. Kaundun SS, Downes J, Jackson LV, Hutchings SJ, Mcindoe E (2021) Impact of a novel W2027L mutation and non-target site resistance on acetyl-coA carboxylase-inhibiting herbicides in a French *Lolium multiflorum* population. Genes 12:1838. https://doi.org/10.3390/genes12111838

Chapter 28

Tubulin-Based Polymorphism (TBP) in Plant Genotyping

Luca Braglia, Floriana Gavazzi, Silvia Gianì, Laura Morello, and Diego Breviario

Abstract

Tubulin-based polymorphism (TBP) is an intron length polymorphism (ILP) method widely applicable to any plant species and particularly suitable for a first and rapid classification of any plant genome. It is based on the selective, polymerase chain reaction (PCR)-based amplification of the two introns present at conserved positions within the coding sequences of plant β-tubulin genes. Amplification releases a simple yet distinctive genomic profile.

Key words DNA polymorphism, DNA fingerprinting, Molecular markers, Introns, Tubulin

1 Introduction

The tubulin-based polymorphism (TBP) method for genotyping can be applied to any plant and at any taxonomic level with no need for a priori sequence information and posteriori deoxyribonucleic acid (DNA) nucleotide sequencing. Applied to any plant DNA preparation, using the same primer pairs targeting exons of the different members of the β-tubulin gene family at their intron boundaries, TBP will always produce a genomic fingerprinting that reflects the number and the length of the different allelic forms of the β-tubulin introns present in the analyzed sample [1–3]. Starting from variety/ecotype level, the more the samples will differ for their taxonomy the more distinct will be the TBP genomic profile and this may be very useful in the genetic characterization of orphan and wild species, landraces and ecotypes, crop-related wild species, hybrids and related parentage, in addition to crops and cultivated species. Similarities between TBP genomic patterns indicate a close genetic relationship that could be further investigated with the use of more penetrating markers or by DNA sequencing. Since TBP is a codominant and nuclear-based marker, it can be of help in the determination of genetic inheritance and the

Yuri Shavrukov (ed.), *Plant Genotyping: Methods and Protocols*, Methods in Molecular Biology, vol. 2638,
https://doi.org/10.1007/978-1-0716-3024-2_28,

reconstruction of lineages. At present, TBP has been successfully applied to more than 100 species belonging to many different genera of seed plants, such as grapes and duckweeds [4–6].

Why is TBP so effective and versatile in its application? This is because TBP is a method whose origin is intimately related to one of the most elegant and evolutionarily conserved biological mechanisms, that is cell division. Starting from unicellular organisms like yeast, but there are tubulin analogs also in bacteria, the cell division process that lead to the production of two daughter cells from one motherrelies on the mitotic spindle and the mitotic spindle is substantially made up by microtubules (MT). In turn, microtubules represent a highly ordered superstructure made up by filaments of tubulin constituted by head to tail addition of two moieties: the alpha and the beta polypeptides. Their amino acid sequence has been selected by evolution for better interaction with the chromosomes, assisting their pairing and transport at the two cell poles [7, 8]. This fundamental role, and related MTs-chromosome binding, reflects in a high conservation of the tubulin coding sequences with a strict and conserved position of the two β-tubulin introns normally present in vertebrates and plants. Thus, these two introns act as a convenient source of DNA polymorphisms that is revealed by an EPIC (Exon-Primed Intron-Crossing) reaction (*see* Fig. 1). In few words: TBP comes from cell biology not from hard nucleic acid sequencing.

Starting from genomic DNA (gDNA) purified by any standard protocol, the list of operations that are typically involved in the application of the TBP method in plant genotyping, includes: (1) DNA quality and quantity assessment; (2) polymerase chain reaction (PCR) amplification of the β-tubulin introns with two primer pairs, one for each of the two introns; (3) separation of the amplicons by Capillary Electrophoresis (CE); (4) recording of the genomic profile, comparison of the newly generated profile to those of reference samples, identity assignment (*see* Fig. 2).

2 Materials

2.1 gDNA Qualitative, Quantitative Evaluation, and Dilution

1. A microvolume ultraviolet-visible (UV-Vis) spectrophotometer for the assay of double-stranded DNA (dsDNA) (*see* **Note 1**).
2. 96-well thin wall PCR plates, 0.2 mL, ribonuclease (RNase)/deoxyribonuclease (DNase)-free and plate-sealing films (DNA plate).
3. Nuclease-free water, for molecular biology.
4. PCR plate centrifuge.

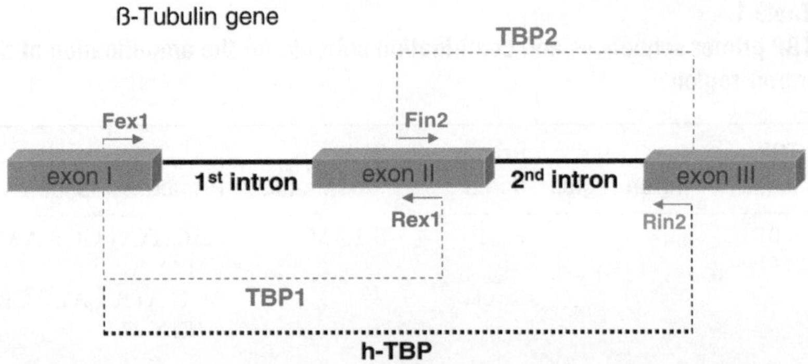

Fig. 1 TBP method. Diagram depicting multiple TBP-based genotyping strategies. Components consist of three different combinations of primer pairs (Fex1/Rex1, Fin2/Rin2, Fex1/Rin2) used to amplify, by touchdown PCR, the first intron (TBP1), the second intron (TBP2), or both introns (h-TBP), present in a conserved position of plant β-tubulin genes

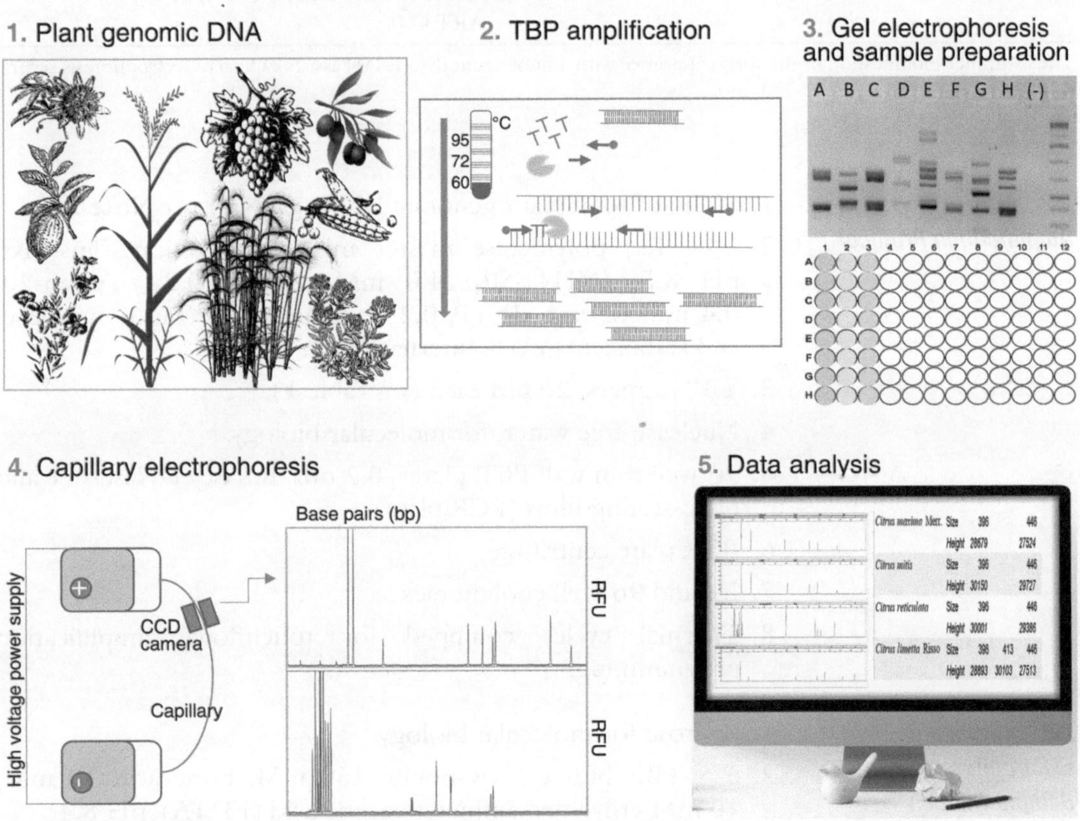

Fig. 2 TBP workflow. The five different steps common to all the TBP versions (TBP1, TBP2, or hTBP) are illustrated

Table 1
TBP primer sequences and combination suitable for the amplification of the respective gene β-tubulin intron region

TBP name	Intron region	Primer name	5′ modification	Primer sequence	Cycle name
TBP1	First	Fex1	6-FAM	AAC TGG GCB AAR GGN CAY TAY AC	TBPTD1i
		Rex1	–	ACC ATR CAY TCR TCD GCR TTY TC	
TBP2	Second	Fin2	6-FAM	GAR AAY GCH GAY GAR TGY ATG	TBPTD2i
		Rin2	–	CRA AVC CBA CCA TGA ARA ART G	
h-TBP	First and second	Fex1	6-FAM	AAC TGG GCB AAR GGN CAY TAY AC	TBPTD2i
		Rin2	–	CRA AVC CBA CCA TGA ARA ART G	

The 5′-primer modification of the forward primer with a fluorescent dye (FAM) assists capillary electrophoresis separation and size estimation of the TBP products

2.2 TBP Amplification Protocol

1. 30 ng of total plant genomic DNA (gDNA) (*see* **Note 2**).
2. 2 × Taq polymerase master mix (composition: Tris–HCl pH 8.5, $(NH4)_2SO_4$, 4.0 mM $MgCl_2$, 0.2% Tween-20, 0.4 mM of each dNTP, 0.2 units/μL VWR Taq polymerase, and stabilizer) (VWR International).
3. TBP primers, 20 μM each (*see* Table 1).
4. Nuclease-free water, for molecular biology.
5. 96-well thin wall PCR plates, 0.2 mL, RNase/DNase-free and plate-sealing films (PCR plate).
6. PCR plate centrifuge.
7. Ice and 96-well cooling rack.
8. Thermal cycler equipped for touchdown amplification programming.

2.3 Agarose Gel Electrophoresis and Sample Dilution

1. Agarose for molecular biology.
2. 5 × TBE buffer (Tris–borate 450 mM, boric acid 10 mM, 10 mM ethylenediaminetetraacetic acid (EDTA), pH 8.4).
3. 6 × DNA gel loading dye, ready to use (10 mM Tris–HCl, pH 8.0; 60 mM EDTA; 0.03% bromophenol blue; 0.03% xylene cyanol FF; and 60% glycerol).
4. 1 Kb DNA ladder (0.2–10 Kb range).

5. Fluorescent dye for dsDNA suitable for agarose gel electrophoresis.
6. Nuclease-free water, for molecular biology.
7. 96-well thin wall PCR plates, 0.2 mL, RNase-/DNase-free and plate sealing films (dilution plate).
8. PCR plate centrifuge.
9. Fluorescence gel imaging acquisition system.

2.4 Sample Preparation and Capillary Electrophoresis Separation

1. Genetic analyzer (8-capillary) 3500 series.
2. Capillary array 8-CAP, 50 cm.
3. Cathode buffer container 3500 series.
4. Anode buffer container 3500 series.
5. GeneScan installation DS-33 Liz 600 kit.
6. Matrix Standard DS-33, Dye Set G5.
7. HI-DI formamide.
8. MicroAmp optical 96-well reaction plates (running plate).
9. Septa for 96-well plates, Flex.
10. POP-7™ (384) polymer 3500 series.
11. GeneScan 1200 LIZ size standard.
12. Thermal cycler.
13. Nuclease-free water.

2.5 Data Analysis

1. 3500 Series Data Collection Software.
2. GeneMapper software v. 5, full upgrade.

3 Methods

The TBP method relies on two independent PCR amplification reactions, each based on a universal plant primer pair, one of which—the forward primer—is labeled with a fluorophore, usually 6-carboxyfluorescein (6-FAM) [9] (*see* Fig. 1). The four primers cannot be combined into one multiplex PCR reaction because the reverse primer (Rex1) for TBP1 is complementary to the forward primer of TBP2 (Fin2). Therefore, the two primers would form dimers during amplification. In addition, a combined version of the technique, the horse-TBP (h-TBP), can be set by combining the forward primer of TBP1 (Fex1) with the reversed primer of TBP2 (Rin2), leading to the simultaneous amplification of both introns (first and second) and the intervening exon [10] (*see* **Note 3**). The workflow of the present protocol, summarized in Fig. 2, includes the following: evaluation of the quality and quantity of the gDNA extracted from plant tissues and its dilution, PCR amplification of the target intron regions carried out with the required primer pair, sample preparation designed to balance the fluorescence signal, capillary electrophoresis allowing a high-resolution TBP amplicon separation, and the final collection and elaboration of data.

3.1 gDNA Qualitative, Quantitative Evaluation, and Dilution

1. Measure through UV spectrophotometry the concentration of 2 μL of gDNA with reference to a dsDNA molecule (*see* **Note 4**).
2. Using a 96-well plate (DNA plate), dilute samples to 5 ng/μL in water and spin briefly.
3. Store samples at −20 °C until further analysis.

3.2 TBP Amplification Protocol

1. Choose a suitable primer pair according to the targeted intron region (*see* Table 1).
2. Allow the reagents to equilibrate on ice, avoiding the direct exposure of the FAM-labeled primers to UV radiation, then mix gently all reagents and spin briefly the tubes.
3. Using a dispenser, transfer 6 μL of each diluted gDNA from the DNA plate into each well of a new 96-well plate (PCR plate) so as to provide 30 ng of template to the TBP reaction. Include in every experiment a negative control reaction, with 6 μL nuclease-free water dH2O (no DNA template).
4. Prepare the PCR reaction cocktail for the desired number of samples, mixing for each sample 15 μL 2 × Taq polymerase master mix, 6 μL nuclease-free water, and, only at the end, 1.5 μL of each 20 μM primer (*see* **Note 5**). Always carefully keep reagents on ice.
5. Select the appropriate touchdown thermal profile according to the targeted intron region (first, second, or the combined version) as reported in Table 2.
6. Preheat the thermal block of the PCR thermal cycler, allowing for more efficient warming of the samples so as to minimize the formation of nonspecific annealing and primer dimers.
7. Vortex gently the PCR reaction cocktail, then distribute 24 μL in each well of the PCR plate. Seal the plate and spin briefly (*see* **Note 6**).
8. Thermal cycle the TBP reaction.

3.3 Agarose Gel Electrophoresis and TBP Amplicons Dilution

1. Cast a 2% (w/v) agarose gel in 0.5 × TBE, including the appropriate volume of fluorescent dye for dsDNA.
2. Prepare samples for loading: add 1 μL 6 × DNA gel loading dye to 5 μL TBP product.
3. Load samples on the agarose gel, including also 0.2 μg of the 1 Kb Plus DNA Ladder.

Table 2
TBP amplification protocols and thermal profiles

	T (°C)	Time		Cycle N°
TBPTD1i	94	3 min		
	94	30 s	−0.5 °C touchdown	
	65	45 s		×14
	72	2 min		
	94	30 s		
	57	45 s		×15
	72	2 min		
	72	30 min		
TBPTD2i	94	3 min		
	94	30 s	−0.7 °C touchdown	
	65	45 s		×14
	72	2 min		
	94	30 s		
	55	45 s		×15
	72	2 min		
	72	30 min		

The use of a thermal cycler equipped for touchdown amplification programming is mandatory for TBP product achievement

4. Run the agarose gel electrophoresis at 80–120 V in 0.5 × TBE buffer until the slower light-blue dye line (xylene cyanol) is approximately 1.5 cm far from the gel wells (*see* **Note** 7).
5. Acquire the imagine of the separated DNA fragments with the available gel-documentation system.
6. Compare the TBP profile's intensity to that of the DNA ladder to determine the proper dilution rate applicable to the samples that will be prepared for loading on the capillary electrophoresis (CE) apparatus (*see* **Note 8**).
7. Prepare the dilution plate, transferring at least 2 μL of each TBP product into a new 0.2 mL 96-well PCR plate, diluting with the required amount of nuclease-free water. Leave the negative controls undiluted. Gently vortex and spin the dilution plate.

3.4 Capillary Electrophoresis Separation

1. Prepare the running plate by transferring 3 μL of each diluted TBP amplification product from the dilution plate. Load the same volume for the undiluted negative control.
2. Prepare the CE running mixture, adding 0.18 μL of 1200 LIZ size standard to 16.82 μL of formamide for each sample (20 μL final volume). Multiply these volumes by the number of samples. Gently vortex and spin the mixture.

3. Dispense 17 μL of CE running mixture into each well of the running plate, and then carefully cover the plate with the dedicated septa (*see* **Note 9**).
4. Denature samples using the thermal cycler, incubating the plate at 95 °C for 5 min, followed by 3 min incubation on ice. Spin the plate (*see* **Note 10**).
5. Prepare the genetic analyzer (3500 series, in our case) according to the instrument's manufacturer instructions and procedures (*see* **Note 11**).
6. Load the capillary array and consumables (running buffers and polymer) as recommended by the manufacturer. Preheat the instrument oven to 60 °C and fill the capillary array with fresh polymer before any series of injections (*see* **Note 12**).
7. Set up the instrument parameters, choosing, from among those provided, the Long Fragment Analysis running module (*see* **Note 13**).
8. Set up the CE assay protocol as follows: dye set, G5 (including 6-FAM, VIC, NED, PET, LIZ size standard); injection and run voltage, 10 kV and 8.5 kV, respectively; injection time, 3 s; and run time, 5100 s (85 min).
9. Fill in the run plate module, choosing the proper Assay Protocol, File Conventional Name, and Results Group in the software plate view, for correct data elaboration and storage.
10. Load the running plate on the instrument and start the CE run.
11. At the end of the run, remove the plate from the instrument and store it at +4 °C until data elaboration has been performed.

3.5 Data Analysis

1. Collect the generated raw data (.fsa data files) from the Data Collection Software.
2. Define the GeneMapper software analysis method as follows: in the Peak Detection algorithm, choose the Advanced mode and fix the peak amplitude threshold at 50 relative fluorescence units (RFU) for all the six dye colors; in the Size-Matching/Calling algorithm, choose the Local Southern method, including the 1200 LIZ size standard definition in the list of size standards used for the sizing of the detected TBP peaks.
3. Upload the .fsa data file to the GeneMapper software and apply the Analysis Method, choosing the LIZ 1200 size standard for sizing the TBP fragments by estimating the peak size, in base pairs (bp); the peak height, in RFU; and their subtended area (*see* **Note 14**).

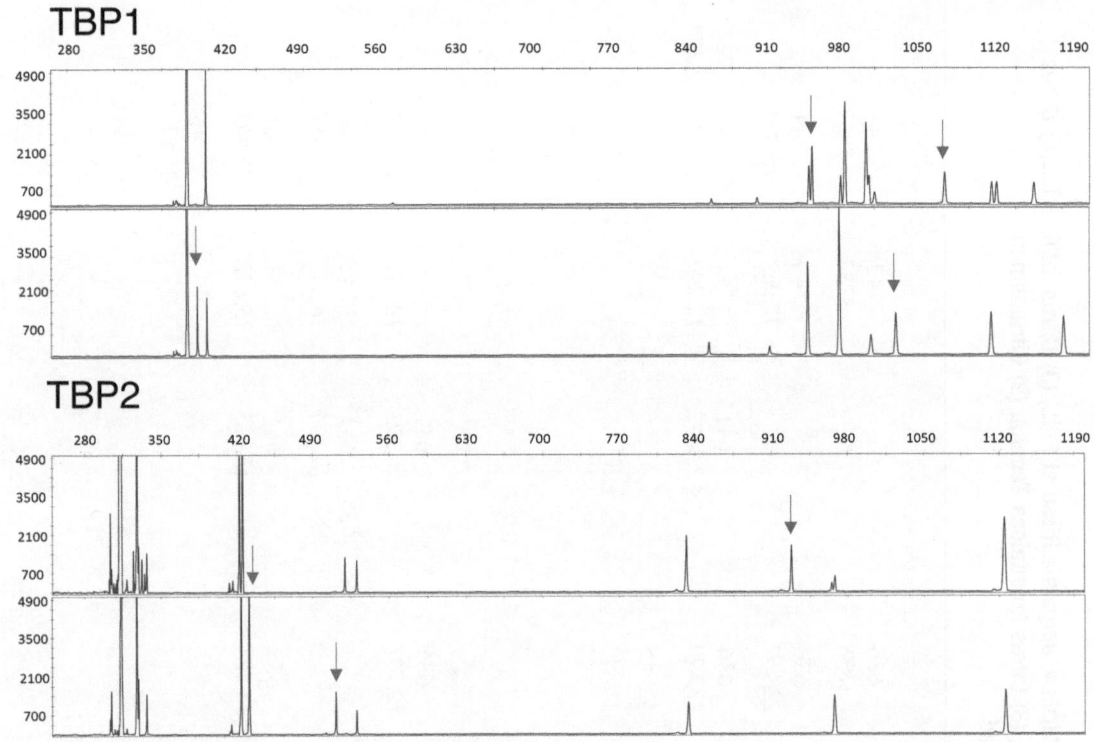

Fig. 3 TBP output. Representative TBP profiles (first-TBP1 and second-TBP2 intron regions) of two *Ranunculus asiaticus* cultivars. Arrows point to some of the most evident intraspecific allelic length variations (peak size) visible in both intron regions

4. After sizing, the CE-TBP data output can be visualized as an electropherogram (fluorescence peak profile) and exported as a .txt data table (*see* Fig. 3).
5. The exported data can be analyzed for subsequent sample comparison by using either statistical software packages or other downstream analysis software (*see* Table 3 and **Note 15**).

4 Notes

1. Any method for the extraction of the total gDNA from plant cells and tissues based on standard laboratory protocols (e.g., cetyltrimethylammonium bromide (CTAB) based) or the use of commercial kits is valid for TBP fingerprinting, provided that it can guarantee a sufficient and suitable template amount for subsequent amplification steps. In fact, the isolation of the total gDNA from plant cells and tissues amenable to molecular analyses can be challenging due to the extreme differences in metabolites and biomolecule production that characterize divergent plant species. Polysaccharides and polyphenols,

Table 3

CE-TBP data output for different *Citrus* species: (1) *Citrus* × *myrtifolia* Raf., (2) *Citrus* × *bergamia* Risso et Poit., (3) *Citrus latifolia* L., (4) *Citrus maxima* Merr., (5) Citrus mitis L., (6) Citrus reticulata L., (7) Citrus limetta Risso, (8) Citrus limettoides Tanaka, (9) Citrus limon Burm. f. "Canaliculata", (10) Citrus sinensis Osbeck, (11) Citrus × paradisi Macfad

TBP1																			
1	Size	396		446			517	593			636					742		744	
	Height	*31,449*		*31,270*			*27,136*	*14,321*			*5878*					*8783*		*8929*	
2	Size	396		446			517				636					743		744	
	Height	*28,377*		*27,673*			*30,503*				*23,967*					*17,963*		*18,521*	
3	Size	396	413	446	474		517				636			641		742		744	747
	Height	*30,837*	*20,663*	*32,300*	*15,691*		*28,997*				*12,590*			*22,582*		*12,299*		*12,419*	*12,270*
4	Size	396		446			517		634		635				722		743		
	Height	*31,149*		*30,978*			*29,995*		*8818*		*5503*				*9948*		*9128*		
5	Size	396		446			517	593	633									745	
	Height	*30,150*		*29,727*			*31,395*	*20,072*	*12,919*									*19,382*	
6	Size	396		446		516	517	593										744	
	Height	*30,001*		*29,386*		*26,081*	*25,499*	*31,775*										*24,998*	
7	Size	396	413	446			517				636			641		742		744	
	Height	*28,893*	*30,103*	*27,513*			*30,391*				*12,707*			*20,193*		*18,434*		*18,841*	
8	Size	396	413	446			517					637	638	641		742		745	
	Height	*29,205*	*30,001*	*27,444*			*30,501*					*4620*	*7396*	*14,226*		*16,337*		*16,588*	
9	Size	396	413	446			517				636			641		742		744	
	Height	*29,615*	*30,756*	*28,195*			*31,287*				*11,405*			*17,443*		*15,801*		*15,903*	
10	Size	396		446			517	593		634								745	
	Height	*29,207*		*28,337*			*30,923*	*24,966*		*14,326*								*19,819*	
11	Size	396		447			517	593				637	638					744	
	Height	*33,300*		*24,796*			*31,656*	*25,727*				*6228*	*7293*					*18,183*	

TBP2																				
1	Size	313	318	323			331		521			687		696		798			1150	1204
	Height	*5690*	*9911*	*5392*			*4428*		*1098*			*2560*		*1805*		*2830*			*8058*	*7607*
2	Size	313	318	323			331	519	521					696		798	800			1203
	Height	*15,414*	*29,230*	*14,151*			*7440*	*2966*	*3070*					*10,698*		*4118*	*3197*			*30,871*
3	Size		318	323	324	325	331	519	522	679				696		798	801	842		1203
	Height		*12,713*	*4714*	*4186*	*2681*	*5198*	*1572*	*749*	*2003*				*1438*		*1476*	*2506*	*1126*		*15,054*
4	Size		318	323			331		521		682			696		798				1203
	Height		*7273*	*6822*			*5021*		*1420*		*1542*			*1570*		*2053*				*9161*
5	Size	313	318	323			331		521			687	691		797					1203
	Height	*6398*	*11,089*	*5987*			*8035*		*1289*			*2790*	*2793*		*3477*					*16,229*
6	Size	313	318				331					687			796	799				1204
	Height	*17,078*	*16,010*				*8703*					*7038*			*3090*	*2211*				*18,350*
7	Size	313	318		324		331	519	521	679				696		799			1150	1202
	Height	*12,818*	*22,595*		*9746*		*6101*	*2224*	*2615*	*5994*				*4315*		*6199*			*15,640*	*16,818*
8	Size	313	318		324		331	519		679				696	796	798			1150	1203
	Height	*11,812*	*22,421*		*8962*		*10,995*	*1990*		*5492*				*5275*	*4788*	*3978*			*17,437*	*16,453*
9	Size	313	318			325	331	519		679				696		799				1203
	Height	*14,243*	*29,502*			*9362*	*7572*	*2797*		*6069*				*5209*		*6089*				*25,708*
10	Size	313	318	323			331					687		696		798	801			1203
	Height	*22,138*	*31,720*	*20,329*			*31,583*					*12,251*		*9890*		*5159*	*6873*			*30,781*
11	Size		318	323			331		521			687		696		798	800			1204
	Height		*26,662*	*17,727*			*18,603*		*4289*			*6190*		*5199*		*6130*	*5135*			*1796*

Each sample is defined by two rows reporting the size, in base pairs, and the height of TBP peaks, in relative fluorescent units (*RFU*), respectively varying in nature and content among plant species and tissues, may copurify with gDNA during its isolation, thus interfering with subsequent gDNA processing steps.

2. The minimum gDNA concentration required is 5 ng/μL, regardless of the type of protocol and the kind of tissue (leaves, seeds, roots, etc.). The choice of method used for evaluating the quantity, and sometimes also quality, of gDNA depends on convenience, practicability, and lab instrument availability.
3. The h-TBP version allows the simultaneous amplification of the genomic region, which encompasses both introns (first and second) of each β-tubulin gene, comprising part of exon1 and exon3 and the whole sequence of exon2. Due to the large size of some of the amplified products, the separation by capillary electrophoresis could result difficult, and variable as a function of analyzed plant species. It is advisable to always carry out a preliminary evaluation of the amplicon size on an agarose gel to verify if this falls within the CE detection range (100–1200 bp).
4. UV spectrophotometry estimates, in addition to concentration, gDNA purity by measuring both the 260/280 nm and the 260/230 nm ratios. A value of ~1.8 for the first ratio is considered acceptable for a "pure" preparation of dsDNA, while 2.0–2.2 values should be obtained by measuring the 260/230 ratio. Values that considerably differ from those just reported denote the presence of contaminants that could inhibit TBP amplification. In this case, changing the gDNA extraction protocol may be advisable, as well as evaluating the introduction of some procedural improvements in the chosen protocol.
5. To prepare the PCR cocktail, multiply the single reagent volumes by the number of experimental samples, and add a negative control (no template). As a rule, it is always advisable to make a little extra of PCR reaction cocktail calculating an additional sample every 12 in order to mitigate errors due to the pipetting of viscous solutions.
6. While dispensing the PCR master mix from the DNA plate, pay attention to keeping the plate constantly on ice or on a 96-well cooling rack to reduce primer dimer formation. In addition, it is recommended to carefully avoid touching the well's edge to prevent contamination. Change the pipette tip for each sample or dispense the mix using an electronic multidispenser.
7. A running time that allows the complete separation of the DNA ladder, and the concurrent separation of the multiple TBP fragments (amplicon size ranges between 0.2–2 Kbp) must be set.
8. The FAM fluorescent intensity signal, that must fall within the detection range capability of the CE apparatus, is dependent on the concentration of the amplified TBP products. Since a wide

range of signal intensity can be obtained when comparing TBP amplicons loaded on the agarose gel from different plant species or tissues, different dilution rates must be applied to each sample to ensure that any will meet the FAM fluorescence detection range required by a successful CE separation. At the first time, to become familiar with the selection of the proper dilution rates, it is recommended to choose a couple of samples that result, by visual inspection on the agarose gel, in a clearly contrasting intensity band profile when compared to the DNA ladder, used as a reference. Then two different dilutions (i.e., 1: 10 and 1:30), at least, of the chosen samples should be performed, followed by the CE separation of the fragments and the corresponding evaluation of the FAM fluorescence intensity of the CE-TBP profile. The CE-TBP profile comparison performed on these two selected samples should be useful for defining more suitable dilution rates for the analysis of the whole samples batch. Notably, both a too low fluorescence signal must be avoided, as it leads to a loss of information (less represented amplicons will be missed), as well as a too high fluorecence signal because it leads to an incorrect or impossible sizing of the peaks and to an increase in background signal.

9. Incorrect sample dilution and preparation would affect the recording of fluorescence signal; thus ensure these are done correctly to allow for efficient samples comparison and a clear detection of the intron size variation.
10. The denaturation step is mandatory in order to prevent the formation of secondary structures in the TBP amplified products, which are capable of altering fragment migration. In addition, an incomplete denaturation of both the size standard and sample TBP fragments would lead to artifacts, increasing background noise and reducing signal intensity. Load samples immediately after denaturation, or store them on ice until ready for loading.
11. Instrument calibration (spatial and spectral), including the performance check, should be regularly performed to assess that the instrument conforms to fragment analysis sizing precision, sizing range, and peak height specifications.
12. Using the same pouch of polymer when running the same batch of samples is highly recommended, even when runs are performed on different days, to carry out sample comparison.
13. The run module is provided by the instrument manufacturer with a Data Collection Software, and optimized according to the instrument model, polymer features, and capillary size. Some settings must be adapted to the analyzed samples, which, in the case of TBP, means the choice of a run module able to guarantee the widest separation range and capable of

maximizing fragment sizing precision. For the TBP analysis, performed trough the supplied CE apparatus, the recommended module is the LongFragAnalysis50_POP7 50 cm (POP-7™) providing a sizing precision as here detailed: range 50–400 bp <0.15; range 401–600 bp <0.30 and range 601–1200 bp <0.45.

14. During sizing, the supplied software performs an internal standard control by comparing the size of the standard peaks of reference included in the TBP sample, and recorded by their relative height and distance from the neighbors, with those of the expected fragments. Sample analysis and sizing fail if some of the observed standard peaks do not meet the expected pattern. Sometimes, high fluorescence intensity emitted from unincorporated primers or dimers can saturate the capability of the charge-coupled device (CCD) camera to such an extent that the correct sizing and control of the standard reference are invalidated. In such a case, the sample cannot be analyzed by the software because the standard cannot be correctly validated. A size-standard, more suitable for TBP amplicon sizing, can also be made by modifying the original length range of reference: 20–1200 bp, 68 single-stranded labeled fragments/peaks into a new range 120–1200 bp, 61 reference peaks, by removing the first seven peaks.

    The lack of any out-of-scale values is mandatory because their presence would result in peak misassignment, a frequent cause of diagnostic errors. If a single out-of-scale value is present in a CE-TBP run, it is recommended to discard the result, go back and increase sample dilution, and then repeat the CE separation. Alternatively, a direct dilution of the out-of-scale sample present in the running plate can be applied. TBP profiles must show similar height of their main peaks to favor a realiable identity assignment and correct allele detection.

15. A comparison of the TBP profiles of different samples can be obtained after aligning the peak size values, organized in ascending order of magnitude, considering a tolerance of ±0.5 bp between each other. This can be performed either manually using a conventional spreadsheet or through a custom-made software tool.

Acknowledgments

The authors wish to thank the following colleagues who, during the last two decades, have worked in our lab contributing to the constant improvement of the technique: Prisca Campanoni, Mauro Bardini, Pietro Blumetti, Marcello Toschi, Andrea Altana, Claudia De Pedro, Elena Ponzoni, Anna Paola Casazza and Francesco Matromauro.

References

1. Bardini M, Lee D, Donini P, Mariani A, Giani S, Toschi M et al (2004) Tubulin-based polymorphism (TBP): a new tool, based on functionally relevant sequences, to assess genetic diversity in plant species. Genome 47: 281–291. https://doi.org/10.1139/g03-132
2. Breviario D, Baird W, Sangoi S, Hilu K, Blumetti P, Giani S (2007) High polymorphism and resolution in targeted fingerprinting with combined β-tubulin introns. Mol Breed 20:249–259. https://doi.org/10.1007/s11032-007-9087-9
3. Gavazzi F, Casazza A, Depedro C, Mastromauro F, Breviario D (2012) Technical improvement of the TBP (tubulin-based polymorphism) method for plant species detection, based on capillary electrophoresis. Electrophoresis 33:2840–2851. https://doi.org/10.1002/elps.201200144
4. Gavazzi F, Braglia L, Mastromauro F, Gianì S, Morello L, Breviario D (2016) The tubulin-based-polymorphism method provides a simple and effective alternative to the genomic profiling of grape. PLoS One 11:e0163335. https://doi.org/10.1371/journal.pone.0163335
5. Braglia L, Breviario D, Gianì S, Gavazzi F, De Gregori J, Morello L (2021) New insights into interspecific hybridization in *Lemna* L. sect. *Lemna* (Lemnaceae Martinov). Plants 10: 2767. https://doi.org/10.3390/plants10122767
6. Braglia L, Gavazzi F, Giovannini A, Nicoletti F, De Benedetti L, Breviario D (2014) TBP-assisted species and hybrid identification in the genus Passiflora. Mol Breed 33:209–219. https://doi.org/10.1007/s11032-013-9945-6
7. Masoud K, Herzog E, Chaboutè ME, Schmit AC (2013) Microtubule nucleation and establishment of the mitotic spindle in vascular plant cells. Plant J 75:245–257. https://doi.org/10.1111/tpj.12179
8. Rasmussen CG, Wright AJ, Muller S (2013) The role of the cytoskeleton and associated proteins in determination of the plant cell division plane. Plant J 75:258–269. https://doi.org/10.1111/tpj.12177
9. Braglia L, Gavazzi L, Morello L, Gianì S, Nick P, Breviario D (2020) On the applicability of the tubulin-based polymorphism (TBP) genotyping method: a comprehensive guide illustrated through the application on different genetic resources in the legume family. Plant Methods 16:86. https://doi.org/10.1186/s13007-020-00627-z
10. Galasso I, Manca A, Braglia L, Martinelli T, Morello L, Breviario D (2011) h-TBP: an approach based on intron-length polymorphism for the rapid isolation and characterization of the multiple members of the β-tubulin gene family in *Camelina sativa* (L.) Crantz. Mol Breed 28:635–645. https://doi.org/10.1007/s11032-010-9515-0

Chapter 29

Multiplexed ISSR Genotyping by Sequencing (MIG-Seq)

Satoshi Nanami

Abstract

Multiplexed inter-simple sequence repeat (ISSR) genotyping by sequencing (MIG-seq) is a simple, rapid, and inexpensive method for detecting single-nucleotide polymorphisms (SNPs) using next-generation sequencing (NGS). The advantages of MIG-seq include easy application to various species without prior genetic information. In addition, this method opens the door to genome-wide nucleotide sequence analyses of low-quality and trace-level deoxyribonucleic acid (DNA) samples, which have previously been difficult to analyze. Another advantage is that the procedure is simple, time-saving, and inexpensive. Recently, MIG-seq has been applied to wild and cultivated plants and has produced novel results. Using invisible DNA information, questions related to gene flow through pollination and seed dispersal, the genetic structure and diversity of populations, clonality, and the hybridization of wild and cultivated plants are being rapidly answered. In this chapter, I present the results of plant research based on MIG-seq and describe the procedure for this method as a user of MIG-seq.

Key words Inter-simple sequence repeat (ISSR), Next-generation sequencing (NGS), Single-nucleotide polymorphisms (SNPs)

1 Introduction

Research on wild plants using genetic information has revealed a great deal about the plant world that was previously invisible to us. To elucidate the mechanisms that sustain plant populations, understanding their reproductive patterns is essential. However, direct observation of pollination and seed dispersal events is constrained by the low visibility of pollen and seeds and the high mobility of pollinators and seed dispersers. Therefore, deoxyribonucleic acid (DNA) markers are used to estimate gene flow based on pollination [1] or seed dispersal distances [2].

In natural plant populations, gene flow via pollination and seed dispersal generally occurs within a limited area. This proximity leads to seedling recruitment near the mother plants and a clumped distribution of consanguineously related individuals with similar genotypes. The nonrandom distribution of genotypes or alleles is

Yuri Shavrukov (ed.), *Plant Genotyping: Methods and Protocols*, Methods in Molecular Biology, vol. 2638, https://doi.org/10.1007/978-1-0716-3024-2_29,

called spatial genetic structure. Fine-scale spatial genetic structure (FSGS; $\lesssim$10 km) is generally characterized by neighboring individuals that are closely related to each other [3]. The intensity of FSGS is closely related to species-specific mating systems, pollination patterns, and seed-dispersal modes. Thus, FSGS is a useful tool for the indirect estimation of gene flow via pollen, seeds, and bulbs within plant populations. In plant populations with clear spatial genetic structures, individuals arising from pollination between neighboring parent plants may be removed from the population due to inbreeding depression [4]. As a result, long-distance pollination makes a relatively large contribution to the recruitment of the next generation [2]. Thus, the spatial genetic structure can affect effective pollination distances through inbreeding depression.

Spatial genetic structure is also observed at geographic scales. Geographic genetic structures may be shaped by various evolutionary forces, including genetic drift, migration, demographic history, and natural selection [5, 6]. Clarifying the geographic genetic structure of natural habitats is an important step in assessing revegetation efforts [7]. First, the evolutionary history of each plant species should be considered. Second, plant populations possess a variety of genes that vary among regions, enabling persistence under changing climatic conditions [8], facilitating range expansion [9], and supporting adaptation to biotic environments [10]. This variability must be considered when attempting natural restoration [11]. Planted plants may be unable to adapt to the local environment, and the locally adapted genetic composition may be disrupted by an influx of genes that were not originally present in the area and that are poorly adapted to the local environment.

When assessing the conservation and restoration of plant populations, attention should be paid to the number of alleles and the proportion of each allele. Genetic diversity has long been noted as a measure of population health [12–14]. Maintaining high genetic diversity is advantageous to survival as individuals with genotypes adapted to environmental change can survive in a fluctuating environment. On the other hand, the loss of genetic diversity makes a population vulnerable to environmental changes and the spread of disease [15]. One example of a genetically uniform plant population that collapsed due to disease is the mid-twentieth-century destruction of a banana variety, Gros Michel [16, 17]. While research on useful cultivated plant species has preceded the study of wild plants [15], interest in the genetic diversity of wild plants has increased in recent years [18].

High genetic diversity within a population is important for the survival of the population; this is the unanimous view of many researchers. However, some plant species increase in number asexually without relying on the combination of female and male gametes [19, 20]. Asexual reproduction results in a genetically identical population of clones. Some plant clones have a tremendous spatial expansion, spreading over 50 km [21], 180 km [22], 1100 km [23], and 1500 km [24]. The existence of these "megaclone" populations in nature suggests that genetic diversity is not required to maintain natural plant populations at least in the short term as long as environmental conditions remain suitable for the extant genotype.

Of the three elements of biodiversity (genetic, species, and ecosystem diversity), species diversity has many open questions. New species have sometimes been discovered through the introduction of DNA information. For example, morphospecies are the basis of plant classification systems. However, in reality, cryptic species, which are two or more divergent species classified as a single species due to their high morphological similarity, may exist, and therefore the use of morphospecies for classification will lead to the underestimation of the number of species. Research on cryptic species has increased recently, supported in large part by the increasing availability of DNA sequences [25].

Increasingly, molecular markers are being used to address species delimitation problems where morphological methods are unreliable or inconclusive, and such markers have often provided great resolution between closely related taxa and hybrids [26]. The advent of relatively inexpensive and rapid DNA sequencing has provided biologists with a new tool for differentiating morphologically similar species [25]. When discussing the number of species, the question of how species are classified inevitably arises.

The biological species concept proposed by Mayr [27] is most commonly used among many others. This model defines a species as a group of interbreeding natural populations that are reproductively isolated from other such groups. In plants, however, reproductive isolation between closely related species is often incomplete, leading to the formation of interspecific hybrids and thus gene introgression [28–30]. Hybridization and associated gene introgression can have a major impact on the fate of a species or population. When gene introgression brings new genes into a species, the genetic diversity of that species may increase, allowing it to adapt better to the new environment [30, 31]. In addition, hybrids can become reproductively isolated from their parent species, leading to speciation [32, 33]. On the other hand, the introgression of genes from other species can threaten the endemism of a species (genetic contamination), especially if that species is rare, thereby reducing species diversity [34].

These findings regarding gene flow, genetic structure, genetic diversity, species diversity, and hybridization have been revealed from invisible DNA information. DNA analysis can transform a conceptual model into a reality. Technological innovations in DNA sequencers, the analytical instruments used to read DNA sequences, have greatly benefited wild plant research. In particular, next-generation sequencing (NGS) supports massively parallel sequencing and has brought revolutionary changes to diverse research fields.

One method based on NGS is multiplexed inter-simple sequence repeat (ISSR) genotyping by sequencing (MIG-seq) [35, 36]. MIG-seq is a simple, rapid, and inexpensive method for detecting single-nucleotide polymorphisms (SNPs) using NGS. In MIG-seq, DNA libraries are constructed through two rounds of polymerase chain reaction (PCR). The first PCR step is conducted using eight pairs of universal multiplex ISSR primers to amplify the ISSR regions of genomic DNA. After the first PCR, products are diluted in deionized water; the second PCR step is performed independently to add individual indices to each sample. Then the second PCR products are pooled into a single mixed library. After purification and fragment size selection, the DNA library is sequenced on an Illumina MiSeq Sequencer (San Diego, CA, USA). The advantages of MIG-seq are as follows.

First, this method is easily applied to various species without prior genetic information, including plants as well as animals [37] and fungi [38]. Species-independent availability contrasts with simple sequence repeat (SSR) DNA markers (e.g., ref. 39), which require development on a species-by-species basis.

Second, this method does not require restriction enzyme treatment of the genomic DNA that serves as the sample. Therefore, the method can be initiated using PCR products. Therefore, MIG-seq allows for the genome-wide nucleotide sequence analysis of DNA samples characterized by low-quality and trace amounts, which have been difficult to analyze in the past. For example, Eguchi et al. [40] applied MIG-seq to aged, millimeter-sized, and dry-mounted specimens of ants and successfully constructed a phylogenetic tree.

In addition, this method is rapid (3 days), simple (two PCR steps), and inexpensive. MIG-seq has already been applied to many studies and is producing novel results (*see* Table 1). It must hold great promise for future application. In this chapter, I present the results of plant research based on MIG-seq and describe the procedure for this method as a user of MIG-seq.

2 Materials

2.1 Reagents and Kits

1. DNA extraction kit for plant tissues.
2. Assay kit for DNA fluorometry.
3. Multiplex PCR kit.
4. AMPureXP beads (Beckman Coulter, Brea, CA, USA).

Table 1
Examples of MIG-seq application to wild plant research and its results

Authors	Subjects	Main results
Hirota et al. (2021) [42]	Phylogeny, species diversity	The phylogenetic relationships of several *Hydrangea* species (Hydrangeaceae) were reviewed. As a result, a plant identified as *Hydrangea serrata*, but morphologically distinct from previously known varieties, was described as a subspecies of *Hydrangea acuminata*, a newly discovered lineage in their study site.
Nakahama et al. (2021) [43]	Gene flow, hybridization	No spatial genetic structure was found within local populations of an endangered species, *Pecteilis radiata* (Orchidaceae), indicating that its pollen or seeds migrated freely at the local scale. On the other hand, genetic disturbance of wild populations occurred through the human transportation of cultivars.
Nakamura et al. (2021) [23]	Clonality, genetic diversity	Genetic diversity was compared among four *Lindera* species (Lauraceae) in Japan, both within local populations and at geographic spatial scales. In the case of *Lindera glauca*, producing seeds through apomixes, it has only female plants present in Japan. Genetic diversity within and among populations was extremely low. The *L. glauca* population in Japan could be considered a "mega-clone" spread over at least 1100 km of distance. *Lindera praecox*, *L. umbellate*, and *L. obtusiloba*, which are dioecious and obligately outcrossing species, had greater heterozygosity than previously published estimates for cosexual plants. This finding might indicate an advantage of dioecy, which avoids inbreeding.
Nota et al. (2022) [11]	Geographic spatial genetic structure	Norway spruce, *Picea abies* (Pinaceae), was found to have been present in southern Fennoscandia as early as 14.7 ± 0.1 cal. kyr BP, and millennia-old clonal spruce trees present today in central Sweden likely arrived through an early Holocene migration from the east. The results are based on ancient sedimentary DNA from multiple European sites.
Okabe et al. (2021) [45]	Phylogeny, species diversity	An unknown plant of *Actinodaphne* was discovered during a field survey, and it was described as a new species. The phylogenetic position of the species among other related ones was determined from morphological data and DNA analysis.
Prasetyo et al. (2020) [46]	Genetic diversity, geographic spatial genetic structure	A weak genetic structure was found in teak, *Tectona grandis* (Lamiaceae), plantations at the geographic scale, indicating that most plantations were established using plant materials from a specific part of the species' natural distribution.

(continued)

Table 1 (continued)

Authors	Subjects	Main results
Sakaguchi et al. (2020) [47]	Phylogeny, species diversity	The phylogenetic origin of an endangered plant species was revealed using DNA molecular data. *Magnolia pseudokobus* (Magnoliaceae) was considered a ploidy variant of *Magnolia kobus* generated through spontaneous triploidization. *Magnolia pseudokobus* appears to be conspecific and is better recognized as a form of *M. kobus*.
Suetsugu et al. (2021) [48]	Hybridization	*Goodyera* × *tanakae* (Orchidaceae) was revealed as a natural hybrid between two related species, *Goodyera foliosa* and *G. velutina*, based on molecular DNA data. The morphology of *Goodyera* × *tanakae* is intermediate between the two parent species.
Wagatsuma et al. (2021) [11]	Geographic spatial genetic structure	*Artemisia indica* var. *maximowiczii* is a native species commonly used for revegetation in Japan. Individuals collected from natural habitats were genetically differentiated between northeastern and southwestern regions in Japan. Most individuals collected from revegetation sites in southwestern Japan were genetically similar to the northeastern populations, suggesting that seeds derived from northeastern populations were used for revegetation throughout Japan.

2.2 Equipment

1. PCR thermal cycler.
2. Agarose gel electrophoresis apparatus.
3. Magnet stand.
4. DNA size selection system.
5. Electrophoresis apparatus for DNA sizing assays.
6. Next-generation sequencer, the Illumina MiSeq sequencing platform.

3 Methods

3.1 DNA Extraction

DNA is extracted from ≤100 mg wet weight or ≤20 mg of lyophilized plant tissue (e.g., leaves, seeds, or cotyledons) using a DNeasy Plant Mini Kit (Qiagen, Hilden, Germany), etc. Alternatively, UniversAll™ Extraction Buffer II (Yeastern Biotech, New Taipei City, Taiwan) can be used for this process. In that case, only 1 mm^3 of plant tissue is needed. One advantage of MIG-seq is that it can be used with low-content and low-quality DNA samples (*see* **Note 1**).

3.2 PCR Amplification

1. The concentration of the DNA solution is measured using a *fluorometer and assay kit for DNA* quantification (Qubit 4 fluorometer and Qubit dsDNA HS Assay Kit; Thermo Fisher Scientific, Waltham, MA, USA). As template DNA for the first PCR step, the necessary DNA concentration is approximately 2–100 ng/μL.
2. The first PCR step is performed to amplify the ISSR regions of genomic DNA using eight forward and reverse primers developed as a MIG-seq primer set [35, 36]. When using a Multiplex PCR Assay Kit Ver. 2 (TaKaRa Bio Inc., Shiga, Japan), the 7 μL PCR mixture contains 1.0 μL of template DNA, 1.4 μL of PCR primers (mixed primers, each at a concentration of 2.0 μM), 3.5 μL of 2× Multiplex PCR Buffer, 0.035 μL of Multiplex PCR Enzyme Mix, and 1.065 μL of water. The PCR cycling conditions are as follows: 94 °C for 1 min, followed by 25 cycles of 94 °C for 30 s, 38 °C for 1 min, and 72 °C for 1 min, and a final hold at 72 °C for 10 min [35] (*see* **Note 2**).
3. From the first PCR product, 2 μL is electrophoresed through a 2% agarose gel to confirm amplification (*see* **Notes 3** and **4**).
4. The second PCR step uses the product of the first PCR diluted to a concentration of 1:50 through the addition of sterile water. Forward and reverse primers for the second PCR are indexed separately [35, 36]. The combination of x forward primers and y reverse primers allows for $x \times y$ individuals to be distinguished through indexing. The second PCR is performed to add complementary sequences as indexes to both ends of the first PCR products for sample identification. Thus, different combinations of primers are used for each sample so that each sample is indexed uniquely (*see* Fig. 1).

   The fragments are amplified with PrimeSTAR GXL DNA Polymerase (TaKaRa Bio Inc.) or a similar brand using 12 μL reaction volumes containing 2.5 μL of the diluted first PCR product as template DNA, 2.4 μL of the second PCR primers (mixed primers, each at a concentration of 2.0 μM), 2.4 μL of 5 × PrimeSTAR Buffer, 0.96 μL of a dNTP mixture, 0.24 μL of PrimeSTAR GXL Polymerase, and 3.5 μL of water. The PCR cycling conditions are as follows: 12 cycles of 98 °C for 10 s, 54 °C for 15 s, and 68 °C for 1 min.

3.3 Purification and Size Selection of the PCR Products

1. From the second PCR product of each sample, 2.0 μL is collected and pooled in a 1.5 mL tube. The mixture is purified with AMPureXP Beads (Beckman Coulter) or a similar product from another manufacturer. DNA fragments in the solution bind to the magnetic beads, and the DNA is recovered by attaching the magnetic beads to a magnetic stand (for example, NGS MagnaStand; Nippon Genetics, Tokyo, Japan).

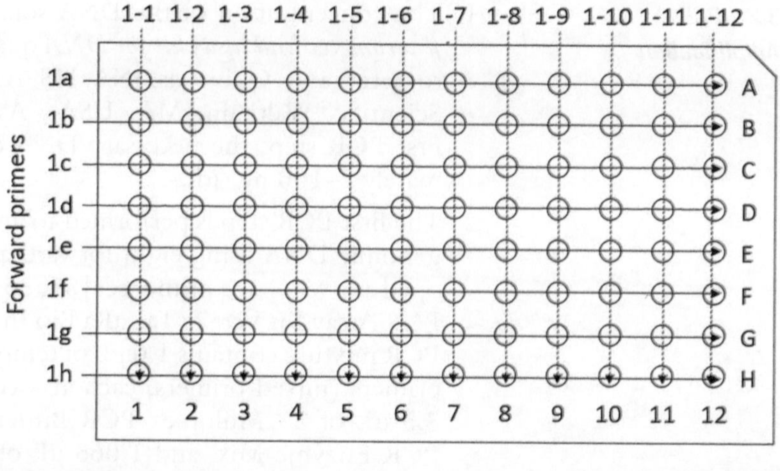

Fig. 1 Application of dual-indexed forward and reverse primers for the second PCR of MIG-seq. In this figure, the combination of 8 forward primers and 12 reverse primers allows for 8 × 12 individuals to be distinguished through indexing. The names of the forward and reverse primers placed on the left and top sides of a 96-well PCR plate followed the Supporting Information S2 in Suyama et al. [36]

2. After purification of the second PCR product, fragments of a size suitable for sequencing (300–800 bp) are selected using a DNA size selection system (for example, Blue Pippin DNA Size Selection System; Sage Science, USA).
3. The DNA library of size-selected fragments is purified again with AMPureXP Beads.

3.4 Estimation of DNA Concentration and Preparation of the Size-Selected Library

1. A TapeStation 4200 electrophoresis system (Agilent Technologies, USA) or a similar brand and a fluorometer are used to measure the average fragment length in the DNA library and the quantity of fragments. Based on the results, the molar concentration of the library (nM) is calculated.
2. The molecular weight of a single base pair of DNA is assumed to be 660. If the average fragment length and the quantity of DNA fragments in the library are L (bp) and X ng/μL, the molar concentration of the library (nM) can be calculated using the following formula:

$$\text{Concentration [nM]} = X/(660 \times L) \times 10^6$$

3. The library is denatured using fresh NaOH (0.2 N) and mixed with 10% Illumina-generated PhiX control libraries according to Illumina's protocol (*see* **Note 5**).
4. Dilute to a concentration of 12 pM the DNA library and submit for sequencing on the Illumina MiSeq sequencing platform using MiSeq Reagent Kit v3.

3.5 Detection of SNPs

1. The sequencing results have to be input into Stacks v.1.47 software for the identification of loci and the detection of SNPs. Following Catchen [41], "ustacks" is used to build loci, "cstacks" to create a locus catalog for all samples, and "sstacks" to check the results against the catalog. Finally, the "population" module is used to estimate a "population genetics statistic" from the population of individual samples.

4 Notes

1. Based on the current experience, the first PCR was successful even with a DNA concentration of 0.1 ng/μL.
2. For a regular concentration of DNA (>10 ng/μL), 25 cycles of PCR is appropriate. For low-concentration DNA samples (<5 ng/μL), 27 cycles is recommended [35].
3. In contrast to general electrophoresis, it is preferable for DNA bands to appear as a "smear-like" pattern rather than distinctly separated. This statement is also based on the Supporting Information S2 in Suyama et al. [36] (*see* Fig. 2).

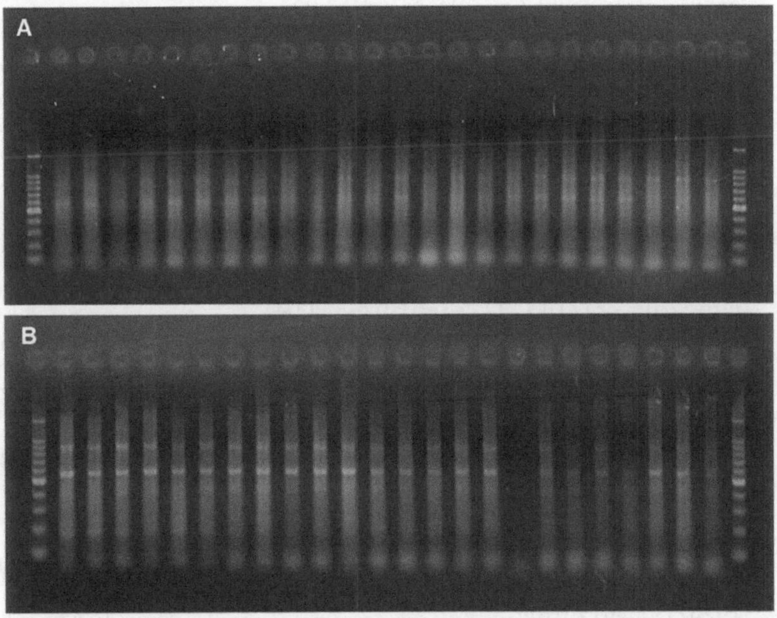

Fig. 2 Agarose gel electrophoresis photograph showing the PCR products with MIG-seq primer set-1 for the first PCR for samples of Japanese *Lindera* species (Lauraceae), and 100 bp DNA Ladder Marker (Gene Ladder 100, Nippon Gene, Tokyo, Japan) are shown on the both sides. Result in (**a**) was better than that in (**b**), because its band pattern was more smear, while (**b**) showed a clear band around 600 bp. PCR error occurred in the ninth sample from the right in (**b**)

4. The first PCR products can be stored in a freezer for up to 2 months.
5. The concentration of PhiX may be adjusted. Higher concentrations yield higher-quality data but reduce the amount of data available. However, the concentration should not be extremely high (e.g., 70%). The concentration of PhiX I can be increased successfully up to 30%, based on the current experience in the laboratory.

References

1. Bai W, Zeng Y, Zhang D (2007) Mating patterns and pollen dispersal in a heterodichogamous tree, *Juglans mandshurica* (Juglandaceae). New Phytol 176:699–707. https://doi.org/10.1111/j.1469-8137.2007.02202.x
2. Isagi Y, Saito D, Kawaguchi H, Tateno R, Watanabe S (2007) Effective pollen dispersal is enhanced by the genetic structure of an *Aesculus turbinata* population. J Ecol 95:983–999. https://doi.org/10.1111/j.1365-2745.2007.01272.x
3. Vekemans X, Hardy OJ (2004) New insights from fine-scale spatial genetic structure analyses in plant populations. Mol Ecol 13:921–935. https://doi.org/10.1046/j.1365-294X.2004.02076.x
4. Nagamitsu T, Ichikawa S, Ozawa M, Shimamura R, Kachi N, Tsumura Y et al (2002) Microsatellite analysis of the breeding system and seed dispersal in *Shorea leprosula* (Dipterocarpaceae). Int J Plant Sci 162:155–159. https://doi.org/10.1086/317902
5. Kawecki TJ, Ebert D (2004) Conceptual issues in local adaptation. Ecol Lett 7:1225–1241. https://doi.org/10.1111/j.1461-0248.2004.00684.x
6. Olson-Manning CF, Wagner MR, Mitchell-Olds T (2012) Adaptive evolution: evaluating empirical support for theoretical predictions. Nat Rev Genet 13:867–877. https://doi.org/10.1038/nrg3322
7. Jordan R, Breed MF, Prober SM, Miller AD, Hoffmann AA (2019) How well do revegetation plantings capture genetic diversity? Biol Lett 15:20190460. https://doi.org/10.1098/rsbl.2019.0460
8. Franks SJ, Hoffmann AA (2012) Genetics of climate change adaptation. Annu Rev Genet 46:185–208. https://doi.org/10.1146/annurev-genet-110711-155511
9. Davis MB, Shaw RG (2001) Range shifts and adaptive responses to quaternary climate change. Science 292:673–679. https://doi.org/10.1126/science.292.5517.673
10. Kaltz O, Shykoff JA (1998) Local adaptation in host-parasite systems. Heredity 81:361–370. https://doi.org/10.1046/j.1365-2540.1998.00435.x
11. Wagatsuma S, Imanishi J, Suyama Y, Matsuo A, Sato MP, Mitsuyuk C et al (2021) Revegetation in Japan overlooks geographical genetic structure of native *Artemisia indica* var. *maximowiczii* populations. Restor Ecol 30:e13567. https://doi.org/10.1111/rec.13567
12. Ellstrand NC, Elam DR (1993) Population genetic consequences of small population size: implications for plant conservation. Annu Rev Ecol Syst 24:217–242. https://doi.org/10.1146/annurev.es.24.110193.001245
13. Lowe AJ, Boshier D, Ward M, Bacles CFE, Navarro C (2005) Genetic resource impacts of habitat loss and degradation; reconciling empirical evidence and predicted theory for neotropical trees. Heredity 95:255–273. https://doi.org/10.1038/sj.hdy.6800725
14. Weeks AR, Sgro CM, Young AG, Frankham R, Mitchell NJ, Miller KM et al (2011) Assessing the benefits and risks of translocations in changing environments: a genetic perspective. Evol Appl 4:709–725. https://doi.org/10.1111/j.1752-4571.2011.00192.x
15. Jump AS, Marchant R, Peñuelas J (2009) Environmental change and the option value of genetic diversity. Trends Plant Sci 14:51–58. https://doi.org/10.1016/j.tplants.2008.10.002
16. Ordonez N, Seidl MF, Waalwijk C, Drenth A, Kilian A, Thomma BPHJ et al (2015) Worse comes to worst: Bananas and Panama disease—when plant and pathogen clones meet. PLoS Pathog 11:e1005197. https://doi.org/10.1371/journal.ppat.1005197
17. Ploetz RC (1994) Panama disease: return of the first banana menace. Int J Pest Manag 40:326–336. https://doi.org/10.1080/09670879409371908

18. Monteiro WP, Dalapicolla J, Carvalho CS, Veiga JC, Vasconcelos S, Ramos SJ (2022) Genetic diversity and structure of an endangered medicinal plant species (*Pilocarpus microphyllus*) in eastern Amazon: implications for conservation. Conserv Genet 23:745–758. https://doi.org/10.1007/s10592-022-01454-6
19. Moriya YD, Nanami S, Sumikura J, Yamakura T, Itoh A (2017) Clonal structure, growth pattern and preemptive space occupancy through sprouting of an invasive tree, *Triadica sebifera*. J For Res 22:8–14. https://doi.org/10.1080/13416979.2016.1265757
20. Suyama Y, Obayashi K, Hayashi I (2000) Clonal structure in a dwarf bamboo (*Sasa senanensis*) population inferred from amplified fragment length polymorphism (AFLP) fingerprints. Mol Ecol 9:901–906. https://doi.org/10.1046/j.1365-294x.2000.00943.x
21. Bricker E, Calladine A, Virnstein R, Waycott M (2018) Mega clonality in an aquatic plant – a potential survival strategy in a changing environment. Front Plant Sci 9:435. https://doi.org/10.3389/fpls.2018.00435
22. Edgeloe JM, Severn-Ellis AA, Bayer PE, Mehravi S, Breed MF, Krauss SL et al (2022) Extensive polyploid clonality was a successful strategy for seagrass to expand into a newly submerged environment. Proc R Soc Lond B Biol Sci 289:20220538. https://doi.org/10.1098/rspb.2022.0538
23. Nakamura M, Nanami S, Okuno S, Hirota SK, Matsuo A, Suyama Y et al (2021) Genetic diversity and structure of apomictic and sexually reproducing *Lindera* species (Lauraceae) in Japan. Forests 27:227. https://doi.org/10.3390/f12020227
24. Pfeiffer T, Klahr A, Peterson A, Levichev IG, Schnittler M (2012) No sex at all? Extremely low genetic diversity in *Gagea spathacea* (Liliaceae) across Europe. Flora 207:372–378. https://doi.org/10.1016/j.flora.2012.03.002
25. Bickford D, Lohman DJ, Sodhi NS, Ng PKL, Meier R, Winker K et al (2007) Cryptic species as a window on diversity and conservation. Trends Ecol Evol 22:48–155. https://doi.org/10.1016/j.tree.2006.11.004
26. Henderson SA, Billotte N, Pintaud JC (2006) Genetic isolation of Cape Verde Island *Phoenix atlantica* (Arecaceae) revealed by microsatellite markers. Conserv Genet 7:213–223. https://doi.org/10.1007/s10592-006-9128-7
27. Mayr E (1942) Systematics and the origin of species from the viewpoint of a zoologist. Columbia University Press, New York
28. Norman C, Ellstrand NC, Prentice HC, Hancock JF (1999) Gene flow and introgression from domesticated plants into their wild relatives. Annu Rev Ecol Syst 30:539–563. https://doi.org/10.1146/annurev.ecolsys.30.1.539
29. Mallet J (2005) Hybridization as an invasion of the genome. Trends Ecol Evol 5:229–237. https://doi.org/10.1016/j.tree.2005.02.010
30. Martinsen GD, Whitham TG, Turek RJ, Keim P (2001) Hybrid populations selectively filter gene introgression between species. Evolution 55:1325–1335. https://doi.org/10.1111/j.0014-3820.2001.tb00655.x
31. Rieseberg LH (1995) The role of hybridization in evolution: old wine in new skins. Am J Bot 82:944–953. https://doi.org/10.1002/j.1537-2197.1995.tb15711.x
32. Chapman MA, Burke JM (2007) Genetic divergence and hybrid speciation. Evolution 61:1773–1780. https://doi.org/10.1111/j.1558-5646.2007.00134.x
33. Rieseberg LH (1997) Hybrid origins of plant species. Annu Rev Ecol Syst 28:359–389. https://doi.org/10.1146/annurev.ecolsys.28.1.359
34. Wolf DE, Takebayashi N, Rieseberg LH (2001) Predicting the risk of extinction through hybridization. Conserv Biol 15: 1039–1053. https://doi.org/10.1046/j.1523-1739.2001.0150041039.x
35. Suyama Y, Matsuki Y (2015) MIG-seq: an effective PCR-based method for genome-wide single-nucleotide polymorphism genotyping using the next-generation sequencing platform. Sci Rep 5:16963. https://doi.org/10.1038/srep16963
36. Suyama Y, Hirota SK, Matsuo A, Tsunamoto Y, Mitsuyuki C, Shimura A et al (2022) Complementary combination of multiplex high-throughput DNA sequencing for molecular phylogeny. Ecol Res 37:171–181. https://doi.org/10.1111/1440-1703.12270
37. Mokodongan DF, Taninaka H, Sara L, Kikuchi T, Yuasa H, Suyama Y et al (2021) Spatial autocorrelation analysis using MIG-seq data indirectly estimated the gamete and larval dispersal range of the blue coral, *Heliopora coerulea*, within reefs. Front Mar Sci 8: 702977. https://doi.org/10.3389/fmars.2021.702977
38. Yamato M, Yamada H, Maeda T, Yamamoto K, Kusakabe R, Orihara T (2022) Clonal spore populations in sporocarps of arbuscular mycorrhizal fungi. Mycorrhiza 32:373–385. https://doi.org/10.1007/s00572-022-01086-1

39. Nanami S, Ikeda S, Tani N, Tan S, Diway B, Harada K et al (2007) Development of microsatellite markers for *Dryobalanops aromatica* (Dipterocarpaceae), a tropical emergent tree in Southeast Asia. Mol Ecol Notes 7:623–625. https://doi.org/10.1111/j.1471-8286.2006.01653.x
40. Eguchi K, Oguri E, Sasaki T, Matsuo A, Nguyen DD, Jaitrong W et al (2020) Revisiting museum collections in the genomic era: potential of MIG-seq for retrieving phylogenetic information from aged minute dry specimens of ants (Hymenoptera: Formicidae) and other small organisms. Myrmecol News 30:151–159. https://doi.org/10.25849/myrmecol.news_030:151
41. Catchen J, Hohenlohe PA, Bassham S, Amores A, Cresko WA (2013) Stacks: an analysis tool set for population genomics. Mol Ecol 22:3124–3140. https://doi.org/10.1111/mec.12354
42. Hirota SK, Yahara T, Fuse K, Sato H, Tagane S, Fujii S et al (2021) Molecular phylogeny and taxonomy of the *Hydrangea serrata* complex (Hydrangeaceae) in western Japan, including a new subspecies of *H. acuminata* from Yakushima. PhytoKeys 188:49–71. https://doi.org/10.3897/phytokeys.188.64259
43. Nakahama N, Asai T, Matsumoto S, Suetsugu K, Kurashima O, Matsuo A et al (2021) Detection and dispersal risk of genetically disturbed individuals in endangered wetland plant species *Pecteilis radiata* (Orchidaceae) in Japan. Biodivers Conserv 30:1913–1927. https://doi.org/10.1007/s10531-021-02174-y
44. Nota K, Klaminder J, Milesi P, Bindler R, Nobile A, van Steijn T et al (2022) Norway spruce postglacial recolonization of Fennoscandia. Nat Commun 13:1333. https://doi.org/10.1038/s41467-022-28976-4
45. Okabe N, Yahara T, Tagane S, Mitsuyuki C, Matsuo A, Sasaki T et al (2021) A new species of *Actinodaphne* (Lauraceae), *A. lambirensis* from Sarawak, Malaysia, and an analysis of its phylogenetic position using MIG-seq and ITS sequences. Acta Phytotaxon Geobot 72:43–59. https://doi.org/10.18942/apg.202005
46. Prasetyo E, Widiyatno IS, Na'iem M, Matsui T, Matsuo A et al (2020) Genetic diversity and the origin of commercial plantation of Indonesian teak on Java Island. Tree Genet Genomes 16:34. https://doi.org/10.1007/s11295-020-1427-5
47. Sakaguchi S, Nagasawa K, Umetsu Y, Nagasawa J, Ichikawa S, Kinoshita S et al (2020) Phylogenetic origin of *Magnolia pseudokobus* (Magnoliaceae), a rare Magnolia extinct in the wild, revealed by chloroplast genome sequencing, genome-wide SNP genotyping and microsatellite analysis. J For Res 25:322–328. https://doi.org/10.1080/13416979.2020.1767268
48. Suetsugu K, Hirota SK, Suyama Y (2021) A new natural hybrid, *Goodyera ×tanakae* (Orchidaceae) from Japan with a discussion on the taxonomic identities of *G. foliosa*, *G. sonoharae*, *G. velutina*, *G. ×maximo-velutina* and *G. henryi*, based on morphological and molecular data. Taiwania 66:277–286. https://doi.org/10.6165/tai.2021.66.277

Chapter 30

Application of SolCAP Genotyping in Potato (*Solanum tuberosum* L.) Association Mapping

Muhammad Farhan Yousaf, Muhammad Abu Bakar Zia, and Muhammad Naeem

Abstract

Potato variety development entails a number of breeding steps, as well as testing and, finally, commercialization. Historically, phenotypic assesment were carried out to select and germplasm development. The US Department of Agriculture (USDA) funded the Solanaceae Coordinated Agricultural Project (SolCAP) to decode genomic resources into tools that breeders and geneticists can use. This project resulted in the creation of a genome-wide single-nucleotide polymorphism (SNP) array that can be used to evaluate elite potato-breeding germplasm. This array was used to genotype a diverse panel of *Solanum* species, as well as numerous biparental, diploid, and tetraploid populations. It has high marker density to generate genetic maps that can be used to identify numerous quantitative trait loci (QTLs) for agronomic, quality, biotic, and abiotic resistance traits. Up to now, numerous QTLs for important traits have been identified using new diploid and tetraploid genetic maps. SNP markers were used to assess germplasm relationships and fingerprint varieties and identify candidate genes. The Infinium 8303 SolCAP Potato array offers a common set of SNP markers that can be used for mapping, germplasm assessment, and fingerprinting with confidence. This array has also been helpful in furthering our understanding of the potato genome. Furthermore, some other Infinium potato arrays (i.e., 12 K, 20 K, and 25 K) have been genotyped, and breeders can map quantitative trait loci (QTLs) across multiple populations to improve our understanding of economically important traits and lead to marker-assisted selection (MAS) and breeding and, ultimately, improved varieties.

Key words SolCAP, Potato, Association mapping

1 Introduction

1.1 Evolution in Breeding and Development of SolCAP Array

The potato crop (*Solanum tuberosum* L.) has high nutritional and economic values. It is an autotetraploid ($2n = 4x = 48$) food crop having a medium-sized genome of about 844 Mbps, and it ranks fourth in overall production following rice, wheat, and maize [1].

Breeders and geneticists face challenges during potato breeding due to its heterozygosity, clonal propagation, and autotetraploid nature, and recently, a variety of developers relied on phenotypic

Yuri Shavrukov (ed.), *Plant Genotyping: Methods and Protocols*, Methods in Molecular Biology, vol. 2638,
https://doi.org/10.1007/978-1-0716-3024-2_30,

evaluation for the development of germplasm. Because of recent advances in next-generation sequencing (NGS) techniques, plant breeding has moved from being a completely phenotyping-based process to being a method with an increased reliance on some level of genotype-based selection [2]. (NGS) technology allow for the rapid development of genome wide markers (SNPs) by sequencing the entire genomes (*see* **Note 1**), allowing researchers to investigate the relationship between genetic and phenotypic diversity with previously unattainable resolution. As a result, next-generation breeding approaches such as association mapping (AM) are critical for potato breeding, with the goal of improving yield-related as well as biotic and abiotic resistance traits. Finding novel SNP markers related to these traits in potatoes through AM will not only shorten the breeding cycle to 5–6 years, but it also has the potential to increase tuber yield [3] (*see* **Note 2**). AM identifies specific functional variants (i.e., loci, alleles) that are linked to phenotypic differences in a trait, which is useful for detecting traits caused by DNA sequence polymorphisms and/or for selecting genotypes that are closely related to the phenotype [4] (*see* **Note 3**). AM is also known as linkage disequilibrium (LD) mapping, and it has advantages over quantitative trait locus (QTL) mapping because it allows more variations to be observed; there is no need to develop segregating populations, and the mapping resolution is generally higher than in populations from controlled crosses/biparental crosses [5, 6].

Recently developed high-throughput genotyping assays such as the SolCAP SNP array (Solanaceae Cooperative Agricultural Project, United States Department of Agriculture (USDA)-funded project, 2009–2014) are used for the genotype analysis of large numbers of SNPs [7] (*see* **Note 4**). The Infinium 8303 SNP array, also known as the SolCAP potato array, is a recent development in this field and was created by using the sequence data from five North American potato varieties (Atlantic, Premier Russet, Snowden, Kennebec, and Shepody) and one European potato variety (Bintje) [7] (*see* **Note 5**). Next-generation transcriptome sequencing was used to generate sequence data for Atlantic, Premier Russet, and Snowden, and public databases were used for Bintje, Kennebec, and Shepody [7]. The array's marker loci were chosen in such a way that 36% of the markers are located in the candidate genes of interest, and 6% are previously mapped markers. For maximum coverage of the potato genome, additional markers (57%) in coding regions were chosen [8] (*see* **Note 6**). Moreover, this array contains 3018 SNPs from candidate genes, 576 from known genes, and 4749 from 69,000 high-confidence SNPs for greatest genome scope [7]. Some other Infinium SolCAP SNP arrays (e.g., 12 K, 20 K, and 25 K) have been genotyped, and breeders can map QTLs by utilizing these arrays across multiple populations to improve our understanding of economically important traits, which will lead to marker-assisted selection (MAS) and breeding and, ultimately, improved varieties. So far, only a few papers have been published that use the SolCAP potato SNP array. The array's design, as well as the integration of two diploid linkage maps with the potato genome sequence, has been described earlier [8]. The SolCAP potato array was also used to genotype a tetraploid biparental mapping population [9]. Stich et al. [10] used it for the first time in European potato germplasm, testing population structure and linkage disequilibrium in a set of 36 tetraploid cultivars and eight diploid potato clones. Through this venture (SolCAP), expected to use genome homology to improve applied breeding in potatoes (*see* **Note** 7). SolCAP's perception was to expand translational genomics beyond commodity boundaries, with a focus on taxonomic groups and DNA sequence homology, so that knowledge and resources could be utilized across the species (*see* **Note 8**). In summary, the SolCAP potato array enables the high-density genotyping of potato germplasm at a low cost by genotyping an individual simultaneously with SNP markers and bridging the gap between genomics and breeding by providing infrastructure to link SNP variations in genes to valuable traits and translating genomic resources into tools that breeders and geneticists could use (*see* **Note 9**).

1.2 SolCAP Array: Recent Outcomes and Applications in Potatoes

The Solanaceae Coordinated Agricultural Project (SolCAP), a USDA-funded project (2009–2014), aimed at exploiting DNA-based sequence homology between genomes in order to positively influence the applied breeding efforts in two most important vegetable crops in the Solanaceae family, i.e., potatoes and tomatoes (*see* **Note 10**). It intended to decrease the gap between genomics and breeding by linking SNP variation to the traits of substantial value, momentarily translating the genomic resources into worthy tools for both breeders and geneticists [11] (*see* **Note 11**). The development of the Infinium 8303 potato array was a major outcome of this project, which contains 3018 SNPs from candidate genes, 536 previously identified markers, and 4749 SNPs (selected from 69 K high-confidence SNPs) [7]. These SNPs were recognized with the aid of the transcriptome of six cultivars, found on the website http://solanaceae.plantbiology.msu.edu. The selection of candidate genes for the project was primarily based on consumer preferences, quality, and profitability trait or market value [11]. The SolCAP platform has been utilized extensively in potatoes to uncover genomic regions associated with agro-morphological traits, quality characteristics, and biotic and abiotic tolerance/resistance. Thus, allelic variations in genes can be linked successfully to the above-listed traits. Furthermore, the SolCAP SNP potato array can efficiently and reliably be used for mapping studies, genetic diversity assessment, and fingerprinting.

The development of the Infinium SolCAP 8303 potato array was a major breakthrough in tracking SNPs with complex traits. For instance, the diploid potato ($2n = 2x = 24$) holds central importance in genetic and breeding studies for hybrid potato breeding because of its simplicity compared to highly heterozygous autotetraploid ($2nc = 4x = 48$) potatoes. But the production of inbred lines in diploid potatoes is limited due to gametophytic self-incompatibility. Hence, it is essential to identify the genomic regions associated with self-fertility and fruit sets. The SolCAP 8303 potato SNP array was used for the genotyping of 103 F_1 hybrid and 92 S3 plants resulting from a cross between *S. tuberosum* L. group *Phureja* DM 1-3516 R44 [DM] and *S. tuberosum* L. group *Tuberosum* RH89-039-16 [RH]. The study revealed that genes responsible for the fruit set were restricted to chromosome 12 in the F_1 hybrids, whereas further advancing the generations to S3 showed the segregation of self-fertile genes on several other chromosomes, like 4, 9, 11, and 12. The segregation patterns suggest that several other factors control the seed set and fruit set in potatoes [12]. Biotic factors such as *Phytophthora infestans* and *Tecia solanivora* represent 40% to 100% losses in potato fields. QTL analysis by virtue of the SolCAP 12 K array successfully detected the genomic areas attached to the incidence and severity of these phytosanitary problems. Fifteen novel QTLs were identified for *T. solanivora* through the QTL mapping of the F_1 biparental population (Roja Nariño (RN) × BGVCOL 15062384 landrace) [13].

Quality is the main concern of consumers and breeders all over the globe. Dry matter content indirectly represents the amount of starch in tubers and is a vital component of the potato processing industry. Similarly, potato glycoalkaloids are natural stress metabolites and are toxic to humans [14]. Esnault et al. [15] used SolCAP 8.3 K SNP array to genotype core collection of 350 accessions from INRAE, France and screened them based on genetic diversity pertaining to the 21 agro-morphological, quality, and consumer-oriented traits like tuber starch content, after cooking blackening and glycoalkaloids, etc. It is pertinent to mention here that about 65% of potato production in the USA is meant for processed potato products, i.e., chips, French fries, and frozen and refrigerated items [16]. Hence, the identification of QTLs associated with them holds great value for both consumers and breeders worldwide. QTLs positioned on chromosomes 4 (6.78 cM) and 7 (47.66 cM) were linked to tuber shape, whereas those present on chromosomes 5 (51.63 cM) and 1 (121.23, 122.30, and 141.26 cM) were associated with specific gravity/starch content, identified by using the Infinium 8303 SNP potato array [17]. Wild potatoes contain rich allelic diversity, addressing the food biofortification issue. In lieu of this, SNP genotyping using the SolCAP 12 K array was performed in 94 F_2 individuals obtained by selfing F_1 progenies resulting from a cross between wild-potato-relative *S. boliviense* (high folate/diploid) and low/medium folate possessing diploid *S. tuberosum* clones (USW4self#3). In total, 497 significant SNPs were identified with 18 SNPs located in close proximity to folate-metabolism-related genes. The identified genes/SNPs would be used for enhancing folate content in potato tubers [18]. The 8.3 K SolCAP potato array was used for the genotyping of 151 F_1 tetraploid progenies of a cross between Atlantic (female) and Superior (male) genotypes. The former parent has a high incidence of hollow heart, was susceptible to common scabs, and contained low amounts of calcium; conversely, the latter parent possesses opposite features with respect to these traits. QTL mapping revealed that six QTLs were associated with variation in tuber calcium, positioned on chromosomes 1, 3, 4, 5, 7, and 8. Likewise, pitted scab incidence and severity were explained by QTLs located on chromosomes 1, 2, 3, 5, 6, 10, and 11. The incidence of hollow heart in the potatoes was linked to QTLs located on chromosomes 3, 5, 6, and 12 [19]. This publication suggests that the SolCAP SNP array provides an efficient platform for translating genome-based knowledge to marker-assisted breeding efforts. Recently, 192 tetraploid potato genotypes of diverse origins were genotyped by using the SolCAP 25 K array, and GWAS (genome-wide association studies) was performed to obtain the root and stolon characteristics of potatoes grown under an aeroponic system. The identified SNPs perform critical functions, such as root signaling and signal transduction (protein kinases SAPK2-like and GT-2 factors), transcriptional and posttranscriptional regulation of genes, and synthesis of sucrose and transporter families (Sus3, SuSy, StSUT1, and UGPases). Potato production and profitability can be enhanced at a significant pace in the era of next-generation sequencing and breeding utilizing SolCAP potato arrays [20].

Another major outcome of the SolCAP array is its application in genetic diversity and fingerprinting studies. North American varieties have been distinguished based on market class through the application of the 8.3 K SolCAP potato array [21]. Vos et al. [22] developed the 20 K SolCAP SNP array for the genotyping of 569 potato genotypes of diverse origins. The contemporary breeding material showed fingerprints of breeding history, like founder effects and the selection and identification of introgression segments. The 12 K SNP array was used by Ellis et al. [23] for the fingerprinting and genetic diversity studies of 250 accessions from the International Potato Center (CIP), Peru. Some misclassified accessions have been identified, along with the recognition of genetic redundancy among individual accessions. The same array was used for the genetic fingerprinting of 46 clones originating from the NWPVD (the Northwest Potato Variety Development) program to address issues of genetic diversity, relatedness, ownership, and intellectual property rights [24]. The potato breeding collection (144 genotypes) of Peru and Colombia has been genotyped by the SolCAP 8.3 K SNP array to assess genetic diversity, structure, and linkage disequilibrium (*see* **Note 12**). Multiple errors in the duplicate candidates and the classification of the potato breeding collection according to the level of ploidy were identified and adjusted (*see* **Note 13**). The diploid and tetraploid genotypes originating from Colombia and Peru have been differentiated. The tetraploid genotypes from Peru hold significantly greater diversity in comparison to diploids and Colombian tetraploids. Conversely, diploid genotypes and Colombian tetraploids possess high LD, and 57.2% polymorphism has been revealed among 144 potato genotypes using the 8 K array, depicting the suitability of this array for genetic analysis [25]. The Infinium 22 K V3 potato array was used for the analysis of the genetic diversity and population structure of variable flesh and skin types, including russet, and the chipping quality of a set of 214 potato clones at Texas A&M University (TAMU) Potato Breeding Program [26]. This array includes SNPs from the Infinium 8303 SNP array, along with high-confidence 69 K SNPs from Ref. [7] and SNPs from 20 K arrays developed by Vos et al. [22]. Selection signatures were found in genes controlling potato skin and flesh color, carbohydrate metabolism (chipping quality), the length of the plant cycle, and tuberization. A region (48.6–49.4 Mb) located on chr 3 controls white to yellow flesh color. SUT4 (sucrose transporter gene) positioned on chr 4 (65.8 Mb) controls the accumulation of sucrose and starch in sink organs. Additionally, pedigree and coancestry analysis is also another application of SNP arrays that helps in avoiding the cross of closely related individuals, thus preventing inbreeding depression. It ensures a continued selection of genetically variable traits; 59% heterozygosity has been reported in TAMU, a potato-breeding program [26] similar to Refs. [21, 27], which observed 56% and 60% heterozygosity, respectively. An increase in heterozygosity and a wide genetic base hint at the presence of genetic diversity, which is essential to a successful breeding program, to achieve biotic and abiotic stress resistance/tolerance and higher yields in potatoes. Conclusively, the SolCAP SNP array (8 K, 12 K, 20 K, 22 K, and 25 K) improved our understanding of the potato genome, and breeders can now use this tool to develop potato varieties much more rapidly, precisely, and efficiently. The summary of SolCAP SNP array applications is presented in Table 1.

Table 1
Application of the SolCAP potato array for the evaluation of traits and genotyping of populations (tetraploids, diploids, and diverse collections)

Characteristics/traits examined	Population	Genotyping platform	Source/ploidy	Major outcome	Reference
Evaluation of self-fertility	*S. tuberosum* L. group *Phureja* DM 1–3516 R44 [DM] and *S. tuberosum* L. group *Tuberosum* RH89–039-16 [RH]	Infinium 8303 SNP potato array	F_1 DRH segregating population ($n = 103$)/diploid	Chromosome 12 harbors genes responsible for the segregation of fruit sets resulting from self-pollination.	[12]
Phytophthora infestans and *Tecia solanivora*	Roja Nariño (RN) × BGVCOL 15062384 landrace, F_1 population of 94 full siblings	Illumina SolCAP 12 K array	Tibaitata Research Center of AGROSAVIA, Colombia/tetraploid	Six and 15 QTLs were identified to be associated with resistance against *P. infestans* and *T. solanivora*; first study to detect QTLs for *T. solanivora*, helpful for the selection of resistant breeding material in potatoes.	[13]
Folate content	Wild potato relative *Solanum boliviense* (PI 597736) × *S. tuberosum* clone (USW4self#3)	Infinium 12 K SolCAP SNP array	Minnesota advanced clones/diploid	Significant SNPs were identified on chromosomes 4 and 6 in close proximity to folate-metabolism-related genes.	[18]
Tuber shape and specific gravity (dry matter content)	Rio Grande Russet × Premier Russet	Infinium 8303 SNP potato array	USDA ARS Small Grains and Potato Germplasm Research Uni/tetraploid	Significant QTLs associated with tuber shape were located on chromosomes 4, 7, and 10, while those related to specific gravity were present on chromosomes 1 and 5.	[17]

(continued)

Table 1
(continued)

Characteristics/traits examined	Population	Genotyping platform	Source/ploidy	Major outcome	Reference
Common scabs, tuber defects (hollow heart), and calcium deficiency	Atlantic × Superior (151 F_1 progeny)	Illumina 8.3 K potato array	Hancock Agricultural Research Station of the University of Wisconsin–Madison/tetraploid	Seven QTLs explain the incidence of pitted common scabs; tuber calcium variation was illustrated by 6 QTLs, while 5 QTLs showed an incidence of hollow heart.	[19]
Root and stolon traits under aeroponic growing system	192 potato genotypes	SolCAP 25 K potato array	German and Turkish breeding lines/tetraploids	Significant SNPs linked to stolon traits were detected on chromosomes 4, 6, 7, 9, 11, and 12, while those attached to root characteristics on chromosomes 1, 2, 3, 9, 11, and 12.	[20]
Identification of genetic diversity and founder effects (pre- and post-1945) via SolSTW SNP potato array development	569 genotypes	SolCAP 20 K SNP array	Dutch commercial genotypes and breeding lines Wageningen UR plant breeding/537 tetraploids and 32 diploids	Nonredundant 15,138 previously identified SNPs and 4454 SNPs from the SolCAP project [7, 8] were used to develop 20 K SNP potato arrays for the assessment of footprints of breeding history in modern-day cultivars, like selection, drift, and inclusion of introgression segments. The founder effect was studied by using this array contributing towards contemporary potato germplasm.	[22]

21 morphological, agronomic, and consumer-oriented quality traits, like tuber dry matter and glycoalkaloids	Core collection of 350 accessions	Infinium 8303 SNP potato array	INRA BrACySol Biological Resource Center, France/ tetraploid	The screening of genetic variation and diversity in the core collection of potato gene banks was maintained at INRAE for various agro-morphological traits, including insect/pest resistance and quality traits. A high level of SNP heterozygosity (= 0.63) value also confirmed high heterozygosity.	[15]
Genetic fingerprinting	46 clones	SolCAP 12 K array	The Northwest Potato Variety Development (NWPVD) Program/tetraploid	The application of the SolCAP 12 K array for genetic fingerprinting of clones originated from the NWPVD program to address issues of genetic diversity, relatedness, ownership, and intellectual property rights.	[24]
Genetic fingerprinting and diversity analysis	250 accessions	Infinium 12 K V2 SolCAP array	International Potato Center (CIP), Peru/diploid, triploid, and tetraploid accessions	There were reports of some putative misclassified accessions and genetic redundancies among individual accessions.	[23]
Genetic analysis (diversity, structure, and LD)	144 genotypes	SolCAP 8 K array	Universidad de Nariño in Colombia/tetraploids and diploids	Genotypes from Columbia and Peru (diploids and tetraploids) were differentiated. Tetraploid genotypes from Peru hold significant diversity compared to diploids.	[25]

(continued)

Table 1 (continued)

Characteristics/traits examined	Population	Genotyping platform	Source/ploidy	Major outcome	Reference
				Adjustments and corrections of redundancy and duplication in potato breeding collections using the SolCAP SNP potato array were made.	
Genetic diversity and population structure of variable flesh and skin types, including russet and chipping quality	214 potato clones	Infinium SolCAP 22 K V3 array	Texas A&M University Potato Breeding Program/ tetraploid	Selection signatures were found in genes controlling potato skin and flesh color, carbohydrate metabolism (chipping quality), length of the plant cycle, and tuberization. A region (48.6–49.4 Mb) located on chromosome 3 controls white to yellow flesh color. *SUT4* (sucrose transporter gene) positioned on chromosome 4 (65.8 Mb) controls the accumulation of sucrose and starch in sink organs.	[26]

2 Materials

2.1 Plant Material and DNA Extraction

1. Genetic material (potato genotypes, including diploids, triploids, and tetraploids).
2. DNA extraction kit (cetyltrimethylammonium bromide (CTAB)).
3. 1% agarose gel stained with ethidium bromide (0.5 ng/μL).
4. NanoDrop 2000 spectrophotometer, for quantifying DNA concentration.
5. Freezer −20 °C, for DNA storage until genotyping.

2.2 Genotyping Platforms and Software

1. Illumina HiScan SQ and iScan, for genotyping.
2. Illumina's GenomeStudio (v 1.0, and 2010.3) software, for SNP calling.
3. Custom Perl script, used to call tetraploids.
4. JMP (version 10.0.2) software, used to determine allelic distribution and heterozygosity.
5. R package *adegenet* or similar, to perform PCA (principal components analysis).
6. R package dendextend version 1.1.2 and RcolorBrewer version 1.1-2, for color labels and palettes for the species tree. The iTOL program can be used to annotate and color tree segments.
7. MareyMap package can be used to plot genetic versus physical distances.
8. MareyMapGUI, for graphical interference.
9. STRUCTURE software.

3 Methods

3.1 SolCAP 8 K SNP Array

3.1.1 Plant Material and DNA Extraction

DNA can be extracted from immature tissues of all parents and offspring using DNeasy Plant Mini Kits, and the material is quantified using the Quant-iT PicoGreen dsDNA Assay Kit and taken to a specific absorption.

3.1.2 Cluster Development and SNP Genotyping

1. DNA can be genotyped from a set of different potato duplicates on an Illumina HiScan Reader using the Infinium HD assay. An Ultra array is used to test the utility of Illumina GenomeStudio software's autoclustering function, and the results are analyzed. Because of the high quality of each SNP and the variable cluster position, no autoclustering software is needed. GenomeStudio is available at http://solcap.msu.edu/Potato_infinium.shtml, for manually positioning clusters for

marker classes. The quality of each SNP is determined manually, and in our study, for example, 637 SNPs were removed from future analyses due to low signal intensity, loose clustering, or other assay failures.

2. SNP genotyping can be carried out on GenomeStudio by generating genotype scores from custom cluster files for D84 and DRH mapping populations.

3.1.3 Data Analysis

1. Prior to mapping, the following data have to be filtered and removed: low-quality SNPs or those that aligned to multiple positions in the super scaffolds generated by the Potato Genome Sequencing Consortium (PGSC), SNPs for which the DM appeared to be heterozygous, SNPs for which both parents were homozygous (*AA6AA* or *AA6BB*), SNPs with minor allele frequency (MAF) of 10%, SNPs with no or greatly malformed segregation (based on chi-square essays with Bonferroni correction for multiple comparisons), and SNPs that cosegregated with other SNPs based on pair-wise comparisons.
2. JoinMap 4 can be used to map the remaining SNPs.
3. In our presented case, the data were entered as a cross-pollinated population type with "lmxll" isolation, and markers were grouped by retrogression mapping using Haldane's function. Markers assigned to association groups had the lowest logarithm of the odds (LOD) score of 3 and the highest LOD score of 10.
4. Estimate the rate of genome-wide recombination using association maps and potato pseudomolecules.
5. Calculate the rate of recombination by fitting a polynomial curve to a plot of genetic versus physical distances using R base packages such as MareyMap packages.
6. To generate the plots, obtain the slope of the curve in the graphical interface MareyMapGUI using the "loess" (or "lowess" for locally weighted scatterplot smoothing) interpolation method.

3.1.4 Managing the Results

1. In total, 69,011 SNPs with high confidence were used from six salable potato cultivars, Atlantic, Premier Russet, Snowden, Bintje, Kennebec, and Shepody, to select SNPs for the Infinium 8303 potato array described in Ref. [7], and approximately 3856 SNPs were chosen based on the position of the candidate genes of interest defined by the potato community or potato genetic markers.
2. A custom Perl script is used to select the remaining 5324 SNPs on the array from a set of 69,011 high-confidence SNPs to provide genome content. The Infinium 8303 potato array is manufactured by Illumina.

3.2 SolCAP 12 K SNP Array

The Infinium 12 K V2 potato array is robust enough to genetically distinguish and give unique fingerprints from a diverse subset of potato landrace accessions (*see* **Note 14**).

3.2.1 Plant Material and DNA Extraction

1. The genetic material was received from the International Potato Center (CIP) in Lima, Peru. Two hundred fifty CIP accessions had been evaluated, which included diploids, triploids, and tetraploids belonging to different taxa. CIP accessions were estimated, and through yearly tubers, regeneration maternal plants were maintained as clones (*see* **Note 13**).
2. The number of accessions included per taxon was calculated [29].
3. To analyze reproducibility, diploid and tetraploid samples had been replicated on the assay following DNA extraction from sample plants. Tetraploid genotypes had been selected for examination [28] (*see* **Note 8**).
4. For DNA extraction, younger leaves of plants were collected from each genotype via in vitro plantlets using the CTAB protocol.
5. Agarose gel (1%) stained with ethidium bromide (0.5 ng/mL) and its concentration are determined by the spectrophotometer NanoDrop 2000. Finally, DNA was diluted to an appropriate concentration of 100 ng/μL and stored at −20 °C until genotyping.

3.2.2 SNP Genotyping

1. The potato material was genotyped using Infinium generation, and the bead heaps were studied using an Illumina iScan SQ.
2. Samples were genotyped as diploids (three-cluster calling, i.e., *AA*, *AB*, and *BB*) and tetraploids (five-cluster calling, i.e., nulliplex = *AAAA*, simplex = *AAAB*, duplex = *AABB*, triplex = *ABBB*, and quadruplex = *BBBB* [8]).
3. Diploids (three-cluster calling, i.e., *AA*, *AB*, and *BB*) were genotyped by using the autoclustering feature of Illumina's GenomeStudio (v 1.0) software and the SolCAP custom three-cluster-calling file [8].
4. Tetraploids (five-cluster calling, i.e., nulliplex = *AAAA*, simplex = *AAAB*, duplex = *AABB*, triplex = *ABBB*, and quadruplex = *BBBB* [8]) were genotyped by using a custom Perl script to call tetraploid genotypes based on theta value thresholds [21]. Due to the lack of a triploid-calling algorithm, the triploid accessions were referred to as diploids and tetraploids.
5. Predetermined low-quality SNPs (http://solcap.msu.edu/potato_Infinium.shtml) and any loci with ≥10 missing data were removed from the SNP genotype data.

3.2.3 Data Analysis

1. The allele distribution of overall SNP genotype data was calculated using JMP (version 10.0.2), a program used for the genetic identification and determination of heterozygotes. The percentage of heterozygous genotypes compared to the total number of genotypes indicates heterozygosity.
2. To reduce the computational time of the data set, all samples with the same fingerprints were reduced to a single sample. The R package *adegenet* was used to perform PCA (principal components analysis). The species tree was constructed in R model 3.2.2 using hierarchical cluster evaluation, the unweighted pair group approach with arithmetic mean (UPGMA) technique, and a common approach to calculating the gap matrix.
3. The color labels and palettes for the species tree were created in R using the dendextend version 1.1.2 and RcolorBrewer version 1.1-2 packages. iTOL was used to annotate and color tree segments [30].
4. After that, filter the sample data set for unique accessions, as well as any samples where the in vitro plantlets did not match their corresponding samples from mother plants (not true-to-type accessions), and then perform STRUCTURE analysis.

3.2.4 Population Structure

Assigning individual to population and multiple population rested on diploid (SNP) genotypes to estimate population structure through the program STRUCTURE version 2.3.4 [31, 32] and through Bayesian approach it interfere population structure that identifies clusters and assigns entities to specific clusters based on Hardy – Weinberg equilibrium and linkage equilibrium.

3.2.5 Managing the Results

1. The Infinium 12 K V2 potato array contains 12 720 SNPs, including those from the original SolCAP Infinium 8303 potato array (*see* **Note 15**), as well as new markers derived from Infinium High-Confidence SNPs (69,000) [7], which were tagged for the identification of genome content, candidate genes, and resistance gene regions.

3.3 SolCAP 20 K SNP Array

3.3.1 Development of SolSTW

1. Subsets of DNA sequence variants are generally used to develop a 20 K SNP array as described in Ref. [33].
2. A diverse design criterion was assumed that reduces the chances of assay failure because of flanking polymorphisms and increases the capability of capturing haplotypes across the diversity of potato germplasm.
3. To create an optimal array, manufacturers recommended that there be no flanking SNPs in the Illumina SNP assay over a region of 60 bp at one side of the SNP. The SNPs taken from Ref. [33] were chosen according to the following criteria: (1) no short insertions and deletions (InDels) and tri- or

quad-SNPs, (2) no InDels in the flanking sequence, (3) only SNPs genotyped with a read depth of ≥15× in at least 25 cultivars, (4) minimum flanking SNP free distance of five nucleotide positions, (5) maximum flanking SNPs at position 6–10 = 1, (6) maximum flanking SNPs at position 6–50 = 5, and (7) no Infinium type I assays. However, if both the left and right flanking sequences met these criteria, the side with the fewest SNPs in the first 10, 25, or 50 bp was chosen.

4. A Kendall tau test was performed to check the technical suitability of SNPs and reduce the redundancy in and maximize the number of SNPs. Eighty-three cultivars from Ref. [33] were taken to make cluster SNPs; correlated SNPs ($r^2 > 0.5$) were placed in one cluster.
5. Following that, the clusters were ordered by gene, and two SNPs were chosen from each cluster. One SNP was chosen for each cluster that lacked two SNPs within a gene.
6. We attempted to achieve a uniform distribution of SNPs across the length of the genome and across the depth of haplotypes by selecting SNPs over the entire length of large introgression segments.
7. Moreover, a subset of 4179 SNPs from the 8303 SolCAP array [7] were included. To improve genome coverage even further, we looked at which PGSC superscaffolds of the potato genome were missing or underrepresented. As a result, additional 284 markers were chosen from the 69,011 SolCAP SNPs discovered by Hamilton et al. [7] (*see* Fig. 1).

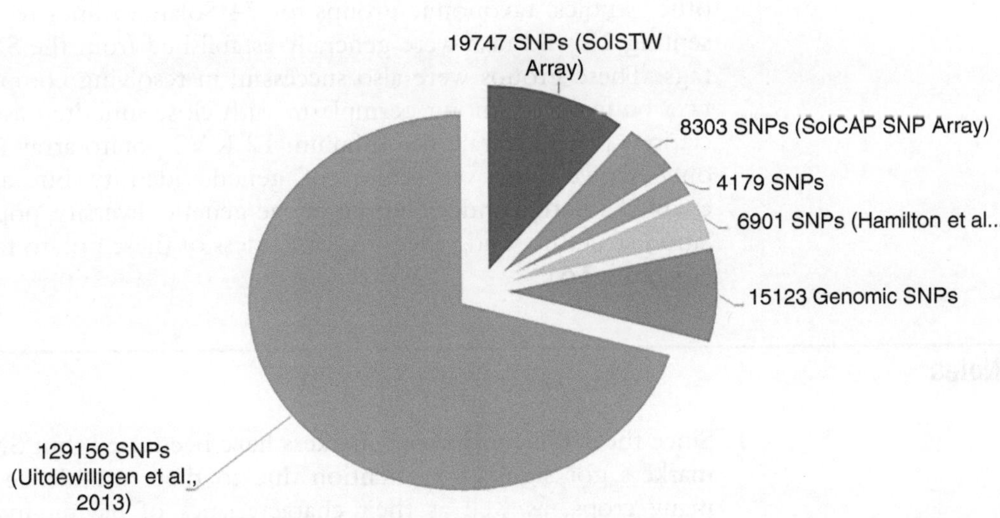

Fig. 1 Pie chart depicting the number of SNPs from several sources used to develop the SolSTW array. The 8303 SNPs originating from the SolCAP array are indicated in red, SNPs from Ref. [33] are indicated in green, and SNPs originating from Ref. [7] are indicated in yellow. This figure does not show the 87 SNPs in the SolSTW array that have been described in both SNP discovery studies [7, 33]: 37 chloroplast SNPs, 124 additional SNPs, and 284 SNPs

3.3.2 Managing the Results

1. The SolSTW SNP array combines SNPs described in two studies [7, 33]. SNPs derived from Ref. [7] were chosen based on previous trial results using the SolCAP SNP array without any new selection criteria. In contrast, the large set of SNPs (129,156) derived from Ref. [33] required stringent selection criteria because only a small subset could be chosen. Because of potatoes' high SNP density, we were able to narrow down the number of potential SNP assays to 59,279 SNPs.
2. Finally, as described by Ref. [33], redundancy among SNPs was reduced by clustering all SNPs according to SNP dose. This resulted in 7019 clusters and 5334 individual SNPs (singletons).
3. Two or more SNP per cluster and gene were chosen for approximately 5200 clusters. One SNP was elected and rounded with 2738 singletons from the remaining roughly 1800 clusters, performing in an aggregate of elected SNPs. SNPs derived from Ref. [33] will be referred to as PotVar SNPs.
4. Potato SNP arrays have been used as a genomic tool in a number of studies to improve cultivated potatoes or gain perception of heritable attributes. These SNP labels were used to measure linkage disequilibrium for genome-wide association (GWA) mapping and population structure in European diploid and tetraploid germplasm [10] (*see* **Note 12**). The SolCAP Infinium 8303 potato array has been used to develop relation charts [8] and genotype populations for QTL analysis [11] and to assess variations in glycoalkaloid biosynthesis [34, 35]. In other studies, taxonomic groups for 74 Solanum lines representing 25 wild taxa were generally established from the SNP tags. These groups were also successful in resolving complex taxa boundaries among germplasm with close inherited associations [36]. Overall, the Infinium 12 K V2 potato array not only allowed the verification of genetic identity but also enabled a better understanding of the genetic diversity, population structure, and inherited relatedness of these potato taxa (*see* **Note 16**).

4 Notes

1. Since the 1980s, numerous markers have been used, but SNP markers got significant attention due to their abundance in many crops, as well as their characteristics of having more repetitive segments and multiple ploidy levels.
2. SNP detection is easier in simple genomes than in complex genomes such as potatoes [37]. A method for reducing genetic complexity and that paves the way for the detection of novel

SNPs has been developed. Nonetheless, identifying (SNP) markers in polyploids, such as potatoes, tobacco, cotton, canola, and wheat can be difficult due to homoeologous loci from the individual subgenomes (homoeologous SNPs) or paralogous loci from duplicated regions of the genome, which increases the rate of false positives [37–40].

3. Marker systems have been used in crops to determine genetic diversity and population structure, detect QTLs, and develop linkage maps that aid in the selection of specific traits, in understanding the influence of genotypes on phenotypes, and in improving or understanding crop plants.
4. SNP markers are powerful tools for the genetic analysis of plant species and are highly useful for breeding because of their low cost and high throughput [38].
5. The Infinium SolCAP SNP array was developed by utilizing potato genotypes from Europe and America to detect allelic variation in genes that are responsible for valuable traits in various crops, particularly potatoes.
6. It is important to note that all SNP arrays contain some level of ascertainment bias, which must be considered when drawing conclusions from population data [41].
7. Flow cytometry has been considered the long-term gold standard for determining ploidy levels in plants. However, the advantages of NGS techniques, including the SolCAP array, make them effective tools for accurately determining ploidy levels without the use of toxic stains or the need for additional equipment.
8. Using SNP-genotype-based selection increases the probability of elite breeding of exceptional potato germplasm.
9. SolCAP's extension and education components have integrated genomics and plant-breeding training so that current breeders can make better use of sequence and genome-wide SNP data for improving crop breeding. By providing integrated marker tools and breeder-friendly databases, SolCAP will enable the international Solanaceae community to pursue marker-based breeding that helps in the development of potato varieties more rapidly, precisely, and efficiently.
10. The SolCAP array has a standard set of markers that may be utilized with confidence by breeders and geneticists to generate mapping, fingerprinting, and genotype assessments.
11. This array has proven to be a significant tool for the advanced comprehension of the potato genome. Furthermore, QTLs are detected by breeders after mapping across multiple populations, which allows us to better understand economically

significant traits and paves the way for marker-assisted selection and breeding.

12. Furthermore, it is used to assess linkage disequilibrium (LD) in diploid and tetraploid germplasm for genome-wide association (GWA) mapping and population structure [10]. CIP's cultivated potato collection began in the early 1970s, with material collected or donated from all over the world. In the early 1980s, research was begun to culture the collection in vitro. The primary reason for the long-term in vitro maintenance of a global genetic resource collection, such as the cultivated potato collection at CIP, is to protect the collection from biotic and abiotic factors that could jeopardize its survival. However, putting a collection in vitro ensures that the large investment (both financial and time) in cleaning pathogen-free accessions is preserved, allowing disease-free germplasm to be distributed around the world for use in research, training, and breeding. Furthermore, an in vitro culture helps maintain the allelic combinations in individual clones in a fixed state, as opposed to keeping accessions as seed collections. Preserving allelic combinations in a fixed state, in conjunction with the complete characterization of data, is a critical component for accelerating genetic gains in breeding programs. The female parental plants from the originally collected or donated samples were kept at CIP, where they were grown annually; the tubers were regenerated, and the tubers that came from plants were matched with the original descriptor data.
13. Some cases of mixed identity were easily distinguished, and the correct sample could be deduced.
14. The longer an accession is handled and manipulated, the higher is the rate of or possibility for mixing or mislabeling. By comparing individual accessions to one another and to the original field-maintained material, the Infinium Potato SolCAP SNP array could generate unique genetic fingerprints in a collection of diverse potato landraces and identify accessions where errors occurred.
15. The Infinium 12 K V2 potato array contains 12,720 SNPs, including those from the original SolCAP Infinium 8303 potato array, as well as markers derived from the Infinium High-Confidence SNPs (69,000) [7], which were chosen for the identification of genome coverage, candidate genes, and resistance gene regions.
16. These arrays have been used as a genomic tool in a number of studies to improve cultivated potatoes or gain insights into their genetic attributes.

References

1. FAOSTAT. https://www.fao.org/faostat/en/#home. Accessed 17 June 2022
2. Varshney RK, Terauchi R, McCouch SR (2014) Harvesting the promising fruits of genomics: applying genome sequencing technologies to crop breeding. PLoS Biol 12: e1001883. https://doi.org/10.1371/journal.pbio.1001883
3. Spindel J, Begum H, Akdemir D, Virk P, Collard B, Redoña E et al (2015) Genomic selection and association mapping in rice (*Oryza sativa*): effect of trait genetic architecture, training population composition, marker number and statistical model on accuracy of rice genomic selection in elite, tropical rice breeding lines. PLoS Genet 11:e1004982. https://doi.org/10.1371/journal.pgen.1004982
4. Oraguzie NC, Wilcox PL, Rikkerink EH, De Silva HN (2007) Linkage disequilibrium. In: Oraguzie NC et al (eds) Association mapping in plants. Springer, New York, pp 11–39. https://doi.org/10.1007/978-0-387-36011-9_2
5. Gupta PK, Rustgi S, Kulwal PL (2005) Linkage disequilibrium and association studies in higher plants: present status and future prospects. Plant Mol Biol 57:461–485. https://doi.org/10.1007/s11103-005-0257-z
6. Myles S, Peiffer J, Brown PJ, Ersoz ES, Zhang Z, Costich DE et al (2009) Association mapping: critical considerations shift from genotyping to experimental design. Plant Cell 21: 2194–2202. https://doi.org/10.1105/tpc.109.068437
7. Hamilton JP, Hansey CN, Whitty BR, Stoffel K, Massa AN, Deynze AV et al (2011) Single nucleotide polymorphism discovery in elite North American potato germplasm. BMC Genomics 12:302. https://doi.org/10.1186/1471-2164-12-302
8. Felcher KJ, Coombs JJ, Massa AN, Hansey CN, Hamilton JP, Veilleux RE et al (2012) Integration of two diploid potato linkage maps with the potato genome sequence. PLoS One 7:e36347. https://doi.org/10.1371/journal.pone.0036347
9. Hackett CA, McLean K, Bryan GJ (2013) Linkage analysis and QTL mapping using SNP dosage data in a tetraploid potato mapping population. PLoS One 8:e63939. https://doi.org/10.1371/journal.pone.0063939
10. Stich B, Urbany C, Hoffmann P, Gebhardt C (2013) Population structure and linkage disequilibrium in diploid and tetraploid potato revealed by genome-wide high-density genotyping using the SolCAP SNP array. Plant Breed 132:718–724. https://doi.org/10.1111/pbr.12102
11. Douches D, Hirsch CN, Manrique-Carpintero NC, Massa AN, Coombs J, Hardigan M et al (2014) The contribution of the Solanaceae coordinated agricultural project to potato breeding. Potato Res 57:215–224. https://doi.org/10.1007/s11540-014-9267-z
12. Peterson BA, Holt SH, Laimbeer FPE, Doulis AG, Coombs J, Douches DS et al (2016) Self-fertility in a cultivated diploid potato population examined with the Infinium 8303 potato single-nucleotide polymorphism array. Plant Genome 9:3. https://doi.org/10.3835/plantgenome2016.01.0003
13. Santa JD, Berdugo-Cely J, Cely-Pardo L, Soto-Suárez M, Mosquera T, Galeano MCH (2018) QTL analysis reveals quantitative resistant loci for *Phytophthora infestans* and *Tecia solanivora* in tetraploid potato (*Solanum tuberosum* L.). PLoS One 13:e0199716. https://doi.org/10.1371/journal.pone.0199716
14. Omayio DG, Abong GO, Okoth MW (2016) A review of occurrence of glycoalkaloids in potato and potato products. Curr Res Nutr Food Sci 4:3. https://doi.org/10.12944/CRNFSJ.4.3.05
15. Esnault F, Pellé R, Dantec JP, Bérard A, Le Paslier MC, Chauvin JE (2016) Development of a potato cultivar (*Solanum tuberosum* L.) core collection, a valuable tool to prospect genetic variation for novel traits. Potato Res 59:329–343. https://doi.org/10.1007/s11540-016-9332-x
16. Potatoes Annual Summary. USDA Economics, Statistics and Market Information System. ID: fx719m44h. https://usda.library.cornell.edu/concern/publications/fx719m44h. Accessed 17 June 2022
17. Park J, Massa AN, Douches D, Coombs J, Akdemir D, Yencho GC et al (2021) Linkage and QTL mapping for tuber shape and specific gravity in a tetraploid mapping population of potato representing the russet market class. BMC Plant Biol 21:507. https://doi.org/10.1186/s12870-021-03265-2
18. Bali S, Robinson BR, Sathuvalli V, Bamberg J, Goyer A (2018) Single nucleotide polymorphism (SNP) markers associated with high folate content in wild potato species. PLoS One 13:e0193415. https://doi.org/10.1371/journal.pone.0193415

19. Zorrilla C, Navarro F, Vega-Semorile S, Palta J (2021) QTL for pitted scab, hollow heart, and tuber calcium identified in a tetraploid population of potato derived from an Atlantic × Superior cross. Crop Sci 61:1630–1651. https://doi.org/10.1002/csc2.20388
20. Yousaf MF, Demirel U, Naeem M, Çalışkan ME (2021) Association mapping reveals novel genomic regions controlling some root and stolon traits in tetraploid potato (*Solanum tuberosum* L.). 3 Biotech 11:174. https://doi.org/10.1007/s13205-021-02727-6
21. Hirsch CN, Hirsch CD, Felcher K, Coombs J, Zarka D, van Deynze A et al (2013) Retrospective view of North American potato (*Solanum tuberosum* L.) breeding in the 20th and 21st centuries. G3 Genes Genomes Genet 3:1003–1013. https://doi.org/10.1534/g3.113.005595
22. Vos PG, Uitdewilligen JGAML, Voorrips RE, Visser RGF, van Eck HJ (2015) Development and analysis of a 20K SNP array for potato (*Solanum tuberosum*): an insight into the breeding history. Theor Appl Genet 128: 2387–2401. https://doi.org/10.1007/s00122-015-2593-y
23. Ellis D, Chavez O, Coombs J, Soto J, Gomez R, Douches D et al (2018) Genetic identity in genebanks: application of the SolCAP 12K SNP array in fingerprinting and diversity analysis in the global in trust potato collection. Genome 61:523–537. https://doi.org/10.1139/gen-2017-0201
24. Bali S, Sathuvalli V, Brown C, Novy R, Ewing L, Debons J et al (2017) Genetic fingerprinting of potato varieties from the northwest potato variety development program. Am J Potato Res 94:54–63. https://doi.org/10.1007/s12230-016-9547-z
25. Berdugo-Cely JA, Martínez-Moncayo C, Lagos-Burbano TC (2021) Genetic analysis of a potato (*Solanum tuberosum* L.) breeding collection for southern Colombia using single nucleotide polymorphism (SNP) markers. PLoS One 16:e0248787. https://doi.org/10.1371/journal.pone.0248787
26. Pandey J, Scheuring DC, Koym JW, Coombs J, Novy RG, Thompson AL et al (2021) Genetic diversity and population structure of advanced clones selected over forty years by a potato breeding program in the USA. Sci Rep 11: 8344. https://doi.org/10.1038/s41598-021-87284-x
27. Igarashi T, Tsuyama M, Ogawa K, Koizumi E, Sanetomo R, Hosaka K (2018) Evaluation of Japanese potatoes using single nucleotide polymorphisms (SNPs). Mol Breed 39:9. https://doi.org/10.1007/s11032-018-0917-8
28. Björn B, Paulo MJ, Mank RA, van Eck HJ, van Eeuwijk FA (2008) Association mapping of quality traits in potato (*Solanum tuberosum* L.). Euphytica 161:47–60. https://doi.org/10.1007/s10681-007-9565-5
29. Hawkes JG (1990) The potato: evolution, biodiversity and genetic resources. Belhaven Press, London, 259 pp
30. Letunic I, Bork P (2016) Interactive tree of life (iTOL) v3: an online tool for the display and annotation of phylogenetic and other trees. Nucleic Acids Res 44:W242–W245. https://doi.org/10.1093/nar/gkw290
31. Pritchard JK, Stephens M, Donnelly P (2000) Inference of population structure using multilocus genotype data. Genetics 155:945–959. https://doi.org/10.1093/genetics/155.2.945
32. Falush D, Stephens M, Pritchard JK (2003) Inference of population structure using multilocus genotype data: linked loci and correlated allele frequencies. Genetics 164:1567–1587. https://doi.org/10.1093/genetics/164.4.1567
33. Uitdewilligen JGAML, Wolters AMA, D'hoop BB, Borm TJA, RGF V, van Eck HJ (2013) A next-generation sequencing method for genotyping-by-sequencing of highly heterozygous autotetraploid potato. PLoS One 8: e62355. https://doi.org/10.1371/journal.pone.0062355
34. Manrique-Carpintero NC, Tokuhisa JG, Ginzberg I, Holliday JA, Veilleux RE (2013) Sequence diversity in coding regions of candidate genes in the glycoalkaloid biosynthetic pathway of wild potato species. G3 Genes Genomes Genet 3:1467–1479. https://doi.org/10.1534/g3.113.007146
35. Manrique-Carpintero NC, Tokuhisa JG, Ginzberg I, Veilleux RE (2014) Allelic variation in genes contributing to glycoalkaloid biosynthesis in a diploid interspecific population of potato. Theor Appl Genet 127:391–405. https://doi.org/10.1007/s00122-013-2226-2
36. Hardigan MA, Bamberg J, Buell CR, Douches DS (2015) Taxonomy and genetic differentiation among wild and cultivated germplasm of *Solanum* sect. Petota Plant Genome 8:25. https://doi.org/10.3835/plantgenome2014.06.0025
37. Mammadov J, Aggarwal R, Buyyarapu R, Kumpatla S (2012) SNP markers and their impact on plant breeding. Int J Plant Genomics

2012:e728398. https://doi.org/10.1155/2012/728398

38. Bertioli DJ, Ozias-Akins P, Chu Y, Dantas KM, Santos SP, Gouvea E et al (2014) The use of SNP markers for linkage mapping in diploid and tetraploid peanuts. G3 Genes Genomes Genet 4:89–96. https://doi.org/10.1534/g3.113.007617
39. Logan-Young CJ, Yu JZ, Verma SK, Percy RG, Pepper AE (2015) SNP discovery in complex allotetraploid genomes (*Gossypium* spp., Malvaceae) using genotyping by sequencing. Appl. Plant Sci 3:1400077. https://doi.org/10.3732/apps.1400077
40. Clevenger JP, Ozias-Akins P (2015) SWEEP: a tool for filtering high-quality SNPs in polyploid crops. G3 Genes Genomes Genet 5:1797–1803. https://doi.org/10.1534/g3.115.019703
41. Lachance J, Tishkoff SA (2013) SNP ascertainment bias in population genetic analyses: why it is important, and how to correct it. BioEssays 35:780–786. https://doi.org/10.1002/bies.201300014

Chapter 31

Fluorescence In Situ Hybridization (FISH) for the Genotyping of Triticeae Tribe Species and Hybrids

Irina Adonina

Abstract

This chapter is dedicated to using fluorescence in situ hybridization (FISH) for the genotyping of Triticeae tribe species and hybrids. The basic method of FISH on metaphase chromosomes is presented with a discussion on its modifications, and deoxyribonucleic acid (DNA) probes that can be useful for genotyping are proposed.

Key words Genotyping, Fluorescence in situ hybridization (FISH), Genomic in situ hybridization (GISH), Triticeae

1 Introduction

Hybridization in situ is the direct method of localizing deoxyribonucleic acid (DNA) sequences on chromosomes when denatured DNA molecules form duplexes with homologous DNA sequences of chromosomes on a slide. Hybridization in situ on plant chromosomes was first carried out in the 1970s [1, 2]. This method has changed significantly over its 50-year history. First of all, this is due to the development of simpler and more efficient labeling systems for DNA probes, with the replacement of radioactive labels by fluorescent labels (fluorescence in situ hybridization (FISH)) [3]. Most often, various repetitive DNA sequences, which average 45.49% of the plant genome [4], are used as probes for FISH. Groups of repeating sequences that have a similar structure and common origin are called families. Repeat families differ in the size of the motif, the number of copies, and the features of their alternation in the genome. Tandem repeats are divided, depending on the length of the repeating unit (monomer), into microsatellites (monomer length of 1–9 bp), minisatellites (10–100 bp), and macrosatellites (>100 bp). Other repeats are scattered along the chromosomes; this category includes mobile elements. Repeat

Yuri Shavrukov (ed.), *Plant Genotyping: Methods and Protocols*, Methods in Molecular Biology, vol. 2638,
https://doi.org/10.1007/978-1-0716-3024-2_31,

sequences can be classified into centromeric, subtelomeric, or intercalary, based on their distribution on chromosomes. Each chromosome has an individual "pattern" of repeats. A special type of FISH—called genomic in situ hybridization (GISH)—is applied when labeled genomic DNA from one of the parent species is used as a probe [5].

FISH and GISH, in particular, are methods that make it possible to study the process of the formation of the genomes of polyploid species, determine the degree of their relationship, analyze the introgressions of foreign genetic material, and localize breakpoints during intergenomic translocations in distant hybrids. These methods of cytogenetics are especially valuable for application in species of tribe Triticeae, which includes such important crops as common and durum wheat, rye, and barley. Many wild members of the tribe may be sources of genetic diversity for cultivated relatives, especially for wheat. Interspecific hybridization is impossible without an analysis of the obtained forms in order to identify alien chromatin and study the features of its preservation and inheritance in a number of generations. In addition to its economic importance, this tribe is characterized by a wide distribution of allopolyploid taxa, which are formed in the course of interspecific and intergeneric hybridization with subsequent transformation of the diploid genomes involved in hybridization.

This chapter presents the basic FISH technique on plant metaphase chromosomes with a discussion of its modifications. The focus is on the application of the technique for genotyping species from the Triticeae tribe and their hybrids (*see* Fig. 1).

2 Materials

2.1 Reagents and Solutions

1. Clark's fixative ("acetic alcohol" or simplified Carnoy): 96% ethyl alcohol—three parts, glacial acetic acid—one part (vol/vol).
2. Acetocarmine, 1–2%. Store in the dark (*see* **Note 1**).
3. 10 mg/mL ribonuclease A (RNase A) in 10 mM Tris–Cl, pH 7.5. Incubate in a boiling water bath for 10–15 min, cool, and then store in aliquots at −20 °C.
4. Deionized formamide (FA), molecular biology grade.
5. 20 × SSC pH 7.0: 3M NaCl, 0.3M Na-citrate.
6. 70% FA in 2 × SSC (70% FA/2 × SSC) per 100 mL of solution: 70 mL FA, 10 mL 20 × SSC, 20 mL H_2O.
7. 50% FA in 2 × SSC (50% FA/2 × SSC) per 100 mL of solution: 50 mL FA, 10 mL 20 × SSC, 40 mL H_2O.
8. Fragmented salmon sperm DNA (10 mg/mL).

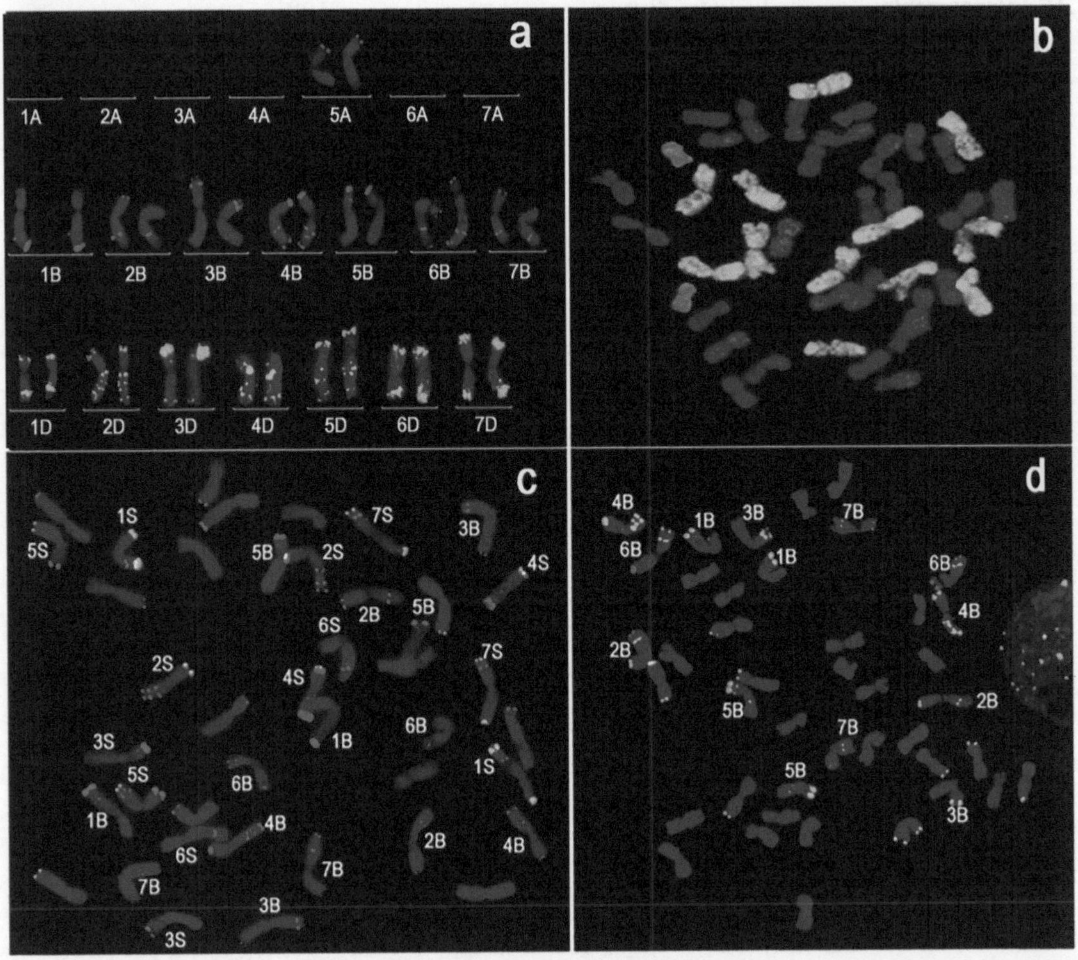

Fig. 1 Examples of using FISH for the genotyping of Triticeae tribe species and hybrids. (**a**) Karyotype of *Triticum aestivum* cv. Saratovskaya 29. FISH with pSc119.2 (red) and pAs1 (green). (**b**) The triticale line with introgressions from *Aegilops umbellulata* Zhuk. GISH with *Ae. umbellulata* DNA (red) and rye DNA (green). (**c**) Amphiploid *T. dicoccum* – *Ae. speltoides*. FISH with pSc119.2 (red) and Spelt1 (green). (**d**) *T. aestivum* – *Ae. speltoides* line. FISH with pSc119.2 (green) and Spelt1 (red). Signals Spelt1 indicate possible translocations from *Ae. speltoides*

9. Blocking DNA (for GISH).
10. Nick-translation labeling kit, dUTP: fluorochrome-, biotin-, or digoxigenin-x-dUTP.
11. Blocking solution: 4 × SSC, 0.1% Tween 20, 3% bovine serum albumin (BSA).
12. Detection buffer: 4 × SSC, 0.1% Tween 20, 1% BSA.
13. Solution C: 4 × SSC, 0.1% Tween 20.
14. Antifade solution, DAPI (4′,6-diamidino-2- phenylindol dihydrochloride) or PI (propidium iodide).

2.2 Consumable, Instruments, and Equipment

1. Petri dishes.
2. Filter paper.
3. Scalpels.
4. Razor blades.
5. Forceps.
6. Pipettes and disposable tips (0.5–20, 20–200, 200–1000 μL).
7. Toothpicks or sharpened wooden sticks.
8. Glass vials (10 mL).
9. Microtubes (volume 1.5–2 mL).
10. Ice container and ice.
11. Glass Coplin staining jars with lids.
12. Parafilm.
13. Nail polish or rubber glue.
14. Microscopic slides.
15. Coverslips: 18 × 18 mm, 22 × 22 mm, 24 × 24 mm, 24 × 32 mm.
16. Slide storage box.
17. Ethanol burner.
18. Liquid nitrogen and metal thermos.
19. Light and epifluorescence microscopes.
20. Water bath.
21. Air thermostat.
22. Thermal cycler.
23. Benchtop centrifuge for microcentrifuge tubes.

3 Methods

3.1 Preparation of Slides with Mitotic Metaphase Chromosomes

1. Seeds are germinated on wet filter paper in Petri dishes. During the night, the dishes with seeds are kept in the refrigerator (temperature +4–5 °C); during the day, the dishes with seeds are kept at a temperature (+23–25 °C). For the preparation of specimens, roots 1–2 cm long are used (*see* **Note 2**).
2. Pretreatment: for the accumulation of dividing cells at the stage of metaphase and the shortening of chromosomes, pretreatment of the plant material is carried out. The easiest way is low-temperature treatment. To do this, the cut roots are placed in vials or Eppendorf Tubes with cold tap water. Keep vials or tubes with roots in ice in the refrigerator for 20–26 h.
3. Fixation is the next step. Freshly prepared Clark's fixative is the most commonly used. Only one rule should be observed

here—about ten times more fixative is needed than plant material. Fixation in the refrigerator at +4 °C is carried out within 2–4 days. If the roots need to be stored for more than a week, you can place the vials in the freezer.

4. For staining, place the roots in 1–2% acetocarmine (15 min to 1 h).
5. Prepare the chromosome specimens (*see* **Note 3**).
6. Take the root out and place it on a glass slide. Using a blade, remove the tip with the root cap and then carefully squeeze the meristematic tissue from the root onto the slide.
7. The meristematic tissue is carefully distributed in a drop of 45% acetic acid and covered with a coverslip (*see* **Note 4**). Place a blade under one of the edges of the coverslip and, holding the glass, gently break up the tissue with careful blows of a toothpick. Tapping with a toothpick should not be on the tissue but next to it, from the side of the blade. The specimen is slightly heated over the flame of an ethanol burner and squashed.
8. The quality of the specimen is assessed by viewing it under a microscope in transmitted light or using phase-contrast microscopy at 10–40× objective magnification.
9. Immerse the end of the slide with the preparation in liquid nitrogen for about 30 s (until the "boiling" stops). Remove the coverslip with a razor blade. The slides are passed through a series of alcohols: 70%, 90%, and 96%, for 5 min each, and then dried in air.
10. The quality of the dry specimen is assessed again by viewing it under a microscope in transmitted light. The preparation should be well spread and flat and should have plenty of chromosomes with good morphology.

3.2 DNA Probes for FISH Genotyping

The main DNA probes currently used for the genotyping of Triticeae tribe species and hybrids are presented in Table 1. GISH is also used for the genotyping of hybrids.

3.3 Plasmid DNA and Genomic Plant DNA Isolation

Plasmid DNA to obtain labeled probes for FISH can be isolated, for example, according to the alkaline lysis method [35]. Currently, many manufacturers of reagents for molecular biology offer a wide range of kits for the isolation and purification of plasmid DNA. Genomic DNA for probe (GISH) or blocking can be isolated using traditional cetyltrimethylammonium bromide (CTAB) [36] or sodium dodecyl sulfate (SDS) methods [37] or by using commercial kits. Therefore, the aspect of DNA isolation is not considered in detail here.

3.4 Probe Labeling

DNA probes are labeled through PCR or nick translation in the presence of modified nucleotides (for example, fluorescein-,

Table 1
Probes for the genotyping of Triticeae tribe species and hybrids

Probe	Characteristic	Origin (reference)	Occurrence in other taxa (reference)
pAs1	Macrosatellite, 1 kb fragment, belonging to the Afa family, localized predominantly in the subtelomeric and intercalary chromosome regions	*Aegilops tauschii* Coss. [6]	D-genome of *Triticum aestivum* L. [7–9], species of the genera *Aegilops* L. spp. [10–13], and *Hordeum* L. [14], *Thinopyrum* A. Löve [15].
pSc119.2	Macrosatellite, 120 bp long sequence, localized in subtelomeric and intercalary regions of chromosomes	*Secale cereale* L. [16]	Widely distributed in many species of the tribe Triticeae and in the closely related tribe Avenae [7–10, 12, 13, 15, 17–19]
Spelt1	Macrosatellite with a monomer length of 150 bp, subtelomeric regions of chromosomes	*Aegilops speltoides* Tausch [20]	Wheat of the Timopheevi group (chromosomes: $2A^tL$, $6A^tS$, 1GL, 2GL, 3GL, 4GL, 5GL, 6GS, and 7GS), the Emmer wheat group (1BS, 2AS, 2AL, 2BL, 3BL, 6BS, and 6BL) [13, 21]
Spelt52 (homolog of pAesKB52)	Macrosatellite with a monomer length of 276 bp, subtelomeric regions of chromosomes	*Ae. speltoides* [22, 23]	*Aegilops* species of the Sitopsis section (*Ae. speltoides*, *Ae. longissima* Schweinf, and Musch L., *Ae. sharonensis* Eig), wheats of the Timopheevi group ($2A^tS$, $4A^tS$, $6A^tS$, 1GS, 6GL) [13, 21]
pTa-535	Macrosatellite, 342 bp tandem repeat localized predominantly in the subtelomeric and intercalary chromosome regions	*T. aestivum* [24]	*Aegilops* [13], *Hordeum* [14], and *Thinopyrum* species [15]
pAesp_SAT86	Minisatellite with a monomer length of 86 bp, 91–94% similarity to the wheat repeat pTa-713 [24]	*Ae. speltoides* [25]	Diploid and polyploid wheats [25], *Thinopyrum* species [15], *Aegilops* species [13]
Fat	Macrosatellite, 460 bp long sequence	*T. aestivum* [26]	*Aegilops* species [27]
pTa71	9 kb long sequence encoding 18S, 5.8S, and 26S rRNA genes, including spacers	*T. aestivum* [28]	Universal probe [13, 29]

(continued)

Table 1 (continued)

Probe	Characteristic	Origin (reference)	Occurrence in other taxa (reference)
pTa794	A 420 bp long sequence containing the 5S rRNA gene and intergenic spacer	*T. aestivum* [30]	Universal probe [13, 29]
$(GAA)_n$ or complementary $(CTT)_n$, $(GTT)_n$, $(AGG)_n$, etc.	Microsatellites	Synthetic oligonucleotides [31]	Different taxa [13, 25, 32–34]

rhodamine-, biotin-, or digoxigenin- deoxyuridine triphosphate (dUTP)). Nick translation is based on the introduction of random, single-stranded nicks in double-stranded DNA by pancreatic deoxyribonuclease I (DNase I). Then *Escherichia coli* DNA polymerase I catalyze the addition of nucleotide residues to the 3′-hydroxyl terminus of a nick (5′ → 3′ polymerase activity) with the simultaneous elimination of nucleotides from the 5′-phosphoryl terminus (5′ → 3′ exonuclease activity). The low reaction temperature (15 °C) is optimal for nick translation. Optimal probe size should be about 200–500 bp. A distinction is made between the direct and indirect methods of labeling probes for FISH. In the direct method, the fluorescent label has been incorporated into the probe, and this probe can be visualized directly once in situ hybridization is completed. The direct method is best for detecting repetitive sequences. However, it is not sensitive enough to detect low-copy sequences; the indirect method is more capable in this regard. The probe cannot be visualized directly in the case of indirect labeling; instead, the reporter molecule is attached to the probe label after hybridization and enables the visualization of probe hybridization sites. Biotin and digoxigenin are most commonly used as labels. Currently, oligonucleotide probes [38, 39] often replace probes based on cloned repeat sequences.

3.5 In Situ Hybridization Procedure

The preparations intended for hybridization with ribosomal gene probes or for GISH are treated with RNase (ribonuclease A) to prevent binding of the probe to cytoplasmic ribonucleic acid (RNA). The preparations intended for hybridization with ribosomal gene probes or for GISH are treated with RNase (ribonuclease A) to prevent binding of the probe to cytoplasmic ribonucleic acid (RNA). Initial solution of RNase (10 mg/mL) is diluted with 2 × SSC in 10 times. 30–50 μL of this solution is applied to each slide, cover with coverslips or Parafilm (square pieces, the size of a coverslip), and incubate in a humid chamber at 37 °C for 1 h. Then rinse the slides three times for 5 min in 2 × SSC, run them through a series of alcohol (70%, 90%, and 96% ethanol, 5 min each), and air-dry.

3.5.1 Denaturing of Slides

Denaturation of chromosomes is a very important step (*see* **Note 5**).

1. The denaturation of chromosome preparations is carried out in a solution of 70% FA/2 × SSC (in glass Coplin staining jars with lids (*see* **Note 6**)) for 2 min at 69–70 °C in a water bath (*see* **Note 7**). An increase in the duration of denaturation can lead to poor chromosome morphology.
2. Slides after chromosome denaturation should be passed through a series of alcohol (70%, 90%, and 96% ethanol, 3–5 min each), cooled to −20 °C, and dried in the air.

3.5.2 Preparation of the Hybridization Mixture and Hybridization

1. The hybridization mixture is prepared on ice in 0.5 mL tubes. The hybridization mixture (volume, 12–20 μL on slide) contains 50% FA, 2 × SSC, 10% dextran sodium sulfate (*see* **Note 8**), 0.5–0.8 μg/μL fragmented salmon sperm DNA or blocking DNA (in the case of GISH) (*see* **Note 9**), and labeled DNA probes (30–50 ng for repeat sequences and 100–150 ng for genomic DNA (GISH)). After adding all components, the hybridization mixture is thoroughly mixed and collected at the bottom of the tube by centrifugation (0.5–1 min).
2. Denature the mixture for 10 min at 80 °C, cool in ice, apply to preparations (avoiding the formation of air bubbles), and cover with coverslips 22 × 22 mm. Place the slides in a humid chamber (plastic box with a lid and humid filter paper at the bottom) overnight at 37 °C (*see* **Note 10**).

3.5.3 Posthybridization Washes

After hybridization, the slides are washed. Posthybridization washes are carried out to remove unbound and nonspecifically bound probes. There are many options for posthybridization washing. We provide the following protocol:

1. Place the slides in a 2 × SSC solution to gently remove the coverslips.
2. Wash the slides one time for 5 min in a 50% FA/2 × SSC solution at 42 °C.
3. Wash preparations for 5 min in solutions:
   - 1 × SSC at room temperature
   - 0.2 × SSC at room temperature
   - 0.1 × SSC at 42 °C

3.6 Detection and Amplification of Hybridization Signals

This step is only required for probes labeled with biotin or digoxigenin (indirect method), not for those labeled with fluorochromes. Slides must not dry out at any stage during detection, or the background will increase significantly.

The detection of biotinylated probes is based on the ability of biotin to attach an avidin molecule to itself. Avidin or streptavidin is covalently bound to a fluorochrome, such as fluorescein (FITC), rhodamine, etc. For the fluorescent detection of digoxigenin, antibodies linked to a fluorochrome molecule are used. In order to distinguish between the hybridization sites of different probes, bio- or dig-probes are detected by various fluorochromes.

We present a protocol in which the detection of biotinylated probes was carried out using fluorescein avidin (FITC-avidin). The hybridization signal was enhanced using fluorescein anti-avidin (anti-avidin-FITC). Digoxigenin-labeled probes were detected using antibodies to digoxigenin associated with rhodamine (anti-digoxigenin-rhodamine Fab fragments). Dilution of antibody solutions was carried out in accordance with the recommendations of the manufacturer.

1. Apply 30–50 μL of blocking solution to the preparations, cover with Parafilm (*see* **Note 11**), put in a humid chamber, and incubate in an air thermostat for 30 min at 37 °C (time is not strictly limited; the samples can be incubated longer).
2. Carefully remove Parafilm, drop 30–50 μL of the detection buffer with FITC-avidin, cover with Parafilm, put in a humid chamber, and incubate for 30–60 min at 37 °C (*see* **Note 12**).
3. Wash the slides three times for 5 min in solution C at 42 °C.
4. Drop anti-avidin-FITC + rhodamine as in **step 2**, cover with Parafilm, put in a humid chamber, and incubate for 30 min at 37 °C.
5. Wash the slides as in **step 3**.
6. Dehydrate the slides sequentially in 70%, 90%, and 96% ethanol, 3–5 min each and air-dry.
7. Mount the slides in a Vectashield mounting medium (Vectashield mounting medium, Vector Laboratories) or other antifade containing 0.5 μg/mL DAPI (4′,6-diamidino-2-phenylindol dihydrochloride) (*see* **Note 13**) for chromosome staining: 12 μL (cover slips 24 × 24 mm or 24 × 32 mm). The coverslips should be rinsed well.
8. It is desirable to seal the preparations along the edge, for example, with nail polish or rubber glue.
9. Store the slides in the dark at 4 °C in a dry, opaque box. Analyze the chromosomal preparations using a fluorescence microscope.

4 Notes

1. To obtain 1–2% acetocarmine, 1–2 g of carmine is dissolved in 45 mL of glacial acetic acid and 55 mL of distilled water. The dissolution is carried out in a flask with a reflux condenser in a

water bath for 30–60 min. After cooling, the carmine solution is filtered and stored in a glass flask or bottle.

2. "Meristem cells can be obtained from newly formed roots of plants growing in soil or hydroponics".
3. Chromosome preparations can be made using an enzyme mixture, for example, the squashing of meristems after enzyme treatment or the air-dry dropping chromosome preparation method [40].
4. The coverslips should be prewashed, for example, in mixture: hydrochloric acid (HCl) — one part, 96% ethyl alcohol — ten parts (vol/vol). Then the coverslips are rinsed with distilled water and with 96% ethanol. Coverslips can be stored in 96% ethanol.
5. Recently, the nondenaturing FISH (ND-FISH) method [41] with oligonucleotide probes has become popular. ND-FISH is performed without using formamide, resulting in reduced damage to chromosome morphology, therefore allowing for multiple hybridization experiments on the same chromosome preparation.
6. The slides can be washed in beakers, but you need to make sure that they do not stick together and that the chromosome preparations are not damaged. Therefore, it is more convenient to use special vessels—Glass Coplin staining jars with lids.
7. For simultaneous denaturation of chromosomes DNA and probes, a hot plate can be used. Hybridization solution with probes add per slide, cover with a coverslip. Slides place on a hot plate [40, 42].
8. The dextran sodium sulfate (DS) solution is very viscous. It can lead to errors in taking the exact amount of the substance. To avoid this, DS can be preheated to 65–80 °C to reduce the viscosity of the solution. Another approach is also possible: first, visually determine what length in the tip corresponds to an equal volume of distilled water, mark with a marker, and it is then found at which measurement the DS solution occupies the same volume as distilled water.
9. The DNA used for blocking needs to be fragmented to about 100–200 bp, for example, by sonication.
10. The hybridization chamber should be sufficiently moist to prevent the slides from drying out but not too wet to prevent the coverslips from slipping off.
11. At the stage of detection-amplification, preparations can be covered either with coverslips (22 × 22 mm or 24 × 24 mm) or with a Parafilm.
12. Do not work with fluorochromes in bright light.

13. DAPI: excitation—$\lambda = 344$ nm (ultraviolet), emission: $\lambda = 449$ nm (blue) + 488 nm (blue). PI (propidium iodide): excitation—$\lambda = 340$ nm (ultraviolet) + 530 nm (green), emission: $\lambda = 615$ nm (red). If the probe is labeled by FITC (excitation: $\lambda = 494$ nm, emission $\lambda = 523$ nm), counterstain propidium iodide (PI) or DAPI can be used for counterstain. If the probe is labeled by rhodamine (excitation: $\lambda = 550$–560 nm, emission: $\lambda = 568$–575 nm) or two probes are used, chromosomes are stained by DAPI.

References

1. Brady T, Clutter ME (1972) Cytolocalization of ribosomal cistrons in plant polytene chromosomes. J Cell Biol 53:827–832. https://doi.org/10.1083/jcb.53.3.827
2. Wimber DE, Duffey PA, Steffensen DM, Prensky W (1974) Localization of the 5S RNA genes in *Zea mays* by RNA-DNA hybridization *in situ*. Chromosoma 47:353–359. https://doi.org/10.1007/BF00326358
3. Jiang J, Gill BS (1994) Nonisotopic *in situ* hybridization and plant genome mapping: the first 10 years. Genome 37:717–725. https://doi.org/10.1139/g94-102
4. Luo X, Chen S, Zhang Y (2022) PlantRep: a database of plant repetitive elements. Plant Cell Rep 41:1163–1166. https://doi.org/10.1007/s00299-021-02817-y
5. Schwarzacher T, Leitch AR, Bennett MD, Heslop-Harrison JS (1989) *In situ* localization of parental genomes in a wide hybrid. Ann Bot 64:315–324. https://doi.org/10.1093/oxfordjournals.aob.a087847
6. Rayburn AL, Gill BS (1986) Isolation of a D-genome specific repeated DNA sequence from *Aegilops squarrosa*. Plant Mol Biol Rep 4:102–109. https://doi.org/10.1007/BF02732107
7. Mukai Y, Nakahara Y, Yamamoto M (1993) Simultaneous discrimination of the three genomes in hexaploid wheat by multicolor fluorescence *in situ* hybridization using total genomic and highly repeated DNA probes. Genome 36: 489–494. https://doi.org/10.1139/g93-067
8. Schneider A, Linc G, Molnar-Lang M (2003) Fluorescence *in situ* hybridization polymorphism using two repetitive DNA clones in different cultivars of wheat. Plant Breed 122:396–400. https://doi.org/10.1046/j.1439-0523.2003.00891.x
9. Adonina IG, Leonova IN, Badaeva ED, Salina EA (2017) Genotyping of hexaploid wheat varieties from different Russian regions. Russ J Genet Appl Res 7:6–13. https://doi.org/10.1134/S2079059717010014
10. Badaeva ED, Friebe B, Gill BS (1996) Genome differentiation in *Aegilops*. 1. Distribution of highly repetitive DNA sequences on chromosome of diploid species. Genome 39:293–306. https://doi.org/10.1139/g96-040
11. Badaeva ED, Amosova AV, Muravenko OV, Samatadze TE, Chikida NN, Zelenin AV et al (2002) Genome differentiation in *Aegilops*. 3. Evolution of the D-genome cluster. Plant Syst Evol 231:163–190. https://doi.org/10.1007/s006060200018
12. Badaeva ED, Amosova AV, Samatadze TE, Zoshchuk SA, Shostak NG, Chikida NN et al (2004) Genome differentiation in *Aegilops*. 4. Evolution of the U-genome cluster. Plant Syst Evol 246:45–76. https://doi.org/10.1007/s00606-003-0072-4
13. Ruban AS, Badaeva ED (2018) Evolution of the S-genomes in *Triticum-Aegilops* alliance: evidences from chromosome analysis. Front Plant Sci 9:1756. https://doi.org/10.3389/fpls.2018.01756
14. Delgado A, Carvalho A, Martın AC, Martın A, Lima-Brito J (2016) Use of the synthetic Oligo-pTa535 and Oligo-pAs1 probes for identification of *Hordeum chilense*-origin chromosomes in hexaploid tritordeum. Genet Resour Crop Evol 63:945–951. https://doi.org/10.1007/s10722-016-0402-3
15. Badaeva ED, Surzhikov SA, Agafonov AV (2019) Molecular-cytogenetic analysis of diploid wheatgrass *Thinopyrum bessarabicum* (Savul. and Rayss) A. Löve. Comp Cytogenet 13:389–402. https://doi.org/10.3897/CompCytogen.v13i4.36879
16. Bedbrook JR, Jones J, O'Dell M, Thompson RD, Flavell RB (1980) A molecular description of telomeric heterochromatin in *Secale* species. Cell 19:545–560. https://doi.org/10.1016/0092-8674(80)90529-2

17. Katsiotis A, Hagidimitriou M, Heslop-Harrison JS (1997) The close relationship between the A and B genomes in *Avena* L. (Poaceae) determined by molecular cytogenetic analysis of total genomic, tandemly and dispersed repetitive DNA. Ann Bot 79: 103–109. https://doi.org/10.1006/anbo.1996.0312
18. Taketa S, Ando H, Takeda K, Harrison GE, Heslop-Harrison JS (2000) The distribution, organization and evolution of two abundant and widespread repetitive DNA sequences in the genus *Hordeum*. Theor Appl Genet 100: 169–176. https://doi.org/10.1007/s001220050023
19. Contento A, Heslop-Harrison JS, Schwarzacher T (2005) Diversity of a major repetitive DNA sequence in diploid and polyploid Triticeae. Cytogenet Genome Res 109:34–42. https://doi.org/10.1159/000082379
20. Salina EA, Pestsova EG, Vershinin AV (1997) "Spelt-1" – a new family of tandem repeats. Rus J Genet 33:437–442
21. Salina EA, Lim YK, Badaeva ED, Shcherban AB, Adonina IG, Amosova AV et al (2006) Phylogenetic reconstruction of *Aegilops* section Sitopsis and the evolution of tandem repeats in the diploids and derived wheat polyploids. Genome 49:1023–1035. https://doi.org/10.1139/g06-050
22. Anamthawat-Jonsson K, Heslop-Harrison JS (1993) Isolation and characterization of genome-specific DNA sequences in Triticeae species. Mol Gen Genet 240:151–158. https://doi.org/10.1007/BF00277052
23. Salina E, Adonina I, Vatolina T, Kurata N (2004) A comparative analysis of the composition and organization of two subtelomeric repeat families in *Aegilops speltoides* Tausch and related species. Genetica 122:227–237. https://doi.org/10.1007/s10709-004-5602-7
24. Komuro S, Endo R, Shikata K, Kato A (2013) Genomic and chromosomal distribution patterns of various repeated DNA sequences in wheat revealed by a fluorescence *in situ* hybridization procedure. Genome 56:131–137. https://doi.org/10.1139/gen-2013-0003
25. Badaeva ED, Amosova AV, Goncharov NP, Macas J, Ruban AS, Grechishnikova IV et al (2015) A set of cytogenetic markers allows the precise identification of all A-genome chromosomes in diploid and polyploid wheat. Cytogenet Genome Res 146:71–79. https://doi.org/10.1159/000433458
26. Paux E, Roger D, Badaeva E, Gay G, Bernard M, Sourdille P et al (2006) Characterizing the composition and evolution of homoeologous genomes in hexaploid wheat through BAC-end sequencing on chromosome 3B. Plant J 48:463–474. https://doi.org/10.1111/j.1365-313X.2006.02891.x
27. Badaeva ED, Zoshchuk SA, Paux E, Gay G, Zoshchuk NV, Roger D et al (2010) Fat element – a new marker for chromosome and genome analysis in the Triticeae. Chromosom Res 18:697–709. https://doi.org/10.1007/s10577-010-9151-x
28. Gerlach WL, Bedbrook JR (1979) Cloning and characterization of ribosomal RNA genes from wheat and barley. Nucleic Acid Res 7:1869–1885. https://doi.org/10.1093/nar/7.7.1869
29. Badaeva ED, Friebe B, Gill BS (1996) Genome differentiation in *Aegilops*. 2. Physical mapping of 5S and 18S-26S ribosomal RNA gene families in diploid species. Genome 39:1150–1158. https://doi.org/10.1139/g96-145
30. Gerlach WL, Dyer TA (1980) Sequence organization of the repeated units in the nucleus of wheat which contains 5S-rRNA genes. Nucleic Acid Res 8:4851–4865. https://doi.org/10.1093/nar/8.21.4851
31. Cuadrado A, Schwarzacher T, Jouve N (2000) Identification of different chromatin classes in wheat using *in situ* hybridization with simple sequence repeat oligonucleotides. Theor Appl Genet 101:711–717. https://doi.org/10.1007/s001220051535
32. Adonina IG, Goncharov NP, Badaeva ED, Sergeeva EM, Petrash NV, Salina EA (2015) (GAA)n microsatellite as an indicator of the A genome reorganization during wheat evolution and domestication. Comp Cytogenet 9: 533–547. https://doi.org/10.3897/CompCytogen.v9i4.5120
33. Hu D, Kang L, Liu Y, Ma J, Tang X, Zeng J et al (2018) A simple and effective ND-FISH probe design for identifying barley (*Hordeum vulgare*) chromosomes. Genet Resour Crop Evol 65:2189–2198. https://doi.org/10.1007/s10722-018-0684-8
34. Zhang Y, Fan C, Chen Y, Wang RRC, Zhang X, Han F et al (2021) Genome evolution during bread wheat formation unveiled by the distribution dynamics of SSR sequences on chromosomes using FISH. BMC Genomics 22:55. https://doi.org/10.1186/s12864-020-07364-6
35. Birnboim HC, Doly J (1975) A rapid alkaline extraction procedure for screening recombinant plasmid DNA. Nucleic Acid Res 7:1513–1523. https://doi.org/10.1093/nar/7.6.1513

36. Kidwell KK, Osborn TC (1992) Simple plant DNA isolation procedures. In: Beckman JS, Osborn TC (eds) Plant genomes: methods for genetic and physical mapping. Kluwer, Dordrecht, pp 1–13
37. Plaschke J, Ganal MW, Röder MS (1995) Detection of genetic diversity in closely related bread wheat using microsatellite markers. Theor Appl Genet 91:1001–1007. https://doi.org/10.1007/BF00223912
38. Tang Z, Yang Z, Fu S (2014) Oligonucleotides replacing the roles of repetitive sequences pAs1, pSc119.2, pTa-535, pTa71, CCS1, and pAWRC.1 for FISH analysis. J Appl Genet 55: 313–318. https://doi.org/10.1007/s13353-014-0215-z
39. Tang S, Tang Z, Qiu L, Yang Z, Li G, Lang T et al (2018) Developing new oligo probes to distinguish specific chromosomal segments and the A, B, D genomes of wheat (*Triticum aestivum* L.) using ND-FISH. Front Plant Sci 9: 1104. https://doi.org/10.3389/fpls.2018.01104
40. Badaeva ED, Ruban AS, Aliyeva-Schnorr L, Municio C, Hesse S, Houben A (2017) *In situ* hybridization to plant chromosomes. In: Liehr T (ed) Fluorescence *in situ* hybridization (FISH). Application guide. Springer protocols handbooks. Springer-Verlag, Berlin Heidelberg, pp 477–494. https://doi.org/10.1007/978-3-662-52959-1_49
41. Cuadrado A, Golczyk H, Jouve N (2009) A novel, simple and rapid nondenaturing FISH (ND-FISH) technique for the detection of plant telomeres. Potential used and possible target structures detected. Chromosom Res 17:755–762. https://doi.org/10.1007/s10577-009-9060-z
42. Zhang P, Friebe B (2009) FISH on plant chromosomes. In: Liehr T (ed) Fluorescence *in situ* hybridization (FISH) – application guide. Springer-Verlag, Berlin Heidelberg, pp 365–394. https://doi.org/10.1007/978-3-540-70581-9_32

Chapter 32

Innovative Advances in Plant Genotyping

William J. W. Thomas, Yueqi Zhang, Junrey C. Amas, Aldrin Y. Cantila, Jaco D. Zandberg, Samantha L. Harvie, and Jacqueline Batley

Abstract

Over the past decade, advances in plant genotyping have been critical in enabling the identification of genetic diversity, in understanding evolution, and in dissecting important traits in both crops and native plants. The widespread popularity of single-nucleotide polymorphisms (SNPs) has prompted significant improvements to SNP-based genotyping, including SNP arrays, genotyping by sequencing, and whole-genome resequencing. More recent approaches, including genotyping structural variants, utilizing pan-genomes to capture species-wide genetic diversity and exploiting machine learning to analyze genotypic data sets, are pushing the boundaries of what plant genotyping can offer. In this chapter, we highlight these innovations and discuss how they will accelerate and advance future genotyping efforts.

Key words Plant genotyping, Single-nucleotide polymorphisms (SNPs), Structural variants (SVs), Pangenomes, Machine learning

1 Introduction

Characterizing the genetic diversity between plant species, populations, cultivars, and even individuals has been one of the biggest overarching goals of genetic research in plants. Due to the seemingly never-ending improvement of sequencing technologies, both in terms of cost and accuracy, genotyping assays have also greatly improved. Plant genotyping has been a critical component of agronomic trait dissection for crop improvement, in gaining insight into the evolutionary histories of many model species, and in classifying and managing the genetic diversity in threatened populations for better conservation outcomes [1, 2]. When studying functional genomics, collecting phenotype data often serves as a bottleneck because of its laborious, subjective, and destructive nature, meaning phenotyping has often lagged behind the advances seen in genotyping [3]. However, with the recent breakthrough of high-throughput phenotyping (HTP), which utilizes remote sensors to

Yuri Shavrukov (ed.), *Plant Genotyping: Methods and Protocols*, Methods in Molecular Biology, vol. 2638,
https://doi.org/10.1007/978-1-0716-3024-2_32,

provide massive amounts of phenotyping data, combined with the application of machine learning to analyze it [4], phenomic information has never been more readily available. Therefore, the ongoing development of better genotyping assays, methodologies, and technologies will be vital in connecting phenotype to genotype and facilitating groundbreaking research in plant genetics.

After the emergence of next-generation sequencing (NGS) in the late 1990s and its subsequent popularity in the 2000s, several marker types were utilized for routine genotyping, including restriction or amplified fragment length polymorphisms (RFLPs/AFLPs), simple sequence repeats (SSRs), and single-nucleotide polymorphisms (SNPs) [5]. More recently, genotyping arrays, especially SNP arrays that target thousands of characterized SNPs, and genotyping-by-sequencing approaches have made cost-effective and reduced representation genotyping more accessible. In the last couple of years alone, however, a number of innovative approaches have been designed to push the boundaries of genotyping. In this chapter, we outline some of these innovations and discuss their potential role in accelerating genotyping efforts in plants.

2 SNP-Based Genotyping

SNPs are widely distributed throughout the genome, including in both exons and introns, in the form of transitions, transversions, insertions, or deletions. SNP-based genotyping is currently the most popular genotyping method because SNPs are abundant, codominant, and reproducible; have a low mutation rate; and can be efficiently processed with SNP arrays and/or genotyping-by-sequencing (GBS) approaches. Three main advantages recently recognized as the driving forces behind the continued popularity of SNP-based genotyping using SNP arrays or GBS are the rapid scanning of high-density genomic SNPs, the high call rate of allelic SNPs, and the cost-effectiveness per data point [6–8].

2.1 SNP Arrays

SNP arrays are based on DNA microarray technology that utilizes probes designed for individual SNP positions and that identifies specific alleles (A/T/C/G) at each position [5, 9]. Recently, improvements have been made to several already established SNP arrays for major crops, so that they have an increased capacity, in some cases containing hundreds of thousands of additional SNPs. For example, the addition of 30 K SNPs to the Brassica 60 K Illumina Infinium™ SNP array enabled the detection of introgressed regions from the *Brassica* B genome [10], as well as candidates for the blackleg resistance gene *Rlm6* [11]. The Axiom®Apple480K SNP array for cultivated apples was developed using high-depth resequencing data from 63 different cultivars and proved to be a better platform for marker-trait association studies in apples [12] compared to other studies using significantly fewer SNP markers [13, 14]. A recent review evaluated the performance of several wheat SNP arrays and showed that the Wheat 35 K, 55 K, 90 K, 820 K, and 660 K SNP arrays were able to detect 89, 22, 1933, 4726, and 17,525 unique SNP–gene associations, respectively. Therefore, the Wheat 660 K SNP array was determined as the most cost-effective and accurate genotyping platform for downstream applications in polyploid wheat, including association studies and candidate gene identification [15]. One recent SNP array innovation is the development of multispecies arrays by Illumina. The Illumina Infinium Wheat-Barley 40 K SNP array is one such example that can be used to genotype both wheat and barley and hybrids of the two [16]. Furthermore, a multispecies Illumina SNP array for pulses, which includes chickpeas, lentils, field peas, faba beans, and lupins, is being developed in Australia [17].

2.2 Genotyping by Sequencing

GBS is a reduced representation genotyping approach that inherently lowers the cost per sample and limits the ascertainment bias from the fixed SNPs in array-based platforms [6]. Using GBS, a total of 14,188 high-quality and nonredundant SNPs were recently identified, providing the first well-resolved phylogeny for 283 *Triticum* accessions [18]. When compared to SNP arrays, GBS was able to detect rare alleles in a diverse collection of barley germplasm [19] and obtain unbiased estimates of the genetic diversity in wheat populations [20]. GBS is particularly suitable for underutilized crops that have less genomic resources as it does not require a reference genome [7]. Since the first GBS was applied in maize and barley using methylation-sensitive restriction enzymes (REs) [21], several methods for GBS have been developed. Some of these include using two REs in double-digest RE-associated DNA sequencing (ddRADSeq) [22], using proprietary kits (standard Illumina TruSeq library preparation kits) for adaptor ligation (ezRAD) [23], shortening the adaptor REs and designing a new barcode-adaptor system (MiddRAD) [24], separating RAD tag isolation and sequencing library preparation (Rapture) [25], using biotinylated RAD fragments as baits for capturing homologous fragments from genomic shotgun sequencing libraries (hyRAD) [26], and utilizing bait–probe loci as target capture GBS (DArT-cap) [27]. A comprehensive list of the different GBS methods currently available is summarized in Table 6 [6].

2.3 Specific-Locus Amplified Fragment Sequencing

Another kind of reduced representation sequencing that has risen in popularity after its inception is specific-locus amplified fragment sequencing (SLAF-Seq). SLAF-Seq is an enhanced version of ddRADSeq, which is optimized for large-scale experiments, taking advantage of reduced representation libraries (RRLs) combined with high-throughput sequencing [28]. Due to its ability to efficiently genotype many samples, SLAF-Seq has facilitated wide-scale marker discovery across a number of plants while at the same time remaining cost-effective. Such markers have been utilized for a range of purposes; for example, SLAF-Seq was used to identify candidate genes for seed weight in *Brassica napus* [29]; to construct a genetic map in eggplants, which will help underpin future trait discovery [30]; and to characterize the genetic relationship of three silvergrass species using SLAF-Seq-derived SNPs [31].

2.4 Whole-Genome Resequencing

Although GBS was originally implemented as a type of reduced representation sequencing, whole-genome resequencing is now becoming a more viable option for genotyping. Skim genotyping by sequencing (SkimGBS) is a low-coverage whole genome approach that begins with sequence alignment, followed by SNP calling, filtering, and imputation, and then downstream analysis [32]. Only low-coverage reads are obtainable through SkimGBS to minimize sequencing costs; therefore, it is often most effective when genotyping recombinant inbred lines (RILs) or doubled haploid (DH) populations, where heterozygous alleles have been eliminated [6, 32, 33]. SkimGBS was effectively applied in soybean to identify a *PP2C-1* allele associated with enhancing 100-seed weight [34]. Furthermore, the distribution of crossovers and gene conversions in both chickpea and canola was revealed through SkimGBS [32], while the unbalanced recombination rate in melon was also observed using a SkimGBS approach [35]. If increased sequence coverage is within budget constraints, then massive amounts of genetic variation can be detected through whole-genome resequencing (WGRS). For example, a large study sequenced a global collection of 588 *B. napus* lines, with an average coverage of ~5×, and obtained nearly 5.3 million SNPs [36].

2.5 Kompetitive Allele-Specific PCR

Kompetitive allele-specific PCR (KASP) is a method that uses fluorescent-based genotyping to discriminate tagged alleles [37], either a favorable or an unfavorable allele. KASP is an extremely cost-effective and flexible approach when genotyping specific markers that have already been characterized (i.e., have known SNPs). It differs from SNP arrays and GBS in that it is not a genome-wide approach. However, KASP is well suited for marker-assisted selection as it can detect both alleles of a specific marker in a single reaction [38]. For example, KASP markers were successfully used to rapidly genotype the SNPs in *AhFAD2*, a gene that controls the conversion of oleic acid into linoleic acid in peanuts, across a large number of seeds [39]. Drought-tolerant KASP markers were consistent with gel-based markers and phenotypes in identifying favorable (tolerant) wheat cultivars [40]. In canola, highly optimized KASP markers were developed for the cloned blackleg resistance genes *Rlm2*, *LepR3*, *Rlm4*, *Rlm7*, and *Rlm9* as cheap and accessible diagnostic tools to discriminate resistant and susceptible lines for improved canola breeding [41].

3 Genotyping Structural Variants

Owing to the advances in long-read sequencing, the two most popular platforms currently being Oxford Nanopore Technologies and PacBio single-molecule real-time sequencing, the large-scale discovery of other types of genetic variation is now possible. Structural variants (SVs) are defined as genomic alterations of at least 50 bp in size [42], including insertions-deletions (InDels), duplications, inversions, translocations, and copy number variations (CNVs). SVs have been increasingly linked to phenotypic outcomes, explaining the missing heritability of key traits in plants [43–45]. The ability to analyze SVs within and between genomes is made possible through advances in computational analysis and the development of several bioinformatics tools designed for analyzing large amounts of sequencing data, including SVs [46]. InDels that are completely absent in the genome are referred to as presence–absence variation (PAV) and can often contain a whole gene [47]. There are two general approaches for identifying SVs in sequencing data: comparing de novo assembled genomes (whole-genome comparison) or mapping sequence reads to a reference genome. These strategies have been applied for genotyping SVs across a range of plant species with different ploidy levels [46]. Recent advances in methods for predicting the three-dimensional structures of chromosomes, such as the high-throughput chromosome conformation capture (Hi-C) [48] and the development of the optical mapping technology [49], have also complemented the assembly of complex genomes, further enhancing the ability to identify and categorize genome-wide SVs.

Once discovered, SVs can also be assayed using array-based approaches, potentially making SV genotyping more high throughput. For instance, the Affymetrix® Axiom array can be customized to detect small to large-scale InDels, as has been demonstrated in maize with the successful genotyping of 86,648 polymorphic InDels [47]. These InDels can be targeted by designing external probes specific to the InDel sequence breakpoints (BP probes) and internal probes targeting the presence/absence regions (PARs) within the internal sequence of the InDels. The internal probe is composed of both polymorphic (OTV probes) and monomorphic sites (MONO probes). Breakpoint probes behave like classical SNP probes where one allele corresponds to the presence and the other to the absence of the InDel, while the internal probes behave like "off-target" variants and are useful in exploring the genetic diversity within InDel sequences [47]. In terms of throughput and accuracy, the Affymetrix® Axiom array is a significant improvement to previous SV genotyping attempts based on comparative genomic hybridization (CGH) technology [50] and SNP arrays [51]. However, it should be emphasized that with this approach, InDel discovery for probe design is influenced by the robustness of the mapping strategies and the quality of the genome assemblies that are used.

While SVs have become increasingly important for crop breeding, they can also be utilized in evolutionary studies for both crop and noncrop species, with potential applications in conservation genetics [52]. Genome-wide SVs complement other genetic markers, including SNPs, and together, they provide a more comprehensive understanding of the full spectrum of genomic variation present. Characterizing SVs will improve the estimation of genetic diversity and population structure and the deciphering of mechanisms of trait differentiation within species [53]. For example, the analysis of SVs in plastid genomes resolved the deep phylogenetic relationship of 50 fern families [54], while SVs identified in comparative genomic analysis between two *Medicago truncatula* ecotypes have been linked to differential responses to metal toxicity, anthocyanin production, and nitrogen-fixation-related characteristics [55]. In yellow monkeyflower (*Mimulus guttatus*), CNV in genes encoding for a transfer RNA (tRNA) ligase was associated with shorter flowering time, affecting drought tolerance and seed production [56]. Integrating SV information for evolutionary or population genetic analysis will help inform management strategies, including the effective maintenance of population viability and the preservation of their adaptive potential to respond to stress conditions [53]. Indeed, the availability of SV genotyping platforms presents promising opportunities for advancing the study of SVs and their role in determining fitness traits for crop improvement, the study of plant evolution, and managing genetic diversity for maximum adaptability.

4 A Pangenome Approach to Genotyping

As our knowledge of genomic variation has exponentially increased, it is now evident that a single reference genome is unable to fully capture the complete genomic landscape of a species [57]. A single reference that does not represent the genomic diversity within a species, population, or cultivar can bias trait association studies, for example, conventional genome-wide associated studies that rely on SNP markers [58]. This approach has limitations to the mapping of traits that are associated with PAV and not SNPs. In addition, no SNPs will be called if they are located in InDel regions that are absent from a reference genome constructed from one individual. On the other hand, if there are no SNPs in linkage disequilibrium with the causative PAV, the correct association will not be identified. Luckily, surrogate measures have been taken to overcome this limitation in GWAS; for example, three reference genome assemblies of maize (B73, PH207, and PHJ89), combined with the whole-seedling gene expression profiles and SNP patterns from 942 diverse inbred lines, were used to map sugarcane mosaic virus resistance, which was influenced by PAV [58].

To capture the complete genetic variations of a species, the concept of "pangenomes" was introduced and was quickly popularized [59]. A pangenome is composed of a core genome, which is present in all individuals, and a variable genome, which is present in only some individuals of a species [60, 61]. Plant pangenomes have now been constructed for a number of plants, including but not limited to bread wheat, *Brachypodium distachyon*, *Brassica napus*, *Brassica oleracea*, *Medicago*, capsicum, pigeon pea, poplar, rice, soybean, sesame, sorghum, sunflower, and tomato [62–73]. The core genome harbors genes that are essential for the life of the organism [69, 74]. A comparison of plant pangenomes shows that the size of the variable genome is influenced by the ploidy level, the rate of outcrossing versus selfing, and the history of artificial selection [75]. The variable genes in the plant pangenome are predominantly involved in signaling and disease resistance, as well as abiotic stress response [67, 71, 74, 76].

Pangenomes that represent a more complete overview of genetic variation within a species are a vital improvement, which has aided trait association studies. For example, the SNP- and PAV-based GWAS using the first pigeon pea pangenome identified stable association with nine agronomically important traits over two cropping times and locations [8, 73]. In another study, PAVs identified from a pangenome constructed from eight *B. napus* accessions were used to successfully map the causal variation for silique length, seed weight, and flowering time in a nested association population based on a single reference assembly ZS11 [77]. Similarly, the sorghum pangenome built using a reference genome and 354 genetically diverse accessions showed that 53% of genes were affected by PAV, among which some were associated with drought adaptation [66]. In addition, the high-quality graph pangenome for cucumber which integrates genetic variants with the reference genome sequences gave insight into the evolution and domestication of cucumber [78]. Utilizing pangenome resources provides a clearer genetic picture and, as they become more available, will accelerate the rapid identification of functional genes through association mapping and improve our understanding of plant evolutionary histories.

5 Genotype-Phenotype Prediction Using Machine Learning

Machine learning (ML) offers a powerful method for data analysis that improves predictive accuracies by developing computer algorithms with statistical techniques [79, 80]. ML algorithms learn patterns from training models and current data and subsequently improve methods and make predictions on new data without explicit programming [80, 81]. ML is particularly powerful when coupled with large data sets such as genotype data [79]. However, the type of ML used is dependent on the data set and the desired outcome. There are three main categories of machine learning that are being applied to genotyping data: (1) supervised, (2) unsupervised, and (3) semi-supervised methods.

Supervised ML models learn from labeled example data sets in order to make predictions on new data sets [79]. For example, supervised ML has been used for sequence mining from large, labeled (annotated) databases for genome selection in plant breeding [82]. Supervised models such as "deepGS" have been shown to outperform traditional genomic selection (GS) models when utilized for genome-wide predictions [80]. The two commonly used models are the random forest algorithms and the support vector machines. Random forest algorithms are an ensemble of ML tools used to predict the output of averaged results from an extensive collection of identically distributed decision trees derived from training data [81]. In contrast, support vector machine learning is a more flexible model as it allows for a specific error threshold when considering fits to the best regression line [83]. Despite innate differences, both models have been successfully used by plant breeders as a genome selection tool by interpreting plant genotypes and phenotypes [81, 84]. *Unsupervised ML* methods learn through patterns in the data by clustering, association, and dimensionality reduction, unlike supervised ML, which requires historical data [79, 85]. Typically, unsupervised ML has been applied where little historical data and low computational resources are available, as well as limited man-hours [86]. Although unsupervised ML is preferred for discovery, semi-supervised machine learning techniques have begun to surpass it in many aspects.

Semi-supervised ML models, also known as deep learning (DL), are a mixture of supervised and unsupervised methods where only some data points have assigned labels [81, 85]. DL integrates other ML categories by enabling the design of a flexible architecture that can optimize feature representation and classification while incorporating domain knowledge into its learning procedure [87]. Furthermore, the performance of DL models is further owed to their ability to explore the relationship between output and input variables by using a combination of neural and hidden layers, forming a network that is similar to the human brain [88]. In the context of plant biology, the two most common models of DL used by plant researchers and breeders are multilayer perception and convolutional neural networks [81]. These models have accelerated our ability to associate plant genotype to phenotype, for example, predicting agronomic traits associated with yield, thousand kernel weight, and days until heading [89]. It is expected that DL models will continue to evolve as emerging deep-learning technology could serve as a powerful machine-learning tool to predict quantitative phenotypes without imputation and also to discover potential associated genotype markers efficiently. For example, in Arabidopsis, an ML model was developed to identify candidate stress-related genes by comparing whole-genome expression data between the control and stress samples [90]. In soybean, an ML model was developed to predict agronomically important traits, including yield, protein, oil, moisture, and height, using SNP markers [91]. Korani et al. [92] showed that a trained ML model can predict true SNP markers twice as effectively as generic prediction methods in difficult plant genomes such as polyploid plants (peanut), and Liu et al. [91] were able to show that by using a DL framework, they could bypass the missing values required for genotype-phenotype predictions by generic methods. These examples showcase the aptitude of ML/DL models for genome analysis and the genotype marker prediction of easy and difficult genome architectures.

Although emerging DL models have predicted phenotypes without input, the pace and accuracy of phenotype prediction can be accelerated further by overcoming the lack of phenotypic data by integrating high-throughput phenotyping (HTP). HTP platforms are tools used to overcome issues with manually screening phenotypic traits, such as labor requirements, human error, and sampling method errors [93–96]. Furthermore, machine/deep learning has the advantage of automatically extracting features from HTP data sets and building an abstract representation of their relationship, depending on a prediction target [88]. For example, the recent use of an ML/DL artificial network was able to identify and predict pests and diseases in plants from HTP image/hyperspectral data sets [80]. By integrating HTP into ML-curated genotype/phenotype models, breeders are able to increase the turnover rate of new varieties as well as identify new emergent qualities [89]. Recent studies have applied machine/deep learning to integrate HTP and environmental and genetic data for selecting varieties under field trial conditions. For example, a study on wheat used generalized Poisson regression, a statistical machine learning method, to merge hyperspectral images with environmental and genetic data to predict count phenotypes for GS [97]. Another study used canopy temperature and vegetation indices for the GS of wheat, highlighting that adding the HTP features increased the yield prediction by 70% in genomic models [98]. In maize, it was shown that a multimodal deep-learning model using multispectral images and genomic data could accurately identify 75% high-yield plots at an early developmental stage [99]. Over the last decade, the adoption and application of ML/DL have improved the predictive ability of plant genotypes, particularly in difficult genome architectures such as polyploid genomes. Furthermore, with the integration of rapid phenotyping platforms such as HTP, ML/DL models are able to drastically increase yield and yield capacity for a number of agricultural crops.

6 Conclusion

Advances in genotyping efforts have been and will continue to be fundamental in identifying, understanding, and exploiting genetic diversity within plants. The development of improved SNP-based genotyping approaches, in the form of SNP arrays and GBS, has given rise to affordable yet powerful data sets that have a range of applications. WGRS is now a feasible genotyping option that can produce massive data sets. Innovative avenues toward genotyping, including using structural variants as an alternative source of variation and incorporating pangenomes to capture species-wide genetic diversity, will continue to expand the limits of what genotyping can provide. In addition, harnessing machine learning to analyze genotypic data will accelerate the timeline for linking genotypes to phenotypes.

Acknowledgments

W.J.W.T would like to acknowledge the support of the Grains Research and Development Corporation.

References

1. Thomas WJW, Anthony JM, Dobrowolski MP, Krauss SL (2021) Optimising the conservation of genetic diversity of the last remaining population of a critically endangered shrub. AoB Plants 13:5. https://doi.org/10.1093/aobpla/plab005
2. Scheben A, Batley J, Edwards D (2018) Revolution in genotyping platforms for crop improvement. In: Advances in biochemical engineering/biotechnology. Springer Science and Business Media Deutschland GmbH, pp 37–52
3. Yang W, Feng H, Zhang X, Zhang J, Doonan JH, Batchelor WD et al (2020) Crop phenomics and high-throughput phenotyping: past decades, current challenges, and future perspectives. Mol Plant 13:187–214. https://doi.org/10.1016/j.molp.2020.01.008
4. Bayer PE, Edwards D (2021) Machine learning in agriculture: from silos to marketplaces. Plant Biotechnol J 19:648–650. https://doi.org/10.1111/pbi.13521
5. Patel DA, Zander M, Dalton-Morgan J, Batley J (2015) Advances in plant genotyping: where the future will take us. In: Methods in molecular biology. Humana, pp 1–11
6. Scheben A, Batley J, Edwards D (2017) Genotyping-by-sequencing approaches to characterize crop genomes: choosing the right tool for the right application. Plant Biotechnol

J 15:149–161. https://doi.org/10.1111/PBI.12645

7. Rasheed A, Hao Y, Xia X, Khan A, Xu Y, Varshney RK et al (2017) Crop breeding chips and genotyping platforms: Progress, challenges, and perspectives. Mol Plant 10:1047–1064. https://doi.org/10.1016/j.molp.2017.06.008
8. Varshney RK, Pandey MK, Bohra A, Singh VK, Thudi M, Saxena RK (2019) Toward the sequence-based breeding in legumes in the post-genome sequencing era. Theor Appl Genet 132:797–816. https://doi.org/10.1007/S00122-018-3252-x
9. LaFramboise T (2009) Single nucleotide polymorphism arrays: a decade of biological, computational and technological advances. Nucleic Acids Res 37:4181–4193. https://doi.org/10.1093/nar/gkp552
10. Quezada-Martinez D, Zou J, Zhang W, Meng J, Batley J, Mason AS (2022) Allele segregation analysis of F_1 hybrids between independent *Brassica* allohexaploid lineages. Chromosoma 131:147–161. https://doi.org/10.1007/S00412-022-00774-3
11. Yang H, Mohd Saad NS, Ibrahim MI, Bayer PE, Neik TX, Severn-Ellis AA et al (2021) Candidate *Rlm6* resistance genes against *Leptosphaeria. maculans* identified through a genome-wide association study in *Brassica juncea* (L.) Czern. Theor Appl Genet 1:3. https://doi.org/10.1007/s00122-021-03803-4
12. Bianco L, Cestaro A, Linsmith G, Muranty H, Denancé C, Théron A et al (2016) Development and validation of the Axiom®Apple480K SNP genotyping array. Plant J 86:62–74. https://doi.org/10.1111/tpj.13145
13. Allard A, Bink MCAM, Martinez S, Kelner JJ, Legave JM, di Guardo M et al (2016) Detecting QTLs and putative candidate genes involved in budbreak and flowering time in an apple multiparental population. J Exp Bot 67: 2875–2888. https://doi.org/10.1093/jxb/erw130
14. Gross BL, Henk AD, Richards CM, Fazio G, Volk GM (2014) Genetic diversity in *Malus* × *domestica* (Rosaceae) through time in response to domestication. Am J Bot 101:1770–1779. https://doi.org/10.3732/ajb.1400297
15. Sun C, Dong Z, Zhao L, Ren Y, Zhang N, Chen F (2020) The Wheat 660K SNP array demonstrates great potential for marker-assisted selection in polyploid wheat. Plant Biotechnol J 18:1354–1360. https://doi.org/10.1111/pbi.13361
16. Keeble-Gagnère G, Pasam R, Forrest KL, Wong D, Robinson H, Godoy J et al (2021) Novel design of imputation-enabled SNP arrays for breeding and research applications supporting multi-species hybridization. Front Plant Sci 12:756877. https://doi.org/10.3389/fpls.2021.756877
17. Kaur S, Keeble-Gagnere G, Pasak RK, Hayden M (2020) Pulse pipeline integration to become more efficient. GRDC, Groundcover
18. Hyun DY, Sebastin R, Lee KJ, Lee GA, Shin MJ, Kim SH et al (2020) Genotyping-by-sequencing derived single nucleotide polymorphisms provide the first well-resolved phylogeny for the genus *Triticum* (Poaceae). Front Plant Sci 11:688. https://doi.org/10.3389/fpls.2020.00688
19. Darrier B, Russell J, Milner SG, Hedley PE, Shaw PD, Macaulay M et al (2019) A comparison of mainstream genotyping platforms for the evaluation and use of barley genetic resources. Front Plant Sci 10:544. https://doi.org/10.3389/fpls.2019.00544
20. Chu J, Zhao Y, Beier S, Schulthess AW, Stein N, Philipp N et al (2020) Suitability of single-nucleotide polymorphism arrays versus genotyping-by-sequencing for genebank genomics in wheat. Front Plant Sci 11:42. https://doi.org/10.3389/fpls.2020.00042
21. Elshire RJ, Glaubitz JC, Sun Q, Poland JA, Kawamoto K, Buckler ES et al (2011) A robust, simple genotyping-by-sequencing (GBS) approach for high diversity species. PLoS One 6:e19379. https://doi.org/10.1371/journal.pone.0019379
22. Poland JA, Brown PJ, Sorrells ME, Jannink JL (2012) Development of high-density genetic maps for barley and wheat using a novel two-enzyme genotyping-by-sequencing approach. PLoS One 7:e32253. https://doi.org/10.1371/journal.pone.0032253
23. Toonen RJ, Puritz JB, Forsman ZH, Whitney JL, Fernandez-Silva I, Andrews KR et al (2013) ezRAD: a simplified method for genomic genotyping in non-model organisms. PeerJ 2013: e203. https://doi.org/10.7717/peerj.203
24. Yang GQ, Chen YM, Wang JP, Guo C, Zhao L, Wang XY et al (2016) Development of a universal and simplified ddRAD library preparation approach for SNP discovery and genotyping in angiosperm plants. Plant Methods 12:39. https://doi.org/10.1186/S13007-016-0139-1
25. Ali OA, O'Rourke SM, Amish SJ, Meek MH, Luikart G, Jeffres C et al (2016) RAD capture (Rapture): flexible and efficient sequence-based genotyping. Genetics 202:389–400. https://doi.org/10.1534/genetics.115.183665

26. Suchan T, Pitteloud C, Gerasimova NS, Kostikova A, Schmid S, Arrigo N et al (2016) Hybridization capture using RAD probes (hyRAD), a new tool for performing genomic analyses on collection specimens. PLoS One 11:e0151651. https://doi.org/10.1371/journal.pone.0151651
27. Guppy JL, Jones DB, Kjeldsen SR, le Port A, Khatkar MS, Wade NM et al (2020) Development and validation of a RAD-Seq target-capture based genotyping assay for routine application in advanced black tiger shrimp (*Penaeus monodon*) breeding programs. BMC Genomics 21:541. https://doi.org/10.1186/S12864-020-06960-w
28. Sun X, Liu D, Zhang X, Li W, Liu H, Hong W et al (2013) SLAF-seq: an efficient method of large-scale *de novo* SNP discovery and genotyping using high-throughput sequencing. PLoS One 8:e58700. https://doi.org/10.1371/journal.pone.0058700
29. Geng X, Jiang C, Yang J, Li W, Liu H, Hong W et al (2016) Rapid identification of candidate genes for seed weight using the SLAF-Seq method in *Brassica napus*. PLoS One 11: e0147580. https://doi.org/10.1371/journal.pone.0147580
30. Wei Q, Wang W, Hu T, Hu H, Wang J, Bao C (2020) Construction of a SNP-based genetic map using SLAF-Seq and QTL analysis of morphological traits in eggplant. Front Genet 11: 178. https://doi.org/10.3389/fgene.2020.00178
31. Chen Z, He Y, Iqbal Y, Shi Y, Huang H, Yi Z (2022) Investigation of genetic relationships within three *Miscanthus* species using SNP markers identified with SLAF-seq. BMC Genomics 23:43. https://doi.org/10.1186/S12864-021-08277-8
32. Bayer PE, Ruperao P, Mason AS, Stiller J, Chan CKK, Hayashi S et al (2015) High-resolution skim genotyping by sequencing reveals the distribution of crossovers and gene conversions in *Cicer arietinum* and *Brassica napus*. Theor Appl Genet 128:1039–1047. https://doi.org/10.1007/S00122-015-2488-y
33. Malmberg MM, Barbulescu DM, Drayton MC, Shinozuka M, Thakur P, Ogaji YO et al (2018) Evaluation and recommendations for routine genotyping using skim whole genome re-sequencing in canola. Front Plant Sci 9: 1809. https://doi.org/10.3389/fpls.2018.01809
34. Lu X, Xiong Q, Cheng T, Li QT, Liu XL, Bi YD et al (2017) A *PP2C-1* allele underlying a quantitative trait locus enhances soybean 100-seed weight. Mol Plant 10:670–684. https://doi.org/10.1016/j.molp.2017.03.006
35. Hu Z, Deng G, Mou H, Xu Y, Chen L, Yang J et al (2018) A re-sequencing-based ultra-dense genetic map reveals a gummy stem blight resistance-associated gene in *Cucumis melo*. DNA Res 25:1–10. https://doi.org/10.1093/dnares/dsx033
36. Lu K, Wei L, Li X, Wang Y, Wu J, Liu M et al (2019) Whole-genome resequencing reveals *Brassica napus* origin and genetic loci involved in its improvement. Nat Commun 10:1154. https://doi.org/10.1038/s41467-019-09134-9
37. He C, Holme J, Anthony J (2014) SNP genotyping: the KASP assay. In: Methods in molecular biology. Humana, pp 75–86
38. Steele KA, Quinton-Tulloch MJ, Amgai RB, Dhakal R, Khatiwada SP, Vyas D et al (2018) Accelerating public sector rice breeding with high-density KASP markers derived from whole genome sequencing of *indica* rice. Mol Breed 38:38. https://doi.org/10.1007/S11032-018-0777-2
39. Zhao S, Li A, Li C, Xia H, Zhao C, Zhang Y et al (2017) Development and application of KASP marker for high throughput detection of *AhFAD2* mutation in peanut. Electron J Biotechnol 25:9–12. https://doi.org/10.1016/j.ejbt.2016.10.010
40. Rehman SU, Sher MA, Saddique MAB, Ali Z, Khan MA, Mao X et al (2021) Development and exploitation of KASP assays for genes underpinning drought tolerance among wheat cultivars from Pakistan. Front Genet 12: 684702. https://doi.org/10.3389/fgene.2021.684702
41. van de Wouw AP, Zhang Y, Saad NSM, Yang H, Sheedy E, Elliott CE et al (2022) Molecular markers for identifying resistance genes in *Brassica napus*. Agronomy 12:985. https://doi.org/10.3390/agronomy12050985
42. Mahmoud M, Gobet N, Cruz-Dávalos DI, Mounier N, Dessimoz C, Sedlazeck FJ (2019) Structural variant calling: the long and the short of it. Genome Biol 20:246. https://doi.org/10.1186/S13059-019-1828-7
43. Bayer PE, Scheben A, Golicz AA, Yuan Y, Faure S, Lee HT et al (2021) Modelling of gene loss propensity in the pangenomes of three *Brassica* species suggests different mechanisms between polyploids and diploids. Plant Biotechnol J 19:2488–2500. https://doi.org/10.1111/pbi.13674
44. Zhou Y, Zhang Z, Bao Z, Li H, Lyu Y, Zan Y et al (2022) Graph pangenome captures

missing heritability and empowers tomato breeding. Nature 606:527–534. https://doi.org/10.1038/s41586-022-04808-9

45. Edwards D, Batley J (2022) Graph pangenomes find missing heritability. Nat Genet 54: 919–920. https://doi.org/10.1038/s41588-022-01099-8
46. Yuan Y, Bayer PE, Batley J, Edwards D (2021) Current status of structural variation studies in plants. Plant Biotechnol J 19:2153–2163. https://doi.org/10.1111/pbi.13646
47. Mabire C, Duarte J, Darracq A, Pirani A, Rimbert H, Madur D et al (2019) High throughput genotyping of structural variations in a complex plant genome using an original Affymetrix® axiom® array. BMC Genomics 20:848. https://doi.org/10.1186/S12864-019-6136-9
48. Oluwadare O, Highsmith M, Cheng J (2019) An overview of methods for reconstructing 3-D chromosome and genome structures from Hi-C data. Biol Proced Online 21:7. https://doi.org/10.1186/S12575-019-0094-0
49. Yuan Y, Chung CYL, Chan TF (2020) Advances in optical mapping for genomic research. Comput Struct Biotechnol J 18: 2051–2062. https://doi.org/10.1016/j.csbj.2020.07.018
50. Pinkel D, Segraves R, Sudar D, Clark S, Poole I, Kowbel D et al (1998) High resolution analysis of DNA copy number variation using comparative genomic hybridization to microarrays. Nat Genet 20:207–211. https://doi.org/10.1038/2524
51. Dellinger AE, Saw SM, Goh LK, Seielstad M, Young TL, Li YJ (2010) Comparative analyses of seven algorithms for copy number variant identification from single nucleotide polymorphism arrays. Nucleic Acids Res 38:e105. https://doi.org/10.1093/nar/gkq040
52. Mérot C, Oomen RA, Tigano A, Wellenreuther M (2020) A roadmap for understanding the evolutionary significance of structural genomic variation. Trends Ecol Evol 35:561–572. https://doi.org/10.1016/j.tree.2020.03.002
53. Wold J, Koepfli KP, Galla SJ, Eccles D, Hogg CJ, le Lec MF et al (2021) Expanding the conservation genomics toolbox: incorporating structural variants to enhance genomic studies for species of conservation concern. Mol Ecol 30:5949–5965. https://doi.org/10.1111/mec.16141
54. Du XY, Kuo LY, Zuo ZY, Li DZ, Lu JM (2022) Structural variation of plastomes provides key insight into the deep phylogeny of ferns. Front Plant Sci 13:862772. https://doi.org/10.3389/fpls.2022.862772
55. Li A, Liu A, Wu S, Qu K, Hu H, Yang J et al (2022) Comparison of structural variants in the whole genome sequences of two *Medicago truncatula* ecotypes: Jemalong A17 and R108. BMC Plant Biol 22:77. https://doi.org/10.1186/S12870-022-03469-0
56. Nelson TC, Monnahan PJ, McIntosh MK, Anderson K, MacArthur-Waltz E, Finseth FR et al (2019) Extreme copy number variation at a tRNA ligase gene affecting phenology and fitness in yellow monkeyflowers. Mol Ecol 28: 1460–1475. https://doi.org/10.1111/mec.14904
57. Coletta RD, Qiu Y, Ou S, Hufford MB, Hirsch CN (2021) How the pan-genome is changing crop genomics and improvement. Genome Biol 22:3. https://doi.org/10.1186/S13059-020-02224-8
58. Gage JL, Vaillancourt B, Hamilton JP, Manrique-Carpintero NC, Gustafson TJ, Barry K et al (2019) Multiple maize reference genomes impact the identification of variants by genome-wide association study in a diverse inbred panel. Plant Genome 12:180069. https://doi.org/10.3835/plantgenome2018.09.0069
59. Bayer PE, Golicz AA, Scheben A, Batley J, Edwards D (2020) Plant pan-genomes are the new reference. Nat Plants 6:914–920. https://doi.org/10.1038/s41477-020-0733-0
60. Petereit J, Bayer PE, Thomas WJW, Fernandez CGT, Amas J, Zhang Y et al (2022) Pangenomics and crop genome adaptation in a changing climate. Plan Theory 11:1949. https://doi.org/10.3390/plants11151949
61. Golicz AA, Bayer PE, Bhalla PL, Batley J, Edwards D (2020) Pangenomics comes of age: from bacteria to plant and animal applications. Trends Genet 36:132–145. https://doi.org/10.1016/j.tig.2019.11.006
62. Ou L, Li D, Lv J, Chen W, Zhang Z, Li X et al (2018) Pan-genome of cultivated pepper (*Capsicum*) and its use in gene presence–absence variation analyses. New Phytol 220:360–363. https://doi.org/10.1111/nph.15413
63. Zhou P, Silverstein KAT, Ramaraj T, Guhlin J, Denny R, Liu J et al (2017) Exploring structural variation and gene family architecture with *de novo* assemblies of 15 *Medicago* genomes. BMC Genomics 18:261. https://doi.org/10.1186/S12864-017-3654-1
64. Zhao Q, Feng Q, Lu H, Li Y, Wang A, Tian Q et al (2018) Pan-genome analysis highlights the extent of genomic variation in cultivated and wild rice. Nat Genet 50:278–284.

https://doi.org/10.1038/s41588-018-0041-z

65. Zhang B, Zhu W, Diao S, Wu X, Lu J, Ding CJ et al (2019) The poplar pangenome provides insights into the evolutionary history of the genus. Commun Biol 2:215. https://doi.org/10.1038/s42003-019-0474-7
66. Ruperao P, Thirunavukkarasu N, Gandham P, Selvanayagam S, Govindaraj M, Nebie B et al (2021) Sorghum pan-genome explores the functional utility for genomic-assisted breeding to accelerate the genetic gain. Front Plant Sci 12:666342. https://doi.org/10.3389/fpls.2021.666342
67. Montenegro JD, Golicz AA, Bayer PE, Hurgobin B, Lee HT, Chan CKK et al (2017) The pangenome of hexaploid bread wheat. Plant J 90:1007–1013. https://doi.org/10.1111/tpj.13515
68. Hübner S, Bercovich N, Todesco M, Mandel JR, Odenheimer J, Ziegler E et al (2018) Sunflower pan-genome analysis shows that hybridization altered gene content and disease resistance. Nat Plants 5:54–62. https://doi.org/10.1038/s41477-018-0329-0
69. Gordon SP, Contreras-Moreira B, Woods DP, des Marais DL, Burgess D, Shu S et al (2017) Extensive gene content variation in the *Brachypodium distachyon* pan-genome correlates with population structure. Nat Commun 8:2184. https://doi.org/10.1038/s41467-017-02292-8
70. Gao L, Gonda I, Sun H, Ma Q, Bao K, Tieman DM et al (2019) The tomato pan-genome uncovers new genes and a rare allele regulating fruit flavor. Nat Genet 51:1044–1051. https://doi.org/10.1038/s41588-019-0410-2
71. Hurgobin B, Golicz AA, Bayer PE, Chan CKK, Tirnaz S, Dolatabadian A et al (2018) Homoeologous exchange is a major cause of gene presence/absence variation in the amphidiploid *Brassica napus*. Plant Biotechnol J 16: 1265–1274. https://doi.org/10.1111/pbi.12867
72. Li Y, Zhou G, Ma J, Jiang W, Jin L, Zhang Z et al (2014) *De novo* assembly of soybean wild relatives for pan-genome analysis of diversity and agronomic traits. Nat Biotechnol 32: 1045–1052. https://doi.org/10.1038/nbt.2979
73. Zhao J, Bayer PE, Ruperao P, Saxena RK, Khan AW, Golicz AA et al (2020) Trait associations in the pangenome of pigeon pea (*Cajanus cajan*). Plant Biotechnol J 18:1946–1954. https://doi.org/10.1111/pbi.13354
74. Wang W, Mauleon R, Hu Z, Chebotarov D, Tai S, Wu Z et al (2018) Genomic variation in 3,010 diverse accessions of Asian cultivated rice. Nature 557:43–49. https://doi.org/10.1038/s41586-018-0063-9
75. Tao Y, Zhao X, Mace E, Henry R, Jordan D (2019) Exploring and exploiting pan-genomics for crop improvement. Mol Plant 12:156–169. https://doi.org/10.1016/j.molp.2018.12.016
76. Golicz AA, Batley J, Edwards D (2016) Towards plant pangenomics. Plant Biotechnol J 14:1099–1105. https://doi.org/10.1111/pbi.12499
77. Song JM, Guan Z, Hu J, Guo C, Yang Z, Wang S et al (2020) Eight high-quality genomes reveal pan-genome architecture and ecotype differentiation of *Brassica napus*. Nat Plants 6: 34–45. https://doi.org/10.1038/s41477-019-0577-7
78. Li H, Wang S, Chai S, Yang Z, Zhang Q, Xin H et al (2022) Graph-based pan-genome reveals structural and sequence variations related to agronomic traits and domestication in cucumber. Nat Commun 13:682. https://doi.org/10.1038/s41467-022-28362-0
79. Libbrecht MW, Noble WS (2015) Machine learning applications in genetics and genomics. Nat Rev Genet 16:321–332. https://doi.org/10.1038/nrg3920
80. Varshney RK, Bohra A, Roorkiwal M, Barmukh R, Cowling WA, Chitikineni A et al (2021) Fast-forward breeding for a food-secure world. Trends Genet 37:1124–1136. https://doi.org/10.1016/j.tig.2021.08.002
81. Sandhu K, Patil SS, Pumphrey M, Carter A (2021) Multitrait machine- and deep-learning models for genomic selection using spectral information in a wheat breeding program. Plant Genome 14:e20119. https://doi.org/10.1002/tpg2.20119
82. Yang A, Zhang W, Wang J, Yang K, Han Y, Zhang L (2020) Review on the application of machine learning algorithms in the sequence data mining of DNA. Front Bioeng Biotechnol 8:1032. https://doi.org/10.3389/fbioe.2020.01032
83. Vapnik VN (2000) The nature of statistical learning theory. Springer, New York
84. Smith PF, Ganesh S, Liu P (2013) A comparison of random forest regression and multiple linear regression for prediction in neuroscience. J Neurosci Methods 220:85–91. https://doi.org/10.1016/j.jneumeth.2013.08.024
85. van Dijk ADJ, Kootstra G, Kruijer W, de Ridder D (2021) Machine learning in plant science and plant breeding. iScience 24:101890. https://doi.org/10.1016/j.isci.2020.101890

86. Yan J, Wang X (2022) Unsupervised and semi-supervised learning: the next frontier in machine learning for plant systems biology. Plant J 111:1527–1538. https://doi.org/10.1111/tpj.15905

87. Gazestani VH, Lewis NE (2019) From genotype to phenotype: augmenting deep learning with networks and systems biology. Curr Opin Syst Biol 15:68–73. https://doi.org/10.1016/j.coisb.2019.04.001

88. LeCun Y, Bengio Y, Hinton G (2015) Deep learning. Nature 521:436–444. https://doi.org/10.1038/nature14539

89. Danilevicz MF, Gill M, Anderson R, Batley J, Bennamoun M, Bayer PE et al (2022) Plant genotype to phenotype prediction using machine learning. Front Genet 13:822173. https://doi.org/10.3389/fgene.2022.822173

90. Wegrzyn JL, Liechty JD, Stevens KA, Wu LS, Loopstra CA, Vasquez-Gross HA et al (2014) Unique features of the loblolly pine (*Pinus taeda* L.) megagenome revealed through sequence annotation. Genetics 196:891–909. https://doi.org/10.1534/genetics.113.159996

91. Liu Y, Wang D, He F, Wang J, Joshi T, Xu D (2019) Phenotype prediction and genome-wide association study using deep convolutional neural network of soybean. Front Genet 10:1091. https://doi.org/10.3389/fgene.2019.01091

92. Korani W, Clevenger JP, Chu Y, Ozias-Akins P (2019) Machine learning as an effective method for identifying true single nucleotide polymorphisms in polyploid plants. Plant Genome 12:180023. https://doi.org/10.3835/plantgenome2018.05.0023

93. Shakoor N, Lee S, Mockler TC (2017) High throughput phenotyping to accelerate crop breeding and monitoring of diseases in the field. Curr Opin Plant Biol 38:184–192. https://doi.org/10.1016/j.pbi.2017.05.006

94. Araus JL, Kefauver SC, Vergara-Díaz O, Vergara-Díaz O, Gracia-Romero A, Rezzouk FZ et al (2022) Crop phenotyping in a context of global change: what to measure and how to do it. J Integr Plant Biol 64:592–618. https://doi.org/10.1111/jipb.13191

95. Hall RD, D'Auria JC, Silva Ferreira AC, Gibon Y, Kruszka D, Mishra P et al (2022) High-throughput plant phenotyping: a role for metabolomics? Trends Plant Sci 27:549–563. https://doi.org/10.1016/j.tplants.2022.02.001

96. Duddu HSN, Johnson EN, Willenborg CJ, Shirtliffe SJ (2019) High-throughput UAV image-based method is more precise than manual rating of herbicide tolerance. Plant Phenom 2019:6036453. https://doi.org/10.34133/2019/6036453

97. Kismiantini, Montesinos-López OA, Crossa J, Setiawan EP, Wutsqa DU (2021) Prediction of count phenotypes using high-resolution images and genomic data. G3 11:jkab035. https://doi.org/10.1093/g3journal/jkab035

98. Rutkoski J, Poland J, Mondal S, Autrique E, Pérez LG, Crossa J et al (2016) Canopy temperature and vegetation indices from high-throughput phenotyping improve accuracy of pedigree and genomic selection for grain yield in wheat. G3(6):2799–2808. https://doi.org/10.1534/g3.116.032888

99. Danilevicz MF, Bayer PE, Boussaid F, Bennamoun M, Edwards D (2021) Maize yield prediction at an early developmental stage using multispectral images and genotype data for preliminary hybrid selection. Remote Sens 13:3976. https://doi.org/10.3390/rs13193976

Index

Yuri Shavrukov (ed.), *Plant Genotyping: Methods and Protocols*, Methods in Molecular Biology, vol. 2638, https://doi.org/10.1007/978-1-0716-3024-2,

S

T

U